PIPE WELDING

First Edition

Larry Jeffus | Bryan Baker

CENGAGE Learning

Australia • Brazil • Mexico • Singapore • United Kingdom • United States

CENGAGE Learning®

Title: Pipe Welding, First Edition
Larry Jeffus, Bryan Baker

SVP, GM Skills & Global Product Management: Dawn Gerrain

Product Director: Matt Seeley

Product Team Manager: Erin Brennan

Associate Product Manager: Nicole Sgueglia

Senior Director, Development: Marah Bellegarde

Senior Product Development Manager: Larry Main

Senior Content Developer: Sharon Chambliss

Product Assistant: Maria Garguilo

Vice President, Marketing Services: Jennifer Ann Baker

Marketing Director: Michele McTighe

Marketing Manager: Jonathan Sheehan

Senior Production Director: Wendy Troeger

Production Director: Andrew Crouth

Senior Content Project Manager: Betsy Hough

Senior Art Director: Benjamin Gleeksman

Cover image: Larry Jeffus

Library of Congress Control Number: 2015943961

ISBN: 978-1-1336-9184-6

Cengage Learning
20 Channel Center Street
Boston, MA 02210
USA

Cengage Learning is a leading provider of customized learning solutions with employees residing in nearly 40 different countries and sales in more than 125 countries around the world. Find your local representative at **www.cengage.com**.

Cengage Learning products are represented in Canada by Nelson Education, Ltd.

To learn more about Cengage Learning, visit **www.cengage.com**
Purchase any of our products at your local college store or at our preferred online store **www.cengagebrain.com**

Notice to the Reader

Publisher does not warrant or guarantee any of the products described herein or perform any independent analysis in connection with any of the product information contained herein. Publisher does not assume, and expressly disclaims, any obligation to obtain and include information other than that provided to it by the manufacturer. The reader is expressly warned to consider and adopt all safety precautions that might be indicated by the activities described herein and to avoid all potential hazards. By following the instructions contained herein, the reader willingly assumes all risks in connection with such instructions. The publisher makes no representations or warranties of any kind, including but not limited to, the warranties of fitness for particular purpose or merchantability, nor are any such representations implied with respect to the material set forth herein, and the publisher takes no responsibility with respect to such material. The publisher shall not be liable for any special, consequential, or exemplary damages resulting, in whole or part, from the readers' use of, or reliance upon, this material.

Printed in the United States of America
Print Number: 01 Print Year: 2015

Dedication

"This book is dedicated to some very special people—my family."

—Larry Jeffus

"I dedicate this book to my wife, daughter, parents, brothers, and in-laws."

—Bryan Baker

Contents

Preface

Introduction

With the introduction of more and more oil and gas exploration sites around the nation and the world, the need for skilled pipe welders is at an all-time high. This demand for more pipe welders is expected to continue growing for decades. In addition to growth in the energy sector, there has been an expansion of pipe-welding jobs in almost every sector of industry.

Students who are preparing for a career in pipe welding will need to:

- be alert and work safely.
- know the theory and application of the various welding and cutting processes.
- read and interpret welding drawings and sketches.
- work well with tools and equipment.
- be able to resolve basic mathematical problems.
- have excellent eye–hand coordination.
- be able to follow written and verbal instructions.
- work with or without close supervision.
- work well individually and in groups.

Some pipe welding jobs may require welders to:

- work outdoors, in all types of weather.
- travel outside their home area.
- stay at a remote worksite for days or weeks at a time.
- work on elevated platforms or down in trenches and excavations.
- work in confined spaces.

Pipe Welding is a comprehensive classroom/shop textbook designed to turn a skilled plate welder into an employable pipe welder. This textbook focuses on pipe welding as a career. It covers the unique safety issues that pipe welders may encounter, in addition to pipe layout, assembly, and welding using shielded metal arc welding (SMAW), gas metal arc welded (GMAW), flux cored arc welded (FCAW), and gas tungsten arc welded (GTAW) processes. It provides a single source for students to build their basic welding skills so that they can become pipe welders. It is written for students who want to take their skill and career potential to the next level to make them more competitive in the job market and take advantage of the surge in demand for pipe welders. This textbook has

- Chapter objectives
- Key terms
- Review questions
- Practices
- Over 500 figures

It covers

- pipe-welding safety
- practices for pipe layout, assembly, and welding
- tools unique to pipe layout, assembly, and welding
- print reading

Pipe welding offers a number of unique safety concerns, such as welders being expected to work in excavations or on scaffolding. These and other issues are addressed to help make pipe welding safer.

Throughout the textbook, students are provided with opportunities to develop their skills in pipe layout, assembly, and welding through step-by-step instructions progressing from the very basics of pipe welding to advanced skills and American Welding Society (AWS) Schools Excelling through National Skills Education (SENSE) certification testing.

There are a number of specialized tools available to pipe welders that aid in layout, assembly, and welding; most of these are illustrated in this textbook. These items can increase a welder's productivity and increase his or her value to the company. Piping systems have many unique attributes as well, which are presented in mechanical drawings that welders must understand in order to fabricate piping systems properly.

Supplements

Instructor Companion Website The Instructor Companion Website, found on cengagebrain.com, includes the following components to help minimize instructor preparation time and help engage students:

- **Microsoft PowerPoint®** lecture slides, which present the highlights of each chapter.
- An **Image Gallery**, which offers a database of images from the text. These can easily be imported into the PowerPoint presentations.
- An **Answer Key** file, which provides the answers to all end-of-chapter review questions.

Cengage Learning Testing Powered by Cognero is a flexible, online system that allows you to:

- author, edit, and manage test bank content from multiple Cengage Learning solutions.
- create multiple test version in an instant.
- deliver tests from your Learning Management System (LMS), your classroom, or wherever you want.

Acknowledgments

To write and illustrate a textbook requires the assistance of many individuals, and the authors and publisher would like to thank the following for their unique contributions to this edition:

- Marilyn K. Burris, for her years of work on this text and graphics
- The American Welding Society, Inc., whose *Welding Journal* was an invaluable source for many of the special-interest articles
- Dewayne Roy, Welding Department Chairman at Mountain View College, Dallas, Texas, for his many contributions to this text
- The Harris Products Group and Jay Jones, for all the help they gave during the preparation of this text
- Hypertherm, for providing access to their engineering department and cutting-edge technology
- Garland Welding Supply Co., Inc., for the loan of material and supplies for photo shoots
- Atmos Energy, its staff, pipe welders, and inspectors, for allowing me access to many of their pipeline welding projects
- The City of Garland, Texas, and Garland Power and Light, for their contributions to the text
- The following instructors, who reviewed the manuscript and gave invaluable recommendations:
    - Michael R. Allen, Pennsylvania College of Technology, Williamsport, PA
    - Ashley Black and Kevin Gratton, Lexington Technology Center, SC
    - Leonard E. Haddox, BS, MS, MA, welding educator
    - Scott Laslo, MS, Columbus State Community College, Columbus, OH
    - Don Murrell, GCHS/GCCC AWS SENSE Program, Garden City, KS
    - Tyler D. Unruh, Lincoln College of Technology, Denver, CO
- Sam Burris, for his expertise in computer graphics that helped make the illustrations and photographs dynamic
- David DuBois, for the use of his welding shop for many of the photo shootings; and for both David and Amy DuBois, for their editorial assistance in preparing the text
- Bryan Baker would specifically like to acknowledge those who have helped him grow in his knowledge and skill as a welder and teacher: Don Mugg, Larry Jeffus, Dewayne Roy, D. A. Smith, Jay Jones, James King, Kyle Emmons, and Pat N. McLeod, PhD.
- Josh Jeffus for the cover photo of him working on the North Slope of Alaska.

About the Author

Larry Jeffus

Larry Jeffus is a welder with over 55 years of welding experience, and he has his own well-equipped welding shop.

In his welding career, he has passed many welding certification tests in a wide variety of processes, positions, and on many different material types and thicknesses. Larry has provided welding and professional consulting services locally, nationally, and internationally to major corporations, small businesses, government agencies, schools, colleges, and individuals. He is a Life Member of the American Welding Society (AWS).

Larry Jeffus has over 40 years of experience as a dedicated classroom teacher and is the author of several Delmar Cengage Learning welding publications. Prior to retiring from teaching, Professor Jeffus taught at Eastfield College, part of the Dallas County Community College District. Since retiring from full-time teaching, he remains very active in the welding community, especially in the field of education. He serves on several welding program technical advisory committees and has visited high schools, colleges, and technical campuses in more than 40 states and four foreign countries. Professor Jeffus was selected as Outstanding Post-Secondary Technical Educator in the State of Texas by the Texas Technical Society. He has served for 12 years as a board member on the Texas Workforce Investment Council in the Texas governor's office, where he works to develop a skilled workforce and bring economic development to the state. He served as a member of the Apprenticeship Project Leadership Team, where he helped establish apprenticeship training programs for Texas, and he has made numerous trips to Washington, D.C., lobbying for vocational and technical education. Larry Jeffus holds a BS degree and has completed postgraduate studies.

Bryan Baker

Bryan Baker became interested in welding while taking metal shop classes in high school. This sparked a desire in him to learn as much about welding as possible. Upon completing high school, he enrolled in Eastfield College and completed an Associate of Applied Science degree in welding technology. He has worked in various industries, including aerospace, manufacturing, communications, food, and construction. He has worked in both large and small private businesses and has had his own welding business. While working in industry, he started teaching as an adjunct instructor at Eastfield College and Mountain View College. His experiences in teaching encouraged him to continue his education and complete both bachelor's and master's degrees from the University of North Texas.

Bryan has been involved in developing welding programs at Northeast Texas Community College and is currently the department chair of welding/industrial trades at Tyler Junior College. In addition, he has served on the board of the East Texas Section of AWS for many years.

Chapter 1

Introduction to Pipe Welding

OBJECTIVES

After completing this chapter, the student should be able to:

- Describe the history of piping systems.
- List the four different major categories of piping systems.
- Explain the various specifications that may be used to specify pipe and tubing.
- Discuss the difference between the three basic groupings for pipe welds.
- Explain the system of pipe schedules.
- List the three general areas that pipe welding jobs may be divided into.
- Discuss the importance of teamwork for pipe welders.

KEY TERMS

Collection piping systems

Construction pipe-welding jobs

Distribution piping systems

High-pressure piping systems

High-strength service pipes

Light-duty service pipes

Low-pressure piping systems

Manufacturing pipe-welding jobs

Medium-pressure piping systems

Medium structural service pipes

Pipeline-welding jobs

Process piping systems

Seamed pipe

Seamless pipe

Transportation piping systems

INTRODUCTION

The earliest piping systems were built by the Greeks approximately 5,000 years ago to provide water to the residences on the island of Crete, **Figure 1-1**. Later, as the Roman Empire began to develop, the Romans expanded the Greek clay piping system of freshwater to include a separate system for drainpipes throughout the city. These early Roman clay piping systems were gradually replaced with lead pipe. The artisans of the day developed a new process to weld sections of lead pipe together. Over time, much of the ancient city of Rome was plumbed for water through lead pipes. Fortunately for the Romans, their water had a high mineral content that quickly coated the inside of the lead pipes, preventing lead poisoning. Small-diameter lead pipes were used throughout Europe and in many of the older cities in the Americas, including New York

(Continued)

City, Boston, and Montreal. Water mains were initially constructed by hollowing out logs. Later, large water mains were made much like very long wooden barrels. Workers used narrow wood strips held together with metal bands tightly wrapped around the outside, **Figure 1-2**. Over time, the lead pipes were removed; however, a few of the early wooden water mains that were constructed are still in service.

These days, there are millions of miles (kilometers) of pipe used in a wide range of applications throughout the world. The ancient world was limited to only a few materials such as clay, lead, and wood; but today, pipe can be made from many popular materials, such as cast iron, steel, stainless steel, copper, brass, plastic, glass, fiberglass, and concrete. In this textbook, the major concentration will be on fitting and welding steel pipe because the skills learned with steel pipe can be transferred to stainless steel, aluminum, and other metal piping materials.

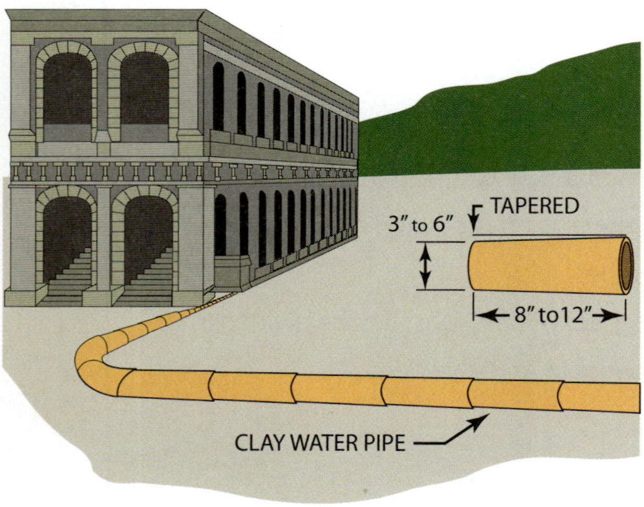

FIGURE 1-1 Clay water pipe systems for fresh water were first used approximately 5,000 years ago.

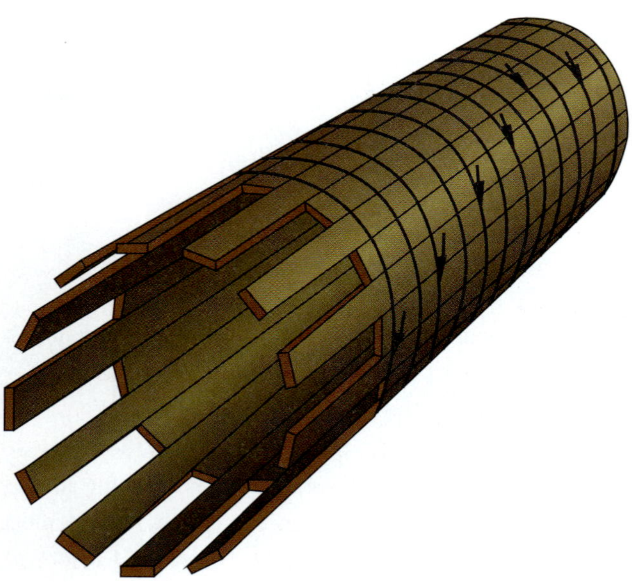

FIGURE 1-2 Some wooden water pipes constructed like long barrels in the 1700s are still in use in some early American cities.

PIPING SYSTEMS

Piping systems can be divided into four major categories: collection, transportation, distribution, and processing.

Collection piping systems have smaller pipes joining larger pipes as the material in the smaller pipes combine, thus requiring a larger pipe to handle the flow. Some examples of collecting piping systems are sewer pipes from various appliances in a home, stormwater drainage pipes in a neighborhood or city, and oilfield piping, where pipes collect product from various wells for storage or processing facilities, **Figure 1-3**.

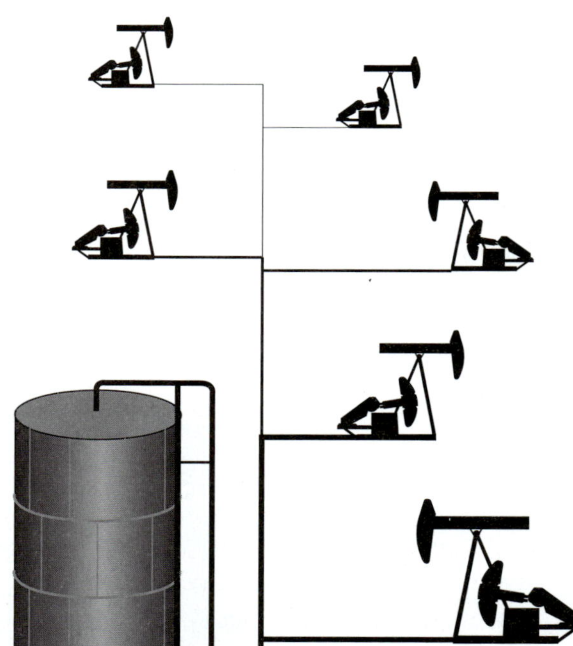

FIGURE 1-3 Collection piping systems.

Transportation piping systems are pipes that move various products throughout the community, state, or nation. Transportation pipelines are typically larger in diameter than collection pipelines and extend for hundreds of miles. They are usually buried beneath the Earth's surface and can be seen only where they surface at pumping or pressure regulator stations, **Figure 1-4**. Some examples of transportation pipelines are water pipes that are used to move raw water from reservoirs to water treatment plants; fuel lines that move gasoline, diesel fuel, and jet fuel from coastal refineries to distribution facilities inland; and the Alaskan pipeline, which moves crude oil 800 miles from Prudhoe Bay to the harbor at Valdez, **Figure 1-5**.

Larry Jeffus

FIGURE 1-4 Natural gas pig insertion and recovery station.

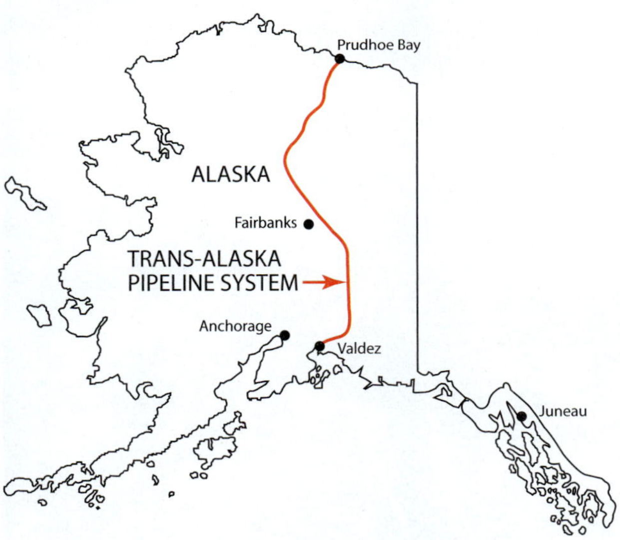

FIGURE 1-5 The trans-Alaskan pipeline went into service in 1977.

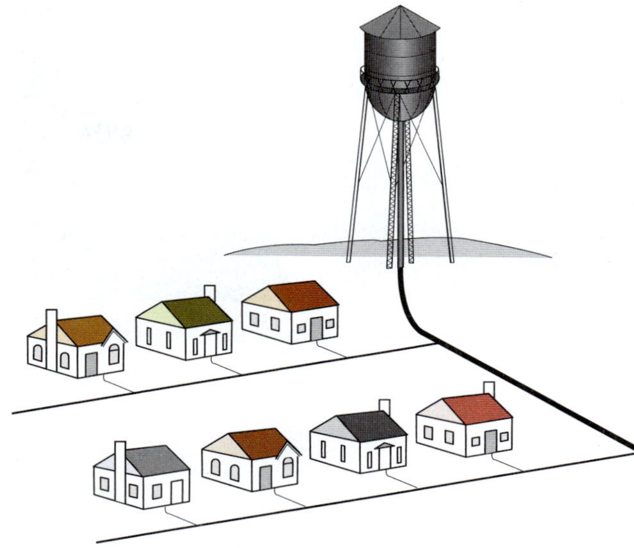

FIGURE 1-6 Distribution piping system.

Distribution piping systems have smaller individual pipes connected to larger mains. They are used to deliver products such as natural gas and water for homes and factories, compressed air for pneumatic tools, shielding gases for welding, and chilled water for air conditioning, **Figure 1-6**. For many years, a brewery in Germany moved beer directly from its plant to a nearby stadium through a distribution piping system.

Process piping systems can be very diverse in size and materials. The diameter of some pipes may be as small as a fraction of an inch (in.) or a few millimeters (mm) all the way through several feet (ft) or meters (m) in diameter. The smaller pipes may be used to supply a chemical agent or other such material from a storage tank to a processing vat. The larger pipes are used to move products through a manufacturing facility or refinery. Examples of processing piping systems can be found in oil refineries, paint manufacturers, cosmetic manufacturers, and many food processing plants, **Figure 1-7**.

Larry Jeffus

FIGURE 1-7 Process piping system.

PIPE APPLICATIONS

Tubing is most often used to build items such as bicycle, motorcycle, race-car, and aircraft frames, handrails, and playground equipment; however, pipe is used for many of these applications, too. Both are used for structural applications, ornamental, or decorative art. Examples of structural applications are a walkway between a hospital and parking garage in Columbia, South Carolina, (**Figure 1-8**), and the framework for pyramid buildings at Moody Gardens in Galveston, TX, **Figure 1-9**. Examples of ornamental applications of pipe include the farm post shown in **Figure 1-10**. Examples of decorative art include the windmill shown in **Figure 1-11**.

Larry Jeffus

FIGURE 1-8 A pedestrian crosswalk over a major city street makes downtown safer.

Larry Jeffus

FIGURE 1-9 Modern pyramids built on frameworks of pipes and tubes.

Larry Jeffus

FIGURE 1-10 Rain gauge/hummingbird feeder pipe garden post.

Larry Jeffus

FIGURE 1-11 A windmill built out of ½-in. (13-mm) electrical mechanical tubing.

HOW PIPE IS MADE

Pipe can be manufactured as seamed or seamless. **Seamed pipe** has a longitudinal weld along its full length. It is formed from flat plates that are first heated then passed through rollers that bend it into the pipe shape. Resistance seam welding machines use rollers to pass the welding current through the pipe to fuse the edges of the plate together, forming the longitudinal weld, **Figure 1-12**. A cutter removes the small amount of weld flash from the outside of the weld. The small weld flash on the inside of the pipe may or may not be removed. The pipe is then cut into sections.

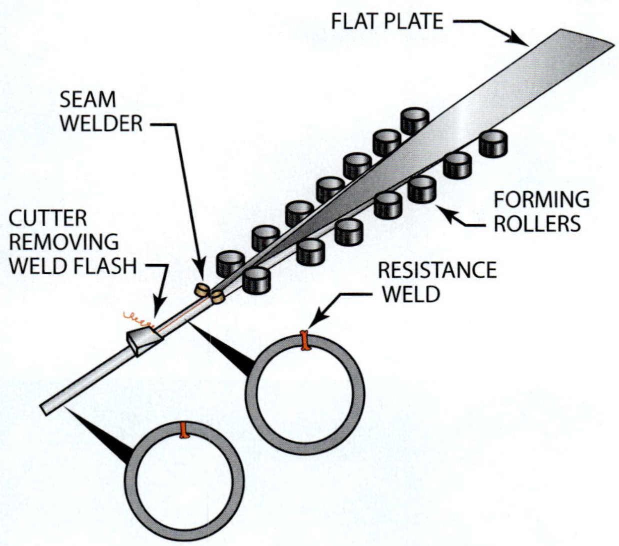

FIGURE 1-12 Seamed pipe manufacturing technique.

Seamless pipe is formed so there is no need to make a longitudinal weld. It is made by forcing heated metal through a forming die that has internal mandrel to form the center opening in the pipe, **Figure 1-13**. Both seamless and seamed pipe sections are thoroughly inspected following manufacturing to ensure that they are sound. Both types of pipes are incredibly strong, and they can provide years of reliable service.

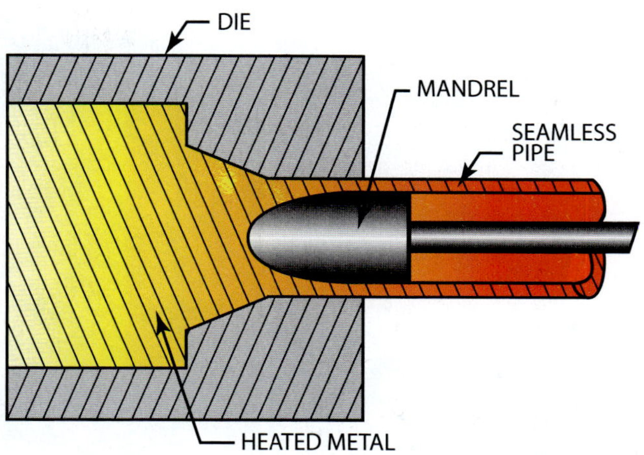

FIGURE 1-13 Seamless pipe manufacturing technique.

Spiral pipe is formed of long coils of hot rolled steel that are wound into a long spiral much like the cardboard tube in the center of a roll of paper towels, **Figure 1-14**. As the metal coil is unwound, its edges are trimmed and beveled so that as it is formed into a round cylinder, the two edges form a double V-groove. The V-groove is first welded on the inside and then on the outside using a process known as *submerged arc welding (SAW)*. Once the pipe is welded, it is cleaned of all flux, and the welds are thoroughly inspected before it is heat treated and coated, **Figure 1-15**.

NOTE

Because pipe and round tubing are very similar in their appearance and applications, the term *pipe* will be used throughout this text, as opposed to *pipe and tubing*. The term *tubing* will be used only when tubing is specifically being addressed. All of the fit-up welding inspections that apply to pipe can also be applied to most tubing welds unless otherwise stated.

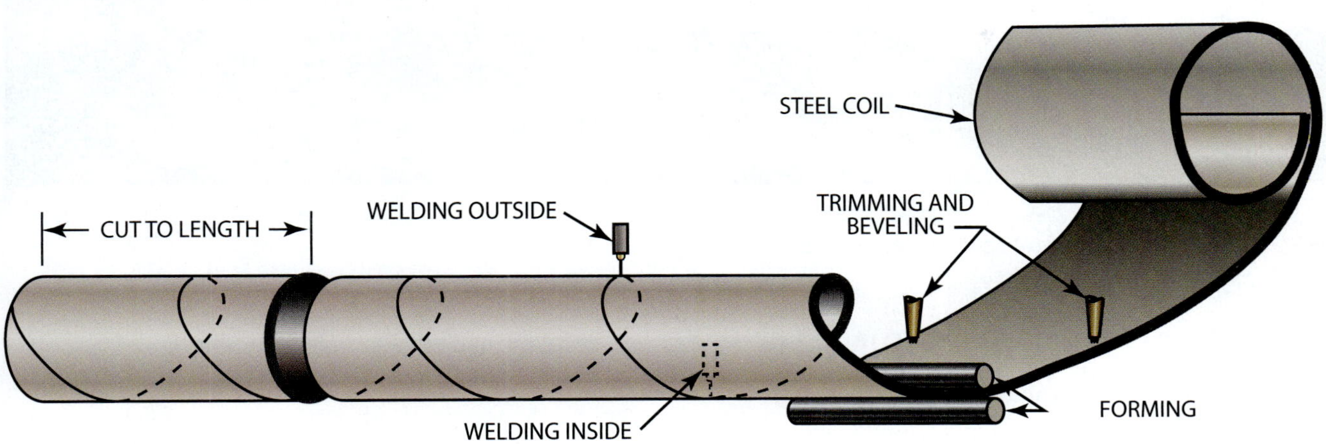

FIGURE 1-14 Spiral pipe manufacturing technique.

(A)

(B)

FIGURE 1-15 (A) 8-ft (2.4-m) spiral water pipe; (B) 6-in (150-mm) spiral pipe.

PIPE VERSUS ROUND TUBING

Pipe and round tubing are not the same, although they can appear the same and may be used for similar applications. Pipe is primarily used to move both liquids like water, gasoline, oil, and other petroleum products; and gases like air, steam, and engine exhaust, **Figure 1-16**. In addition, tens of thousands of other products are safely and efficiently moved through millions of miles of pipe every day.

FIGURE 1-16 Formula 1 racecar exhaust manifold constructed from stainless steel tubing.

Round tubing is used for both structural applications and to move fluids. Tubing is the structural component of many common products such as classroom and office furniture, highway guardrails, playground equipment, and roller coasters.

Pipe and round tubing are different in the way they are specified. All or part of the following pipe and round tubing specifications may be included on a bill of materials or material specification sheet. For example, the weight of pipe or round tubing, usually expressed as pounds per foot (lb/ft) or kilograms per meter (kg/m) of pipe, may be written in the specifications, but it is not often included since it can be found on most pipe charts.

Pipe Standards

One of the major differences between pipe and tubing is how the material is specified. When the American Standards Association, now the American National Standards Institute (ANSI), first standardized the dimensions for steel and iron pipe and tubing, there were few applications for pipe, so only a few different pressure ranges were available. The various pressure ranges determined the thickness of the wall of the pipe, **Figure 1-17**. The original standards only included standard, extra strong, and double extra strong. Today, standard (STD) pipe is often schedule 40, extra strong (XS) pipe is often schedule 80, and double extra strong (XXS) pipe does not have a schedule number. **Table 1-1** shows the various pipe schedules and wall thicknesses for a variety of pipe diameters that are available today. Changes in the marketplace have resulted in a significant increase in the various pressure ranges and wall thicknesses of pipe. Some of the things that have affected the wall thicknesses of pipe are changes in material and application. Since stainless steel pipe, which was first introduced in the 1920s, is much stronger and less likely to have corrosion reduce the thickness of the pipe over time, this allows a much thinner wall section for the same schedule. In addition, thinner-walled sections of pipe were introduced for applications other than critical, such as drainage, ornamental, or decorative work. One drawback to the thinner sections is that they cannot be threaded because of the thinner walls.

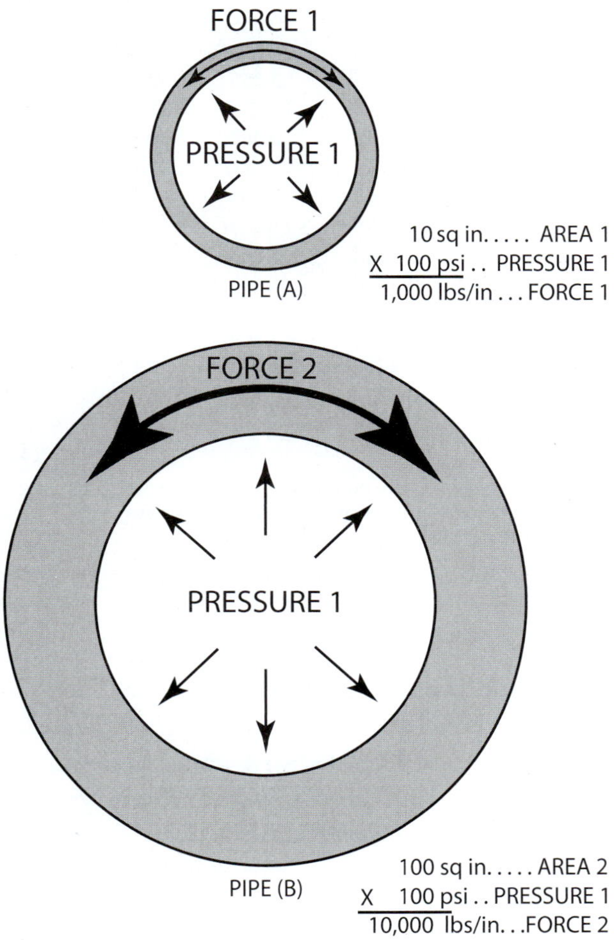

FIGURE 1-17 Diameter and wall thickness versus pressure.

Pipe Specifications Typical pipe specifications include:

- **Length**—This is the overall length of a section or piece of pipe. Common lengths for steel pipe are 10, 20, and 21 ft (3, 6, and 6.5 m).
- **Diameter**—The inside diameter (ID) dimension is given for pipe that is smaller than 12 in. (300 mm) in diameter, and the outside diameter (OD) dimension is given for pipe that is 12 in. (300 mm) or larger. The actual diameter of pipe may vary slightly because of many factors; therefore, the term *nominal pipe size (NPS)* is used for standard pipe, and *Diamétre Nominal (DN)* is used for metric pipe.

			Schedule (SCH) and Wall Thickness decimal inches (mm)							
NPS*	DN**	OD***	SCH 5	SCH 10	SCH 30	SCH 40 STD	SCH 80 XS	SCH 120	SCH 160	XXS
1/8	6	0.405 (10.29)	0.035 (0.889)	0.049 (1.245)	0.057 (1.448)	0.068 (1.727)	0.095 (2.413)	—	—	—
1/4	8	0.540 (13.72)	0.049 (1.245)	0.065 (1.651)	0.073 (1.854)	0.088 (2.235)	0.119 (3.023)	—	—	—
3/8	10	0.675 (17.15)	0.049 (1.245)	0.065 (1.651)	0.073 (1.854)	0.091 (2.311)	0.126 (3.200)	—	—	—
1/2	15	0.840 (21.34)	0.065 (1.651)	0.083 (2.108)	0.095 (2.413)	0.109 (2.769)	0.147 (3.734)	—	0.188 (4.775)	0.294 (7.468)
3/4	20	1.050 (26.67)	0.065 (1.651)	0.083 (2.108)	0.095 (2.413)	0.113 (2.870)	0.154 (3.912)	—	0.219 (5.563)	0.308 (7.823)
1	25	1.315 (33.40)	0.065 (1.651)	0.109 (2.769)	0.114 (2.896)	0.133 (3.378)	0.179 (4.547)	—	0.250 (6.350)	0.358 (9.093)
1 1/4	32	1.660 (42.16)	0.065 (1.651)	0.109 (2.769)	0.117 (2.972)	0.140 (3.556)	0.191 (4.851)	—	0.250 (6.350)	0.382 (9.703)
1 1/2	40	1.900 (48.26)	0.065 (1.651)	0.109 (2.769)	0.125 (3.175)	0.145 (3.683)	0.200 (5.080)	—	0.281 (7.137)	0.400 (10.160)
2	50	2.375 (60.33)	0.065 (1.651)	0.109 (2.769)	0.125 (3.175)	0.154 (3.912)	0.218 (5.537)	0.250 (6.350)	0.343 (8.712)	0.436 (11.074)
2 1/2	65	2.875 (73.02)	0.083 (2.108)	0.120 (3.048)	0.188 (4.775)	0.203 (5.156)	0.276 (7.010)	0.300 (7.620)	0.375 (9.525)	0.552 (14.021)
3	80	3.500 (88.90)	0.083 (2.108)	0.120 (3.048)	0.188 (4.775)	0.216 (5.486)	0.300 (7.620)	0.350 (8.890)	0.438 (11.125)	0.600 (15.240)
3 1/2	90	4.000 (101.60)	0.083 (2.108)	0.120 (3.048)	0.188 (4.775)	0.226 (5.740)	0.318 (8.077)	—	—	0.636 (16.154)

*NPS - Nominal pipe size (given in inches)
**DN - Diamètre nominal/nominal diameter (given in millimeters)
***OD - Outside diameter

TABLE 1-1 Pipe Schedule and Wall Thickness

- **Wall thickness**—Pipe wall thickness was originally standardized back in 1939. The early system used a system of schedules that were related to a pressure range. The basis of that original system grouped pipe into three pressure ranges—STD, XS, and XXS. The current standard for pipe wall thickness is the National Pipe Standard, Table 1-1. Note on the chart that the current system still has some relationship to the earlier STD, XS, and XXS terminology, but it is important to note that now there is not a pressure range relationship (i.e., a Schedule 40 STD steel pipe has a higher pressure range than a Schedule 40 PVC pipe would have).
- **Material**—The American Society of Mechanical Engineers (ASME) has a number of specifications that cover a wide range of materials and methods of manufacturing pipe. The ASME specifications range from AS-135, for electric resistant welded steel pipe, to AS-378, for seamless austenitic high-temperature pipe.
- **Fittings**—The ends of pipe can be joined by different methods such as butt welded, socket welded, or threaded.

Pipe Dimensioning

The ASME standard for pipe scheduling lists the OD of the pipe, and the wall thickness can vary based on the schedule; therefore, the ID of a 6-in. (152-mm) Schedule 40 pipe is slightly larger than the ID of a 6-inch (152-mm) Schedule 80 pipe. This system is used on pipe from 1/8 in up to 12 in. (3–305 mm). Above 12 in (305 mm), the OD of the pipe is given, and the wall thickness changes the ID of the pipe, **Figure 1-18**.

Round Tubing Specifications Generally, tubing has a more rigid set of specifications as compared to most pipe. Tubing specifications may include:

- **Material**—Tubing is available in a number of different types of metal, as well as other materials such as plastic, rubber, silicon, fiberglass, composite, and glass. ASTM International (formally the American Society of Testing and Materials) has established standard specifications that cover most types of metal tubing.
- **Diameter**—The diameter for tubing is always expressed as the OD, **Figure 1-19**.

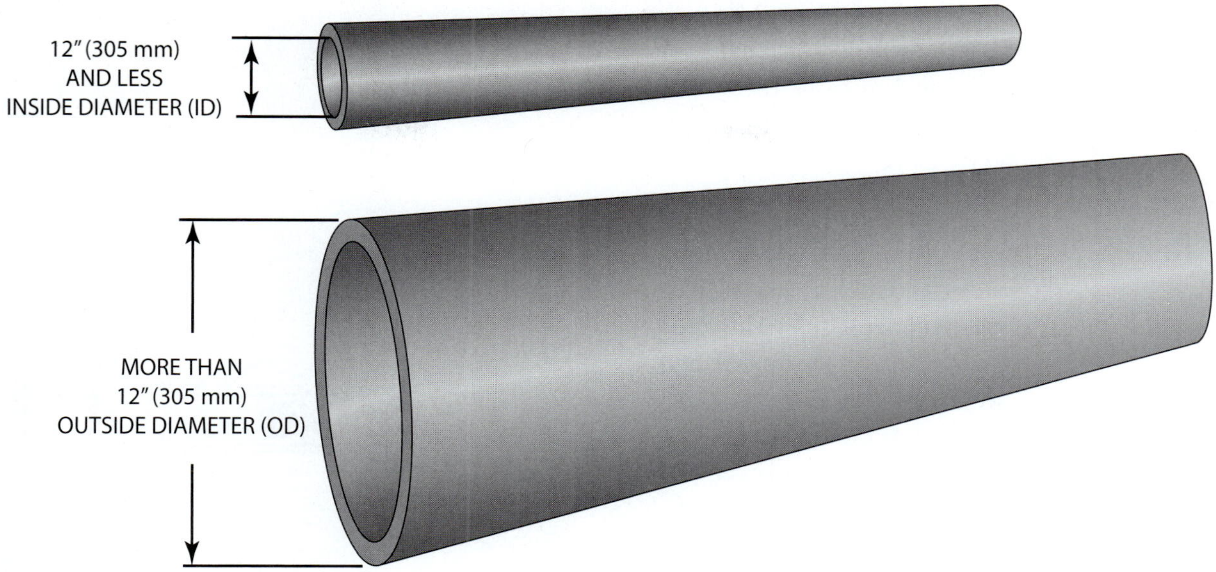

FIGURE 1-18 ID versus OD for pipe diameters.

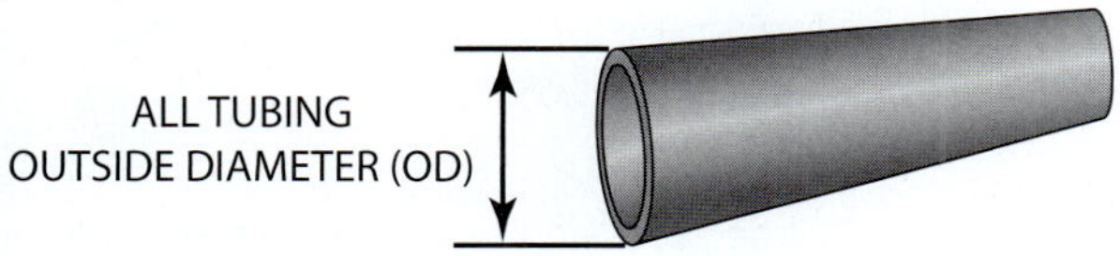

FIGURE 1-19 Tubing diameter measurement.

- **Wall thickness**—The wall thickness of rigid round tubing may be given as a gauge thickness or as a specific dimension.
- **Shape**—The shape of tubing must be specified because it is available in shapes other than round, such as rectangular, square, oval, and a wide range of specialty shapes.
- **Length**—Rigid round tubing can be purchased in lengths ranging from 5–24 ft (1.5–7 m), and flexible round tubing can be purchased in coils ranging from a few feet to several hundred feet. Common lengths for flexible tubing are coils of 25 and 50 ft (7.5 and 15 m).
- **Pressure range**—Because some types of tubing are designed to carry pressurized fluids, a pressure range for these tubes would be included in their specifications.
- **Temperature range**—The operating temperature range or the minimum and/or maximum service temperature are important because both high and low temperatures will affect the performance of tubes.
- **Bending radius**—The minimum-bending radius of a tube can be a major determination in the selection since tubing is often bent to a particular shape before welding.
- **Strength**—Tubing that is designed to be used in structural applications will include the tensile strength, bending strength, compression strength, and other characteristics as needed for particular applications.
- **Finish**—Tubing is available in a variety of surface finishes, including cleaned, galvanized, polished, and coated.
- **Temper**—Annealed tubing is soft and more easily bent or shaped than half-hardened or fully hardened tubing. Half-hardened and fully hardened tubing is often referred to as *rigid tubing*.
- **Application**—The intended application of tubing will affect the available or approved types in some cases. Aerospace and aircraft tubing must have certification paperwork that must be kept with the tubing through the fabrication process all the way to the final customer. Some of the applications that can be specified for tubing include general-purpose, automotive, chemical, cryogenic, food and beverage processing, medical, and oil and fuel, among others.

Tubing Dimensions

Round tubing is specified by the OD and wall thickness, so a 6-in. (152-mm) diameter tube with a quarter-inch wall thickness could possibly fit inside of a 6-in. (152-mm) Schedule 40 pipe. It is possible, then, to have a 14-in.

(355-mm)-diameter pipe and 14-in. (355-mm)-diameter tube that would have the same wall thickness if the tubing wall thickness was specified to be the same as the pipe.

PRESSURE AND STRENGTH RANGES

In this textbook, pipe welded joints are grouped into three general categories based on their strength requirements for both piping systems and structural applications. These groupings are loosely based on the level of critical service that the pipe is performing. Basically, the groups relate to the degree or extent of injury, property damage, or both that might happen if a catastrophic failure of the welded joint were to occur. The three general categories are:

- Low-pressure or light-duty service
- Medium-pressure or medium structural service
- High-pressure or heavy structural service

Low-pressure piping systems may be used for collector systems such as wastewater systems, agricultural irrigation systems, and building sprinkler systems. **Light-duty service pipes** may be used for bicycle stands, agricultural fences and gates, and art sculptures. If a catastrophic failure occurred in a low-pressure piping system or light structural service application, there might be some property damage, but little chance of injuries.

Medium-pressure piping systems may be used for water supplies, compressed air, and residential gas distribution. **Medium structural service pipes** may be used for signposts, railroad crossing signals, **Figure 1-20**, and truck brush guards. If a catastrophic failure occurred in a medium-pressure system or medium structural service application, there would be some property damage, and there could be some minor injuries.

High-pressure piping systems may be used for cross-country transmission of oil, gasoline, or natural gas, oil refinery, and high-pressure steam pipes in power plants. **High-strength service pipes** may be used for aircraft engine mounts, building frameworks, and tank support legs. A catastrophic failure in high-pressure systems and high-strength structures will result in significant property damage and can result in serious injury or possible death.

So the welding requirements for each of the systems varies from a low-pressure system with a visual inspection, medium pressure with a little greater scrutiny, and a high-pressure system that would be thoroughly tested, **Figure 1-21**.

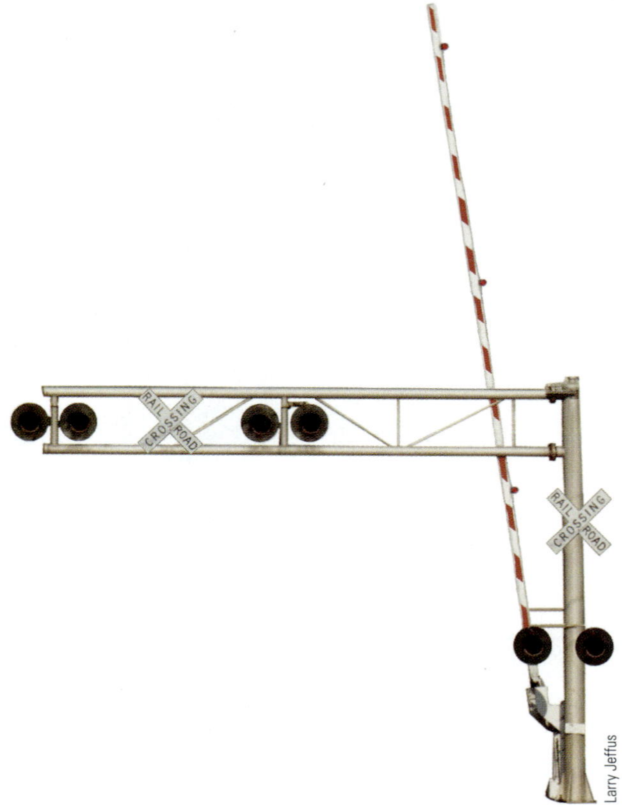

FIGURE 1-20 Railroad crossing signals are an example of medium structural service pipes.

FIGURE 1-21 X-ray testing of pipe welds.

PIPE SYSTEMS CODES AND STANDARDS

The American Petroleum Institute (API) 1104 code is one of the most widely used piping welding standards in the industry. It is used for both structural and piping applications. The American Welding Society D1.1 structural welding code has specifications for structural pipe applications. These applications can vary from a system that is primarily ornamental, such as a piece of art or yard decoration, to something that is more critical, such as a handrail or guardrail, to an application that would be critical if it failed, such as the crosswalk or structural building components.

PIPE WELDING PROCESSES

All of the major welding processes can be used to weld pipe. The most important factor to consider when selecting a welding process is the applicable codes and standards. But within these codes and standards, there are opportunities to make choices between different welding processes. Some of the other factors that should be considered when selecting the welding process include the pipe diameter and schedule, the pipe material, whether the welding will be done in a shop or in the field, and the number of welds that will be required.

Steel pipe that is an inch or less in diameter can be oxyacetylene welded (OAW), gas tungsten arc welded (GTAW), gas metal arc welded (GMAW), or flux core arc welded (FCAW). But with these small-diameter pipes, shielded metal arc welding (SMAW) can be a difficult process. Larger-diameter pipe and thick-walled pipe cannot be oxyacetylene welded OAW efficiently, but almost any other welding process can be used.

Stainless steel pipe can be welded with GTAW, GMAW, and FCAW. Aluminum pipe is usually welded with GTAW.

This textbook will cover SMAW, GTAW, GMAW, and FCAW, which are the most commonly used welding processes for pipe welding.

TYPES OF PIPE WELDING JOBS

Pipe welding jobs can be divided into three general areas—manufacturing, construction, and pipeline. Each of these general job groupings has its own skills, opportunities, and requirements.

Manufacturing Pipe Welding Jobs Examples of these pipe welding jobs include welding many common products, ranging from motorcycle frames to booms for cranes and many other items, **Table 1-2**. When the size and shape of parts being manufactured allow the use of weld positioners, the welds can be made in the rotated flat position (1G). Because the pipe welder does not have to transition between welding positions, 1G pipe welding is considered to be easier than the other welding positions. Many manufacturing companies have climate-controlled work areas, so welding in the summer is not too hot and welding in the winter is not too cold.

For many welders, manufacturing jobs offer the convenience of not having to travel to different job sites. These jobs also offer the stability of working with the same people. Because production requirements may be high, the welders may be required to weld for longer periods of time than welders working in other types of jobs.

Construction Pipe Welding Jobs Examples of these welding jobs include constructing oil refineries, chemical plants, boiler room piping, and many other structures, **Table 1-3**. The size and complexity of construction projects can vary greatly. Some nuclear plant construction projects may last for years, while a pumping station job may take only a few weeks. Construction at a power plant

Pipe Welded Items	
Welding booth tables and coupon stands	Handrails
Racecar frames	Utility trailers
Rails for stage lighting	Sculptures
Boilers	Automotive engine hoist
Food processing equipment	Earth-moving machine
A/C chillers	Dental office equipment
Cooling towers	Exercise equipment
Pickup truck bumpers and racks	Aircraft frames
Gas manifolds for medical and welding gases	Commercial kitchen carts and table frames
Furniture	Farm equipment
Highway signpost	Trailer axles
Railroad crossing guards	Cattle squeeze shoot
Commercial swimming pool equipment	Oil drilling equipment
Gas pumps	Equipment and machines for industries
Hospital equipment	Plastic injection machines
Picnic table frame	Micro beer brewing equipment
Outdoor furniture frames	Oil drilling equipment
Wastewater treatment equipment	Pumping stations

TABLE 1-2 Pipe Items That Are Manufactured with Pipe

Construction Projects	
Coal power plants	Pumping stations
Natural gas power plants	Breweries
Bridges	Distilleries
Nuclear power plants	Bottling plants
Amusement rides	Refrigeration plants
Cooling towers	Fire sprinkler systems
Ships	Water supply systems
Water treatment plants	Wineries

TABLE 1-3 Construction Projects Made with Pipe

or refinery will require a great deal of skill to make quality welds on various types and sizes of pipe, as well as requiring welders to weld in all positions while, in some cases, working around obstructions. In addition to welding skills, welders often must be comfortable working high above the ground.

Almost all construction jobs require traveling, and some may stipulate that welders relocate to a different city or state. There may be little protection from the rain, heat, cold, wind, snow, or other weather conditions. From the first day on the job, construction welders are working themselves out of a job—when the building is finished, they have to look for another job.

Pipeline Welding Jobs Examples of these jobs include all types of cross-country pipelines, **Table 1-4**. Pipelines can range in length from a mile or less to hundreds of miles. They may also range in diameter from an inch or two to several feet in diameter. In most cases, the pipe is welded above grade, but welders may be required to work in open trenches. Much like construction welders, pipeline welders may be required to travel or relocate and work in all types of weather.

Types of Pipelines	
Freshwater pipelines	Gasoline pipelines
Sewer pipelines	Irrigation water pipelines
Natural gas pipelines	Dredge pipelines
Crude oil pipelines	

TABLE 1-4 Pipelines

TEAMWORK

There is a lot of work that needs to be done to make a weld on large-diameter pipe. Often there is more work than a single welder can do in a timely manner to meet the scheduled deadline, so most pipeline welding jobs and many pipe construction welding projects are done by a crew, **Figure 1-22**. The crews are made up of welders and helpers, all of whom take directions from the site supervisor. On large jobs where several welding crews are working, the supervisor may be the project engineer. On smaller crews, the pipe welder is the person in charge of the welding crew.

Larry Jeffus

FIGURE 1-22 It takes teamwork to move large pipe sections into place.

Each member of the crew has specific job responsibilities. The welder acts as the leader and directs the other crew members. The welder helper assists with the preparation and fit-up of the pipe joint. They are also responsible for wire brushing and grinding the weld between weld passes and weld crater cleanup (if needed) before a new electrode is started. The welder helper may also be responsible for making small adjustments in the welding amperage settings as requested by the welder, **Figure 1-23**.

Larry Jeffus

FIGURE 1-23 Welder helpers make a valuable contribution to fashioning quality pipe welds.

A crew member may serve as the fire watch. The fire watch's primary responsibilities are to make sure that the site is safe so no accidental fires get started, alert the welder and sound the alarm if a fire does occur, and be prepared to use the available fire extinguisher to put out the fire. When all of the combustible materials cannot be removed from the welding area, this person may wet them down or cover them up, **Figure 1-24**.

Larry Jeffus

FIGURE 1-24 Plywood used to protect dry grass from the sparks given off from the cutting torch.

Other crew members may operate equipment to lift and place pipe joints and pipe sections in place, as well as performing other tasks as required.

Leadership

A team leader must have knowledge of the Equal Employment Opportunity Laws regarding job discrimination and make sure that these laws are followed. A good team leader should have a positive attitude, think and plan ahead, treat everyone with respect, show confidence in crew members' abilities, and most important, communicate clearly to each member what is expected of him or her. It is important to recognize the various strengths and weaknesses of each crew member so that that person can be used in the most effective manner to get the job done. When necessary, leaders should coach crew members in how they want them to do their jobs. Often, demonstrating the proper way to do something can be helpful.

A crew that works well together will be much more productive than one that is dysfunctional. As crew leader, you need to be able to help resolve conflicts between team members. Identify the key issues that might be causing the conflict, and get the crew to agree on how the problem can be resolved. A work crew's performance is usually evaluated on the productiveness of the team, so hard work and cooperation are rewarded.

METRIC UNITS

Both standard and metric (SI) units are given in this text. The SI units are in parentheses ( ) following the standard unit. When nonspecific values are used—for example, "set the gauge at 2 psig," where 2 is an approximate value—the SI units have been rounded to the nearest whole number. Round-off occurs in these cases to agree with the standard value and because whole numbers are easier to work with. SI units are not rounded off only when the standard unit is an exact measurement.

Often students have difficulty understanding metric units because exact conversions are used even when the standard measurement was an approximation. Rounding off the metric units makes understanding the metric system much easier, **Table 1-5**. By using this approximation method, you can make most standard-to-metric conversions in your head, without needing to use a calculator.

Standard	Metric
1/4 in.	= 6 mm
1/2 in.	= 13 mm
3/4 in.	= 18 mm
1 in.	= 25 mm
2 in.	= 50 mm
1/2 gal	= 2 L
1 gal	= 4 L
1 lb	= 1/2 K
2 lb	= 1 K
1 psig	= 7 kPa
1°F	= 2°C

TABLE 1-5 Approximate Standard to Metric Conversions

Once you have learned to use approximations for metric, you will find it easier to make exact conversions whenever necessary. Conversions must be exact in the shop when a part is dimensioned with one system's units and the other system must be used to fabricate the part. For that reason, you must be able to make those conversions. **Tables 1-6** and **1-7** are set up to be used with or without the aid of a calculator. Many calculators today have built-in standard/metric conversions. It is a good idea to know how to make these conversions with and without these aids, of course. Practice making such conversions whenever the opportunity arises.

TEMPERATURE
Units

°F (each 1° change)	=	0.555°C (change)
°C (each 1°change)	=	1.8°F (change)
32°F (ice freezing)	=	0°Celsius
212°F (boiling water)	=	100°Celsius
–460°F (absolute zero)	=	0°Rankine
–273°C (absolute zero)	=	0°Kelvin

Conversions

°F to °C ______ °F – 32 = ______ × .555 = ______ °C
°C to °F ______ °C × 1.8 = ______ + 32 = ______ °F

LINEAR MEASUREMENT
Units

1 inch	=	25.4 millimeters
1 inch	=	2.54 centimeters
1 millimeter	=	0.0394 inch
1 centimeter	=	0.3937 inch
12 inches	=	1 foot
3 feet	=	1 yard
5280 feet	=	1 mile
10 millimeters	=	1 centimeter
10 centimeters	=	1 decimeter
10 decimeters	=	1 meter
1000 meters	=	1 kilometer

Conversions

in. to mm	______ in.	×	25.4	=	______	mm
in. to cm	______ in.	×	2.54	=	______	cm
ft to mm	______ ft	×	304.8	=	______	mm
ft to m	______ ft	×	0.3048	=	______	m
mm to in.	______ mm	×	0.0394	=	______	in.
cm to in.	______ cm	×	0.3937	=	______	in.
mm to ft	______ mm	×	0.00328	=	______	ft
m to ft	______ m	×	3.28	=	______	ft

AREA MEASUREMENT
Units

1 sq in.	=	0.0069 sq ft
1 sq ft	=	144 sq in.
1 sq ft	=	0.111 sq yd
1 sq yd	=	9 sq ft
1 sq in.	=	645.16 sq mm
1 sq mm	=	0.00155 sq in.
1 sq cm	=	100 sq mm
1 sq m	=	1000 sq cm

Conversions

sq in. to sq mm ______ sq in. × 645.16 = ______ sq mm
sq mm to sq in. ______ sq mm × 0.00155 = ______ sq in.

VOLUME MEASUREMENT
Units

1 cu in.	=	0.000578 cu ft
1 cu ft	=	1728 cu in.
1 cu ft	=	0.03704 cu yd
1 cu ft	=	28.32 L
1 cu ft	=	7.48 gal (U.S.)
1 gal (U.S.)	=	3.737 L
1 cu yd	=	27 cu ft
1 gal	=	0.1336 cu ft
1 cu in.	=	16.39 cu cm
1 L	=	1000 cu cm
1 L	=	61.02 cu in.
1 L	=	0.03531 cu ft
1 L	=	0.2642 gal (U.S.)
1 cu yd	=	0.769 cu m
1 cu m	=	1.3 cu yd

Conversions

cu in. to L	______ cu in.	×	0.01638	=	______	L
L to cu in.	______ L	×	61.02	=	______	cu in.
cu ft to L	______ cu ft	×	28.32	=	______	L
L to cu ft	______ L	×	0.03531	=	______	cu ft
L to gal	______ L	×	0.2642	=	______	gal
gal to L	______ gal	×	3.737	=	______	L

WEIGHT (MASS) MEASUREMENT
Units

1 oz	=	0.0625 lb
1 lb	=	16 oz
1 oz	=	28.35 g
1 g	=	0.03527 oz
1 lb	=	0.0005 ton
1 ton	=	2000 lb
1 oz	=	0.283 kg
1 lb	=	0.4535 kg
1 kg	=	35.27 oz
1 kg	=	2.205 lb
1 kg	=	1,000 g

Conversions

lb to kg	______ lb	×	0.4535	=	______	kg
kg to lb	______ kg	×	2.205	=	______	lb
oz to g	______ oz	×	0.03527	=	______	g
g to oz	______ g	×	28.35	=	______	oz

PRESSURE AND FORCE MEASUREMENTS
Units

1 psig	=	6.8948 kPa
1 kPa	=	0.145 psig
1 psig	=	0.000703 kg/sq mm
1 kg/sq mm	=	6894 psig
1 lb (force)	=	4.448 N
1 N (force)	=	0.2248 lb

Conversions

psig to kPa	______ psig	×	6.8948	=	______	kPa
kPa to psig	______ kPa	×	0.145	=	______	psig
lb to N	______ lb	×	4.448	=	______	N
N to lb	______ N	×	0.2248	=	______	psig

VELOCITY MEASUREMENTS
Units

1 in./sec	=	0.0833 ft/sec
1 ft/sec	=	12 in./sec
1 ft/min	=	720 in./sec
1 in./sec	=	0.4233 mm/sec
1 mm/sec	=	2.362 in./sec
1 cfm	=	0.4719 L/min
1 L/min	=	2.119 cfm

Conversions

ft/min to in./sec ____ ft/min × 720 = ____ in./sec
in./min to mm/sec ____ in./min × .4233 = ____ mm/sec
mm/sec. to in./min ____ mm/sec × 2.362 = ____ in./min
cfm to L/min ____ cfm × 0.4719 = ____ L/min
L/min to cfm ____ L/min × 2.119 = ____ cfm

TABLE 1-6 Formulas for Calculating the Exact Conversion of Standard to Metric

U.S. Customer (Standard) Units

°F	=	degrees Fahrenheit				
°R	=	degrees Rankine				
	=	degrees absolute F				
lb	=	pound				
psi	=	pounds per square inch				
	=	lb per sq in.				
psia	=	pounds per square inch absolute				
	=	psi + atmospheric pressure				
in.	=	inches	=	in.	=	"
ft	=	foot or feet	=	ft	=	'
sq in.	=	square inch	=	in.		
sq ft	=	square foot	=	ft		
cu in.	=	cubic inch	=	in.		
cu ft	=	cubic foot	=	ft		
ft-lb	=	foot-pound				
ton	=	ton of refrigeration effect				
qt	=	quart				

Metric Units (SI)

°C	=	degrees Celsius
°K	=	Kelvin
mm	=	millimeter
cm	=	centimeter
cm^2	=	centimeter squared
cm^3	=	centimeter cubed
dm	=	decimeter
dm^2	=	decimeter squared
dm^3	=	decimeter cubed
m	=	meter
m^2	=	meter squared
m^3	=	meter cubed
L	=	liter
g	=	gram
kg	=	kilogram
J	=	joule
kJ	=	kilojoule
N	=	newton
Pa	=	pascal
kPa	=	kilopascal
W	=	watt
kW	=	kilowatt
MW	=	megawatt

Miscellaneous Abbreviations

P	=	pressure	sec	=	seconds
h	=	hours	r	=	radius of circle
D	=	diameter	π	=	3.1416 (a constant used in determining the area of a circle)
A	=	area			
V	=	volume			
□	=	infinity			

TABLE 1-7 Abbreviations for Standard Units

Summary

Pipe welding offers welders a wide range of opportunities for jobs in a wide range of industries. It also can be one of the most rewarding welding fields, both personally and financially. The high-quality welds required for most pipe welding let welders truly demonstrate their welding abilities. Completing a high-quality pipe weld job can be very personally rewarding. Because of the high level of skill required for many pipe-welding jobs, these welders receive some of the highest rates of pay.

The diversity of the jobs will allow pipe welders to work in a single area for years, or allow them the freedom to travel and experience different parts of the country.

Review Questions

1. What material was used to make water pipes during the early Greek and Roman Empire?
2. List two examples of collection piping systems.
3. Which category of piping system would the Alaskan pipeline be classified in?
4. Which category of piping system would be used to supply compressed air for pneumatic tools in a factory?
5. List four examples of process piping systems.
6. What process is used to weld the longitudinal seam on seamed pipe?
7. What is used to form the center opening in seamless pipe?
8. What is pipe's primary purpose?
9. List two items that round tubing can be used to construct.
10. What was the basis for the original three pipe schedules?
11. List two common lengths that pipe may be purchased in.
12. What do the abbreviations NPS and DN stand for?
13. What does the abbreviation XXS stand for?
14. Above what amount does pipe diameter dimensioning change from OD to ID?
15. What organization has established standard specifications that cover most types of tubing?
16. Why is it important to specify shape when ordering tubing?
17. List two examples of a low-pressure piping system.

(Continued)

18. List two examples of a light-duty structural application.
19. List two examples of a medium-pressure piping system.
20. List two examples of a medium-service structural application.
21. List two examples of a high-pressure piping system.
22. List two examples of a high-strength service structural application.
23. What are the most commonly used pipe-welding processes?
24. Which type of welding job offers the convenience of not having to travel to different job sites?
25. Building piping in a boiler room is most like which type of welding job?
26. List four qualities that a good team leader should have.

Chapter 2

Welding Safety

OBJECTIVES

After completing this chapter, the student should be able to:

- Describe the types of personal protective equipment (PPE) required to keep themselves safe.
- Demonstrate the proper way of storing, handling, and disposing of hazardous and nonhazardous materials.
- Explain the hazards and precautions that must be addressed to prevent falls.
- Explain the hazards and precautions associated with working in or around excavations.
- Describe how to prevent accidental fires when working outside a shop or building.
- Demonstrate how to rig and lift a load safely.

KEY TERMS

American National Standards Institute (ANSI)

Benching

Cave-ins

Excavation

Extension ladder

Fire watch

Ground-fault circuit interrupter (GFCI)

National Institution for Occupational Safety and Health (NIOSH)

Occupational Safety and Health Administration (OSHA)

Personal protective equipment (PPE)

Planned maintenance (PM)

Powered air-purifying respirators (PAPRs)

Safety Data Sheet (SDS)

Self-contained breathing apparatus (SCBA)

Shoring

Sloping

Stepladders

Straight ladders

Supplied-air respirators (SARs)

Trench

Underwriters Laboratories (UL)

INTRODUCTION

Good welding safety practices are extremely important to every work site. By the time most welders have developed the skills required to start learning pipe welding, they have already developed the necessary safety habits. The safety material in this chapter is primarily designed as a review of the major safety issues facing all welders. It is designed to reinforce what should be a firmly based set of safe working habits by now.

Because pipe welding can offer some unique safety challenges, such as working at heights above the ground and working in excavations below the ground, those topics will be covered in depth here.

All welders should periodically review welding and job safety practices. This chapter can serve as a basis for such a review. Additional safety material can be reviewed in textbooks such as *Welding Principles and Applications* and *Safety for Welders*, as well as in the safety guides that manufacturers supply with their equipment.

If an accident does occur on a welding site, it can have consequences far beyond just the person who gets injured. Serious accidents can result in local, state, or national investigations. For example, if the federal office of **Occupational Safety and Health Administration (OSHA)** becomes involved, the job site may be closed for hours, days, weeks, months, or even permanently. While the job site is closed for an investigation, you may be out of work, without pay. If it is determined that your intentional actions contributed to the accident, you may lose your job, be fined, or worse.

BURNS

Pipe welders are continuously exposed to hot welding sparks, hot metal, the arc's light rays, and other burn hazards. They must take proper precautions to prevent these burn hazards from resulting in painful injuries. The key to preventing burns starts with proper clothing. **Figure 2-1** shows the typical clothing and **personal protection equipment (PPE)** that a pipe welder should wear for burn protection. Additional PPE, such as leather aprons, spats, and leggings, may be needed; and in extreme cases, a full fire-resistant suit might be required.

FIGURE 2-1 Personal protection equipment.

NOTE

Pipe welding crew members may not be fully aware of potential burn hazards, so a safety briefing about these should be given to them before work starts. In addition, the crew members should be advised of any actions the welder will be taking that might expose them to possibility of being burned, such as oxy-fuel cutting.

INJURIES

Everyone hopes that there will never be job-related injuries; however, accidents may occur. Trained medical personnel must treat serious injuries immediately. Minor injuries such as small scrapes, cuts, or burns may be treated with the job site first aid kit. Report all injuries to the job site supervisor, site safety official, or company representative as soon as possible. This information can be used to improve job site safety.

EYE, FACE, AND EAR PROTECTION

Eye and Face Protection

Because the possibility exists that you could be struck in the eye at any time by flying welding or grinding sparks or dirt or debris, you must wear eye protection at all times when working. Most of the time, welders wear safety glasses with side shields even under their welding hoods, **Figure 2-2**. But in extremely windy conditions, goggles may be needed. The glasses and face shield can be clear, or they may be slightly shaded. The shading will help protect your eyes from accidental arc flash burns. Welding helmets that have self-darkening lenses, flip-front lenses, or a full face shield can be used to protect your face when grinding.

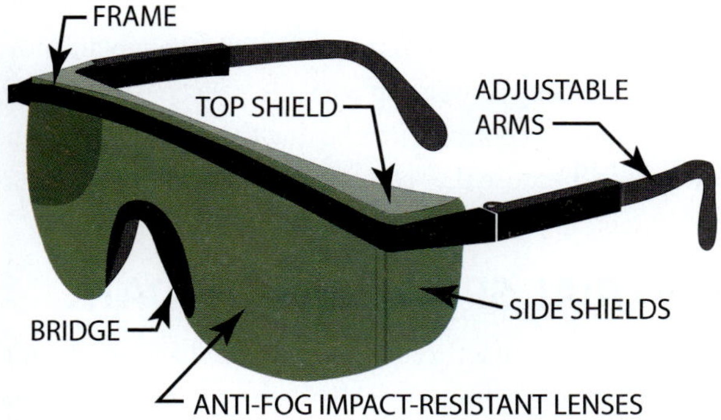

FIGURE 2-2 Safety glasses.

AUTHOR'S NOTE

Recently, I was suddenly struck by shot blast debris at a pipe welding site even though I was more than 75 feet away from the sandblaster. Fortunately, I was wearing my safety glasses.

Ear Protection

Grinding, engine generator welders, excavation equipment, air compressors, and other pieces of equipment can produce dangerous levels of noise, **Table 2-1**. To protect your hearing from possible damage caused by prolonged exposure to high levels of sound, some type of ear protection should be worn, **Table 2-2**. Earplugs fit into the ear canal. They can be disposable or reusable and work well when you need to protect yourself from high sound levels but still want to hear someone speaking to you. Earmuffs fit over the entire ear. They also provide protection to the ears from flying sparks. Some electronic earmuffs have sound canceling features that eliminate undesirable sound but still allow speech to be easily understood.

Typical Sound Pressure Levels	Sound Pressure Level (dB)
Chipping hammer	120
Large air compressor	110
Angle grinder	100
Engine generators	90
Small air compressor	80
Average factory noise	70
Conversation speech	60

TABLE 2-1 Typical Types of Welding Equipment and the Sound Levels They Can Produce

Average Daily Sound Level	Time of Noise
90	8 hours
95	4 hours
100	2 hours
105	1 hour
110	30 minutes
115	15 minutes

TABLE 2-2 Sound Levels and Time Limits of Exposure Without Proper Ear Protection

CAUTION

Damage to your hearing caused by high sound levels may not be detected until later in life, and the resulting loss in hearing is nonrecoverable. Your hearing will not improve with time, and each exposure to high levels of sound will further damage your hearing. So always wear hearing protection when you are working in an area with high sound levels.

RESPIRATORY PROTECTION

Although almost all transmission pipeline welding, and much of construction welding, is done out of doors with good ventilation, all pipe welders need to be aware of potential respiratory hazards and take the necessary precautions to protect themselves and the other crew members. When new pipes are being used, there are no additional hazards beyond other welding types. However, anytime and anywhere repair work is being performed, there can be additional hazards from product residuals left in the pipe or in the area.

CAUTION

Always refer to the Safety Data Sheet (SDS) for the product or products that the pipeline carries before any welding or cutting begins on piping. If a product is flammable, explosive, or in any way dangerous, the area must be completely cleaned before welding begins.

Residual product in the area can be a significant respiratory problem if you are making a repair on a leaky pipe. Do not assume that any residual product is safe. You must make sure that the product itself is absolutely safe. Many materials will decompose and form hazardous vapor or fume byproducts when exposed to the heat of the arc or ultraviolet and infrared arc light. The safest thing is to remove all the material from the welding area.

Always check with the company or job site safety officer before beginning any welding.

Welders are responsible for following the welding shop's established written respiratory protection program. Guidelines for the respiratory protection program are available from the OSHA office in Washington, DC.

Training must be a part of the welding shop's respiratory protection program. This training should include instruction for any or all of the following procedures:

- Proper use of respirators, including techniques for putting them on and removing them
- Schedules for cleaning, disinfecting, storing, inspecting, repairing, discarding, and performing other aspects of maintaining respiratory protection equipment
- Selection of the proper respirators for use in the workplace and any respiratory equipment limitations
- Procedures to test for tight-fitting respirators
- Proper use of respirators in both routine and reasonably foreseeable emergency situations
- Regular evaluation of the effectiveness of the program

All respiratory protection equipment used in a welding shop should be certified by the National Institution for Occupational Safety and Health (NIOSH). Some of the types of respiratory protection equipment that may be used are the following:

- Air-purifying respirators have an air-purifying filter, cartridge, or canister that removes specific air contaminants by passing ambient air through the air-purifying element.
- Atmosphere-supplying respirators supply breathing air from a source independent of the ambient atmosphere; this includes both **supplied-air respirators (SARs),** or airline respirators, which are atmosphere-supplying respirators that have air piped in through a flexible hose from a large central air supply; and **self-contained breathing apparatus (SCBA)** units, for which the breathing air source is designed to be carried by the user.
- Demand respirators are atmosphere-supplying respirators that admit breathing air to the facepiece only when negative pressure is created inside it by inhalation.
- Positive pressure respirators are respirators in which the pressure inside the respiratory inlet covering exceeds the ambient air pressure outside the respirator.
- **Powered air-purifying respirators (PAPRs)** are air-purifying respirators that use a blower to force the ambient air through air-purifying elements to the inlet covering.

VENTILATION

Even though much of pipe welding is done outside, additional ventilation may be required for some jobs. Examples of when additional ventilation can be needed are when you are welding in an excavated area or inside a structure. For the most part, portable fans can be used to provide enough air movement to ventilate the welding area adequately. Care must be taken when using portable fans so that the arc shielding is not blown away, because that can cause weld defects to occur.

MATERIAL SPECIFICATION DATA

All material manufacturers must provide to users detailed information regarding possible hazards resulting from use of their products. These **Safety Data Sheets (SDS)** must be provided to anyone using the products or anyone working in the area where the products are being used. Companies will post these sheets on a bulletin board or put them in a convenient place near the work area. Some states have right-to-know laws that require specific training of all employees who handle or work in areas with hazardous materials.

WASTE MATERIAL RECYCLING AND DISPOSAL

Proper recycling and disposal of waste materials at welding job sites are very important. This is especially true when work is being done on cross-country pipelines. Because these pipelines are being constructed in open areas, many of which are environmentally sensitive, every effort should be made to protect the environment. Many times, there are special regulations in place to protect the environment. It is your responsibility to familiarize yourself and your crew with these regulations so that the regulations can be followed.

All welding generates waste materials. Much of the waste is scrap metal. All scrap metals, including electrode stubs, can be recycled. Recycling is good for the environment, and recycling metal can be a source of revenue for the welding shop.

Recycling of other materials may not be as easy as that of metal. Many local communities have recycling programs, which may differ significantly as to which materials they will accept and how the materials need to be separated for recycling. So check with your local community before starting work in a new area.

Some of the waste generated during welding may be considered hazardous materials, such as burned flux, cleaning solvents, paints, oils, and other chemicals. Check with the material manufacturer or an environmental consultant to determine if any waste material is considered hazardous.

CAUTION

Throwing hazardous waste material into the trash, pouring it on the ground, or dumping it down the drain is illegal. Before you dispose of any welding shop waste that is considered hazardous, you must first consult local, state, and federal regulations. *Protecting our environment from pollution is everyone's responsibility.*

FALLS

Falls are the major cause of workplace injuries and death. Any time that work is being performed on overhead platforms, elevated workstations, scaffolds, or around holes in the floors and walls, appropriate fall protection must be installed and used. In addition to protecting workers from falls, it is important that materials and equipment be secured so it cannot fall onto workers below. OSHA has strict guidelines regarding fall protection, **Table 2-3**. Businesses and industries can enact stricter fall protection regulations, but they may not allow workers to violate the OSHA minimum standards.

It is important to be proactive to reduce the hazards on a work site that might contribute to a fall. Employers are required to provide a workplace that is free from any known hazards. Floors must be clear of obstructions, such as building materials and scrap, and also kept dry when possible. Some of the things that must be done to prevent falls include:

Fall Protection Requirements	
Workplace or Type of Work	Potential Fall Distance
General industry workplaces	4 feet
Shipyards	5 feet
Construction industry	6 feet
Longshoring operations	8 feet
Working over dangerous Equipment and machinery	At any height, no matter how small

TABLE 2-3 When Fall Protection Is Required

- Guard every floor hole into which a worker can accidentally walk with a railing with a toeboard and/or a floor hole cover, **Figure 2-3**.
- Provide a guardrail and toeboard around every elevated floor and open-sided platform or runway.
- Regardless of the elevation in question, if a worker can fall into or onto dangerous machines or equipment (such as a vat of acid or a conveyor belt), employers must provide guardrails and toeboards to prevent workers from getting injured.

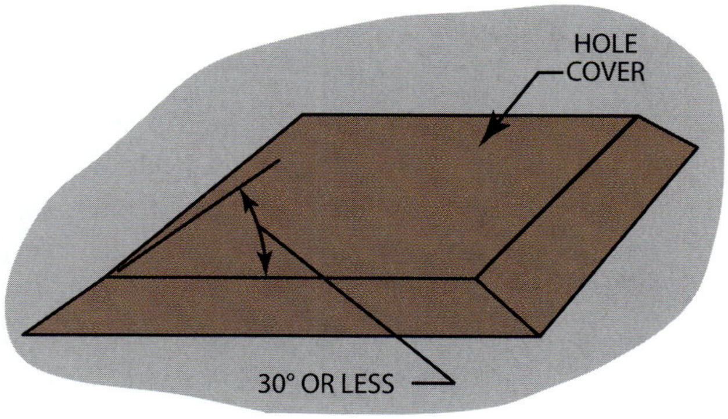

FIGURE 2-3 Specifications for a hole cover.

Guardrails must have a top handrail that is 42 in. (1 m) [plus or minus 3 in. (8 cm)] above the floor, have a midrail that is approximately halfway between the top handrail and floor (around 21 in), and a toeboard, **Figure 2-4**.

NOTE

To protect against falls, workers must be provided any required PPE and fall protection training.

Some of the other ways of providing worker protection from fall-related injuries include safety harnesses and line and safety nets.

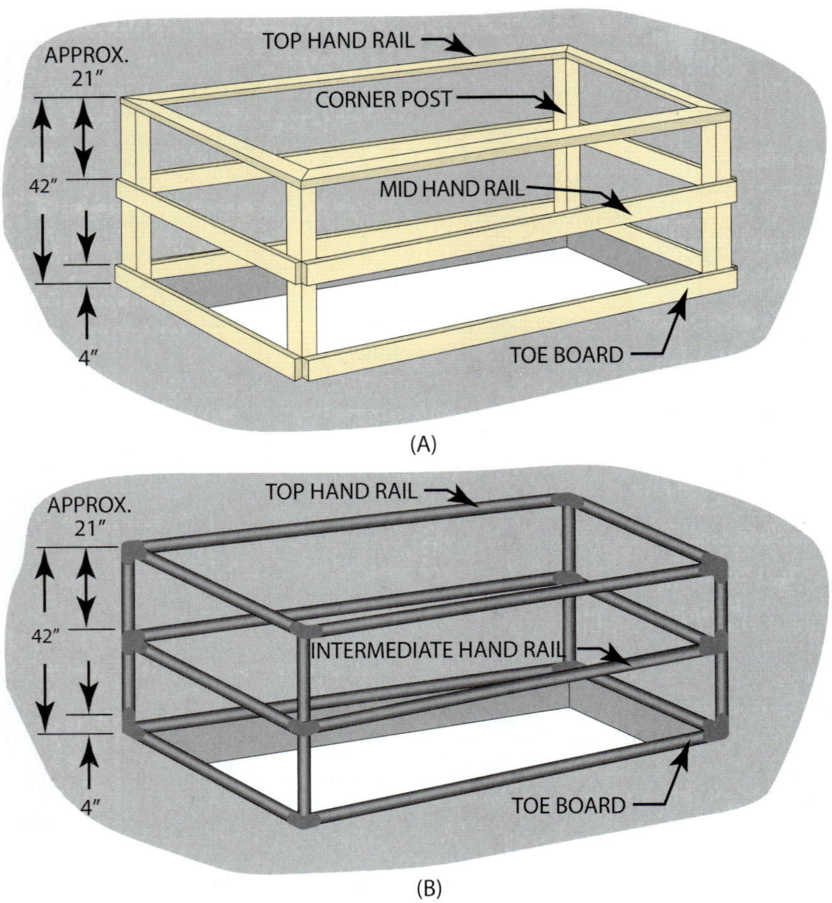

TOP HAND RAIL shall be 39 to 45 in. above the walkway or working level and be able to support at least 200 lbs of force.

MID HAND RAIL (intermediate hand rail) shall be installed between the top rail and walkway or working level around 21 in. and able to support at least 150 lbs of force.

TOE BOARD shall be at least three in. wide and be able to withstand at least 50 lbs of force.

CORNER POST shall be anchored to withstand a force of 200 lbs when applied to the top rail in any direction once the guard rail is completed.

FIGURE 2-4 Specifications for a hole guardrail.

Scaffolding

Scaffolding, sometimes called *staging*, consists of temporary structures (usually constructed of pipe) that are used to support people or material. The most commonly used scaffolding consists of two end frames that are joined together by cross brace tubes, **Figure 2-5**. To make a taller scaffold, the scaffolding frames can be stacked vertically, with one on top of the other, using scaffolding insert pins. The frames may also be joined horizontally with interlocking cross brace tubes to make a longer scaffold, **Figure 2-6**. The legs of the scaffold frame may be set in flat base plates, casters or leveling jacks.

The safety guidelines for fall protection and guardrails as discussed above also apply to scaffolding. Some of the additional safety concerns for scaffolding are:

- **Footing**—It is important that the scaffolding is set on a solid base that is capable of bearing the load of the scaffold.
- **Platforms**—These must be fully supported, capable of supporting the load, and fully cover the deck area.
- **Training**—Anyone working on scaffolding must complete training on the hazards involved.

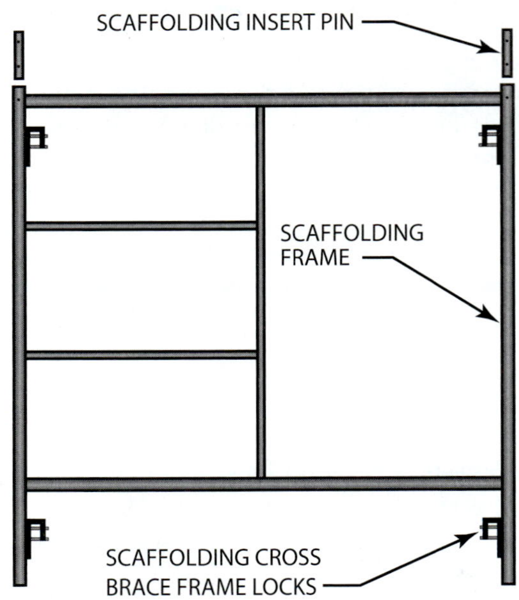

FIGURE 2-5 Scaffolding part identification.

- **Inspection**—All scaffolding must be inspected before each shift begins work.
- **Erecting and dismantling**—All fall protections must be followed as scaffolding is erected and disassembled.

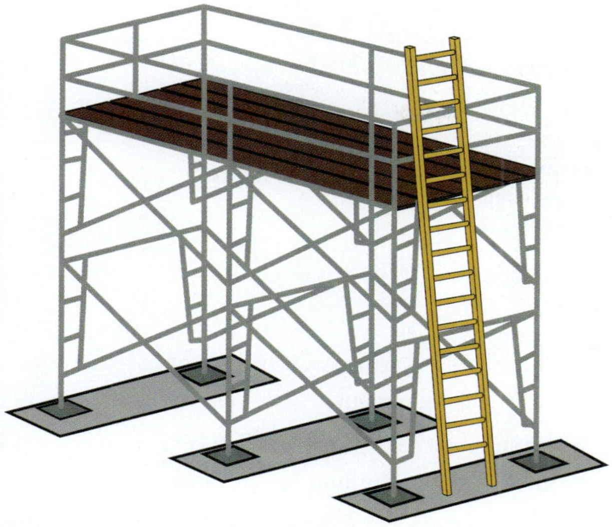

FIGURE 2-6 Suggested way that scaffolding can be set up.

FIGURE 2-7 Stepladder.

Ladder Safety

Falls from portable ladders are major causes of occupational fatalities and injuries, so it is important to know how to use them properly and safely. Often pipe welding is done up on scaffolding or down in ditches, so there are times that a ladder must be used to reach the work area, or even to weld from. Even a fall from a small step stool can cause injury. Often people feel that when a ladder starts to fall, they can just jump free; however, there is nothing solid under your feet, so you cannot do this. If someone falls onto debris or equipment, the injuries can be compounded; so it is important to keep the area around the base of a ladder clear.

Some of the things that must never be done with a ladder include:

- Never place a ladder on anything such as boxes to increase its reach.
- Never move a ladder with anyone on it.
- Never overload a ladder with more weight than it is rated for.
- Never use a stepladder unless it is fully opened, or use it as a straight ladder.
- Never stand on the top two rungs.

Types of Ladders

The two major types of ladders used for welding are stepladders and straight ladders. **Stepladders** prop open and are self-supporting, **Figure 2-7**. **Straight ladders** must be leaned against a stable surface, and they are available as fixed-length ladders or **extension ladders**, **Figure 2-8**.

Ladders can be made from wood, aluminum, or fiberglass. Each type has its advantages and disadvantages,

FIGURE 2-8 Straight (left) and extension ladders.

Table 2-4. When choosing a ladder for a job, consider the height required for the job and how much load (weight) it must carry. Look for the **American National Standards Institute (ANSI)** or **Underwriters Laboratories (UL)** label to ensure that the ladder is constructed to a standard of safety.

Ladder Inspection

Always inspect a ladder prior to using it. Over time, ladders can become worn or damaged. Look for loose or damaged steps, rungs, rails, braces, and safety feet. Check to see that all the hardware is tight, including the hinges, locks, nuts, bolts, screws, and rivets. Wooden ladders must be checked

Ladder Materials		
Materials	**Advantages**	**Disadvantages**
Wood	Electrically nonconductive	Long-term exposure to weather will cause rotting
Aluminum	Lightweight Weather resistant	Electrically conductive Shakier than wood or fiberglass
Fiberglass	Electrically nonconductive Weather resistant	Heavier than aluminum and wood Fiberglass splinters

TABLE 2-4 Types of Ladder Material and Their Advantages and Disadvantages

for cracks, rot, or wood decay. Never use a defective ladder. Make any necessary repairs before it is used; or, if it cannot be repaired, replace it.

Rules for Ladder Use

Read the entire ladder manufacturer's list of safety rules before using the ladder for the first time. Stepladders must be locked in the full-open position with the spreaders. Straight or extension ladders must be used at the proper angle; either too steep or too flat is dangerous, **Figure 2-9**.

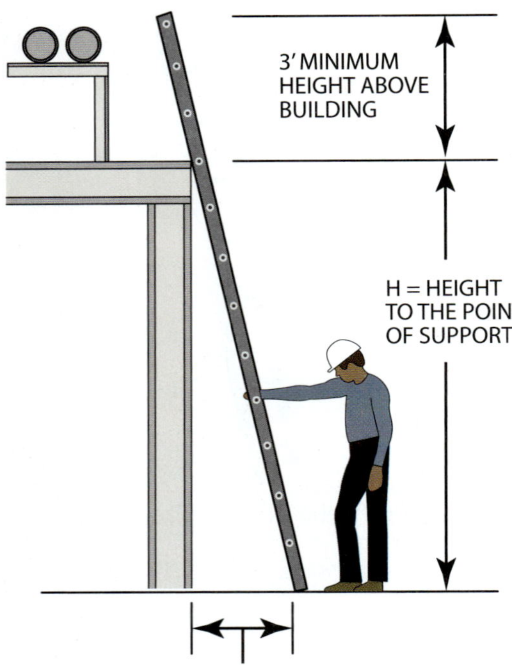

FIGURE 2-9 Proper setup for using a straight ladder.

The following are general safety and usage rules for ladders:

- Follow all recommended practices for safe use and storage.
- Avoid electrical hazards! Look for overhead power lines before handling a ladder. Metal ladders should never be used near power lines or exposed energized electrical equipment.
- Do not exceed the manufacturer's recommended maximum weight limit for the ladder.
- Before setting up a ladder, make certain that it will be erected on a level, solid surface.
- Never use a ladder in a wet or muddy area where water or mud will be tracked up the ladder's steps or rungs. Only climb or descend ladders with clean, dry shoes.
- The proper angle for setting up a straight or extension ladder is to place its base a quarter of the working length of the ladder from the wall or other vertical surface.
- Tie the ladder securely in place.
- Always maintain a three-point (two hands and a foot, or two feet and a hand) contact on the ladder when climbing. Keep your body near the middle of the step and always face the ladder while climbing.
- Do not carry tools and supplies in your hand as you climb or descend a ladder. Use a rope to raise or lower the items once you are safely in place.
- Wear well-fitting shoes or boots.

CAUTION

Never use ladders near live electrical wires. Turn the power off to any wires that will be near your ladder when working. Never move a ladder in the upright position near any electrical wires, even if you think the power on the line is off.

EXCAVATIONS

Because many pipelines are buried beneath the Earth's surface, pipe welders must often work in some type of excavation to access these pipes. All work being done in an excavation poses unique safety issues not normally associated with other work areas. The term *excavation* is very broad; OSHA defines an **excavation** as "any man-made

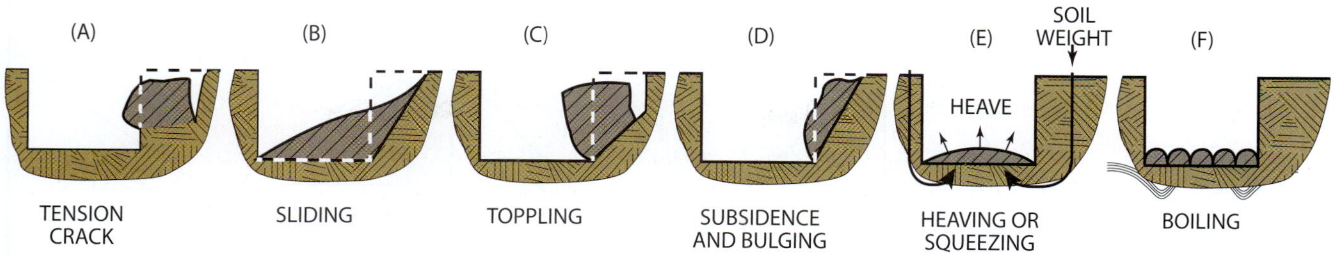

FIGURE 2-10 Types of excavation wall and floor failures.

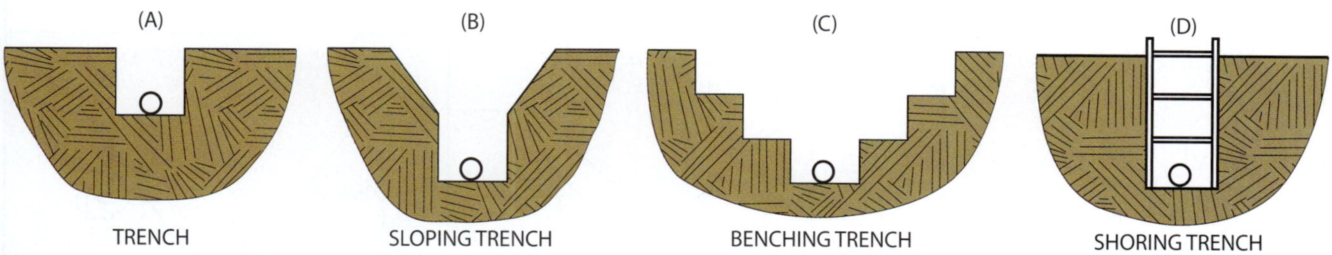

FIGURE 2-11 Recommended methods for opening an excavation.

cut, cavity, trench, or depression in the earth's surface formed by earth removal." OSHA defines the related term **trench** as "a narrow underground excavation that is deeper than it is wide, and no wider than 15 feet (4.5 meters)."

CAUTION

Never begin an excavation until the site has been surveyed for underground utilities. Most states have laws that require contractors to call a published phone number for their "Call Before You Dig" service, which will make sure that all the required utility companies have been contacted prior to beginning the excavation. Failure to do this can result in fines and make you liable for any damage caused by the excavation.

Cave-ins

One of the most serious safety hazards to workers in excavations is **cave-ins**, **Figure 2-10**. Excavation cave-ins result in the most serious injuries and too often result in work site fatalities.

One of the first things that an engineer must do at a proposed excavation site is a soil analysis. There are many different types of soil; some types, such as clay soils, are less prone to cave-ins than sandy soils. In addition, many excavation sites may have more than one soil type, or different soil types layered on top of each other. The analysis is used to determine the type of soil and the appropriate excavation wall design, **Figure 2-11**. The dirt removed from a trench is referred to as *spoil*, **Figure 2-12**.

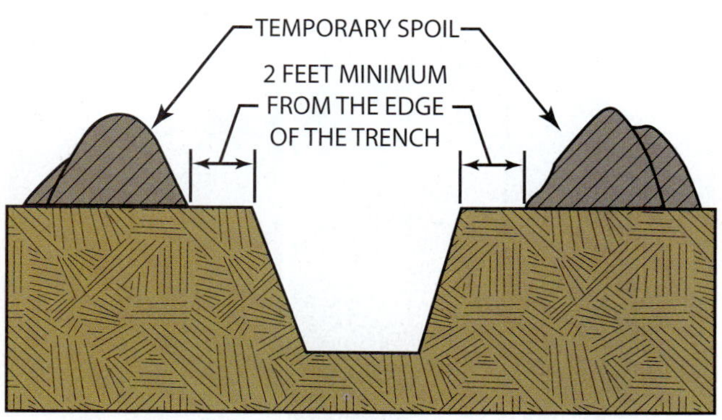

FIGURE 2-12 Safe way to place dirt when digging.

Excavation Walls

There are a number of ways to prevent cave-ins, including:

- **Sloping**—The angle of the sloped wall of an excavation is determined by the soil type, **Figure 2-13**. Less stable soil types and layered soil types require a much shallower slope angle than the more stable soil types.
- **Benching**—Stair steps cut into the wall of an excavation can make it a little easier to get in and out of it, **Figure 2-14**. It's also possible to combine **sloping** and **benching** with some soil types, **Figure 2-15**.

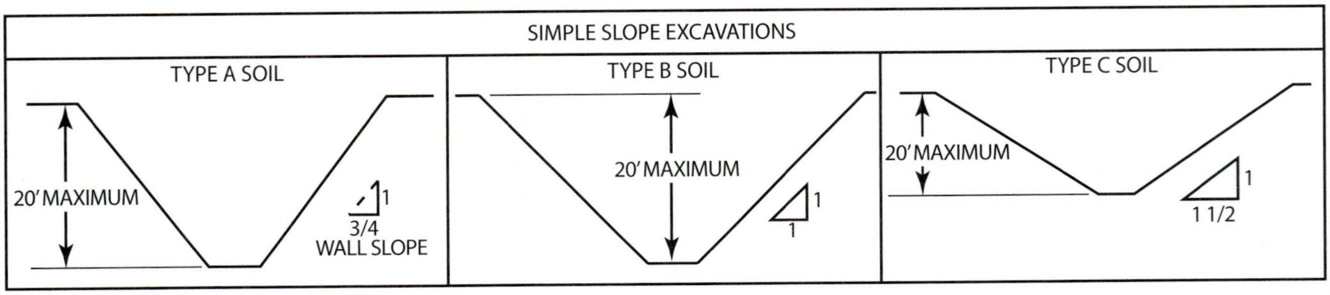

FIGURE 2-13 Sloping for different types of soil.

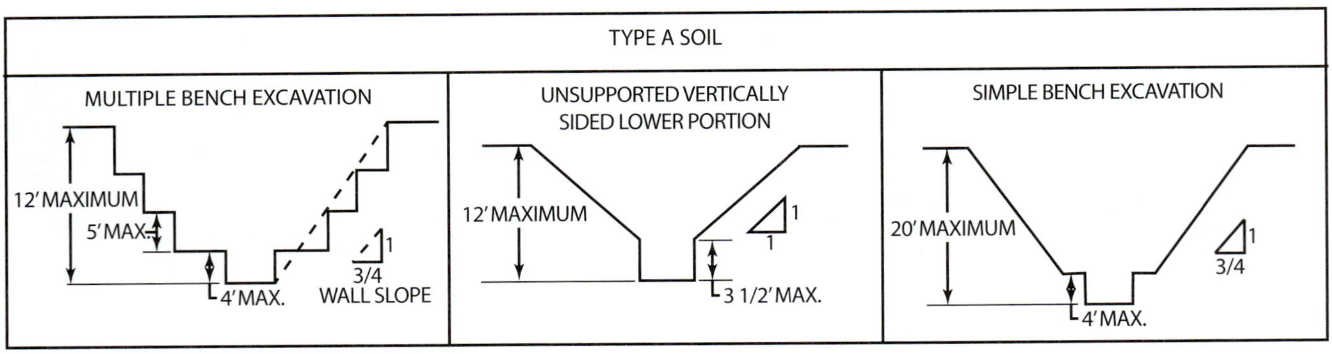

FIGURE 2-14 Benching for different depths of excavations.

FIGURE 2-15 Example of benching (on the left wall) and sloping (on the right wall).

- **Shoring**—Because trenches are a narrower type of excavation, these walls can be secured against cave-ins by bracing them with one of several types of **shoring**, **Figure 2-16**.
- **Shielding**—One of the most commonly used methods of protecting against cave-ins in trenches are trench boxes, **Figure 2-17**. Metal trench boxes can be easily moved along a trench or from one excavation site to another. The space between the trench wall and the wall of the trench box should be kept as small as possible.

Other Excavation Hazards

Welders must be alert for other safety hazards while working in or around an excavation, including:

- **Falls**—Pay very close attention while walking around open excavations to avoid accidentally stepping into them.
- **Slipping**—Often the dirt in and around an excavation can turn into mud, which greatly increases the potential of slipping. Slipping and sliding can cause strained or pulled muscles and can result in a fall.
- **Falling loads**—Be alert to loads that may be moving overhead. Although nothing should ever be moved over workers who are in an excavation, crane or lift operators may not be able to see that someone is working in an excavation.

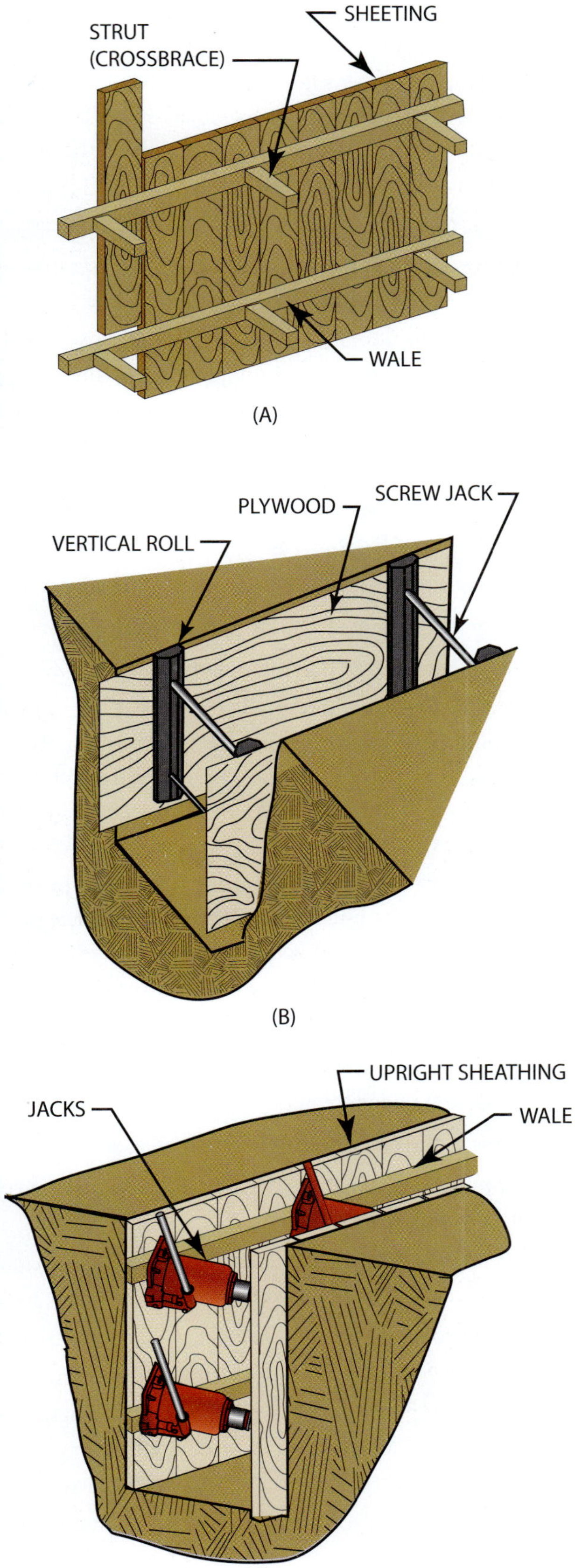

FIGURE 2-16 Methods of securing an excavation wall from cave-ins. (A) Vertical planking, (B) plywood, and (C) hydraulic jacks can be used to replace cross bracing and screw jacks.

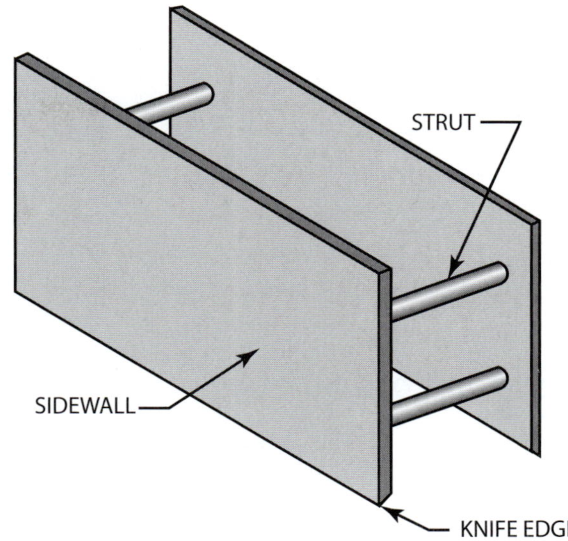

FIGURE 2-17 A metal trench box is a good way of securing the walls of excavations.

- **Hazardous atmosphere**—Many hazardous gases, such as carbon monoxide from equipment exhaust, are heavier than air and can collect inside an excavation.
- **Mobile equipment**—Drivers and equipment operators my accidentally drive into an excavation.
- **Underground utilities**—There is a wide variety of utilities that may be located in a proposed excavation site, such as telecommunication cables, wires, and fiber optics; pipes for water, gas, and sewer; and wires for power. It is the responsibility of the excavation equipment operator to locate these underground utilities before any digging begins. Damaging underground utilities such as electrical cables, gas pipes, and water lines can be deadly.

ELECTRICAL SAFETY

The best way to avoid an electrical shock is not to come in contact with electricity in the first place. There are many factors that affect exactly how much electricity it takes to be fatal. However, it is generally accepted that it only takes around 0.03 amperes (amps) to kill a person. To put this in perspective, a 60-watt incandescent light bulb uses around 0.5 amps—16 times higher than the fatal dose.

These are some general electrical safety steps that can be taken to minimize your risk of an electrical shock and possible death:

- Do not work on equipment that is still plugged into a power source.
- Repair or discard all worn or damaged electric power tools.

- Do not continue working if you think that you are not safe.
- Do not continue working with electrical equipment if there are electrical storms in the area because a lightning strike miles away can be carried by power wires to your job site.
- Complete a cardiopulmonary resuscitation (CPR) course and encourage others to do so too.

Extension Cord Safety

Because much pipe welding is done on location, it is necessary to use extension cords for lighting, grinders, fans, drills, and other portable tools. Because of the potential of electrical shock or electrocution posed by extension cords, they must have three wires and have a **ground-fault circuit interrupter (GFCI)**as part of the cord or be attached to a GFCI at the power source. In fact, it is a violation of federal safety rules not to have a three-wire extension cord and not to use a GFCI.

When a high current draw is required through an extension cord, there will be some degree of drop in the voltage. This voltage drop is much like the drop in water pressure when a long extension hose is used. As the voltage drop increases, you will experience a loss in power in drills, saws, and other motor-driven tools. The real problem is not the loss of power, but that the voltage drop will cause additional heat buildup in the motor. Too much heat buildup will result in the motor burning out. Keeping an extension cord as short as possible and making sure it can safely carry the current will help reduce voltage drop. **Table 2-4** shows the correct size extension cord to use based on cord length and nameplate amperage rating. If in doubt, use the next larger size. The smaller the gauge number of an extension cord, the larger the cord.

Extension cords should be checked frequently while in use to detect unusual heating. Any cable that feels more than slightly warm to a bare hand placed outside the insulation should be checked immediately for overloading.

Safety Rules for Portable Electric Tools

Portable electric tools such as grinders, drills, motorized pipe beveling cutters, fans, and lights are commonly needed on pipe welding jobs. They are often plugged into portable welding machines, and the chance of a fatal electrical shock is just as real as if they were plugged into an electrical outlet in the shop.

Following are a few safety precautions that should be observed. These general rules apply to all power tools, and they should be strictly obeyed to avoid injury to the operator and damage to the power tool:

- Know the tool. Learn the tool's applications and limitations, as well as its specific potential hazards by reading the manufacturer's literature.
- Ground the portable power tool unless it is double insulated. If the tool is equipped with a three-prong plug, it must be plugged into a three-hole electrical receptacle. If an adapter is used to accommodate a two-pronged receptacle, the adapter wire must be attached to a known ground. Never remove the third prong.
- Do not expose the power tool to rain or wet locations.
- Keep the work area well lit.
- Avoid chemical or corrosive environments.
- Because electric tools spark, portable electric tools should never be started or operated in the presence of propane, natural gas, gasoline, paint thinner, acetylene, or other flammable vapors that could cause a fire or explosion.
- Power tools will do the job better and more safely if operated at the rate for which the tool was designed; do not force a tool to operate beyond its limits.
- Use the right tool for the job. Never use a tool for any purpose other than that for which it was designed.
- Wear eye protectors such as safety glasses or goggles while you operate power tools.
- Wear a dust mask if the operation creates dust.
- Never carry a tool by its cord or yank it to disconnect it from the receptacle.
- Secure your work with clamps. It is safer than using your hands, and it frees both hands to operate the tool.
- Keep proper footing and balance at all times. Do not overreach when operating a power tool.
- Maintain power tools. Follow the manufacturer's instructions for lubricating and changing accessories. Replace all worn, broken, or lost parts immediately.
- Disconnect the tools from the power source when they are not in use.
- Form the habit of checking to see that any keys or wrenches are removed from the tool before turning it on.

- Avoid accidental starting. Do not carry a plugged-in tool with your finger on the switch. Be sure that the switch is off when plugging in the tool.
- Be sure that any accessories and cutting bits are attached securely to the tool.
- Do not use tools with cracked or damaged housings.
- When operating a portable power tool, give it your full and undivided attention; avoid dangerous distractions.

GENERAL WORK CLOTHING

Because of the amount and temperature of hot sparks, your general work clothes should be made of 100% cotton. When working in a cold environment, make sure that any jacket or coat is not made with or does not contain synthetic materials, such as nylon, rayon, or polyester. These materials are easily melted, and they produce a hot, sticky ash; some produce poisonous gases to boot. The clothing must also stop ultraviolet light from passing through it. This is accomplished if the material chosen is a dark color, thick, and tightly woven.

The following are some guidelines for selecting work clothing:

- Shirts must be long-sleeved to protect the arms, have a high-buttoned collar to protect the neck, be long enough to protect the waist, and have flaps on the pockets to keep sparks out (or have no pockets), as shown in **Figure 2-18**.

FIGURE 2-18 Welders and welding helpers must wear appropriate general work clothing.

- Pants must have legs long enough to cover the tops of the boots and must be without cuffs that could catch sparks.
- Boots must have high tops to keep out sparks, steel toes to prevent crushed toes, **Figure 2-19**, and smooth tops to prevent sparks from being trapped in seams.

FIGURE 2-19 Steel toe safety boots are required on most job sites.

- Caps should be thick enough to prevent sparks from burning the top of a welder's head.

Clothing must be relatively tight-fitting in order to prevent excessive folds or wrinkles that might trap sparks. Also, clothing must be free of frayed edges, which might easily catch fire from welding sparks.

Butane lighters and matches may catch fire or explode if they are subjected to welding heat or sparks. There is no safe place to carry these items when welding.

Special Protective Clothing

In addition to general work clothing, extra protective clothing is often needed for many welding jobs. Leather is often the best material to use for such garments, as it is lightweight, flexible, resists burning, and is readily available. Ready-to-wear leather protection includes capes, jackets, aprons, sleeves, gloves, caps, pants, kneepads, and spats, among other items. Synthetic insulating materials are also available.

FIRE PROTECTION

Too often, we read in the news that a "welder's torch" started a fire. No one wants to be the center of a lead story about a fire started by a welder, so we need to do everything to prevent accidental fires from being started. Fire is a constant danger in all welding situations, but it is a special problem for pipe welding outdoors, such as welding on cross-country pipelines. Fires can spread quickly in materials such as dried grass, farm crops, underbrush, and woodlands, especially during a drought or in the winter when plants die back and dry out. The possibility of accidental fires being started can be decreased by removing vegetation and other flammable materials 35 ft (10.7 m) or more from the welding area.

It is not always possible to remove all combustible materials when welding outside. If you must work around combustible materials, wet the area first, and then keep a bucket of water and a fire extinguisher handy and use a fire watch, **Figure 2-20**.

FIGURE 2-21 In most areas, it is the welder's responsibility to know if there are fire restrictions, so call before you weld.

FIGURE 2-20 A fire watch has both a water sprayer and a fire extinguisher. A fire safety officer does not have to see a fire happen; the burned grass on a site may be enough to shut down welding.

Fire Watch

A **fire watch** can be provided by any person who knows how to sound the alarm and use a fire extinguisher. The fire extinguisher must be the type required to put out a fire of the type of combustible materials near the welding. Combustible materials that cannot be removed from the welding area should be soaked with water or covered with sand or noncombustible insulating blankets, whichever is available.

CAUTION

Never weld or cut in a dry area or any area that has been posted by the county or state as fire restricted, **Figure 2-21**. You could be held legally responsible if you ignore this warning and a fire starts.

PLANNED MAINTENANCE

Follow the manufacturer's routine schedule of equipment maintenance. Small problems, if fixed in time, can prevent the loss of valuable time due to equipment breakdown or injury.

Any maintenance beyond routine external maintenance should be referred to a trained service technician. In most areas, it is against the law for anyone but a factory-trained repair technician to work on pressure regulators. Electrical shock and exploding regulators can cause serious injury or death.

Hoses

Hoses should be kept out of the direct line of sparks. Any leaking or bad joints must be repaired.

Cables

Welding cables and extension cords can be damaged by welding and cutting sparks, as well as becoming scraped and gouged as they are pulled around a job site. So it is important to inspect them for damage that might create a safety hazard.

Hand Tools

Tools that are used outside and exposed to damp conditions need to be cleaned and well oiled to prevent rusting. Keep knives and other cutting tools sharpened.

Work Area

The work area in welding shops should be kept picked up and swept clean. Collections of steel, welding electrode stubs, wire, hoses, and cables are difficult to work around and easy to trip over, **Figure 2-22**. Keeping the work area clean is important not only for safety, but also because antilitter laws can be broken if food wrappers, Styrofoam™ cups, cardboard, and other trash are not disposed of

Larry Jeffus

FIGURE 2-22 A metal or plastic bucket can make a handy tool and scrap electrode holder.

properly. You must also make sure that this trash is not blown off of the job site.

MATERIAL HANDLING

Proper lifting, moving, and handling of large, heavy, welded assemblies are important to the safety of the workers and the weldment. Improper work habits can cause serious personal injury, as well as cause damage to equipment and materials. When you are lifting a heavy object, the weight of the object should be distributed evenly between both hands, and your legs should be used to lift, not your back, **Figure 2-23**. Do not try to lift a large or bulky object without help if the object is heavier than you can lift with one hand.

(A) CORRECT (B) INCORRECT

FIGURE 2-23 Lift with your legs, not your back.

Hauling

Most pipe welding is done on site, not in a building or shop. This means that the welding and fabrication equipment has to be hauled to the job site. Keeping everything together makes it safer and easier to haul. Depending on the operation's needs, a truck or trailer can be set up for welding, **Figures 2-24a** and **2-24b.**

Larry Jeffus

(A)

Larry Jeffus

(B)

FIGURE 2-24 Pipe welders' trucks are portable workshops.

CAUTION

It is dangerous to store fuel and oil in the same toolbox with regulators and electrodes. If contaminated with fuel or oil, the regulators can explode and the electrodes may not work correctly. Securely store fuel and oil containers in the open vehicle bed, Figure 2-25.

Heavy Equipment

A variety of equipment can be used on a job site to move, support, or hold pipe and pipe sections. Some of the most commonly used pieces of equipment are excavators, backhoes, boom trucks, and cranes. Excavators may have rubber tires or tracks, and both types come in a wide range of sizes or capacities. Excavators with tracks are also called *trackhoes*, **Figure 2-26**. The tracks on an excavator allow it to move easily over rough terrain. A shovel is usually attached to the end of the boom for digging. It can also be used to lift loads. The cab and boom can be rotated in a full 360-degree circle.

Backhoes, also called *diggers*, are specially equipped tractors that usually have a bucket or front loader

Larry Jeffus

FIGURE 2-25 Welding gas cylinders are capped when not in use.

Larry Jeffus

FIGURE 2-26 Always stay well clear of loads that are being moved.

mounted on the front and a boom with a shovel in the back, **Figure 2-27**. The addition of the front bucket lets backhoes move or load dirt, gravel, sand, and other bulk materials around the job site as needed. The rear boom has a little more than a 180-degree side-to-side swing.

Larry Jeffus

FIGURE 2-27 Backhoe outriggers must always be extended before lifting a load.

Boom trucks and cranes have longer booms, which allow them to reach higher and farther than either excavators or backhoes. Their ability to reach farther is a significant advantage when there are obstacles such as creeks or environmentally sensitive areas that excavators or backhoes cannot cross.

It is important not to overload lifting equipment. The capacity of the equipment should be checked before trying to lift a load. Keep any load as close to the ground as possible while it is being moved, **Figure 2-28**. Pushing a load is better than pulling a load, **Figure 2-29**. It is advisable

Larry Jeffus

FIGURE 2-28 Never have any part of your body under a load as it is being held or moved.

Larry Jeffus

FIGURE 2-29 Pushing a load is safer than pulling.

to stand to one side of ropes, chains, and cables that are being used to move or lift a load. If they break and snap back, they will miss you. If it is necessary to pull a load, use a rope.

Rigging

Rigging is the term used to describe the securing of a load to be lifted. Over the years, the most common items used to secure a load have been chains, cables, ropes, and straps. Today, straps are the most common rigging material. There are a number of ways that a load can be rigged so that it will be balanced and stable during the lifting, **Figure 2-30**. When a single sling is used for lifting, the lift operator may pick up the load several times as the sling is moved to find the balance point, **Figure 2-31**. Using a slightly longer strap will result in a slightly larger angle between the strip and the load, **Figure 2-32**.

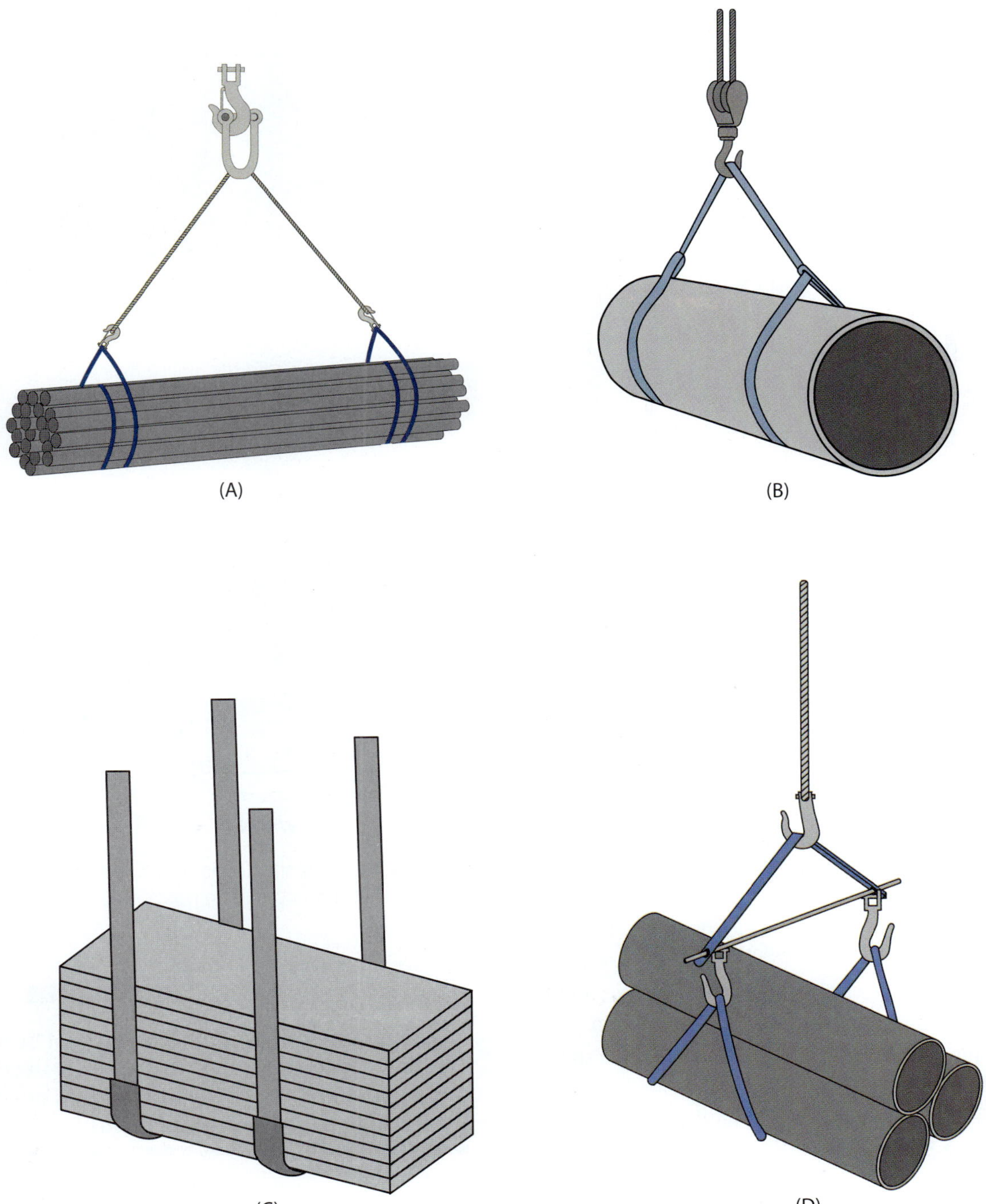

FIGURE 2-30 Loads must be properly slung before lifting.

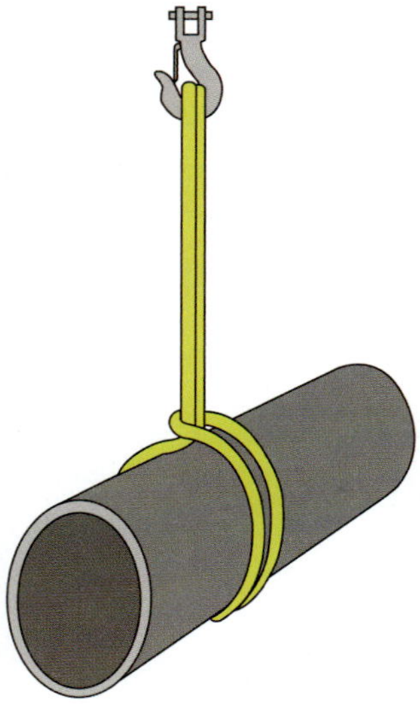

FIGURE 2-31 A properly located choker can be used to lift long pipe sections.

FIGURE 2-33 Pipe hooks.

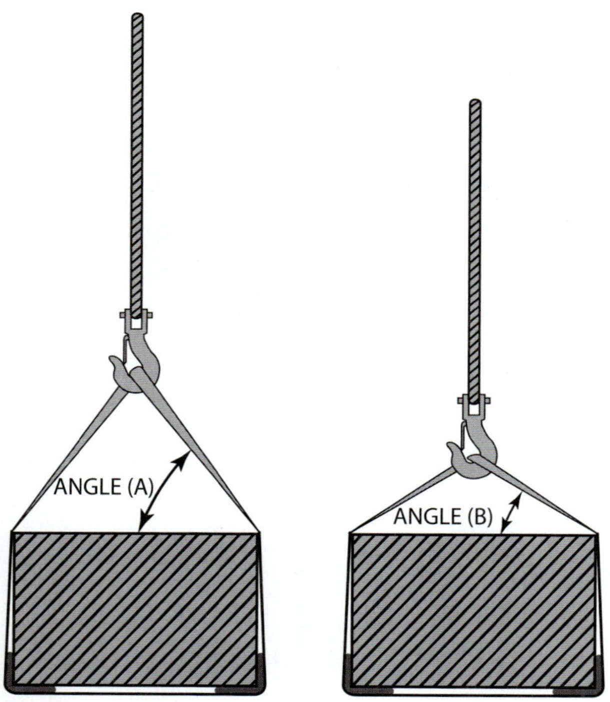

FIGURE 2-32 Using a longer strap puts less pressure on the corners of the load and less strain on the straps, as shown on the left.

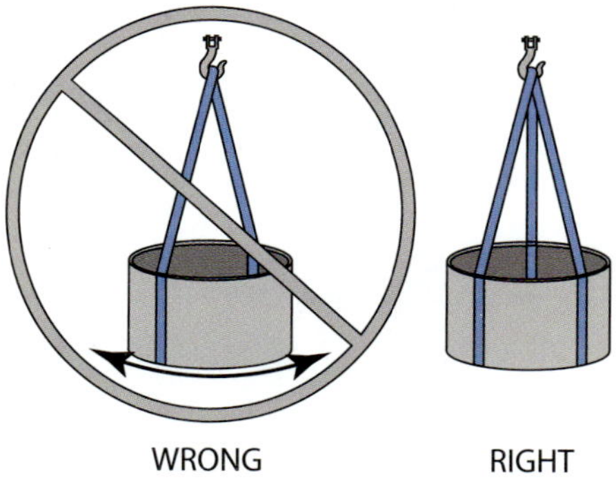

FIGURE 2-34 Smaller loads may need more straps to keep them from tipping and possibly falling.

Pipe hooks can be used to lift and move pipes and fittings, but they can be in the way when trying to fit pieces together, **Figure 2-33**. When possible, connect the rigging to the item being lifted in a manner that will minimize swinging or the possibility of it being dumped out of the sling, **Figure 2-34**.

CAUTION

Chains, cables, ropes, and slings must be inspected for damage each time they are used. Also, make sure that their strength rating is greater that the weight of the load to be lifted.

Summary

Safety in all aspects of welding fabrication is of utmost importance. When done correctly, welding and fabrication are safe. You are responsible for your own safety. If you do not think a job can be done safely, do not even begin it. If you have safety concerns about a job, request assistance.

You must read and follow all of the manufacturer's operation and safety literature for any equipment. Do not assume that any new piece of equipment has the exact same operating instructions or safety rules. As equipment changes, even slightly, manufacturers update the literature; you must read it to know that you are working in the safest manner.

Periodically performing maintenance, servicing, and doing a safety check of equipment will make doing every job safer. In addition to safety, keeping equipment well maintained will reduce the operating cost. Equipment that is in good working order will make the job go better, faster, and safer.

Further safety information is available in *Safety for Welders* by Larry F Jeffus (Delmar Learning, 1980) and other safety material is available from the U.S. Department of Labor.

Review Questions

1. What are some of the consequences that can occur after an accident on the welding site?
2. What precaution can be taken to prevent burns?
3. List three examples of equipment that can produce dangerous levels of noise so that ear protection is necessary.
4. What can cause respiratory problems when performing repair work on pipes?
5. What is the difference between air-purifying respirators and atmosphere-supplying respirators?
6. What are two examples of situations when additional ventilation might be needed?
7. What information does an MSDS supply?
8. What are the benefits of recycling scrap metal?
9. Give three examples of welding waste material that may be considered hazardous.
10. Why is it important to keep the area around the base of a ladder clear?
11. What are two factors that should be considered when choosing an appropriate ladder for a job?
12. What should be inspected on a ladder prior to use?
13. Why are metal ladders never used near power lines?
14. What is the proper angle for setting up a straight or extension ladder?
15. How does OSHA define a trench?
16. What must a site be surveyed for before excavation?
17. Why is a soil analysis one of the first things done at a proposed excavation site?
18. Which type of excavation walls have stair steps cut into the wall of an excavation?
19. List three excavation hazards.
20. Describe two steps that can be taken to minimize the risk of an electrical shock.
21. What are two requirements for extension cords to avoid electrical shock?
22. Why is it important to read the manufacturer's literature for electric tools?
23. Why should electric tools never be used in the presence of propane, natural gas, gasoline, paint thinner, or acetylene?
24. What type of fabric should general work clothes be made of?
25. If it is not possible to remove all combustible material when welding outside, what can be done to decrease the possibility of an accidental fire?
26. How should you lift a heavy object?
27. Why is it dangerous to store fuel and oil in the same toolbox with regulators and electrodes?
28. List three commonly used pieces of equipment that are used on job sites for excavation and to move, support, or hold pipe and pipe sections.
29. When should chains, cables, ropes, and slings be inspected for damage?

Chapter 3

Shop Math

OBJECTIVES

After completing this chapter, the student should be able to:

- Explain the difference between the standard system and the metric system.
- Follow the sequence of operations when solving math problems.
- Add and subtract mixed units.
- Add, subtract, multiply, and divide fractions.
- Find the common denominator of fractions and reduce fractions.
- Convert fractions to decimals and decimals to fractions.
- Determine acceptable measurements given a tolerance.
- Round numbers.
- Calculate the volume of a straight pipe, cylinder, or tank.

KEY TERMS

Decimal fractions

Dimensioning

Dimensioning tolerance

Equation

Fractions

Formula

Metric system

Mixed units

Standard system

Whole numbers

INTRODUCTION

The most common use of math in welding shops is for dimensioning and pricing. **Dimensioning** is the process of defining the size of an object's width, height, and length. For dimensioning, most pipe welding uses the **standard system** (also known as the *English system*), which uses feet, inches, and fractions. A few areas use metric dimensioning. The **metric system**, sometimes abbreviated as *SI* from the French term *Le Système International d' Unités*, is made up of seven base units, including those for length, temperature, and weight, **Table 3-1**. For pricing, dollars and cents calculations can include labor costs in hours and minutes; material costs in pounds, ounces, feet, and inches; fuel and oil in quarts and gallons for welding generators; and overhead costs as a percentage of the company's cost for insurance and other expenses.

(*Continued*)

Although most pipe welders use the standard system, the math functions of addition, subtraction, multiplication, and division of dimensions are easier in the metric system than in the standard system because the metric system is based on decimals. The advantage of a decimal system is that a calculator can be used more easily than with the mixed-numbered standard system.

Almost all pipe welding dimensioning uses mixed units. An example of a mixed unit is feet and inches, based on 12 in. to the foot. So the largest number of inches is 11 because when you add another inch, it becomes 12 in., which is expressed as 1 ft. Other common examples of mixed units are pounds and ounces and hours and minutes. Mixed units present unique problems for addition, subtraction, multiplication, and division because each type of unit must be worked separately. For example, you cannot add feet to inches without first converting the feet to inches.

Most welders use calculators because they eliminate arithmetic errors. There are a few calculators that can add, subtract, multiply, and divide standard fractional dimensions, and some can even work with mixed units of feet and inches. However, most math using fractions must be worked manually.

For pricing and cost estimating, welders must use math to enable their welding business to operate profitably. The owner or manager must be able to make cost-effective welding decisions. A number of factors affect the cost of producing weldments. Some of these factors include the following:

- Material
- Weld design
- Welding processes
- Finishing
- Labor
- Overhead

Abbreviations for Units				
Units	**Standard**		**Metric**	
Temperature	F	Fahrenheit	C	Celsius
Length	yd ft in.	yard feet inches	m cm mm	meter centimeter millimeter
Weight	lb oz	pounds ounces	kg g	kilogram grams
Liquid	gal qt pt oz	gallons quarts pints ounces	l	liters
Pressure	psi	pounds per square in.	kPa	kilopascal

TABLE 3-1 Mathematical Abbreviations for Units

SHOP MATH

Mathematics has many branches—calculus, geometry, trigonometry, and so on; but *arithmetic* is the most basic branch of mathematics; it is primarily involved with combining numbers by addition, subtraction, multiplication, and division. Arithmetic is the math that most welders use on a daily basis. Welding layout, and fit-up to some extent, may involve geometry and trigonometry. Both these branches of mathematics deal with angles and points and are related to laying out and fitting up weldments with complex shapes.

TYPES OF NUMBERS

There are certain types of numbers that are commonly used when figuring calculations for pipe welding. These types of numbers are reviewed next:

- **Whole numbers** are numbers used to express units in increments of 1, so they can be divided evenly by the number 1. Examples of whole numbers are 1, 2, 3, 4, 5, 6, 7, 8, 9, 10. They are the easiest type of numbers to use with all types of mathematical functions.

A **decimal fraction** is a number that uses a decimal point to denote a unit that is smaller than 1 (though the number itself can be greater than 1). Examples of decimal fractions are 0.5, 0.25, 0.8725, 9.3, 6.2, 3.14, and 225.4321. Depending on their place after the decimal point, decimal fractions are expressed in units that are 10 times, 100 times, and 1,000 times smaller than 1, and so on, **Figure 3-1**.

- **Mixed units** are measurements containing numbers that are expressed in two or more different units. An example of a mixed unit is a linear measurement such as 3 ft 6 in., with part of the measurement expressed in feet (3) and the other part in inches (6). Other examples of mixed units are angular measurements such as 90°0', weight measurements such as 12 lb 8 oz, and time measurements such as 2 hours,

1,000,000.0	millions
100,000.0	hundred thousand
10,000.0	ten thousand
1,000.0	thousand
100.0	hundreds
10.0	tens
1.0	ones
tenths	0.1
hundredths	0.01
thousandths	0.001
ten thousandths	0.0001
hundred thousandths	0.00001
	0.000001

FIGURE 3-1 Whole numbers and decimal fraction terminology.

35 min, 20 sec. The most common types of mixed units used in welding fabrication are linear dimensions, angular dimensions, weight, and time. Linear dimensions use units of feet and inches (or meters, centimeters, and millimeters in the metric system); angular dimensions use degrees, minutes, and seconds; weight uses pounds and ounces (or kilograms and grams in the metric system); and time uses hours, minutes, and seconds. The different units may be separated with a space ( ) or dash (–). It is important to keep the different number units straight when performing mathematical functions.

A **fraction** is two or more numbers used to express a unit smaller than 1. Examples of fractions are 1/4, 1/2, 7/16, 4 5/8, and 10 1/2. When whole numbers are combined with fractions, they are called *mixed fractions* (e.g., 9 1/2). A dash (–) or slash (/) is used to separate the top and bottom numbers of a fraction. The denominator is the bottom number of a fraction, and the numerator is the top number, **Figure 3-2**. When fractions are added, subtracted, multiplied, or divided, they often must be converted to a common denominator before the operation can be performed.

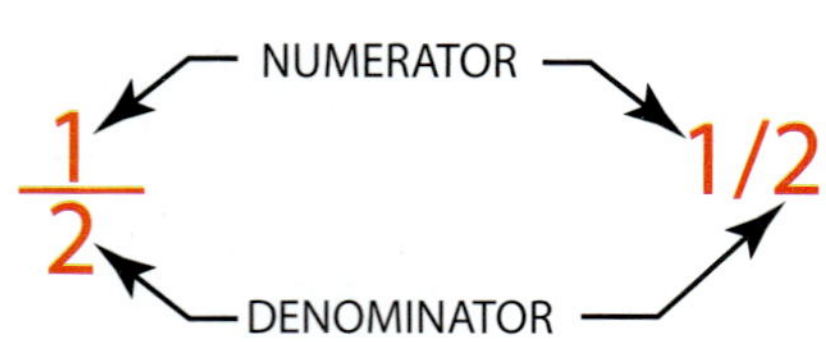

FIGURE 3-2 Fraction terminology.

GENERAL MATH RULES

All of the math problems in this chapter will be set up and worked in the same manner:

First Step	The equation or formula will be stated on the first line.
Second Step	"Where:" explains the meaning of any variables.
Third Step	State the problem's known values and what answer is needed.
Fourth Step	Write the equation or formula.
Fifth Step	Write the known values in place of the variables.
Sixth Step	One mathematical calculation will be performed per line.
Seventh Step	Add as many lines as necessary to complete the problem's calculations.
Eighth Step	Give the answer, including units.
Ninth Step	Explain the answer in a written statement.

When working a math problem by hand, write it down in a vertical series of lines, with each individual step on its own line. Keeping the equal symbol (=) on each line lined up vertically will reduce confusion and make it easier to refer back to the problem later. Working math problems in this very structured way makes it easier for you to look back at the examples in this textbook and work any new problems you encounter in a welding shop. Years later, it can be very frustrating to know the equation or formula but not remember the sequence of mathematical steps.

Sometimes formula expressions will have a superscript. Superscripts can be numbers or letters written to the upper right of a formula expression, such as the 2 in the expression, 3^2. In this case, it means that the number 3 is squared (or multiplied by itself two times). Expressions may also have a subscript. Subscripts can be numbers or letters written to the lower right of an expression. They are often used to define an expression, such as the term *weld* in the expression, CS_{weld}. In this case, it is identifying the CS (cross section) as that of the weld. Some of the formula expressions used in this textbook are shown in **Table 3-2**.

EQUATIONS AND FORMULAS

An **equation** is a mathematical statement in which both sides are equal to each other; for example, $2X = 1Y$. In this equation, the value of X is always going to be 1/2 of the value of Y. An example of an equation used in metal fabrication would be:

$$\text{number of hours worked (hrs)} \times \text{pay per hour (\$)} = \text{total labor bill (T), or hrs} \times \$ = T$$

If either the hours or the pay rate goes up, the total bill goes up as well, and vice versa.

Expression		Meaning
b	=	base dimension of weld
BD	=	bead depth
BOR	=	burn-off rate in pounds per hour
CS_{root}	=	cross-sectional area of the root opening
CS_{weld}	=	cross-sectional area of the weld
DE	=	deposit efficiency
DR	=	deposition rate in pounds per hour
EL	=	electrode length
GV	=	groove volume
h	=	height dimension of weld
LOCH	=	labor and overhead cost per hour
l	=	length dimension of weld
MD	=	metal density (weight of metal in pounds per cubic inch)
OF	=	operating factor in decimal percent
%DE	=	percentage of weld deposition efficiency
PT	=	plate thickness
RG	=	root gap
RO	=	root opening
SL	=	stub loss
TCS	=	total cross-sectional area
TLOC	=	Total Labor and Overhead Cost
$Wt_{Electrode\ Used}$	=	weight of electrode used
$Wt_{Weld\ Metal}$	=	weight of weld metal
LH	=	weld leg height
LW	=	weld leg width
WL	=	weld length

TABLE 3-2 Formulas

The direct labor cost can be determined by the following equation:

Equation 1

$$C = hrs \times rate$$

Where

C = cost of labor
hrs = the hours worked on the project
rate = hourly rate of pay for the welder

Problem 1

Find the total labor bill for 7 hours of work at $25 per hour, **Figure 3-3.**

$$C = hrs \times rate$$
$$C = 7 \times 25$$
$$C = \$175.00$$

Answer 1

The total cost of labor for the 7 hours of work at $25 per hour would be $175.

A **formula** is a mathematical statement of the relationship of items. It also defines how one cell of data relates to another cell of data—for example, wt = [(l" × w" × t") ÷ 1,728] × wt/ft. In this formula, you must first find the volume of material by multiplying length (l) times width (w) times thickness (t) before dividing that number by the number of cubic inches in a cubic foot (1,728), and then multiply that number by the material's weight per cubic foot (wt/ft).

The **sequence of mathematical operations** is important when working formulas and equations. For example, 8 × 3 ÷ 2 × 4 = 48, but (8 × 3) ÷ (2 × 4) = 3. When a formula has more than one mathematical operation, the operations must be performed in the following order:

1. Perform all operations within parentheses.
2. Resolve any exponents.

JOB CARD

Job 3" Pipe Saddle on 8" Manifold Date April 11 Welder lfj

Starting Time	Ending Time	Total Time
7:00 am	11.30 am	4.5 hrs
12:00 pm	2.30 pm	2.5 hrs
	Total Hours	7 hrs
	Hourly Rate	x $25/hr
	Total Labor	$175.00

FIGURE 3-3 Problem 1.

3. Do all multiplication and division, working from left to right.
4. Do all addition and subtraction, working from left to right.

MIXED UNITS

Adding and Subtracting Mixed Units

When adding mixed units such as feet and inches, you have to add each type of number together first. For example, you would add the inches to inches and feet to feet. An example of a mixed unit problem that you might find in pipe welding would be to determine the total length in feet is of two sections of pipe and a valve, as in the problem given next.

Problem 2

One section of pipe is 10' 2" long, another is 8' 3" long, and a valve is 5" long, **Figure 3-4.** The first step would be to write the numbers in columns, with feet over feet and inches over inches. The second step would be to add the inches to the inches and the feet to the feet.

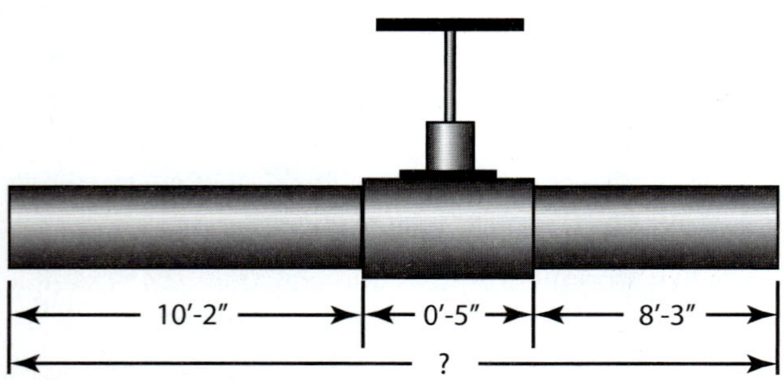

FIGURE 3-4 Problem 2.

Equation 2

$$TL = l_1 + l_2 + l_3$$

Where

TL = Total length of the parts

l_1 = length 1

l_2 = length 2

l_3 = length 3

Add 10' 2" + 5" + 8' 3"

	Feet Column	Inch Column
1st Step	10'	2"
		5"
2nd Step	8'	3"
	18'	10"

Answer 2

The total length of the two lengths of pipe and valve is 18' 10".

Problem 3

When subtracting mixed units, use the same steps as were used for adding. For example, to see how many feet of scrap pipe you have left from a 7' 8" piece when 5' 3", is cut off **(Figure 3-5)**, you would do the following:

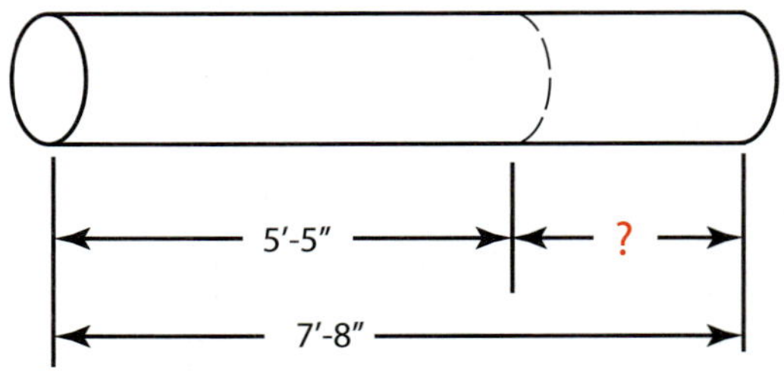

FIGURE 3-5 Problem 3.

Equation 3

$$TL = l_1 - l_2$$

Where

TL = Total length of the parts

l_1 = length 1

l_2 = length 2

Subtract 7' 8" − 5' 3"

		Feet Column	Inch Column
First Step		7'	8"
Second Step	−	5'	3"
		2'	5"

Answer 3

The length of scrap pipe left from the original 7' 8" piece is 2' 5".

When some mixed units are added, the sum can be reduced. For example, if you add 7 in. to 7 in., the total would be 14 in., which is the same as 1' 2". To reduce any inches equal to or larger than 12 in. to feet and inches, you divide the inches by 12. For example, how long will a piece of steel bar need to be if you are going to cut both a 7' 8" piece and a 6' 6" piece from it? Reduce the answer in inches to feet and inches using **Table 3-3**. Pounds and ounces are also mixed units, and they can be reduced by dividing the ounces by 16.

Add 7' 8" + 6' 6"

		Feet Column	Inches Column
First Step		7'	8"
Second Step	+	6'	6"
		13'	14"
Reduce and Add			
Third Step		13'	14 ÷ 12
Fourth Step		13 + 1	2
Fifth Step		14'	2"

Feet and Inches	
Ft Fraction	**Inches**
1/12	1
2/12	2
3/12	3
4/12	4
5/12	5
6/12	6
7/12	7
8/12	8
9/12	9
10/12	10
11/12	11
12/12	12

Pounds and Ounces	
Lb Fraction	**Ounces**
1/16	1
2/16	2
3/16	3
4/16	4
5/16	5
6/16	6
7/16	7
8/16	8
9/16	9
10/16	10
11/16	11
12/16	12
13/16	13
14/16	14
15/16	15
16/16	16

TABLE 3-3 Length and Weight Conversion Table

Problem 4

When subtracting one mixed unit from another, and the small unit being subtracted is larger than the small unit it is being subtracted from (i.e., 1' 4" from 2' 2"), you must make that unit larger. You can increase the smaller unit in a mixed-number problem by subtracting one whole larger unit then dividing it into the smaller units (as demonstrated next). For example, 2' 2", **Figure 3-6A**, has the same dimension as 1' 14", **Figure 3-6B**, and 4 lbs 8 oz, **Figure 3-6C**, is the same weight as 3 lbs 24 oz, **Figure 3-6D**.

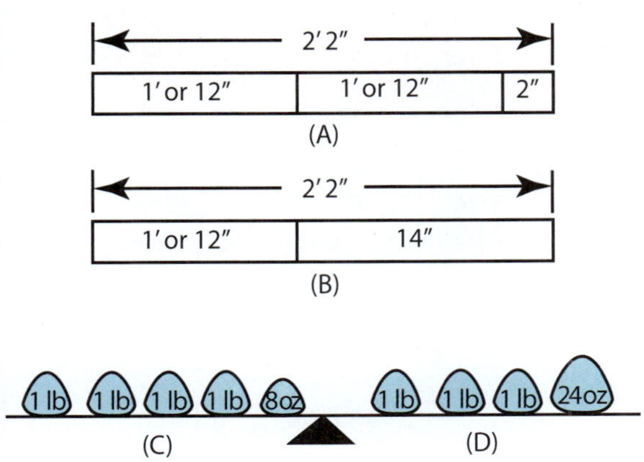

FIGURE 3-6 Problem 4.

Convert 2' 2"

First Step 2' = 1' 12"
Second Step 1' 12" + 2" = 1' 14"

Convert 4 lbs 8 oz

First Step 4 lbs = 3 lbs 16 oz
Second Step 3 lbs 16 oz + 8 oz = 3 lbs 24 oz

With 2' 2" converted to 1' 14", you can now subtract 1' 4" the same way that you subtracted mixed numbers earlier in the chapter.

Subtract 1' 14" − 1' 4"

First Step 1' − 1' = 0'
Second Step 14" − 4" = 10"
Third Step 0' + 10" = 10"

To subtract 2 lbs 10 oz from 4 lbs 8 oz, you have to convert 1 pound to 16 oz and add that to the 8 oz (as shown in Figure 3-6C and D). Now 4 lbs 8 oz has become 3 lbs 24 oz.

Subtract 3 lbs 24 oz − 2 lbs 10oz

First Step 3 lbs − 2 lbs = 1 lb
Second Step 24 oz − 10 oz = 14 oz
Third Step 1 lb + 12 oz = 1 lb 14 oz

When adding multiple mixed numbers, add all of the inches first, and then all of the feet. Then reduce the inches to feet and add the new mixed number to get the final answer. The same process is followed when adding pounds and ounces. Find the total length of angle iron needed if the following pieces are to be cut out.

Add 2' 6" + 5' 6" + 7' 3" + 8' 2" + 1' 1" + 3' 9"

	Feet Column	Inches Column
First Step		
	2'	6"
	5'	6"
	7'	3"
	8'	2"
	1'	1"
Second Step	+ 3'	9"
	26'	27"

Reduce and Add

Third Step	26'	27 ÷ 12
Fourth Step	26 + 2	3
Fifth Step	28'	3"

FRACTIONS

Fractions are commonly used in pipe fabrication for dimensioning a distance that is less than an inch. **Figure 3-7** shows the common inch fractions most often used in pipe fabrication. Because it is difficult for most manual welding to work with fractions smaller than 1/16 in. (such as 1/32 and 1/64 in.), these smaller dimensions are not commonly used in fabrication. When they are used, it is most often with some form of automated or machine welding or cutting process.

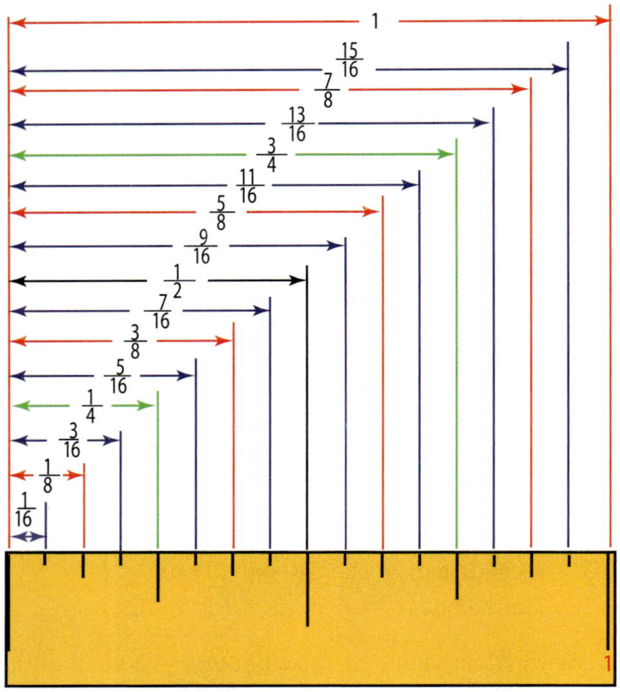

FIGURE 3-7 Fractional measurements of 1 in.

Adding and Subtracting Fractions

The fractional dimensions found on welding fabrication drawings will have a denominator of 2, 4, 8, or 16. If the denominators are not the same, they can easily be made the same, **Table 3-4.** When all of the denominators of the fractions are the same, adding and subtracting can be accomplished quickly by following these simple rules:

1. Add the numerators.
2. Reduce the resulting fraction to the lowest common denominator.

Inch	Half	Fourth	Eighths	Sixteenths
				1/16
			1/8	2/16
				3/16
		1/4	2/8	4/16
				5/16
			3/8	6/16
				7/16
	1/2	2/4	4/8	8/16
				9/16
			5/8	10/16
				11/16
		3/4	6/8	12/16
				13/16
			7/8	14/16
				15/16
1	2/2	4/4	8/8	16/16

TABLE 3-4 Reducing Fractions Conversion Table

Add 1/8 and 3/8. Both denominators are the same, so all you have to do is add the numerators:

$$\text{Add } \frac{1}{8} + \frac{3}{8}$$

$$\frac{1+3}{8}$$

$$\frac{4}{8}$$

Subtract 5/8 from 7/8. Both denominators are the same, so all you have to do is subtract the numerators:

$$\text{Subtract } \frac{7}{8} - \frac{5}{8}$$

$$\frac{7-5}{8}$$

$$\frac{2}{8}$$

Finding the Fractions' Common Denominator

When the denominators of two fractions to be added or subtracted are different, one or both must be converted so that both denominators are the same. To convert a denominator, multiply both the numerator and the denominator of the fraction by the same number. For example, to convert 1/4 to 16ths, you would multiply both the numerator and denominator by 4: $4 \times 1 = 4$ and $4 \times 4 = 16$, which is 4/16. There are conversion tables available to convert the denominators and numerators, **Table 3-5.** To add 1/2 and

Inch Fraction	Inch Decimal	mm
1/16	1.5875	0.062500
1/8	3.1750	0.125000
3/16	4.7625	0.187500
1/4	6.3500	0.250000
5/16	7.9375	0.312500
3/8	9.5250	0.375000
7/16	11.1125	0.437500
1/2	12.7000	0.500000
9/16	14.2875	0.562500
5/8	15.8750	0.625000
11/16	17.4625	0.687500
3/4	19.0500	0.750000
13/16	20.6375	0.812500
7/8	22.2250	0.875000
15/16	23.8125	0.937500
1	25.4000	1.000000

TABLE 3-5 Standard Linear Fractional and Decimal Units to Metric Units

1/16 where both denominators are different, you must first find the common denominator. In this case, it is 16.

Convert the Fraction $\frac{1}{2} = \frac{8}{16}$

Add $\frac{8}{16} + \frac{1}{16}$

$\frac{8+1}{16}$

$\frac{9}{16}$

Reducing Fractions

Some fractions can be reduced to a lower denominator. For example, 4/8 is the same as 1/2. When you're working with a rule or tape measure on the job, you can locate either measurement. Reductions may be necessary only when you are adding or subtracting several different dimensions or various fractional units.

NOTE

It is important to be able to communicate clearly on the job. For example, you would want to reduce the fraction and ask for a 3/8-in.-thick piece of steel, not a 6/16-in.-thick piece of steel.

The normal way to reduce a fraction is to find the largest number that can be divided into both the denominator and numerator. If you are good at math, that can be easily done; but if you are not good at doing math in your head, there is an alternative way of reducing fractions. For reducing fractions in a welding fabrication shop, it is often easiest to divide both the numerator and denominator by 2. This method will simplify the reduction because all the fractional units found on shop rules and tapes are divisible by 2 (e.g., halves, fourths, eighths, 16ths, and 32nds). Using this method may require more than one reduction, but the simplicity of dividing by 2 offsets the time needed to repeat the reduction. For example, both the denominator and numerator of 4/8 can be divided by 2, so 4 ÷ 2 = 2 and 8 ÷ 2 = 4. That would make the new fraction 2/4, which can be reduced again by dividing the denominator and numerator once more by 2. This last division results in 2/4 being reduced to 1/2. Reduction of fractions becomes easier with practice.

Reduce 14/16 and 4/16 to their lowest common denominators:

Reduce $\frac{14}{16}$

$\frac{14 \div 2 = 7}{16 \div 2 = 8}$

$\frac{7}{8}$

The new fraction for 14/16 is 7/8 in., which is the lowest form.

First Reduction $\frac{4}{16}$

$\frac{4 \div 2 = 2}{16 \div 2 = 8}$

Second Reduction $\frac{2}{8}$

$\frac{2 \div 2 = 1}{8 \div 2 = 4}$

$\frac{1}{4}$

The new fraction for 4/16 is 1/4 in., which is the lowest form.

MULTIPLYING AND DIVIDING FRACTIONS

Although fractions can be multiplied and divided directly, the easiest way to work with them is to convert the fraction to a decimal fraction and, if needed, convert it back to a fraction once you have an answer. By converting the fraction to a decimal fraction, you can easily use a calculator to do the math. One slight problem with doing that, however, is that some fractions can result in recurring decimals when they are converted to decimal fractions and then divided; for example, 1/3 becomes 0.333333333. If you add 0.333333333 + 0.333333333 + 0.333333333, you get 0.999999999, which is not equal to 1, but if you add 1/3 + 1/3 + 1/3, you get 3/3, which is 1. But realistically, 0.999999999 is only 0.0000000001 less than 1, and most welding fabrications use a dimensional tolerance of ±1/16 to ±1/8, which is a lot larger than that.

Multiplying Fractions

To multiply fractions without converting, follow these steps:

1. Multiply the two numerators to get the new numerator.
2. Multiply the two denominators to get the new denominator.
3. Simplify the resulting fraction, if possible.

Equation 4

$$A = L \times W$$

where

A = area
L = length
W = width

Problem 4

Find the area of a 3/4" long by 7/8" wide piece of metal:

Multiply A = L × W

$$A = \frac{3}{4} \times \frac{7}{8}$$

$$A = \frac{3 \times 7 = 21}{4 \times 8 = 32}$$

$$A = \frac{21}{32}$$

Answer 4

The area would be 21/32 in.^{2}.

Dividing Fractions

To divide fractions, invert the divisor, and then multiply using steps 2 and 3 as given previously.

To determine how many 1/4-in. pieces could be cut out of 3/4 in., you would do the following:

$$\text{Divide Number} = \frac{3}{4} \div \frac{1}{4}$$

$$\text{Number} = \frac{3}{4} \times \frac{4}{1}$$

$$\text{Number} = \frac{3 \times 4 = 12}{4 \times 1 = 4}$$

$$\text{Number} = \frac{12}{4}$$

The fraction 12/4 can be reduced to 3, so you could cut three pieces from the 3/4-in. piece of metal.

CONVERTING NUMBERS

Dimensions on a drawing are usually given in a consistent format and unit type. For example, everything would be in standard units or metric units. Very seldom will you have to make conversions when reading a drawing. However, you may be asked to install a valve or mount a pump that is dimensioned in metric while everything else is given as standard units. You may be able to simply measure the part using a tape or rule of the same measurement type as your drawing. Sometimes you will be working from the manufacturer's drawing, and then you may have to make conversions of measurements.

Converting Fractions to Decimals

From time to time, it may be necessary to convert fractional numbers to decimal numbers. A fraction-to-decimal conversion is needed before most calculators can be used to solve problems containing fractions. There are some calculators that will allow the inputting of fractions without converting them to decimals.

RULE: To convert a fraction to a decimal, divide the numerator (top number in the fraction) by the denominator (bottom number in the fraction) or use a conversion chart.

To convert 3/4 to a decimal:

$3 \div 4 = 0.75$

To convert 7/8 to a decimal:

$7 \div 8 = 0.875$

Tolerances

All measuring, whether on a part or on the drawing, is essentially an estimate because no matter how accurately the measurement is made, there could always be a more accurate way of making it. The more accurate the measurement, the more time it takes to obtain it. To save time while still making an acceptable measurement, dimensioning tolerances have been established. Most drawings usually state a **dimensioning tolerance**, the amount by which the part can be larger or smaller than the stated dimensions and still be acceptable. Tolerances are usually expressed as plus (+) and minus (−). If the tolerance is the same for both the plus and the minus, it can be written using the symbol ± (pronounced "plus or minus"), **Table 3-6**. In addition to the tolerance for a part, there may be an overall tolerance for the completed weldment. This dimension ensures that if all the parts are either too large or too small, their cumulative effect will not make the completed weldment too large or too small. Most weldments use a tolerance of ±1/16 in. or ±1/8 in., **Figure 3-8**.

Drawing Dimensions		Acceptable Dimensions	
Dimension	Tolerance	Minimum	Maximum
12"	± 1/8"	11 7/8"	12 1/8"
2' 8"	± 1/4"	2' 7 3/4"	2' 8 1/4"
10'	± 1/8"	9' 11 7/8"	10' 1/8"
11"	± 0.125"	10.875"	11.125"
6'	± 0.25"	5' 11.75"	6' 0.25"
250 mm	± 5 mm	245 mm	255 mm
300 mm	+ 5 mm - 0 mm	300 mm	305 mm
175 cm	± 10 mm	174 cm	176 cm

TABLE 3-6 Example of Tolerance Ranges

Converting Decimals to Fractions

The process of converting decimals to fractions is less exact than the conversion of fractions to decimals. Except for specific decimals, the conversion will leave a remainder unless a small enough fraction is selected. For example, the decimal 0.765 is very close to the decimal 0.75, which easily converts to the fraction 3/4. The difference between 0.765 and 0.75 is 0.015 (0.765 − 0.75 = 0.015), which is within the acceptable tolerance for most welding

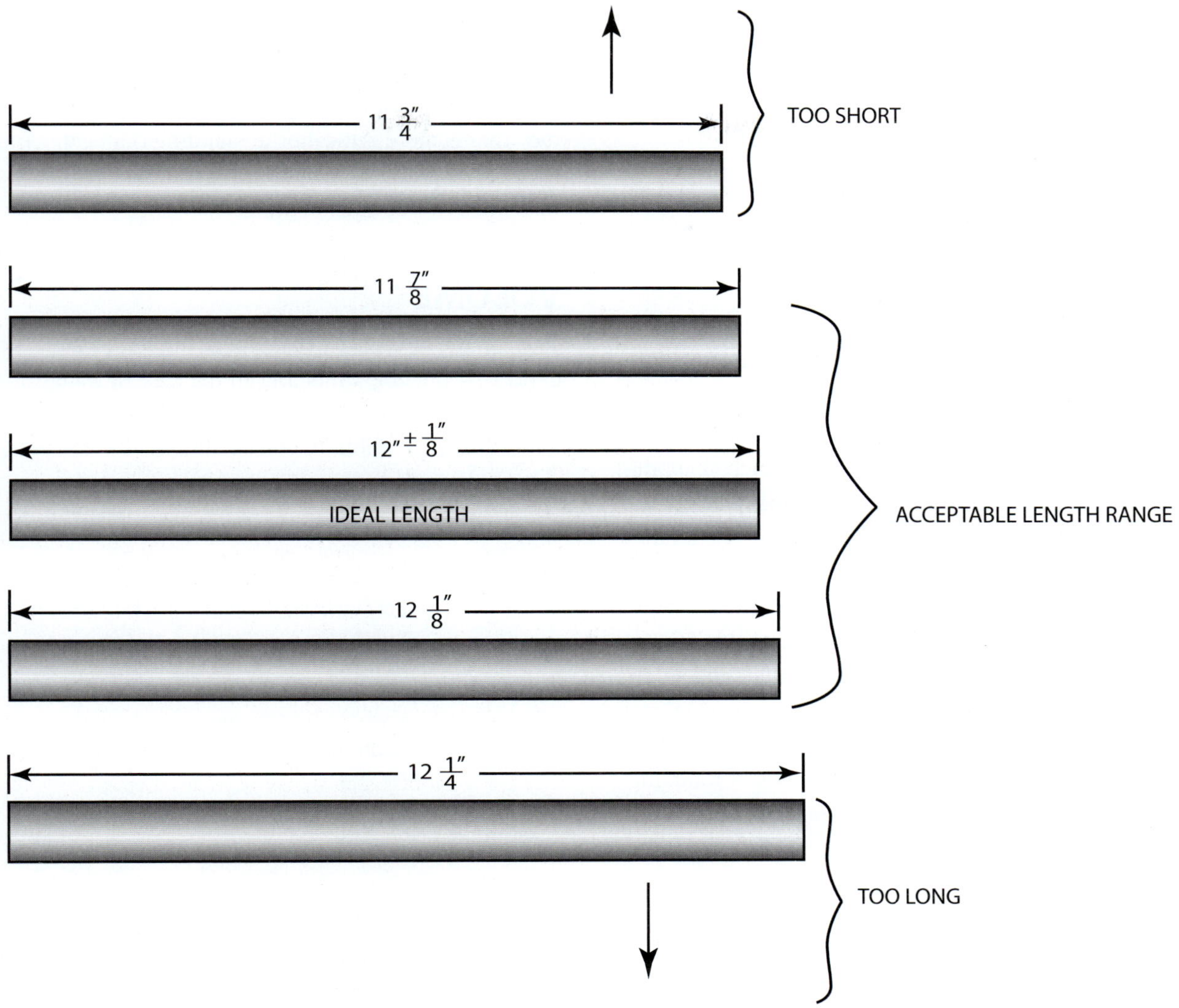

FIGURE 3-8 Tolerances within a range.

applications. If you are working to a ±1/8-in. (3-mm) tolerance that has up to a 1/4-in. (6-mm) difference from the minimum to maximum dimensions, a measurement of 3/4 is acceptable. More accurately, 0.765 can be converted to 49/64 in., a dimension that would be hard to lay out and impossible to cut using a hand torch.

RULE: To convert a decimal to a fraction, multiply the decimal by the denominator of the fractional units desired; that is, for 8ths (1/8), use 8; for 4ths (1/4), use 4; and so on. After multiplying, place the product (dropping or rounding off the decimal remainder) over the fractional denominator used as the multiplier.

To convert 0.75 to 4ths:

0.75 × 4 = 3.0 or 3/4

To convert 0.75 to 8ths:

0.75 × 8 = 6.0 or 6/8, which will reduce to 3/4

To convert 0.51 to 4ths:

0.51 × 4 = 2.04 or 2/4, which will reduce to 1/2

CONVERSION CHARTS

Occasionally, a welder must convert the units used on the drawing to the type of units used on the layout rule or tape. Fortunately, charts are available that can be used to easily convert between fractions, decimals, and metric units. To use these charts, **Table 3-3**, locate the original dimension and then look at the dimension in the adjacent columns of the new units needed.

To convert 1/16 in. to millimeters:

1/16 in. = 1.5875 mm

To convert 0.5 in. to a fraction:

0.5 in. = 1/2 in.

To convert 0.375 in. to millimeters:

0.375 in. = 9.525 mm

To convert 25 mm to a decimal inch:

25 mm = 0.98425 in.

To convert 19 mm to a fractional inch:

19 mm = 3/4 in. (approximately)

Both metric-to-standard conversions and standard-to-metric conversions result in answers that often contain long strings of decimal numbers. Often this new converted number, because of the decimals now attached to it, cannot be easily located on the rule or tape. In addition, most of the layout and fabrication work welders perform does not require such levels of accuracy. These small decimal fractions, in inch or millimeter scales, represent such a small difference that they cannot be laid out with a steel rule or tape. Such small differences can be important to some weldments, but in these cases, some machining is required to obtain that level of accuracy. Because these small units are not normally included in a layout, they can be rounded off. Round off millimeter units to the nearest whole number: for example, 19.050 mm would be 19 mm; 1.5875 mm would be 2 mm; and so on. Round off decimal inch units to the nearest 1/16-in. fractional unit; 0.47244 in. would become 0.5 in. (1/2 in.), and 0.23622 in. would become 0.25 in. (1/4 in.). In both cases of rounding, the whole number obtained is well within most welding layout and fabrication drawing tolerances.

Using the rounding off method of conversions with the conversion chart makes the following converted units easier to locate on rules and tapes:

To convert 1/2 in. to millimeters:

1/2 in. = 13 mm

To convert 0.625 in. to millimeters:

0.625 in. = 16 mm

To convert 2 3/4 in. to millimeters:

$2 \times 25.4 = 50.8$

$3/4 = 19.0$

$50.8 + 19 = 69.8$

69.8 rounded to 70 mm

To convert 5.5 in. to millimeters:

$5 \times 25.4 = 127.0$

$0.5 = 12.7$

$127 + 12.7 = 139.7$

139.7 rounded to 140 mm

To convert 10 mm to fraction inches:

10 mm = 3/8 in.

To convert 14 mm to decimal inches:

14 mm = 0.5625 in.

To convert 300 mm to fraction inches:

$300 \div 25.4 = 11.81$ in., rounded to 11 13/16 in.

To convert 240 mm to decimal inches:

$240 \div 25.4 = 9.44$ in., rounded to 9 7/16 in.

When a weldment's specifications do call for an accuracy that is more critical or demanding than can expected to be found with a steel rule and marker, the parts are often machined to size after welding.

VOLUME

The volume or capacity of a pipe, cylinder, or tank is often expressed in cubic units, but in the case of water or other fluids, volume may be expressed as gallons (liters in the metric system), **Table 3-7**. The volume of a straight pipe, cylinder, or tank can be found with Equation 1.

Volume Units		
cubic inches	cu in	in^3
cubic foot	cu ft	ft^3
cubic yard	cu yd	yd^3
pint	pt	
quart	qt	
gallon	gal	
liter	L	

TABLE 3-7 Volume Units

Equation 5

$$V = 1 \times \pi \times r^2$$

Where

V = volume

1 = length of the pipe

π = Pi, which is 3.14

r^2 = the radius of the pipe squared

Problem 5

The inside diameter (ID) of a 3-in. Schedule 40 pipe, as shown in **Table 3-8**, is 3 in. Find the volume of a 36-in.-long section.

V = volume

1 = 36 in.

$\pi = 3.14$

r = 1.5 (1/2 the 3-in. diameter)

$V = 1 \times \pi \times r^2$

$V = 36 \times 3.14 \times 1.5 \times 1.5$

$V = 36 \times 3.14 \times 2.25$

$V = 36 \times 7.065$

$V = 254.34\ in.^3$

Answer 5

The volume of the 36-in.-long, 3-in.-diameter section of Schedule 40 pipe is $254.34\ in^3$.

Low Carbon Steel

PIPE

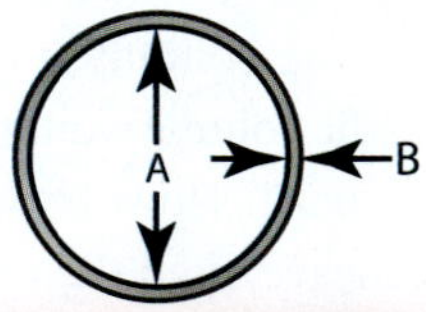

ID 'A' (inch)	Wall Thickness 'B' (inch)	OD (inch)	SCH	Weight per foot (lb)*
1/4	0.08	0.54	40	0.42
3/8	0.091	0.675	40	0.57
1/2	0.109	0.84	40	0.85
3/4	0.113	1.05	40	1.13
1	0.133	1.315	40	1.68
1 1/4	0.14	1.66	40	2.27
1 1/4	0.191	1.66	80	3
1 1/2	0.145	1.9	40	2.72
1 1/2	0.2	1.9	80	3.63
2	0.154	2.375	40	3.65
2	0.218	2.375	80	5.02
2 1/2	0.203	2.875	40	5.79
2 1/2	0.276	2.875	80	7.66
3	0.216	3.5	40	7.58
3	0.3	3.5	80	10.25
3 1/2	0.226	4	40	9.11
3 1/2	0.318	4	80	12.51
4	0.237	4.5	40	10.79
4	0.337	4.5	80	14.98
5	0.258	5.563	40	14.62
5	0.375	5.563	80	20.78
6	0.28	6.625	40	18.97
6	0.432	6.625	80	28.57
8	0.322	8.625	40	28.55
8	0.5	8.625	80	43.39

*Approximate Weight

TABLE 3-8 Standard Unit Pipe Sizing Table

Low Carbon Steel

PIPE

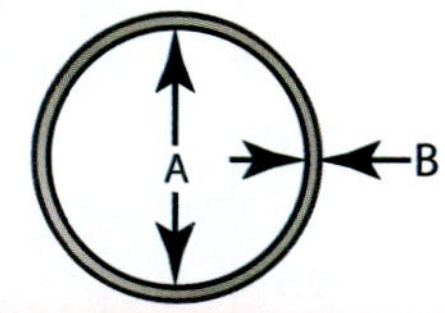

ID 'A' (mm)	Wall Thickness 'B' (mm)	OD (mm)	SCH	Weight per m (kg)*
6.35	2.03	13.72	40	0.63
9.53	2.31	17.15	40	0.85
12.70	2.77	21.34	40	1.27
19.05	2.87	26.67	40	1.68
25.40	3.38	33.40	40	2.50
31.75	3.56	42.16	40	3.38
31.75	4.85	42.16	80	4.47
38.10	3.68	48.26	40	4.05
38.10	5.08	48.26	80	5.41
50.80	3.91	60.33	40	5.44
50.80	5.54	60.33	80	7.48
63.50	5.16	73.03	40	8.62
63.50	7.01	73.03	80	11.41
76.20	5.49	88.90	40	11.29
76.20	7.62	88.90	80	15.27
88.90	5.74	101.60	40	13.57
88.90	8.08	101.60	80	18.63
101.60	6.02	114.30	40	16.07
101.60	8.56	114.30	80	22.31
127.00	6.55	141.30	40	21.78
127.00	9.53	141.30	80	30.95
152.40	7.11	168.28	40	28.26
152.40	10.97	168.28	80	42.56
203.20	8.18	219.08	40	42.53
203.20	12.70	219.08	80	64.63

*Approximate Weight

TABLE 3-9 Metric Unit Pipe Sizing Table

Equation 6

ID = OD − (Wall Thickness + Wall Thickness)

Where

ID = inside diameter
OD = outside diameter
Wall Thickness = the pipe thickness as measured, or from a pipe specification table

Problem 6

To find the volume of a 100-mm-long piece of straight 10 gauge round tubing that has an outside diameter (OD) of 73.03 mm, you must first determine the ID, **Table 3-9**. SI lists the wall thickness of the tubing as 3.43 mm.

ID = OD − (Wall Thickness + Wall Thickness)
ID = 73.03 − (3.43 + 3.43)
ID = 73.03 − 6.86
ID = 66.17 mm

The ID is 66.17 mm.

$r = ID \div 2$
$r = 66.17 \div 2$
$r = 33.08$ mm

The radius is 33.08 mm.

$V = l \times \pi \times r^2$
$V = 100 \times 3.14 \times 33.08 \times 33.08$
$V = 100 \times 3.14 \times 1092$
$V = 100 \times 3429$
$V = 342{,}900 \text{ mm}^3$

Answer 6

The volume of the 100-mm-long 73.03-mm diameter section of Schedule 10 gauge is 342,900 mm^3.

An approximation of the volume of a pipe fitting such as a 90-degree elbow, valve, or reducer can be estimated using the standard volume formula given here. The exact volume of pipe fittings can be obtained from the manufacturer technical data sheet.

Tank Capacities

The calculation of the volume of a tank with a flat head and bottom would use the same formula as the one for finding the volume of a straight pipe. However, many of the tanks connected to a manufacturing, heating, or processing system have convex or concave pressure heads. These curved heads allow the tank to withstand higher pressure that flat-headed tanks. Calculating the volume of the head is calculated the same way whether it is convex (adds to the volume of the tank) or concave (reduces the volume), **Figure 3-9**. The volume of a convex or concave head can be found by using Equation 2.

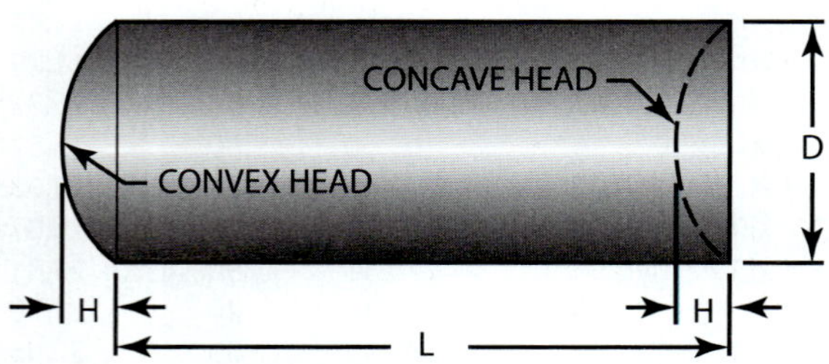

FIGURE 3-9 Problem 7.

Equation 7

$$P = 1.5708 \times H \times \left(\frac{H^2}{3} + \frac{D^2}{4}\right)$$

P = volume of convex or concave head
1.5708 = constant
H = depth of convex or concave head
D = diameter of tank

Problem 7

Find the volume of a convex tank head that is 4 feet in diameter and 1 foot deep.

$$P = 1.5708 \times H \times \left(\frac{H^2}{3} + \frac{D^2}{4}\right)$$

$$P = 1.5708 \times 1 \times \left(\frac{1^2}{3} + \frac{4^2}{4}\right)$$

$$P = 1.5708 \times 1 \times \left(\frac{1}{3} + \frac{16}{4}\right)$$

$$P = 1.5708 \times 1 \times (0.333 + 4)$$

$$P = 1.5708 \times 1 \times (4.333)$$

$$P = 1.5708 \times 4.333$$

$$P = 6.8^3 \text{ ft}$$

Answer 7

The volume of the convex head is 6.8 ft^3. If the tank has two convex heads, then the volume of the tank would be increased by twice the volume of one head, and if both heads are concave, the volume would be decreased by twice the volume of one head. However, if one head is convex and the other is concave, then their volumes cancel out.

Measuring

Measuring for most welded fabrications does not require accuracies greater than what can be obtained with a steel rule or a steel tape, **Figure 3-10**. Tape measures and steel rules are available in standard units, decimal fractions, and metric units. Some may even have two or more different measuring units on the same tape. Using tapes and rules with multiple measuring units can make laying out weldments much easier when you are working with drawings and parts that have different unit measuring systems, **Figure 3-11**.

Larry Jeffus

FIGURE 3-10 Steel tapes with widths of 3/4 in. to 1 in. (19 mm to 25 mm) are more useful than narrower tapes because wider tapes can be extended more without buckling.

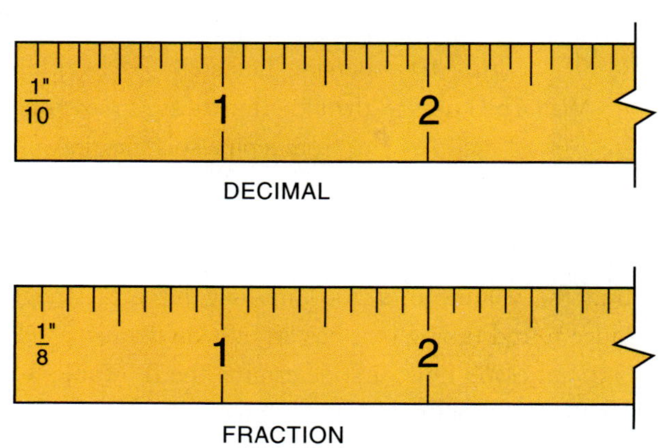

FIGURE 3-11 Some tapes have both decimal and fractional measurements.

Summary

Pipe welders are often required to do math that ranges from simple adding and subtracting dimensions to complex calculations affecting the entire job site. Having a good working knowledge of basic math principles will enable the welder to solve simple problems with pencil and paper or soapstone and scrap metal used as writing implements. The more complex problems can be solved with the aid of a personal computer.

Examples of personal computers available to pipe welders include programmable calculators, smart phones, tablets, and other mobile devices. These devices have significantly improved the pipe welders' access to programs and apps that help them solve math problems. However, it is still important that pipe welders understand the basic math principles outlined in this chapter so that when they enter data on one of these devices, they are more likely to recognize the difference between a valid answer and an incorrect one. For example, if the wrong key is struck, the answer displayed could be slightly off—or even thousands of hours or millions of pounds off. Having a good understanding of math will enable the welder to recognize such errors.

Review Questions

1. List three examples of whole numbers.

2. What are the number and decimal terms for the following:

- **(a)** 10,000.0
- **(b)** 0.0001
- **(c)** 100,000.0
- **(d)** 0.1

3. Give an example of a mixed unit.

4. What is the meaning of a subscript number or letter written at the lower right of an expression?

5. Find the total labor bill for 8 hours of work at $30 per hour.

6. List the order of operations that must be followed when a formula has more than one mathematical operation.

7. What is the sum of the following mixed numbers:

- **(a)** 10' 2" + 1' 8" + 4' 1",
- **(b)** 6' 3" + 7' 2" + 3' 4"

8. What would the total length of two sections of pipe be if one section is 4' 6", another pipe section is 3' 1", and the connecting valve is 4"?

9. What is the remainder if 4' 10" is cut off of a 12' 0" pipe?

10. If you cut a section of pipe 8' 9" out of a 20' length of pipe, how much of the pipe will remain?

11. Add the following lengths of pipe and convert the answer:

- **(a)** 3' 4" + 5' 6" + 6' 7" + 10' 5",
- **(b)** 9' 2" + 7' 8" + 2' 10" + 8' 6"

12. If you cut 9' 1 7/8" out of a section of pipe 11' 6 1/4" long, how much pipe would remain?

13. Reduce the following fractions to the lowest common denominator:

- **(a)** 4/8,
- **(b)** 16/32,
- **(c)** 10/64,
- **(d)** 5/15

14. List the three steps required to multiply fractions.

15. Convert the following fractions to decimal equivalents:

- **(a)** 1/2,
- **(b)** 7/8,
- **(c)** 5/8,
- **(d)** 3/16

16. Based on the tolerance given, what is the minimum and maximum lengths of the following:

- **(a)** 10" ± 1/16 in.
- **(b)** 5 1/2" ± 1/8

17. What is the volume of a 24-in.-long piece of 8" diameter Schedule 80 pipe?

Chapter 4

Blueprint Reading and Welding Symbols

OBJECTIVES

After completing this chapter, the student should be able to:

- Explain the purpose of mechanical drawings (blueprints) and what type of information they can convey.
- Identify the different types of lines on a mechanical drawing and explain what they represent.
- Sketch orthographic and isometric drawings.
- Demonstrate the ability to find dimensions on a drawing.
- Identify piping and welding symbols.
- Explain the purpose of components that might be connected to a piping system.
- Identify parts shown in an orthographic and isometric drawing.
- Produce a takeoff of materials required to produce a system from a drawing.
- Explain the components of a welding symbol.
- Describe the purpose of groove welds and fillet welds.

KEY TERMS

Alphabet of Lines
Arrow side symbol
Blueprint
Drawing scale
Isometric drawings
Material takeoff
Mechanical drawings (blueprints)
Orthographic drawings
Other side symbol
Perspective drawings
Shutoff valves
Takeoff
Throttling valves

INTRODUCTION

Mechanical drawings (blueprints) are used in construction because without them, the amount of written information needed to accurately convey a design engineer's requirements for even the simplest piping system would require volumes of narrative. The expression "A picture is worth a thousand words" is an understatement when it comes to technical mechanical drawings.

(Continued)

The main purpose of mechanical drawings is to accurately convey the requirements and specifications needed to correctly produce the requested system. Mechanical drawings include information such as size, material, joining methods, part location, part orientation, and surface finishing details.

The term "blueprint" is an old term used to refer to any set of mechanical drawings. It dates to a period of time when an ammonia-based copying process was used to make copies of drawings before modern copiers and printers were invented. The process left the lines white, while turning the rest of the paper blue. The smell of ammonia was always strong around these machines and could linger on the prints for days afterward. The term "blueprint" is still used today; however, thankfully, the ammonia-based copying process has long been abandoned.

Most of the lines, dimensioning, and symbols used on earlier drawings remain the same today. Knowing how to read contemporary mechanical drawings, therefore, will help you read mechanical drawings that were made almost 100 years ago. For example, **Figure 4-1** shows the patent drawings for a pipe welder done in August 1931. It is one of several drawings submitted for an automatic, unshielded, bare wire, pipe-welding machine.

After completing this chapter, you should be able to identify five or more different types of lines used in this drawing and explain each of their uses.

Piping systems can be joined by welding, threading, flanges, brazing or soldering (also called "sweating"), and coupling clamps, **Figure 4-2**. This chapter will focus on welding and threading, but bolts, nuts, and screws will also be covered since they are used to join flanges and other components to piping systems.

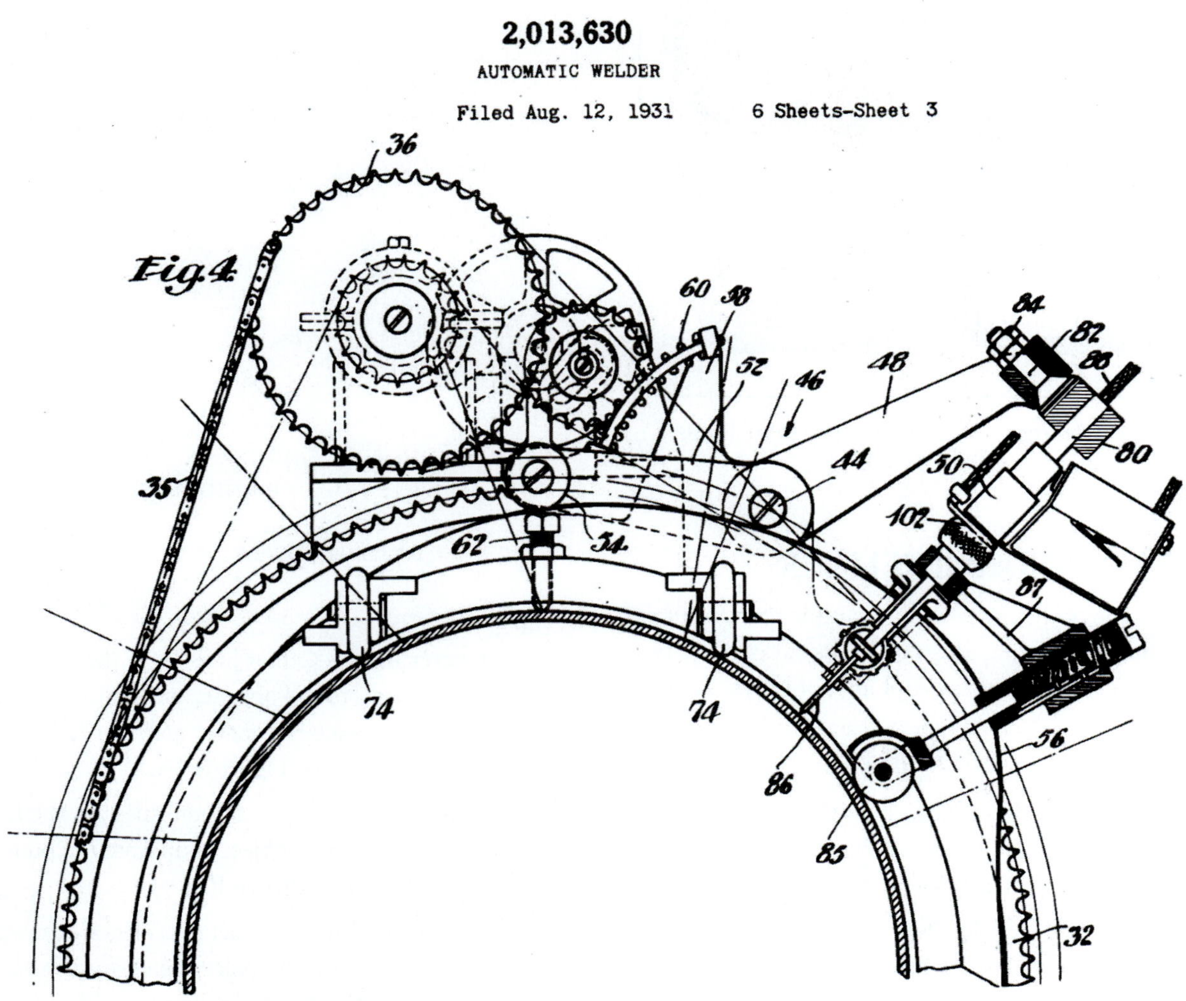

FIGURE 4-1 Patent drawing for an automated welding machine from 1931.

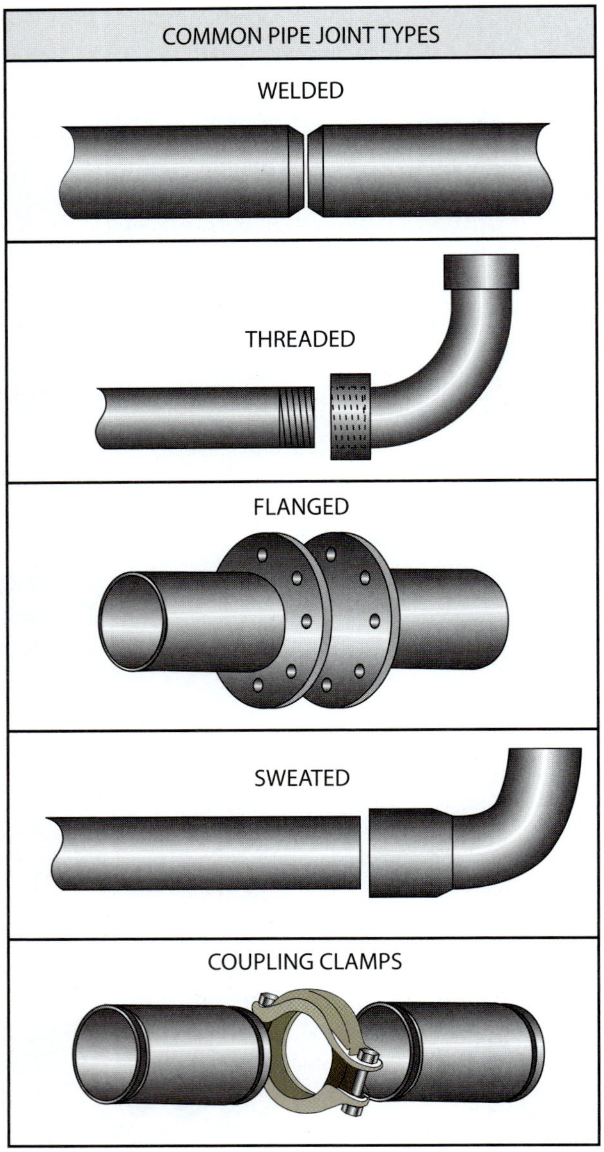

FIGURE 4-2 Common types of pipe-joining methods.

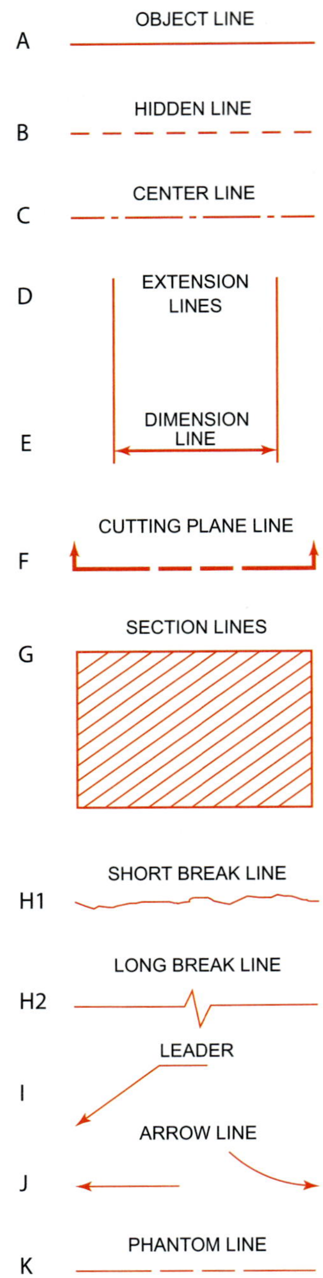

TABLE 4-1 Alphabet of Lines

TYPES OF DRAWING LINES

Different types of lines are used to outline the various surfaces of the object being illustrated. The various line types are collectively known as the **Alphabet of Lines, Table 4-1** and **Figure 4-3**.

The types of lines used on drawings are:

- **Object line**—Object lines show the edge of an object, the intersection of surfaces that form corners or edges, and the extent of a curved surface, such as the sides of a cylinder, **Figure 4-3(A)**.
- **Hidden lines**—Hidden lines show the same features as object lines except that the corners, edges, and curved surfaces cannot be seen because they are hidden behind the surface of the object, **Figure 4-3(B)**.
- **Center lines**—Center lines show the center point of circles, arcs, round objects, and symmetrical objects. They also locate the center point for holes, irregular curves, and bolts, **Figure 4-3(C)**.
- **Extension lines**—Extension lines are the lines that come out from an object and locate the points being dimensioned, **Figure 4-3(D)**.
- **Dimension lines**—Dimension lines are drawn so that their ends touch the object being measured, or they may touch the extension line from the object being measured. Numbers in the dimension line or next to it give the size or length of an object, **Figure 4-3(E).**

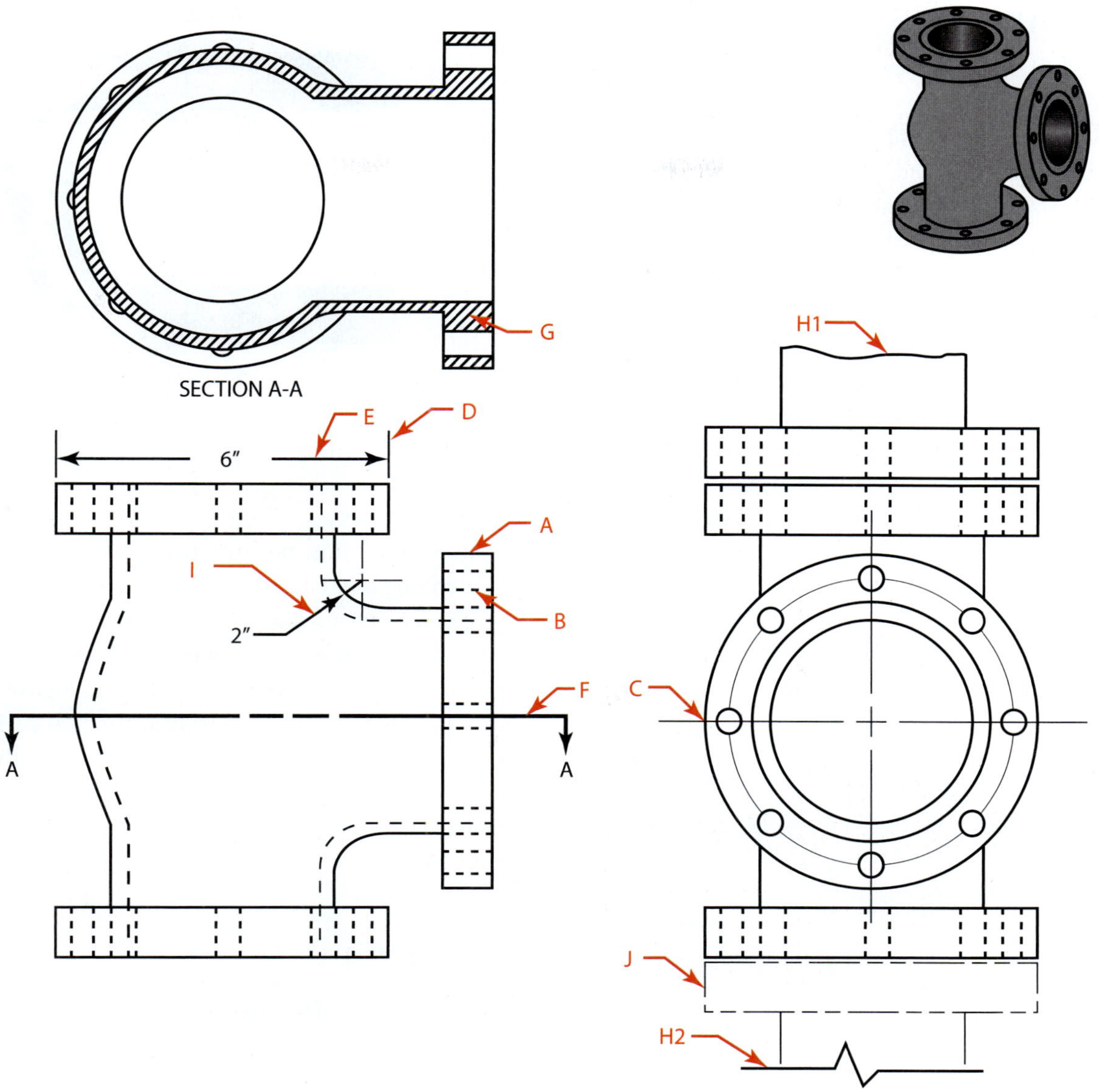

FIGURE 4-3 Examples of the use of the Alphabet of Lines.

- **Cutting plane lines**—Cutting plane lines represent an imaginary cut through the object. They are used to expose the details of internal parts that would not be shown clearly with hidden lines, **Figure 4-3(F)**.
- **Section lines**—Section lines show a surface that has been cut away with a cutting plane line in an imaginary sense to show internal details. Different patterns of section lines are used to show different types of materials. The evenly spaced diagonal lines for cast iron are often used as the default pattern, **Figure 4-3(G)**.
- **Break lines**—There are two types of break lines: long and short. Both show that part of an object has been removed. This is often done when a long, uniform object needs to be shortened to fit the drawing page, **Figure 4-3(H)**.
- **Leaders**—Leaders are used to connect information to the place on the drawing the information applies to. They are also used as the base for welding symbols. Leaders have an angled line with an arrow on one end pointing at the place on the drawing, and a straight horizontal line next to the pertinent information, **Figure 4-3(I)**.
- **Arrow lines**—Arrow lines are used to show the location of a dimension where the dimension was too large to fit in the space, **Figure 4-3(J)**.
- **Phantom lines**—Phantom lines show an alternative position of a moving part or the extent of motion, such as the on/off position of a power switch. They can also be used as a placeholder for a part that will be added later, **Figure 4-3(K)**.

PIPE DRAWINGS

Orthographic drawings are made of one or more two-dimensional (2D) drawings called "views" that are drawn as if you were looking directly at one side of the object (i.e., front, top, right side, etc.). They are often also referred to as "mechanical drawings." The views are much like a photo that looks straight at one side of an object where only two dimensions of the object can be seen. A second view (photo) of the object is taken to show another side so that the actual shape of the object can be determined, **Figure 4-4**. The alignment of the views is very important because that makes it possible to compare the adjacent views so you can find the third dimension from an adjacent view. Since the views are lined up, you can go in a straight line from one view to the other to get more information, **Figure 4-5**. From the front view of the pipe in **Figure 4-6**, it is not possible to determine how long the fitting is, but with the top view, you can see that it is short.

FIGURE 4-4 The front view of a round cylinder and square cube would both be exactly the same.

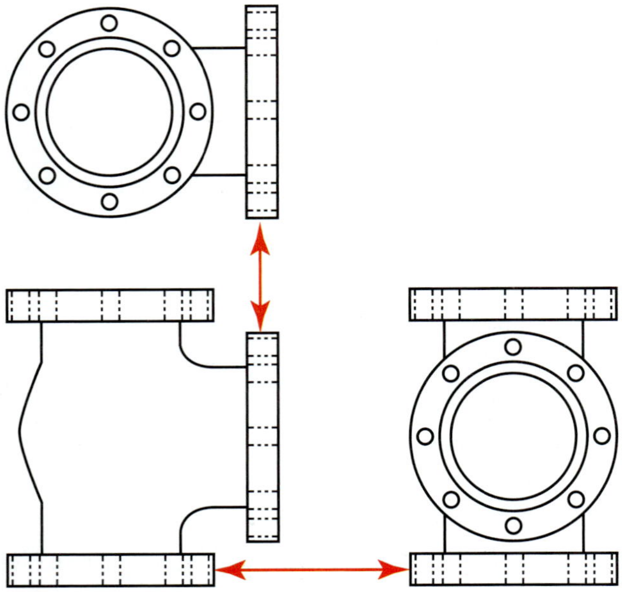

FIGURE 4-5 Parts or items shown in the front-view drawing are directly below the same parts or items shown in the top-view drawing, and directly to the left of the same parts or items in the side-view drawing.

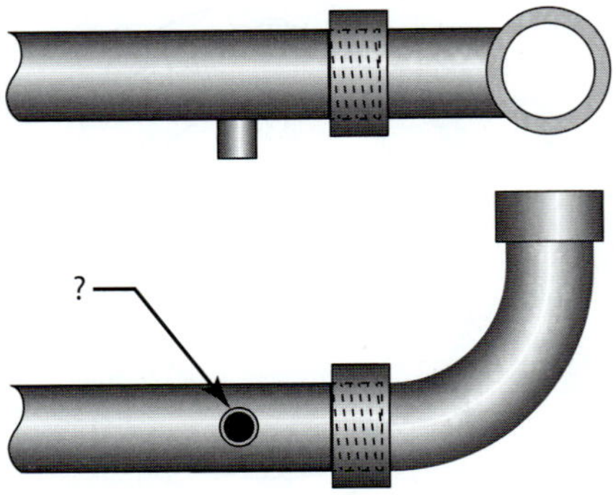

FIGURE 4-6 You have to look at the top view to identify the part shown on the front view.

Although there are six possible views that can be used to make an orthographic drawing, **Figure 4-7**, the most commonly used are the front, top, and right-side views. On a few occasions, a special view, one that looks at the object from an unusual angle, may be used to show details on a surface that is not parallel to one of the six standard views.

PRACTICE 4-1

Sketching Side Views

The following steps will help you sketch the missing right-side view for the 12 various-shaped blocks, **Figure 4-8:**

1. On a sheet of graph paper or a blank sheet of paper, sketch the front and top views of drawing number 1, **Figure 4-9(A)**.
2. Sketch the corners of the right-side view, **Figure 4-9(B).**
3. Referring to the top and front views, sketch the right-side view, **Figure 4-9(C)**.
4. Repeat steps 1–3 with each of the remaining 11 objects, Figure 4-8.
5. List the numbers of the side views of all the objects that are exactly alike, even though the front and top views are not alike. ◆

An advantage of orthographic drawings is that with the use of extension and dimension lines, very detailed dimensions can be shown. A disadvantage of these drawings is that because they are 2D, they require welders to "picture" in their minds what the object would look like in three dimensions (3D).

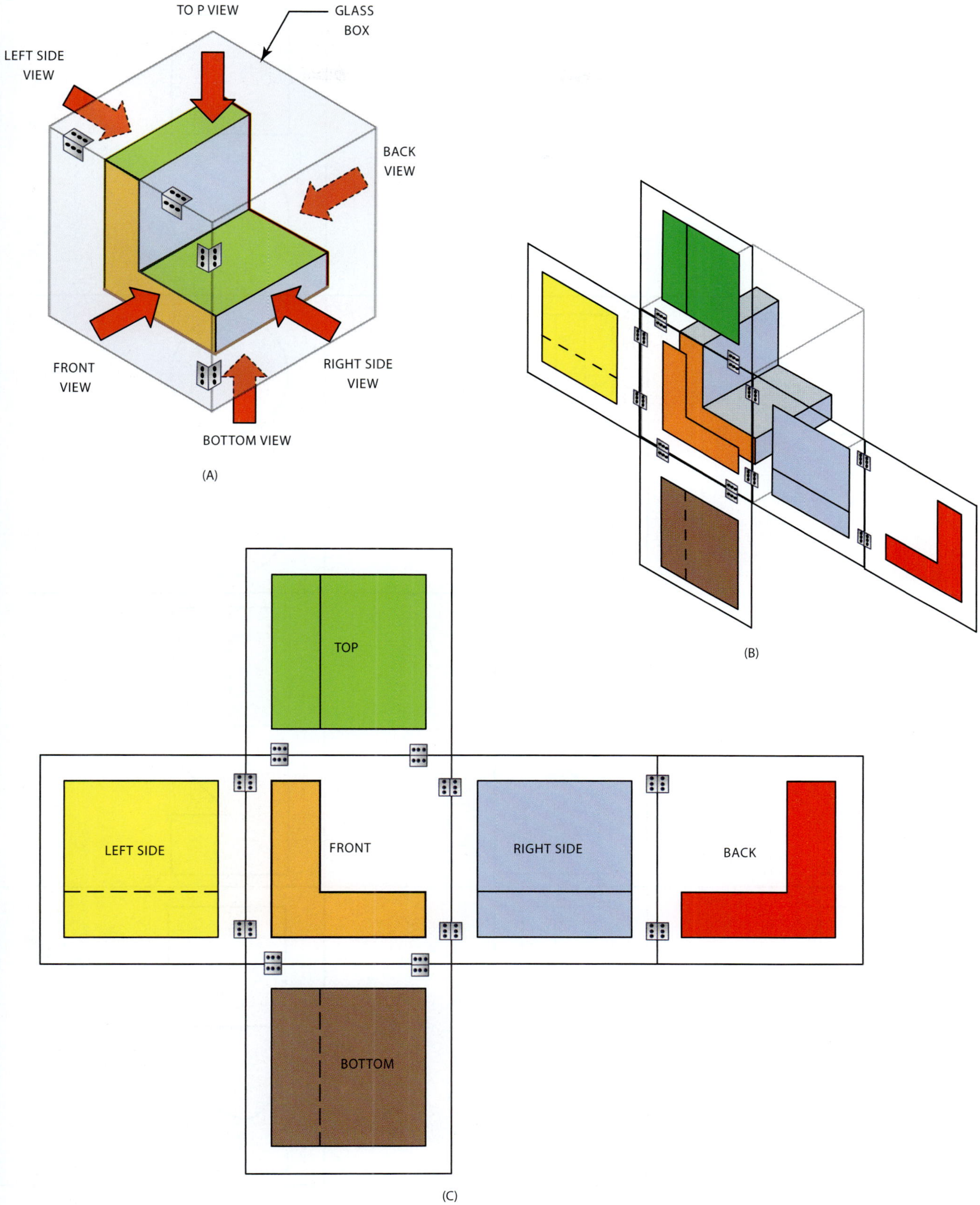

FIGURE 4-7 The orientation of the standard views in a mechanical drawing. (A) The object as seen through a glass box, (B) unfolding the glass box, and (C) laying the glass box flat.

FIGURE 4-8 Practice 4-1.

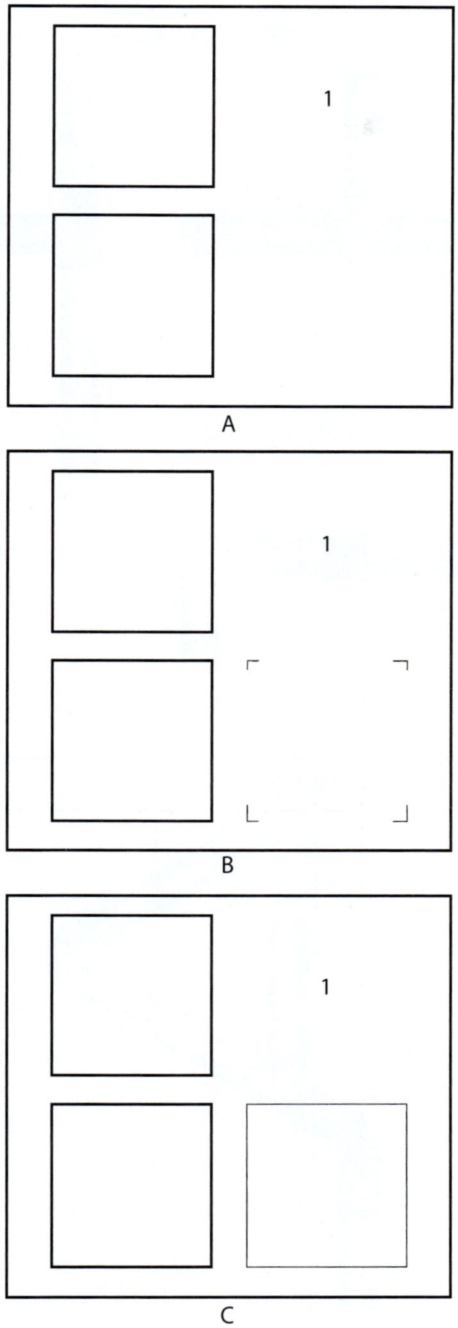

FIGURE 4-9 The steps to follow to sketch the missing view.

Isometric drawings are pictorial (picturelike) drawings that use lines drawn at 30° angles to the right and left of vertical so that the top, right, and left sides can all be shown in a single drawing, **Figure 4-10**. This type of drawing is often easier for a welder to understand than are orthographic drawings. **Perspective drawings** are the most picturelike because the lines converge at points on the horizon, just as railroad tracks appear to converge at the horizon. Isometric drawings are often used for pipe layouts because they can more easily show the relationships of the various components of the system.

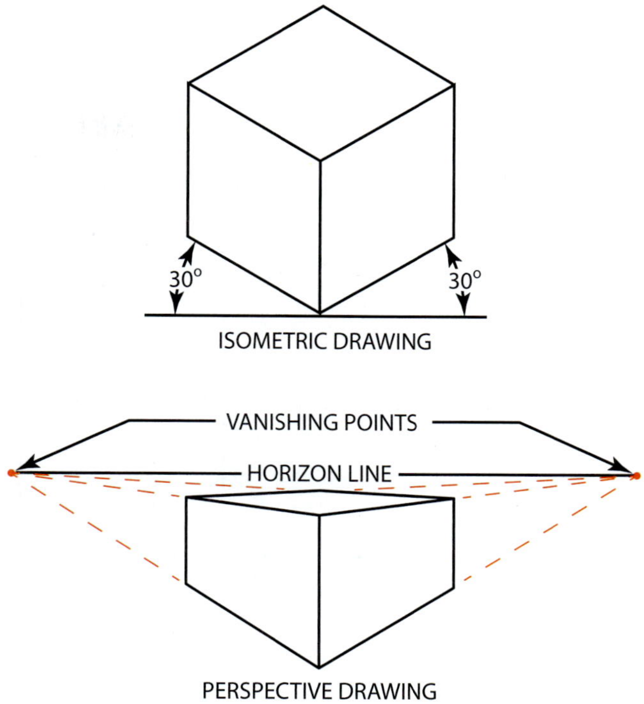

FIGURE 4-10 Two types of pictorial drawings: isometric and perspective.

PRACTICE 4-2

Sketching Isometric Views

The following steps will help you sketch the missing right-side view for the 12 variously shaped blocks, **Figure 4-11:**

1. On a sheet of graph paper or a blank sheet of paper, sketch the isometric corner shown in box 1, **Figure 4-12(A)**.
2. Referring to the block shown in box 1 of Figure 4-8, copy the isometric sketch of the block shown in **Figure 4-12(B).**
3. Sketch each of the remaining 11 objects shown in Figure 4-8, in the same way. ◆

Both orthographic and isometric drawings can be made manually or with a computer program. Once the programs are mastered, computer-aided design (CAD) can create drawings faster and easier than drawing them with pencil and paper. In the past, the pencil-and-paper drawing method was considered more portable than computer-aided drafting; however, with the introduction of touch-screen tablet computers and smart phones, that is no longer true. Sharing drawings between tablets has often eliminated the need to print them. In addition, computer drawings can be more lifelike when the pipes are shaded, **Figure 4-13**. However, both methods can create drawings with enough detail that a skilled welder can follow them to fabricate the piping system.

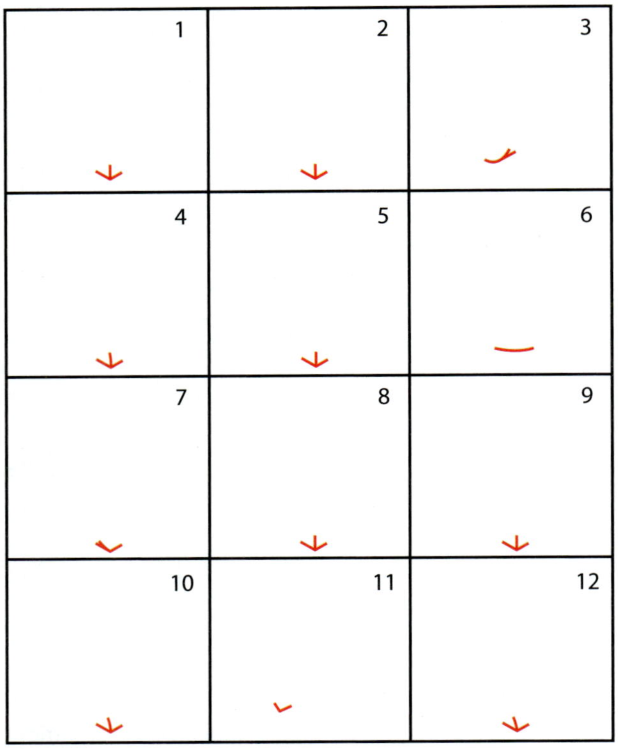

FIGURE 4-11 Practice 4-2.

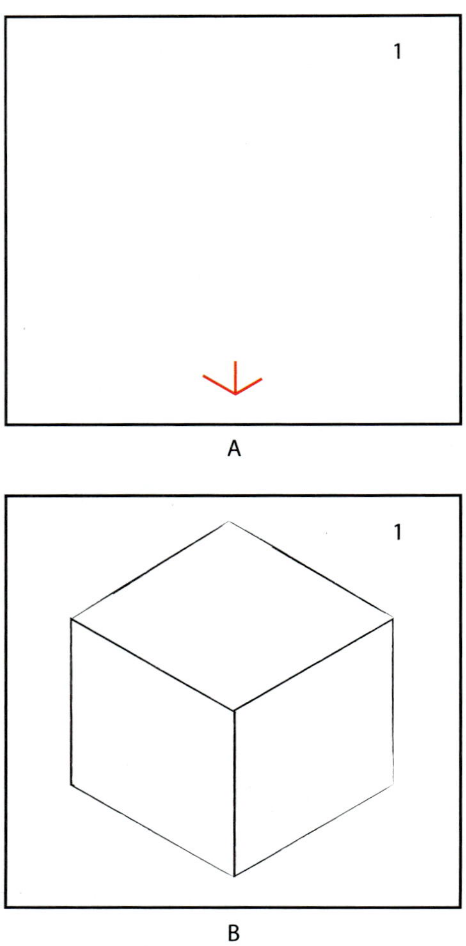

FIGURE 4-12 Example of Practice 4-2.

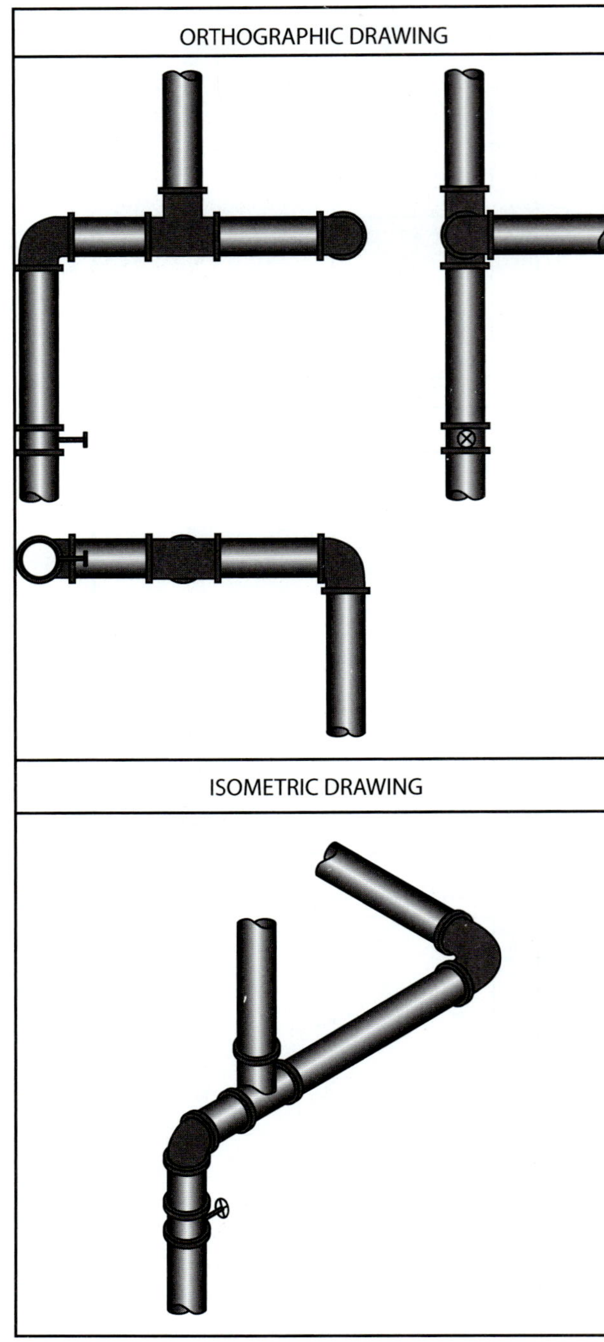

FIGURE 4-13 Computer-generated pipe drawings.

DIMENSIONING

Most of the dimensions on a drawing will be given in standard units, but some systems may use metric units; and parts, such as valves and pumps, may use both systems. Large piping projects typically use feet, inches, and fractions of inches. However, surveyors laying out a cross-country pipeline will typically use feet and decimal fractions of feet rather than inches.

Often it is necessary to look at other views to locate all of the dimensions required to build an object.

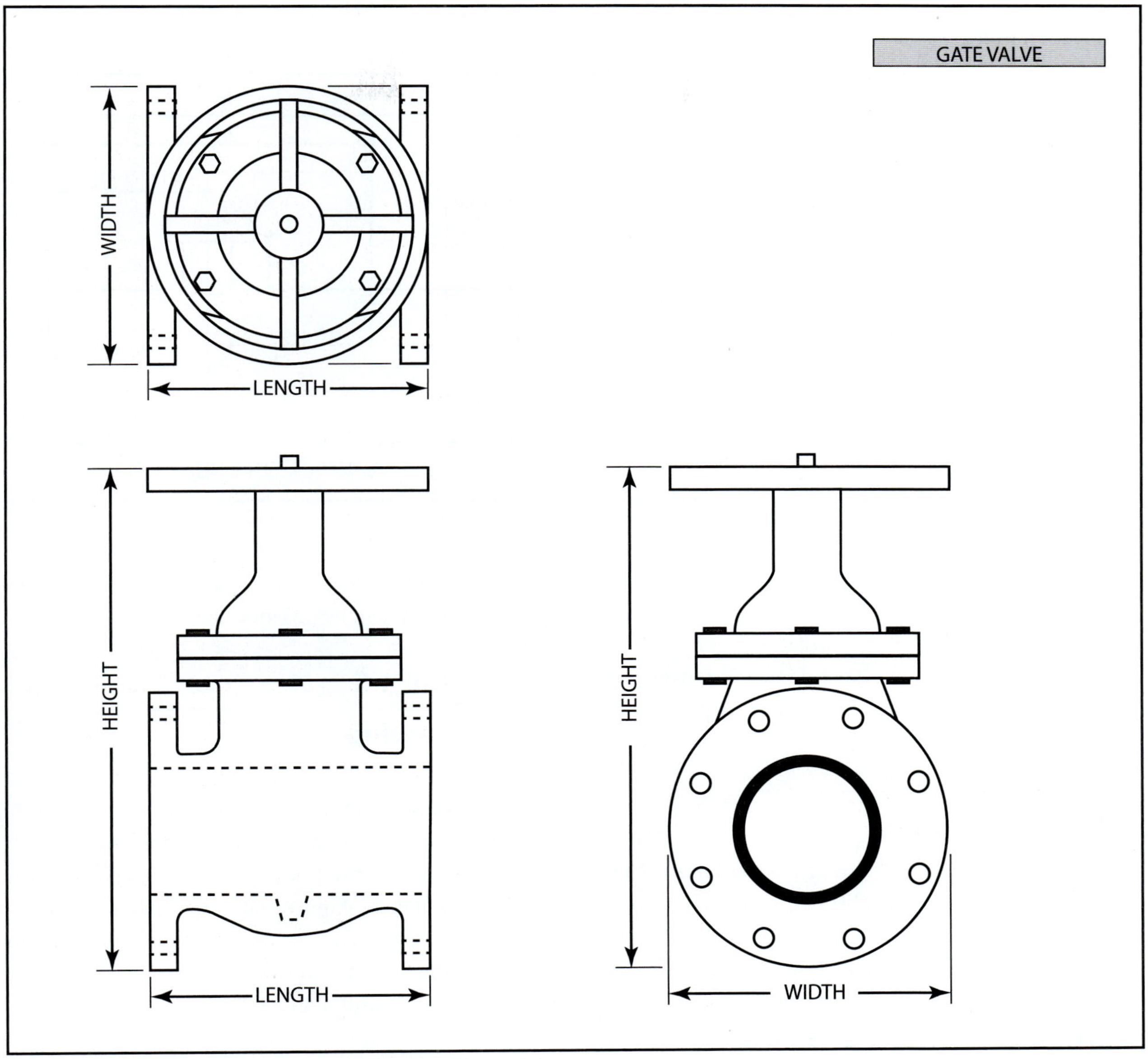

FIGURE 4-14 Where dimensions can be located on drawings.

Knowing how the views are arranged makes it easier to locate dimensions. Length dimensions can be determined from the front and top views, while height dimensions can be found from the front and right-side views. Width dimensions can be found on the top and right-side views, **Figure 4-14**.

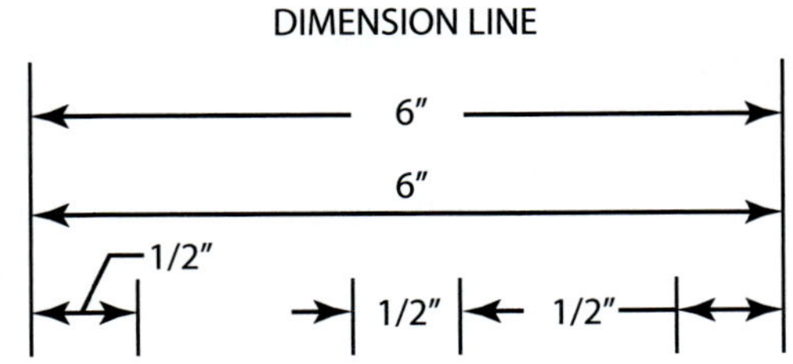

FIGURE 4-15 Methods of dimensioning.

READING DIMENSIONS

The dimensions on a drawing may appear in a gap in the dimension line or above or below the dimension line, or a leader is used to show what the dimension is measuring, **Figure 4-15**. An arrow, slash, or dot may be used at the intersection of the end of a dimension line and an extension line, center line, object line, or hidden line, **Figure 4-16**.

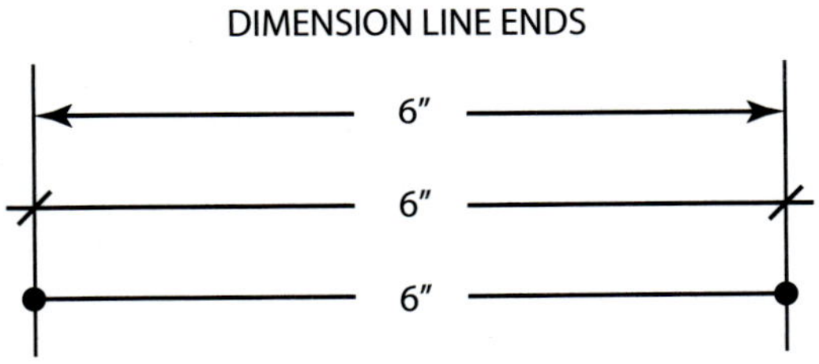

FIGURE 4-16 Types of ends for dimension lines.

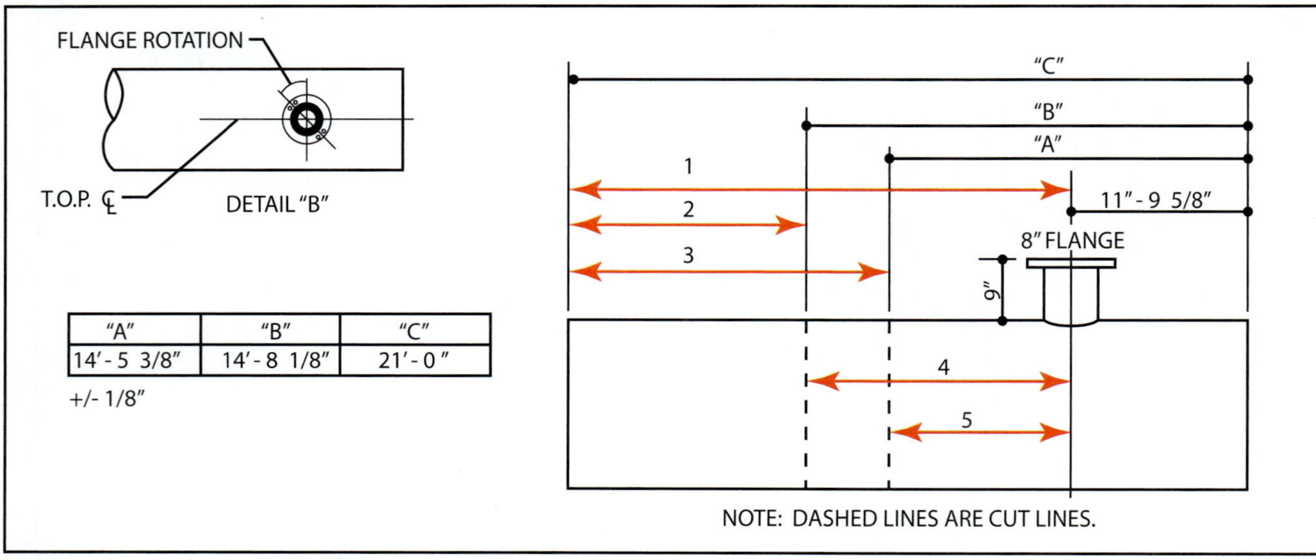

FIGURE 4-17 Practice 4-3.

PRACTICE 4-3

Reading Dimensions

Use basic math to determine dimensions 1 through 5 on **Figure 4-17**. Write the answers on a sheet of paper. ◆

PRACTICE 4-4

Locating Dimensions

Find the missing dimensions 1 through 10 on **Figure 4-18**. On a sheet of paper, write the numbers 1 through 10 vertically. Next to each number, write the standard and metric dimensions represented by the numbered arrows on the drawing. ◆

Finding Missing Dimensions

If the needed dimensions cannot be found on the drawings, try to obtain them by measuring the drawing itself. However, note that even if the original drawing was made very accurately, the paper it is on can change size with changes in humidity, and copies of the original drawing are never the exact same size. The most acceptable way of determining missing dimensions is to contact the person who made the drawing.

Drawing Scale

It is not possible to make every drawing to actual size; for that reason, you must change the drawing size. The **drawing scale** is used to define the ratio between the drawing size and the size of the actual part that the drawing represents. Sometimes the drawing may be made larger than actual size; other times, it may be made smaller. Often the primary factor that controls the scale used for a drawing is the size of the paper being used. It is not necessary for the welder/fabricator to know the scale used to make the drawing. All this person needs to be able to do is read the dimensions and apply them to the piping system being fabricated.

PRACTICE 4-5

Scale Drawing

You are going to make a scale drawing on a blank 8 1/2" by 11" sheet of paper of the gate valve shown in Figure 4-18. Compare the scale shown on the drawing with a standard rule or an architectural or engineering scale to determine the closest match, **Figure 4-19**. Use the scale that you determine is the best match and convert it to a new scale so that your drawing fits on the sheet of paper with no more than a 1-in. (25-mm) margin. ◆

PRACTICE 4-6

Reading Dimensions Without a Common Denominator

Sometimes the fractional dimensions on a drawing will not have a common denominator. In **Figure 4-20**, the dimensions for the sections of the pipe length are given in 4ths, 8ths, and 16ths. Also, there may be missing dimensions on a drawing, so you will have to calculate them yourself from the information that is provided. To calculate the total length of the pipe section in Figure 4-20, perform the following steps:

1. On a blank sheet of paper, draw vertical lines on the page to form four columns.
2. Label the columns: "Dimension," "Inches," "Fractions," and "16ths."
3. Write the numbers 1 through 14 vertically down the left side of the page. Under the last number (14), write the word "Total"; **Figure 4-21**.

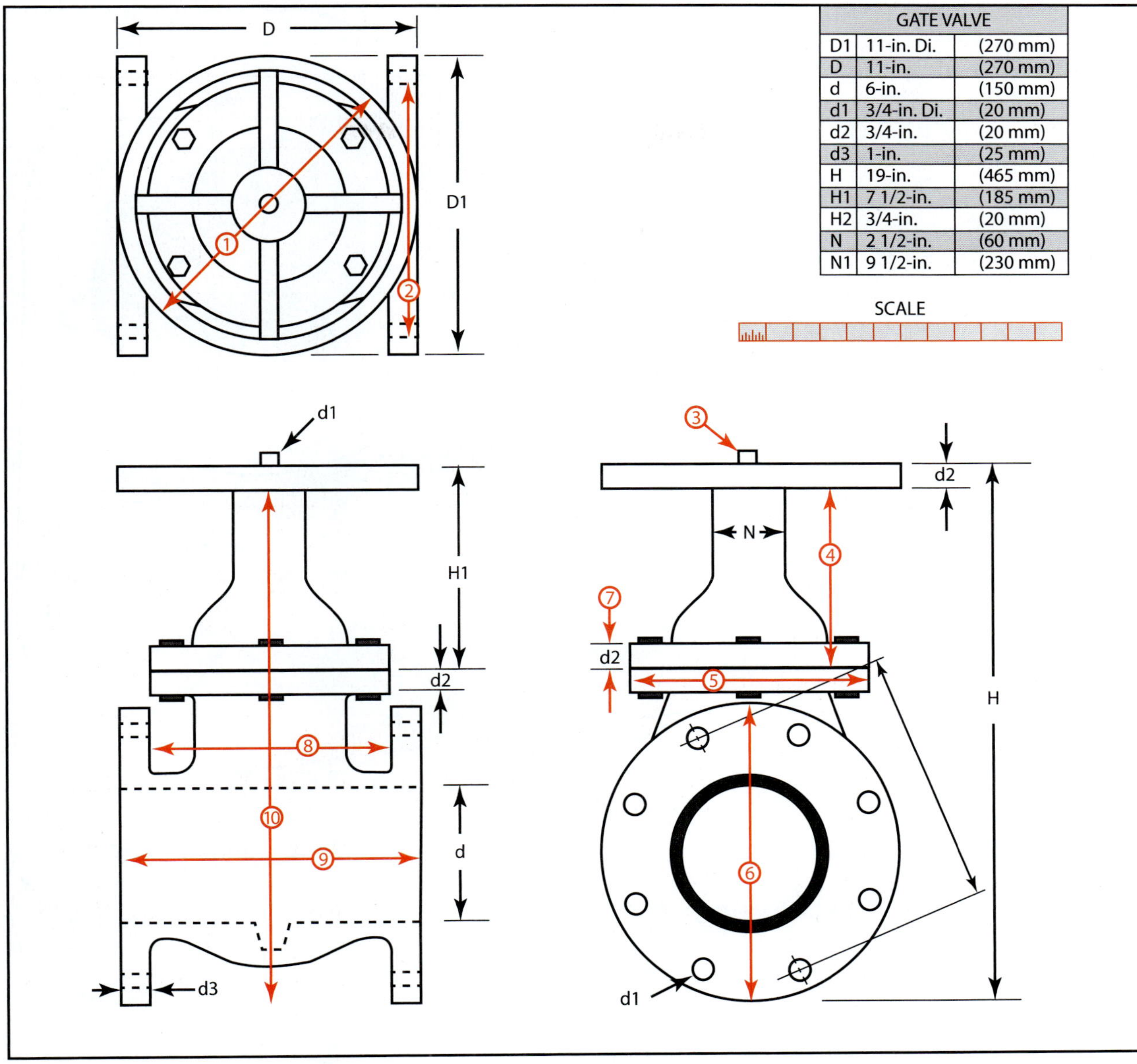

GATE VALVE		
D1	11-in. Di.	(270 mm)
D	11-in.	(270 mm)
d	6-in.	(150 mm)
d1	3/4-in. Di.	(20 mm)
d2	3/4-in.	(20 mm)
d3	1-in.	(25 mm)
H	19-in.	(465 mm)
H1	7 1/2-in.	(185 mm)
H2	3/4-in.	(20 mm)
N	2 1/2-in.	(60 mm)
N1	9 1/2-in.	(230 mm)

FIGURE 4-18 Practice 4-4.

Next, fill in the columns of the table as follows:

4. Enter each of the 14 dimensions given along the bottom edge of the pipe in Figure 4-20 into the "Dimensions" column of your table from Figure 4-21, starting with 30 5/8.
5. Separate each of the dimensions that you listed in the "Dimensions" column into whole numbers and fractions by putting the whole number in the "Inches" column and the fractional amount into the "Fractions" column. (On line 1 in the "Inches" column, write the number "30." On line 1 in the "Fractions" column, write "5/8.")
6. On line 1 in the column labeled "16ths," convert the 5/8 in the previous column to 16ths and write down just the numerator of your answer; 5/8 = 10/16, so write the number 10 in the column. Here, 16ths is used because it is the common denominator for the dimension fractions used in this pipe assembly.
7. Repeat these steps on each line until all of the dimensions have been written on the sheet of paper.
8. Add up all the entries in the "Inches" column and write the total at the bottom on the "Total" line.
9. Add up all the entries in the "16ths" column and write the total at the bottom on the "Total" line.
10. Divide the total of the "Inches" column by 12 to reduce the number to feet and inches.
11. Divide the total of the "16ths" column by 16 to reduce the number to inches and 16ths.
12. Add the results of the last two steps. This will be the total length of the pipe. (Hint: The correct answer is between 35 and 40 feet.) ◆

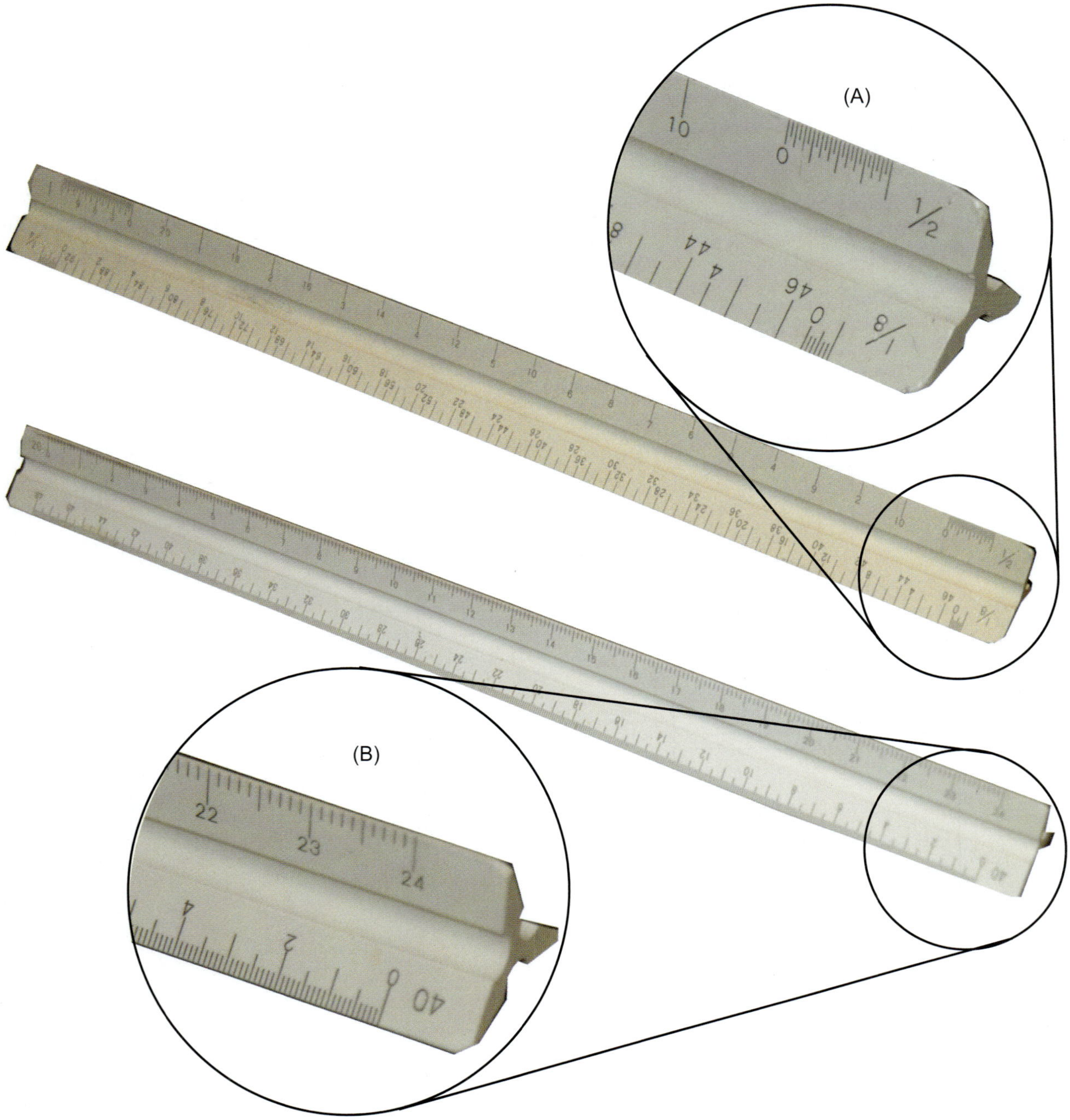

Larry Jeffus

FIGURE 4-19 The two types of scales: standard or architectural.

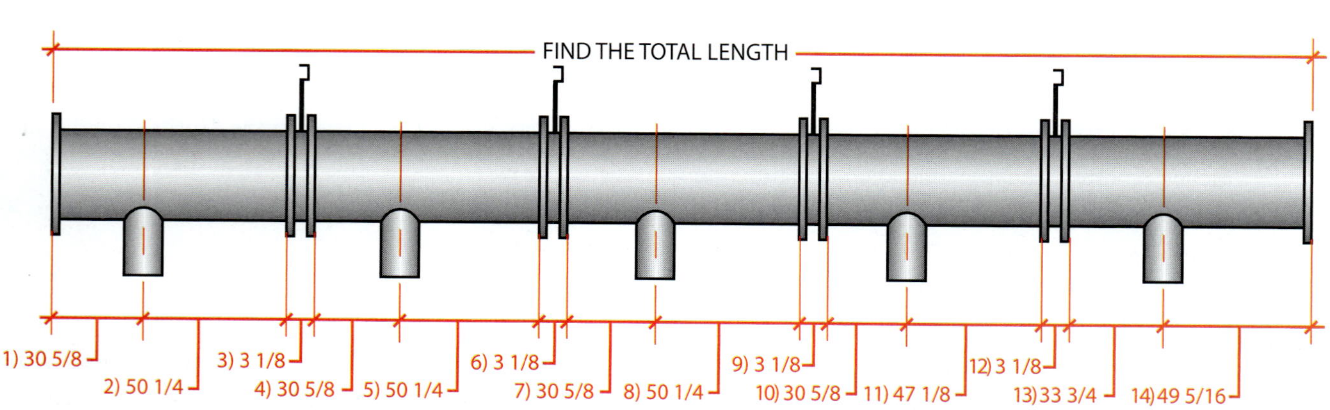

FIGURE 4-20 Practice 4-6.

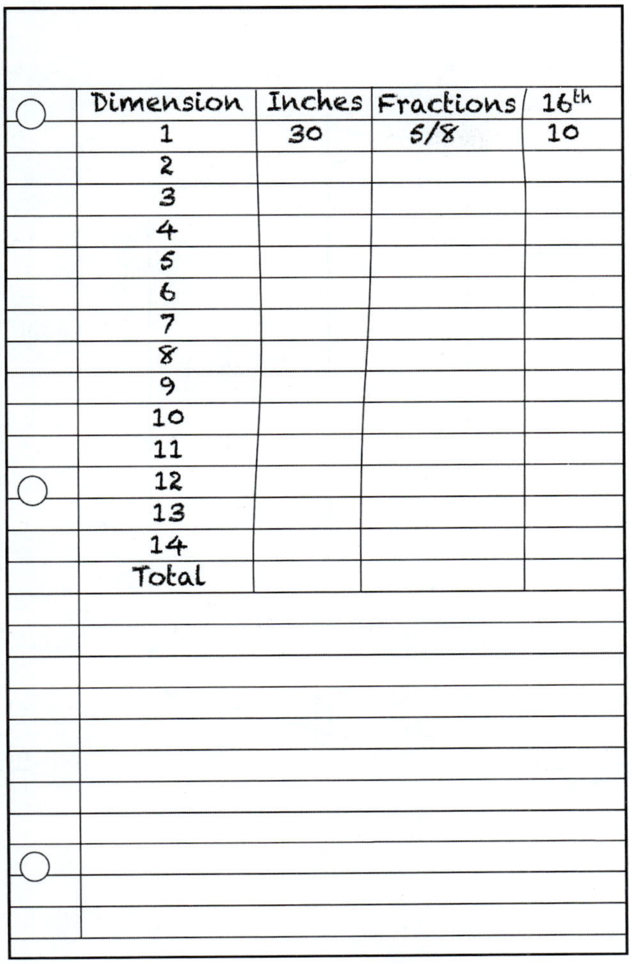

FIGURE 4-21 Worksheet for Practice 4-6.

> **NOTE**
>
> Keep your drawing clean and well away from any welding. Avoid writing or doing calculations on the drawing unless you are noting changes. Sometimes it may be necessary to make a change in the system as it is being fabricated. When these changes are added to the "as drawn" drawing, the drawing is referred to as the "as built" drawing. It is important to keep these as built drawings, so they should be filed following the project for reference at a later date. The better care you take with these drawings, the easier it will be for someone else to use them.

PIPING SYMBOLS

Piping and welding symbols are used to convey specific information. The American Welding Society (AWS) has standardized welding symbols, which makes it easier for everyone involved with the project to work accurately. Several different organizations have created their own sets of piping symbols. Although there are many similarities between the different symbol sets, a component symbol key is usually added to the drawing for clarification.

PRACTICE 4-7

Identifying Piping Symbols

In this practice, you will be using **Table 4-2** to identify the different pipe fittings and valves shown in this tank,

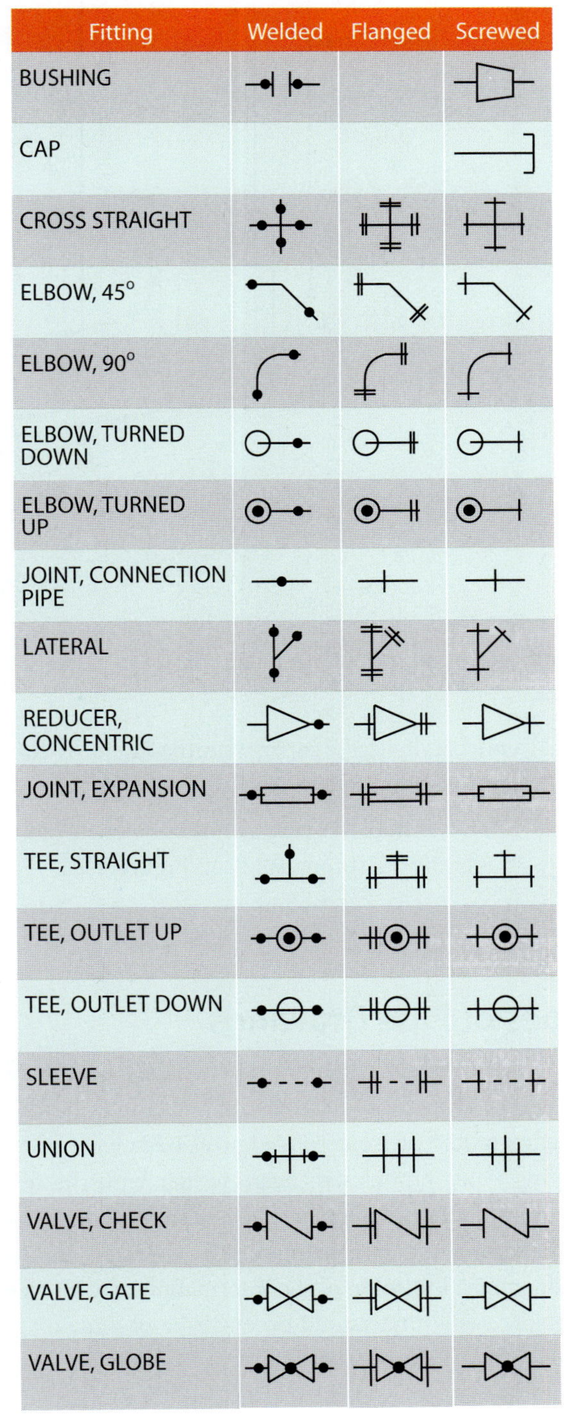

AN "X" MAY REPLACE THE CIRCLE "●" TO DENOTE A WELDED JOINT.

TABLE 4-2 Pipe Symbols

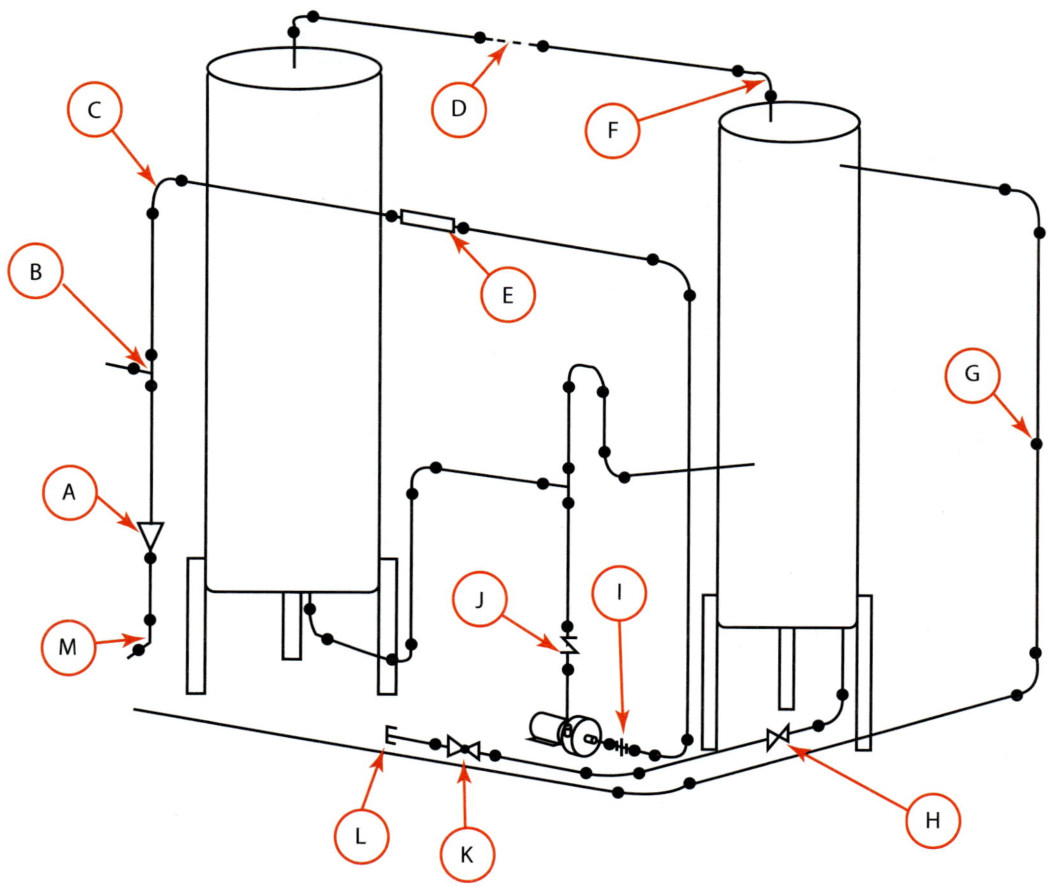

FIGURE 4-22 Practice 4-7.

piping, and pump layout. To begin, perform the following steps:

1. On a blank sheet of paper, write the letters "A" through "M" in a column down the left side of the page.
2. Use the examples of pipe symbols shown in Table 4-2 to identify "A" through "M" on **Figure 4-22**.
3. Write the part name next to the letter corresponding to the part being identified. ◆

Types of Pipe Drawings

Piping systems and pipeline drawings may be single- or double-line drawings, and each of these may be drawn as orthographic or isometric, **Figure 4-23**. Each of these drawing types provides the pipe welder with the necessary information to fabricate the required piping project. Much of the same information can be depicted on any of the drawings, but some of the information may be clearer and more easily understood on one type of drawing than another. Pipe drawings use different symbols for systems and joints that are welded (B), threaded (C), or flanged (D), Figure 4-23.

Welding symbols can be added to any of the types of pipe drawings, along with the written welding procedures to provide the welder with all the information required to fabricate the piping project.

MATERIALS

There are a variety of materials that can be used as part of the fabrication of a piping system. The larger the system, the more diverse the materials can be. This section will cover some of the basic materials that pipe welders may find on job sites.

Pipe

Most stock sections of steel pipe are 21 feet long. Pipe diameter is given as nominal because the actual diameter of the pipe may vary slightly. It can be ordered with one of three standard ends, **Figure 4-24**. Plain-end (PE) pipes are square-cut and are used with socket-welded fittings, **Figure 4-25**. Threaded-end (TE) pipes have pipe threads precut to one of the American National Thread types. The national pipe taper (NPT) has a slight taper so that the further the part is threaded into the fitting, the tighter the fitting gets, **Figure 4-26**. The national pipe straight (NPS) threads are straight. The ends on beveled-end (BE) pipes are prepared for a V-groove weld, **Figure 4-27**.

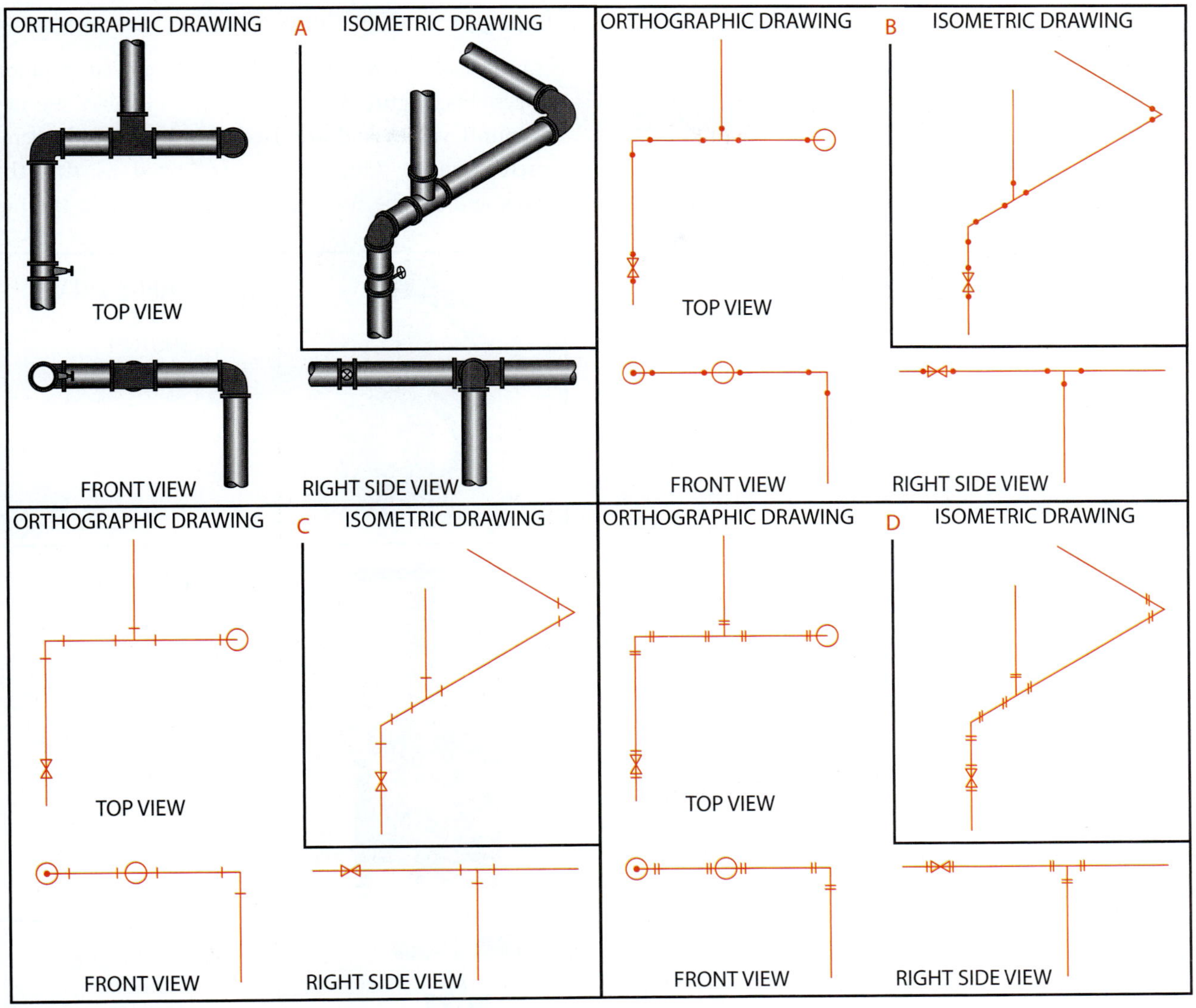

FIGURE 4-23 Types of pipe drawings.

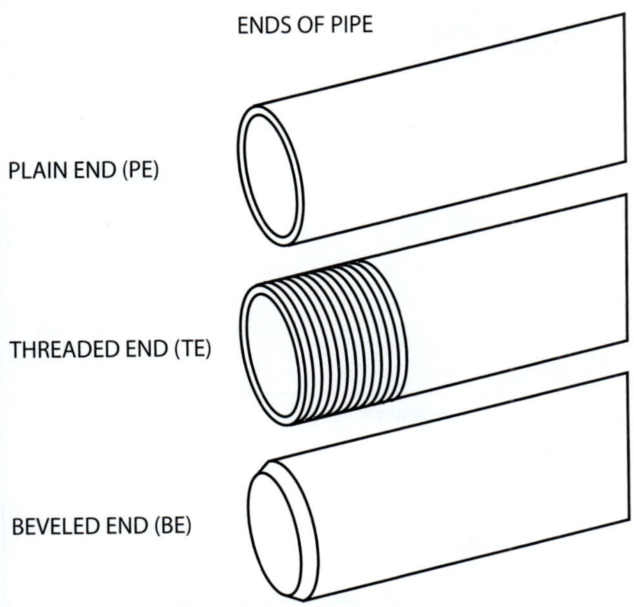

FIGURE 4-24 Types of pipe ends.

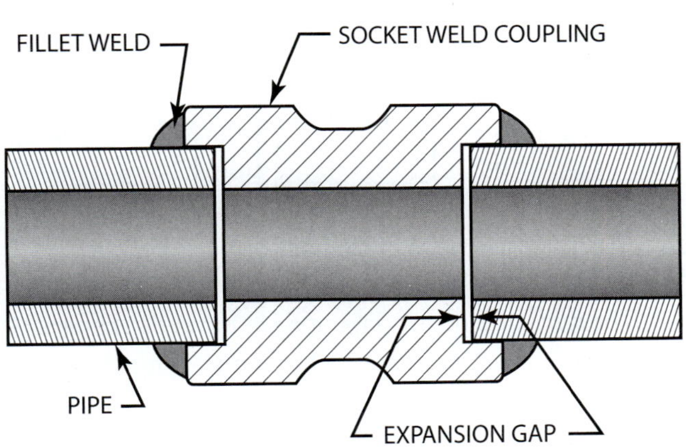

FIGURE 4-25 Socket-welded joint.

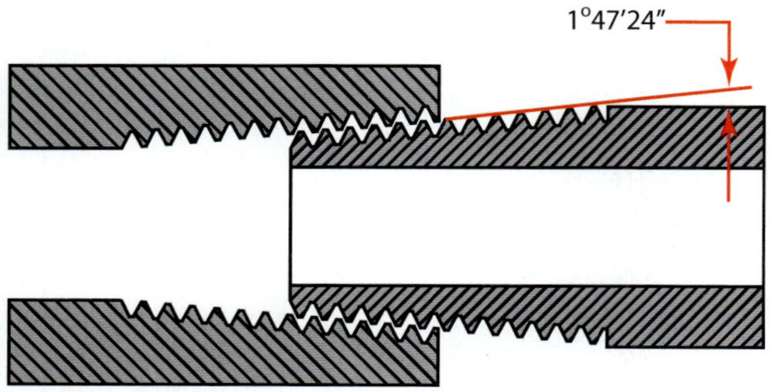

FIGURE 4-26 Pipe threads.

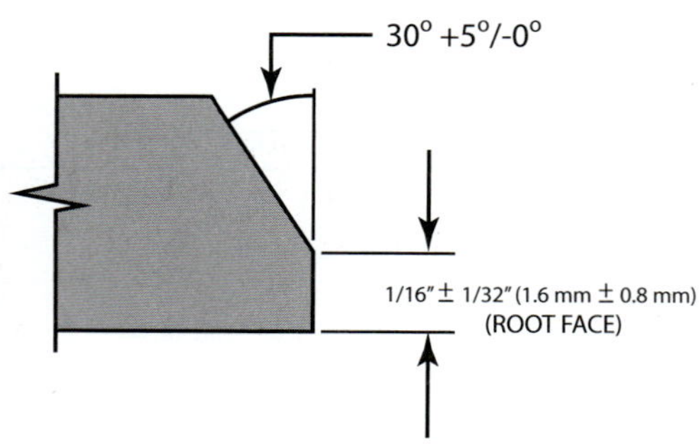

FIGURE 4-27 Pipe beveled end for welding.

> **NOTE**
>
> Teflon thread tape and pipe thread compound are used primarily as lubricants so that the fitting can be threaded further without sealing the joint.

Fittings

Pipe elbow and tee fittings can have the same-size openings on all sides, or they can have different-size openings so that the fitting can also serve as a reducer. If the larger nominal-size opening is the same on all openings, only one dimension needs to be written on the takeoff. If the fitting is being used as a reducer, then the largest opening diameter is listed first for elbows. For tees, the largest opening is listed first, followed by the opening size on the opposite end, and then the size of the center opening, **Figure 4-28**.

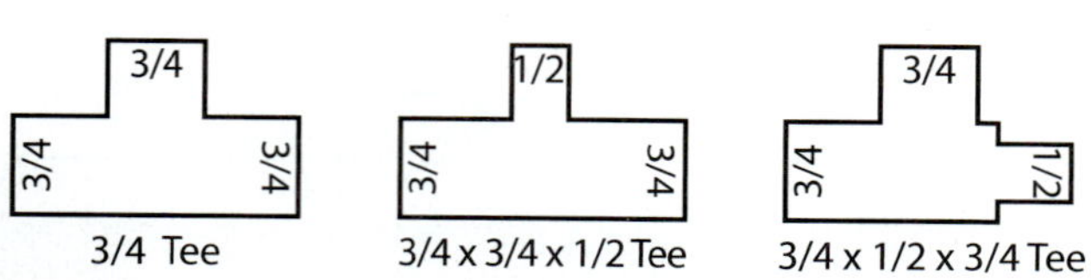

FIGURE 4-28 Methods of dimensioning pipe tee fittings.

The sizes of pipe in a piping system can be changed to maintain uniform pressure and flow through the system. A "reducer" is a fitting that is used when a size change is needed. There is an opening on either end, each of a different size.

Valves

A number of different types of valves are used in plumbing systems, **Figure 4-29**. Some types of valves are rated as **shutoff valves** and are used only in the full open or closed position. Other valves are rated as **throttling valves** and can be used to control the flow in the pipe.

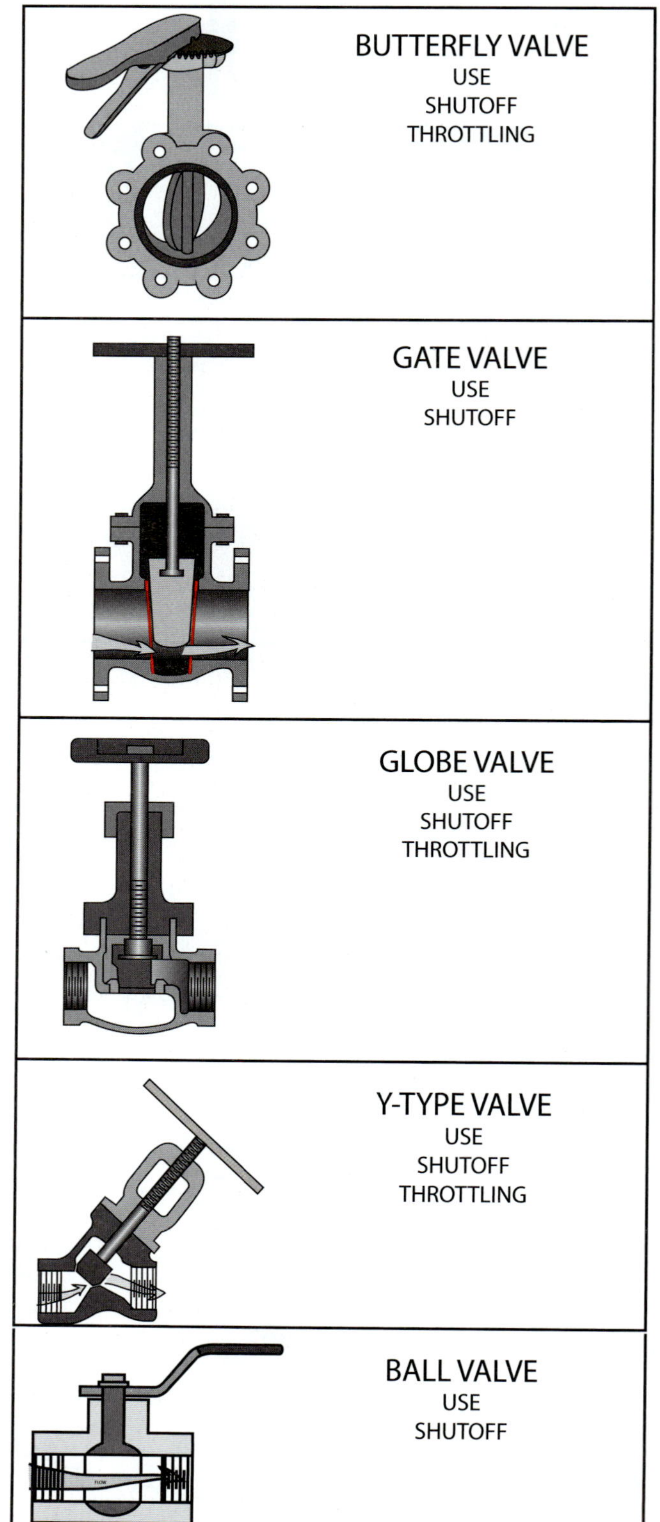

FIGURE 4-29 Common types of valves and their applications.

Throttling valves can also be used as shutoff valves, but if shutoff valves are used for that purpose for long periods of time, their valve seats may be damaged.

"Flanges" and "unions" are devices that are used to connect components in a piping system so that they can be easily removed and replaced if service is needed. Flanges are available in a wide range of sizes, from 1/2 in. up to 24 in., and in several pressure classes. There are a number of ways that a flange can be attached to the pipe, some of the most common of which are threaded, socket weld, and welding neck. Also, the face of the flange may be flat or recessed for a gasket, or it can have a groove for an O-ring. Unions are only available for smaller pipe sizes, ranging from a fraction of an inch to up around 4 in.

Control valves are used to stop and start flow or to modulate flow in a piping system. The most common type of control valves use pneumatic (air), electromagnetic solenoid, or electric motors.

Pumps

Most pumps used in piping systems are of the centrifugal type. Centrifugal pumps work by creating a pressure difference as the result of centrifugal forces caused as an impeller spins in a fluid, **Figure 4-30**. An "impeller" is a round disk with blades that pull some fluid into its center (A) while slinging other fluid outward (B), where it flows out through the discharge opening (C). Pump impellers may have many different designs depending on the fluid they are pumping. Some may look more like a fan, while others may appear to be paddles on a wheel; however, they all work the same way.

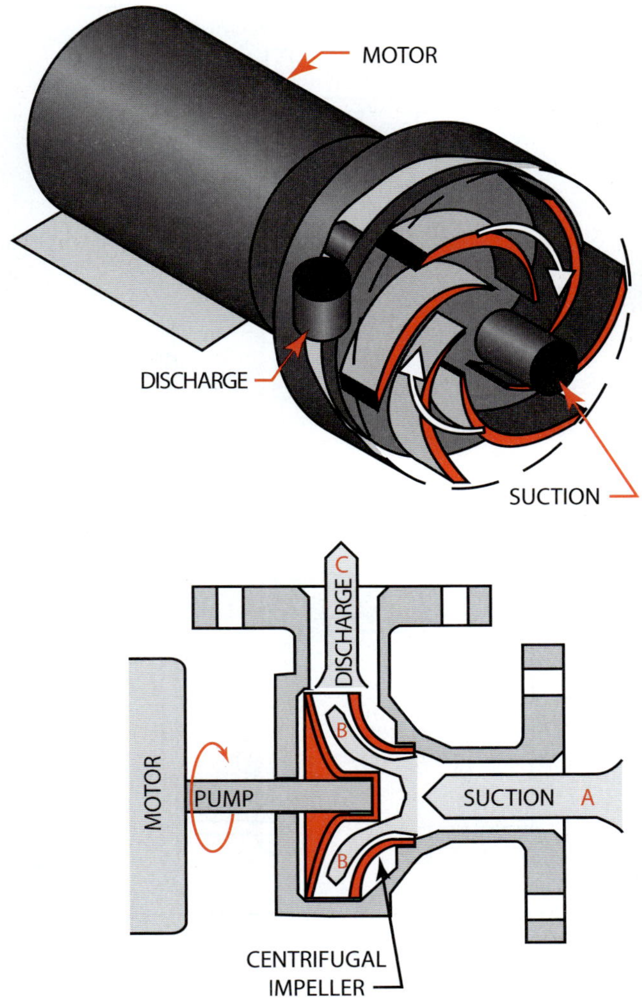

FIGURE 4-30 Centrifugal pump.

NOTE

When you stir a glass of water quickly with a spoon, the water around the sides rises as the center sinks. The water rises because the centrifugal force pushing outward has increased the water pressure around the sides of the glass. At the same time, it reduces the water pressure in the center. This is exactly how a centrifugal pump works.

Centrifugal pumps are not positive displacement pumps, so any resistance to flow in the system affects the flow produced by these pumps. For example, if a pump is connected to very short inlet and outlet pipes, it will pump more gallons of fluid per minute than if the same pump were connected to a very long or very thin pipe. Pump manufacturers provide charts that show the flow rate for their pumps at different inlet and outlet pressures.

The input pressure of a centrifugal pump has a greater effect on the pump's output; for that reason, they work best when there is a straight section of pipe whose length is at least ten times the diameter of the pipe, **Figure 4-31**. But this section can be as short as five times the diameter if space is limited. Under no circumstances should a 90° elbow be attached directly to the pump. A short straight section or 90° elbow will cause turbulence in the water and reduce pumping efficiency and pump life.

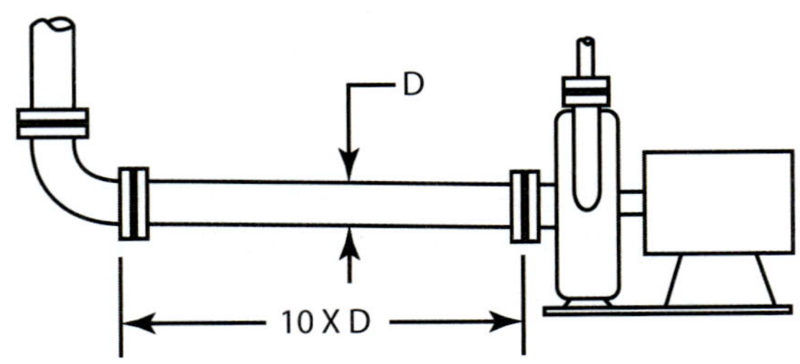

FIGURE 4-31 The correct way to connect the suction pipe to a centrifugal pump.

The circulating pump with the boiler, **Figure 4-32**, is typical of ones that would be found in a small hydronic heating system. The major components shown in addition to the connecting piping are the boiler, the circulating pump, globe valves, and check valves.

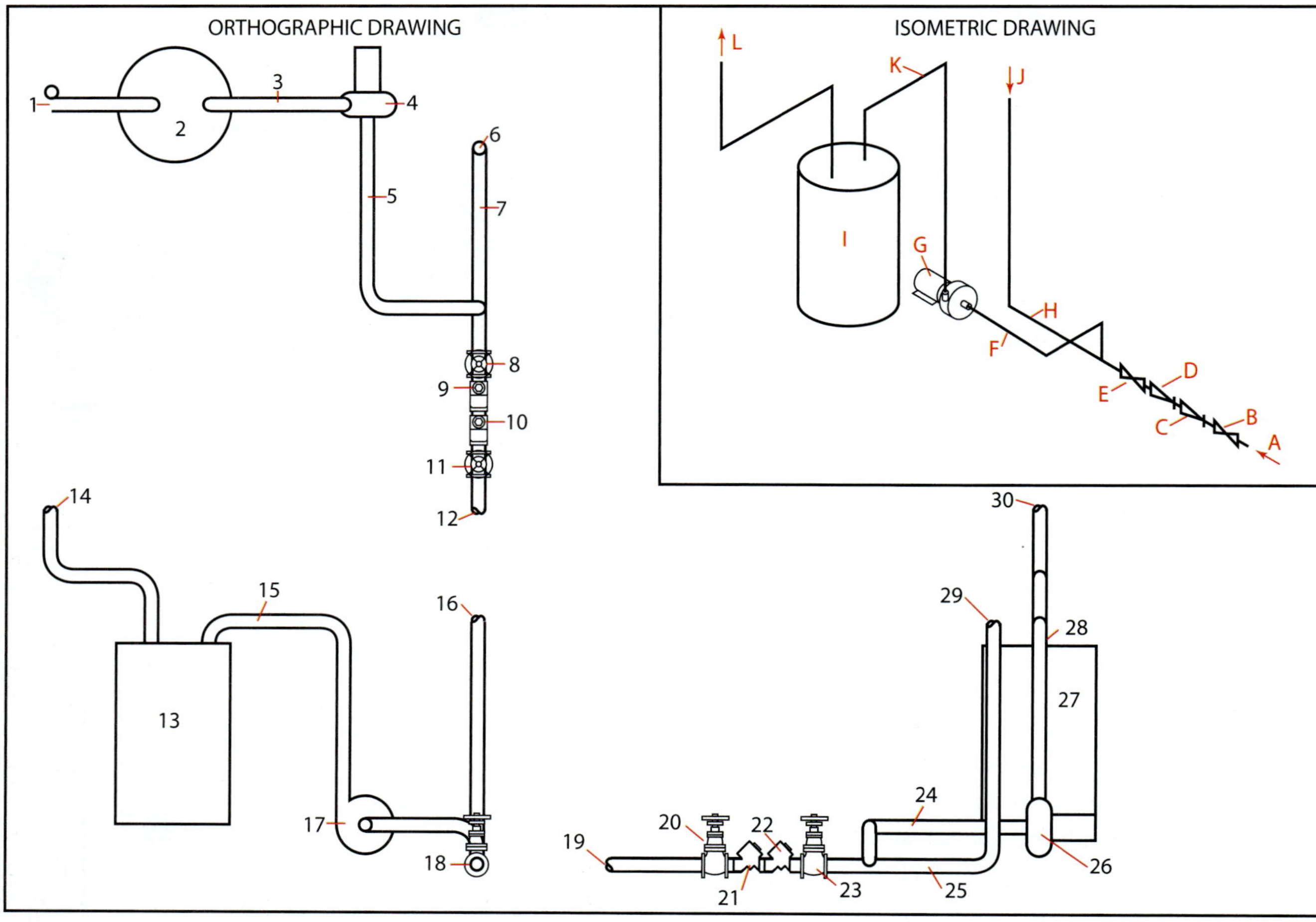

FIGURE 4-32 Hydronic heating piping system.

Check Valves

The system shown in **Figure 4-32(A)** is connected to a public water supply. All hydronic systems lose some water, and by having them attached to a water supply, additional water called "makeup water" can be automatically added to refill the system. When a system is connected to a public water supply, there is a chance that if the city water system loses pressure, the hydronic system could drain back into the potable water supply. To prevent this, codes and local ordinances and laws require that a double-check valve be installed in line with a public water supply that is connected directly to any piping system, **Figure 4-32(C and D).** Often, annual inspections of the double-check valves are required.

PRACTICE 4-8

Part Identification

The following steps will help you identify the parts shown in the orthographic and isometric drawings, Figure 4-32:

1. On a blank, lined sheet of paper, write the letters A through L vertically down the left side.
2. Refer to the letters next to parts in the isometric drawing to identify the numbers next to the same parts that are shown in the orthographic drawing.
3. Write the numbers next to the letters on the sheet of paper.
4. For example, I in the isometric drawing is the same as 2 in the top view, 13 in the front view, and 27 in the right-side view of the orthographic drawing.
5. In this instance, you would write *I, 2, 13,* and *27* on your sheet of paper. ◆

MATERIAL TAKEOFF

The construction term **material takeoff** refers to the process of analyzing drawings, material specifications, and other design documents to determine all the materials that will be needed to fabricate the system. Sometimes the term is shortened to just *takeoff* (as in, "Will you do the takeoff on this job?"). On some smaller construction jobs, the welder will do the takeoff to create a bill of materials.

PRACTICE 4-9

Zone Hydronic Heating System Takeoff

The three-zone hydronic heating system single-line piping layout shown in **Figure 4-33** is typical of what a welder might find in a commercial building. This is the type of heating zone that might be attached to the boiler and circulating pump shown in Figure 4-31. Refer to Figure 4-32 and follow the instructions given here to fill in the takeoff information for the system components for the blanks listed in **Table 4-3:**

1. On a blank, lined sheet of paper, make a copy of Table 4-3. Start by drawing two vertical lines to make the three columns. Now write the item numbers 1 through 20 vertically down the left side of the paper, and then write the material descriptions next to the item numbers.
2. On a sheet of graph paper or a blank sheet of paper, sketch the three-zone hydronic heating system shown in, Figure 4-33.
3. Count the number of gate valves by following the piping circuit starting at the hot water supply line all the way around to the return water line.
4. Repeat step 3 until you have the number of valves for each of the 20 items.

NOTE

Place a small check mark next to each of the items you find on the drawing. This will let you make sure that every item in the piping system has been counted.

Next, you will do a takeoff for the pipe required to fabricate the three-zone hydronic system. Using Figure 4-33, calculate the number of feet of 6-in., 4-in., and 2-in. pipe that will be required to fabricate this system. Other than the heat exchange coils, all of the other symbols are not to scale.

5. On a blank, lined sheet of paper, write "6 inches," "4 inches," and "2 inches" vertically down the left side.
6. Using the drawing scale of one square equals 1 foot, starting at the hot water supply line (1 foot line) and continuing to the center of the concentric reducer (30-foot line), a total of 29 feet of 6-inch pipe would be needed for this section.
7. Repeat this process for the rest of the 6-in. pipe, and write the total number of feet of 6-in. pipe on the 6-in. line.
8. Repeat steps 5 and 6 with the 4-in. and 2-in. piping.
9. BE steel pipe comes in 21-foot lengths. Divide the total length of each size of pipe by 21 to calculate the total number of 21-foot pipe sections that are needed. ◆

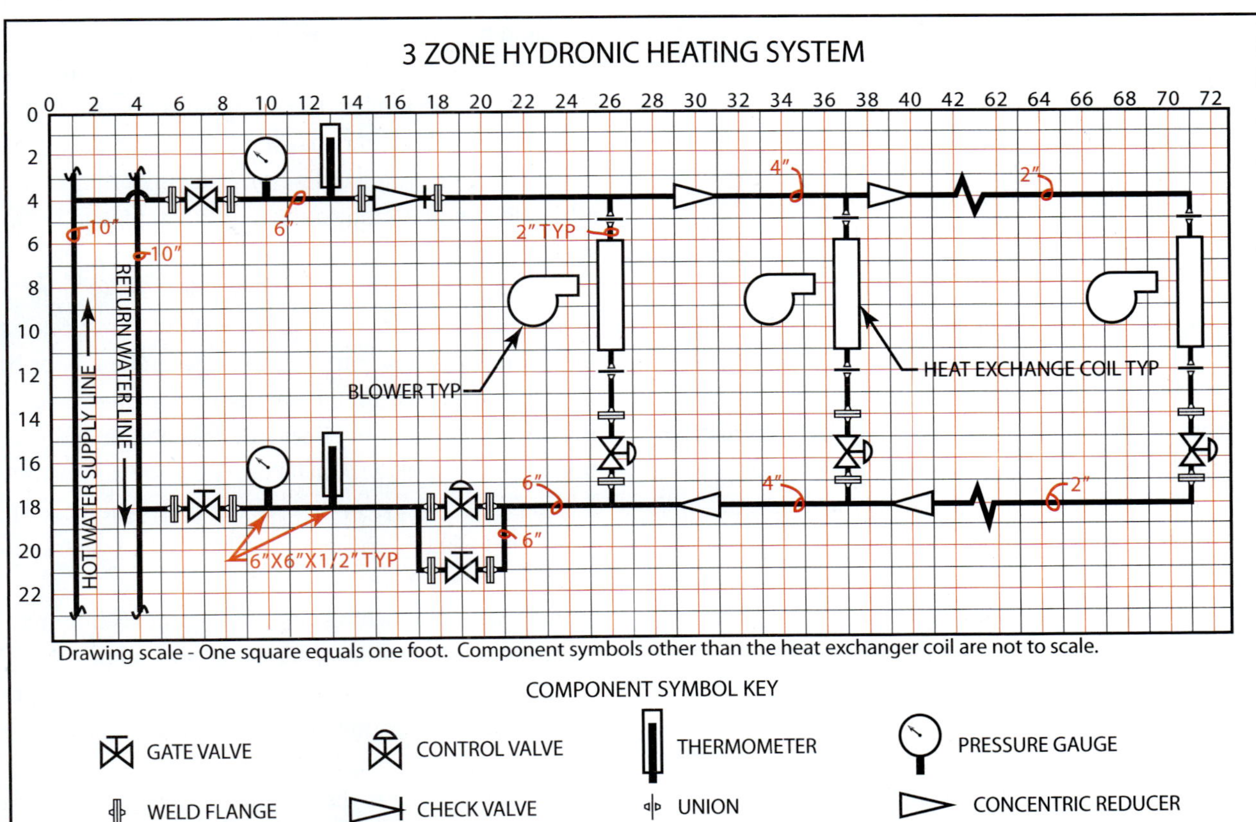

FIGURE 4-33 Practice 4-9.

Three Zone Hydronic Heating System Parts Takeoff		
Item No.	Material Description	Quantity
1.	6" GATE VALVE	
2.	PRESSURE GAUGE	
3.	THERMOMETER	
4.	6" CHECK VALVE	
5.	6" WELD FLANGE	
6.	2" WELD FLANGE	
7.	2" UNIONS	
8.	2" CONTROL VALVES	
9.	6" CONTROL VALVES	
10.	10" X 10" X 6" TEE	
11.	6"X6"X1/2" TEE	
12.	6"X6"X2" TEE	
13.	4"X4"X2" TEE	
14.	6"X6"X6" TEE	
15.	2" 90o ELBOW	
16.	6" 90o ELBOW	
17.	21'6" PIPE	
18.	21'4" PIPE	
19.	21'2" PIPE	
20.	HEAT EXCHANGE COILS	
		(C) Jeffus 2013

TABLE 4-3 Worksheet for Practice 4-7

NOTE

Remember to add another section of pipe if longer lengths of pipe (e.g., 42', 63', etc.) of pipe are needed.

WELDING SYMBOLS

The use of welding symbols enables a designer to indicate clearly to the welder important detailed information regarding the weld. The information in the welding symbol can include details such as length, depth of penetration, height of reinforcement, groove type, groove dimensions, location, process, filler metal, strength, number of welds, weld shape, and surface finishing.

Welding symbols are a shorthand language for the welder. They save time and money and ensure understanding and accuracy. The AWS has a set of standardized welding symbols. Some of the more common symbols for welding are reproduced in this chapter. If more information is desired about symbols or how they apply to all forms of manual and automatic machine welding, these symbols can be found in *Standard Symbols for Welding, Brazing, and Nondestructive Examination,* ANSI/AWS A2.4, a manual published as an American national standards document by the AWS.

Figure 4-34(A) shows the basic components of welding symbols, consisting of a reference line with an arrow on one end. Other information relating to various features of the weld are shown by symbols, abbreviations, and figures located around the reference line. A tail is added to the basic symbol as needed for the placement of specific information.

Types of Welds

Groove welds and fillet welds are the most commonly used types in the piping industry. Some of the other types of welds include flange, plug or slot, spot or projecting, seam, back or backing, and surfacing. All of the basic symbols are shown in **Figure 4-34(B)**.

Weld Location

Welding symbols are applied to a short horizontal line called the "reference line," which serves as a base for the welding symbols, and it has an arrow pointing to the joint where the weld is to be made. The reference line serves as the base for all of the symbols, is always drawn horizontally, and has two sides: the arrow side and the other side. Any welding symbol that appears below the reference line is applied to the same side of the joint the arrow touches and is referred to as the **arrow side symbol**. Any symbol that appears above the reference line is referred to as the **other side symbol,** indicating that the weld is to be made on the other side of the joint.

Accordingly, the terms "arrow side," "other side," and "both sides" are used to locate the weld with respect to the joint. The arrow line can be drawn from one end or both ends of a reference line to the location of the weld. The arrow line can point to either side of the joint and extend either upward or downward. Pipe welds are almost always deposited on the arrow side of the joint (the near side); the desired weld symbol is placed below the reference line, **Figure 4-35(A)**.

As a way of attaching cables to a pipe, a ring may be welded to it using a fillet weld on one side, **Figure 4-35(B)**. The weld may be made on both sides to give the ring more support; in that case, the same weld symbol appears above and below the reference line, **Figure 4-35(C)**.

The tail is added to the basic welding symbol when necessary to designate the welding specifications, procedures, or other supplementary information needed to make the weld,

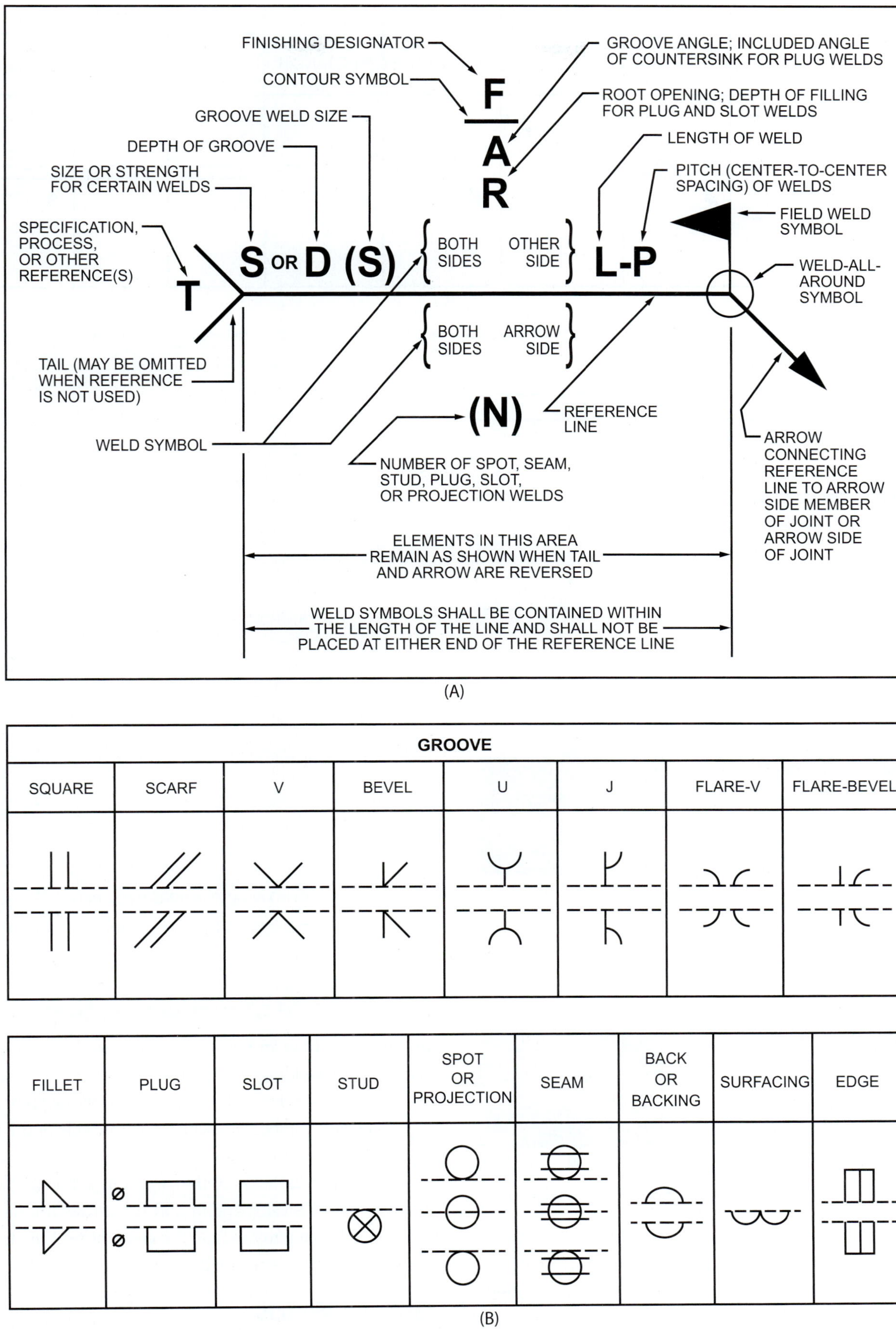

FIGURE 4-34 (A) Standard locations of elements of a welding symbol. (B) Symbols for different types of welds. (C) Supplementary symbols. (*Continued*)

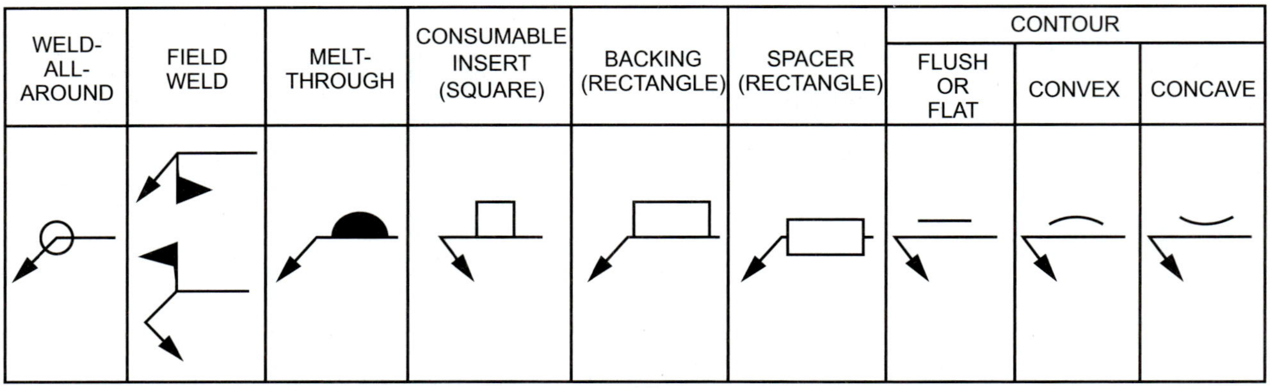

(C)

FIGURE 4-34 (A) Standard locations of elements of a welding symbol. (B) Symbols for different types of welds. (C) Supplementary symbols. (*Continued*)

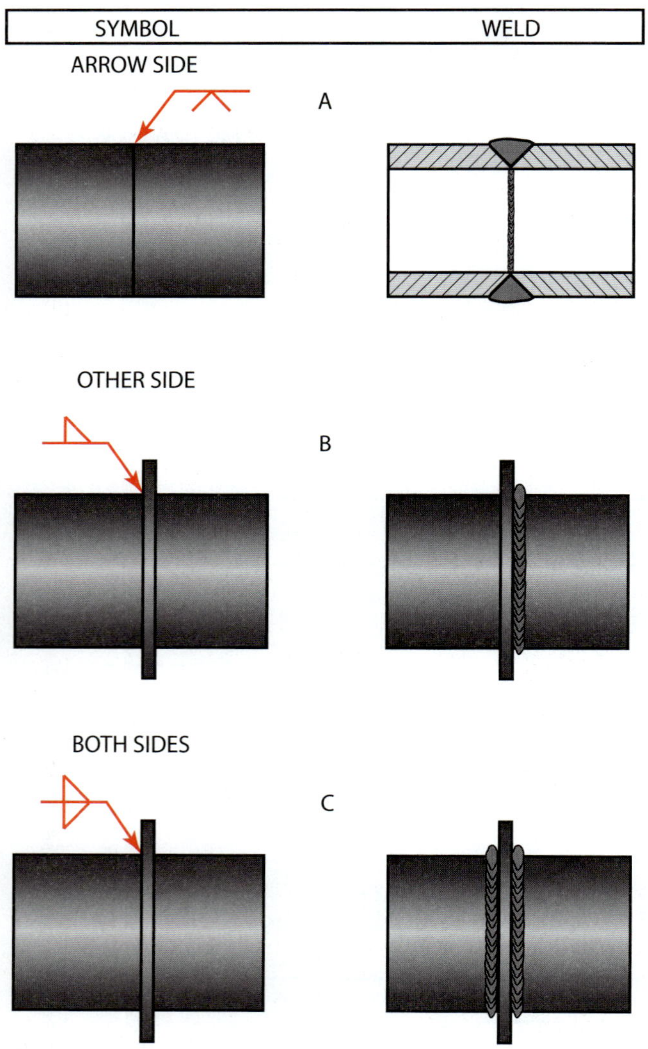

FIGURE 4-35 Arrow side and other side symbol designations.

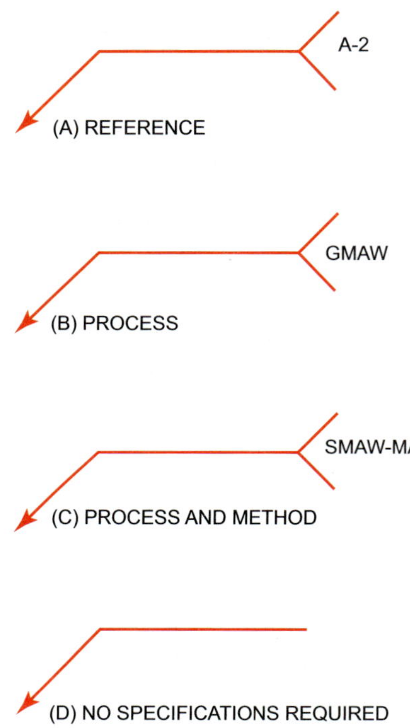

FIGURE 4-36 Information that can be found in the tail of a weld symbol.

Figure 4-36. The notation placed in the tail of the symbol may indicate the welding process to be used, the type of filler metal needed, whether pinging or root chipping is required, and other information pertaining to the weld. If notations are not used, the tail of the symbol is omitted. For joints that are to have more than one weld, a symbol is shown for each weld.

Significance of Arrow Location

In the case of fillet and groove welding symbols, the arrow connects the welding symbol reference line to one side of the joint. The surface of the joint that the arrow point actually touches is considered to be the arrow side of the joint. The side opposite the arrow side of the joint is considered to be the other (far) side of the joint. On a drawing where a single line illustrates a joint and the arrow of a welding symbol is directed to the line, the arrow side of the joint is considered to be the near side of the joint.

Groove Welds

Joint strength is improved by cutting a bevel to form a v-groove on the ends of pipe. This allows the weld to penetrate deeper into the joint, which increases the joint strength without restricting flexibility.

The grooves can be cut into base metal in a number of different ways. They can be cut using an oxy-fuel cutting torch, air carbon arc cutting, or plasma arc cutting, machined, or ground.

The various features of groove welds are as follows:

- Single-groove and symmetrical double-groove welds extend completely through the pipe wall being joined. No size is included on the pipe weld symbol, **Figure 4-37(A, B).**
- The root opening of groove pipe welds is the user's standard unless otherwise indicated. The root opening of groove welds, when not the user's standard, is shown inside the weld symbol, **Figure 4-37(E, F).**

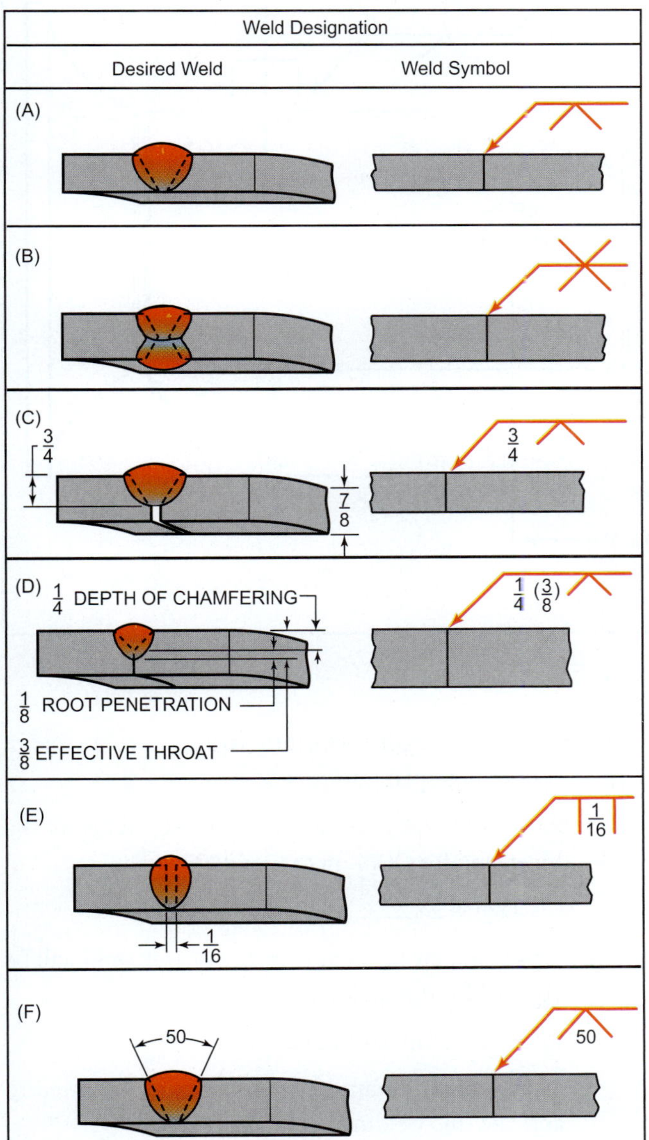

FIGURE 4-37 Common specifications found on groove welds.

- The root face's main purpose is to minimize the burnthrough that can occur with a featheredge. The size of the root face is important to ensure good root fusion, **Figure 4-38**.

PRACTICE 4-10

Weldment Identification

Provide the needed information for the short-flanged coupling to be fabricated. On a blank, lined sheet of paper, write and sketch the answers to the following items for **Figure 4-39**:

- What is dimension B for the 4" pipe?
  - Dimension B ______, minimum ______, and maximum ______
- What is dimension A, which is used to locate the 1/2" hole to drill for the fitting?
  - Dimension A ______, minimum ______, and maximum ______
- What is dimension C, which is used to locate the 1/2" hole to drill for the fitting?
  - Dimension C ______, minimum ______, and maximum ______
- Sketch the weld profile for the E welds, assuming a 30° weld angle.
  - Sketch E
- Sketch the two different cross sections of weld D at points D1 and D2, assuming 0° root opening.
  - Sketch D1
  - Sketch D2 ◆

PRACTICE 4-11

Weld Identification

Provide the following information for the pipe weldment shown in **Figure 4-40:**

- On a blank, lined sheet of paper, write and sketch the answers to the following questions. The four items that make up this weldment are shown in the front, right side, and detail views.
  1. What other letter or letters point to the same part as (E) in the side view? ____________
  2. What other letter or letters point to the same part as (F) in the side view? ____________
  3. What other letter or letters point to the same part as (G) in the side view? ____________
  4. What other letter or letters point to the same part as (H) in the side view? ____________

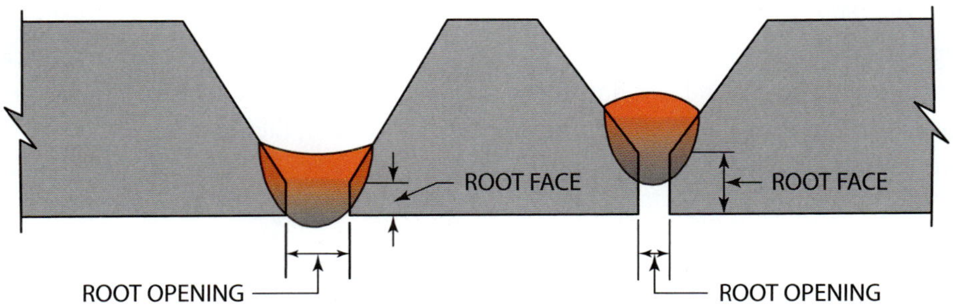

FIGURE 4-38 The root face size affects weld penetration.

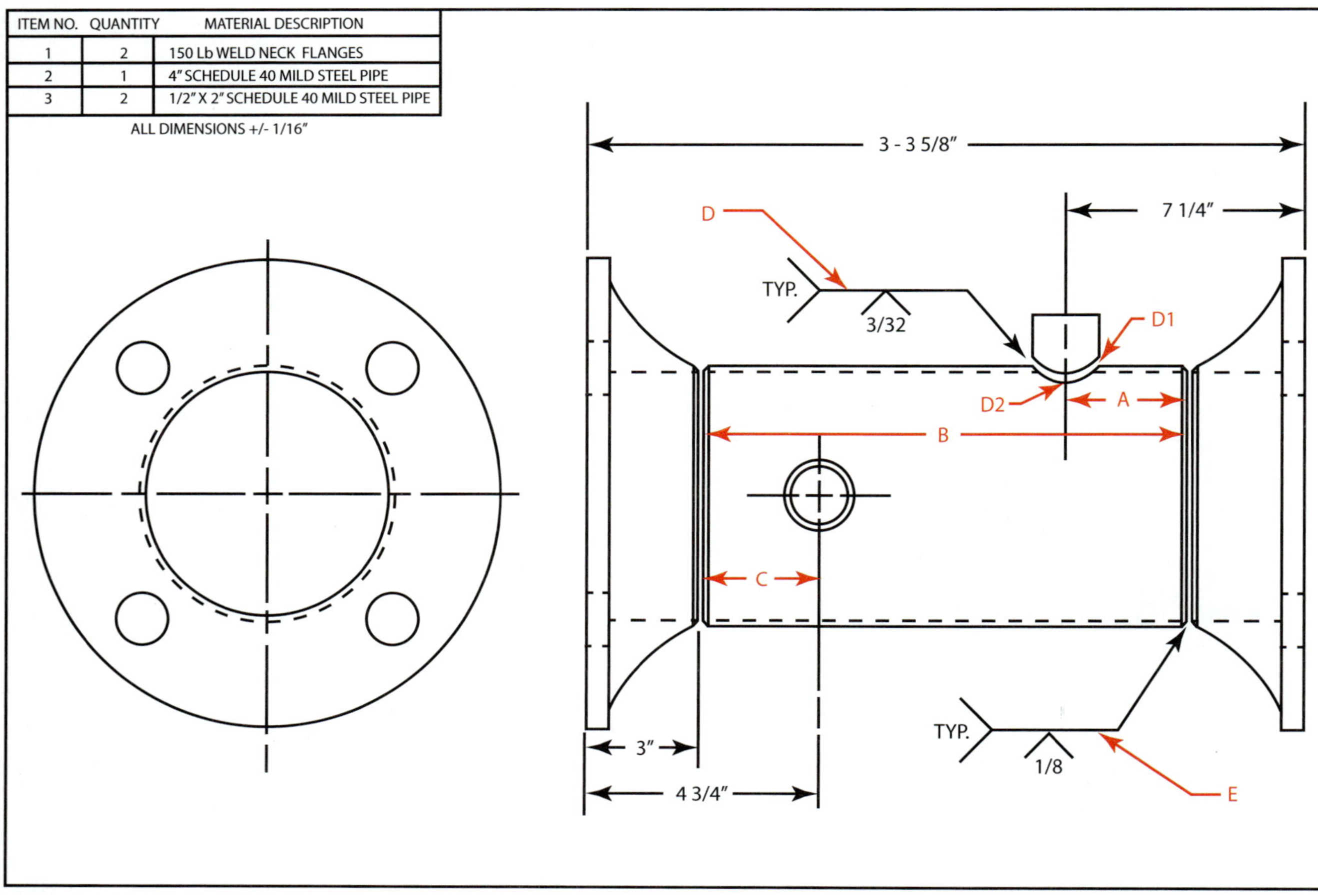

ITEM NO.	QUANTITY	MATERIAL DESCRIPTION
1	2	150 Lb WELD NECK FLANGES
2	1	4" SCHEDULE 40 MILD STEEL PIPE
3	2	1/2" X 2" SCHEDULE 40 MILD STEEL PIPE

ALL DIMENSIONS +/- 1/16"

FIGURE 4-39 Practice 4-10.

- Sketch the weld profile for the (1) weld, assuming a 0° root opening.

  5. Sketch 1

- Sketch the weld profile for the (3) weld, assuming a 0° root opening.

  6. Sketch 3

- According to the hidden lines, does part (J) in the front view open into part (D)?

  7. _______ Yes, it opens into part (D). _______ No, it does not open into part (D).

- According to the hidden lines, does part (B) in the detail A view open into part (C)?

  8. _______ Yes, it opens into part (C). _______ No, it does not open into part (C).

- Both large pipes are 4" Schedule 40. What is the dimension for OD/4 on the saddle?

  9. OD/4 = _______"

- Approximately how many inches of 1/4" weld will be required for this weldment?

  10. _______" of 1/4" weld

- What is the minimum length of 4" pipe required to fabricate this weldment?

  11. _______ of 4-in. pipe.

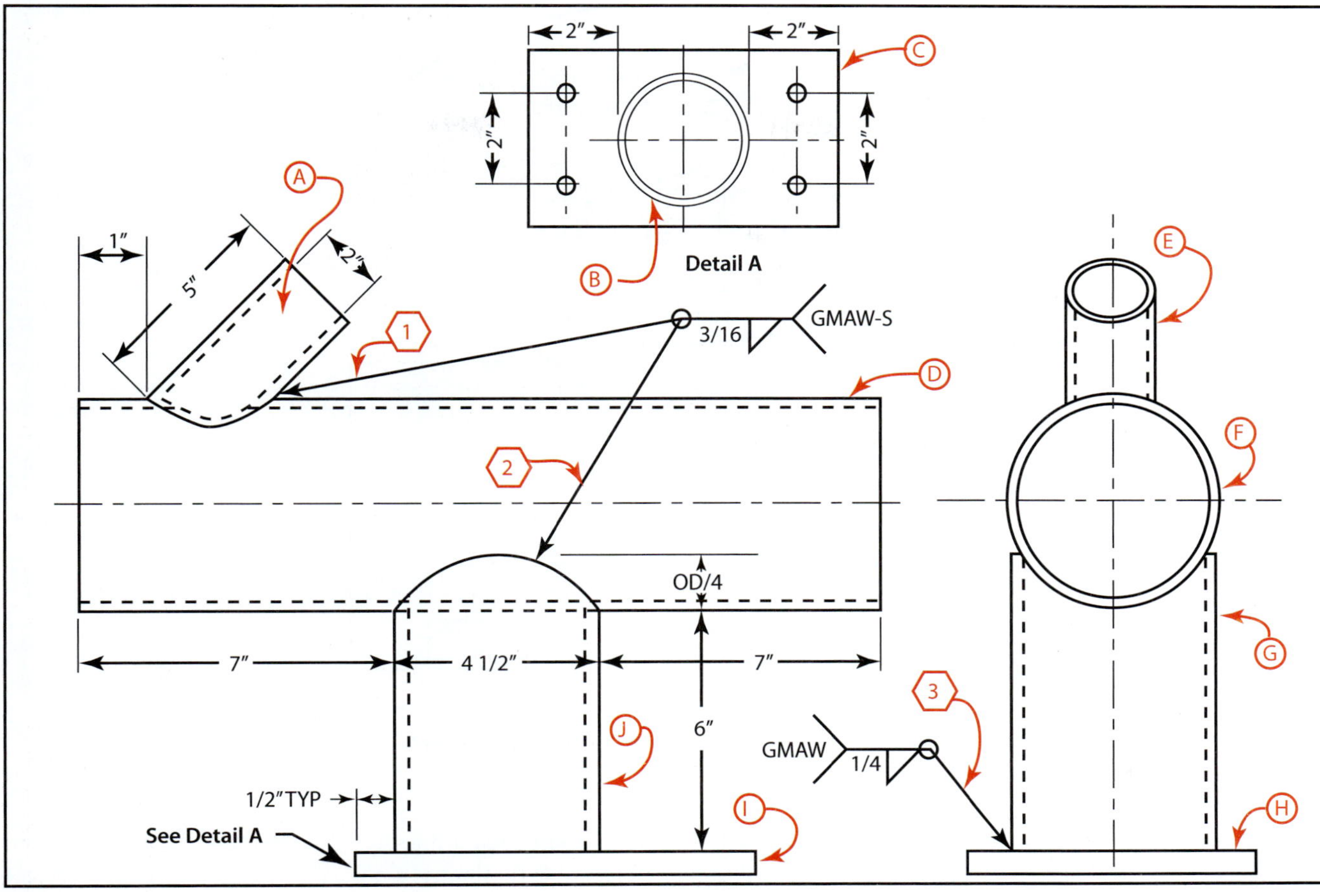

FIGURE 4-40 Practice 4-11.

Consumable Inserts (Backing)

A "consumable insert" is a ring of metal that is placed on the inside of a pipe joint to prevent the molten metal from dripping through the open root, **Figure 4-41**. It helps to ensure that 100% of the base metal's thickness is fused by the pipe weld. The insert must be thick enough to withstand the heat of the root pass as it is burned in. The insert is usually not removed on piping following the completion of the job.

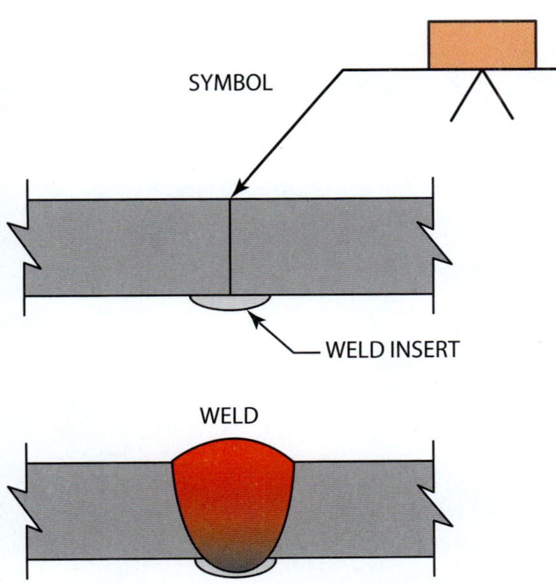

AWS A2.4:2012, Figure AWS Welding Symbol Chart adapted with permission of the American Welding Society (AWS), Miami, FL

FIGURE 4-41 Symbols for consumable inserts.

Fillet Welds

Dimensions of fillet welds are shown on the same side of the reference line as the weld symbol and are shown to the left of the symbol. A fillet weld is approximately triangular, and it is used to join lap joints and tee joints, **Figure 4-42**.

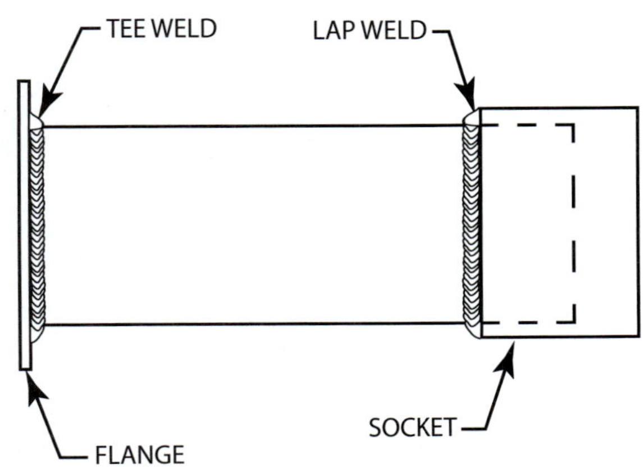

FIGURE 4-42 Fillet welds.

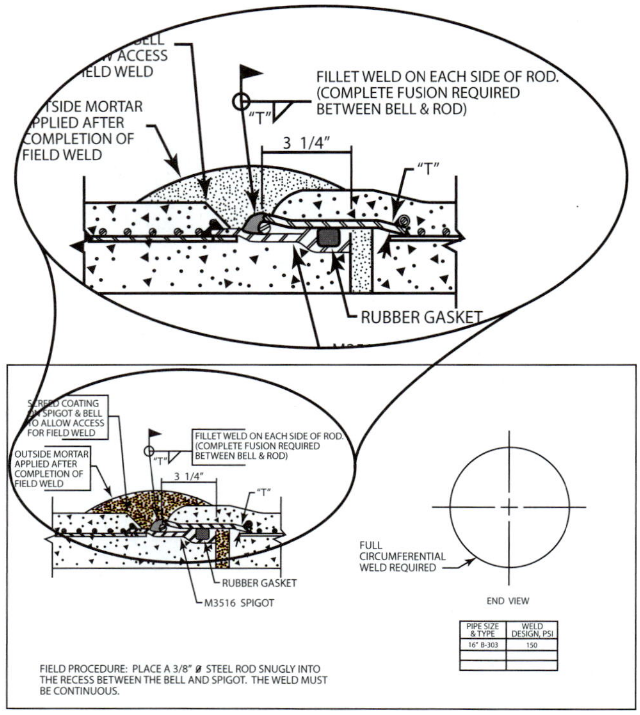

FIGURE 4-43 A fillet weld is used to both connect and seal the joint in a concrete drain pipe.

A fillet weld is used to hold two pieces of concrete wastewater or stormwater pipe together securely to prevent shifting soil from separating the ends of the pipe, which would cause a leak, **Figure 4-43**.

Summary

Pipe fabricating and welding is a very diverse field, and it requires many skills such as threading, blueprint reading, and symbol identification. This chapter has provided the basics in these areas to allow a new pipe fabricator or welder to enter the field. Over time and with field experience, you will develop more skills and knowledge, along with some useful shortcuts. Keeping notes is a good way of reinforcing these new skills. Also, notes can be referred to from time to time as you work.

Review Questions

1. What is the purpose of mechanical drawings (blueprints)?
2. What type of information might mechanical drawings contain?
3. Name five ways that piping systems can be joined together.
4. What is used to join flanges and other components to piping systems?
5. What type of line shows the edge of an object?
6. What type of line would be used to show the size or length of an object?
7. What type of line can show an alternative position of a moving part or the extent of motion of a part?
8. What is another name commonly used to refer to orthographic drawings?
9. What are the three most common views shown in an orthographic drawing?
10. What type of drawings are picturelike and use lines drawn at 30° angles to the right and left of vertical?
11. What are two advantages of using a computer program to make drawings?

(Continued)

12. On what typical views can you find height dimensions?
13. What is an "as drawn" drawing called when changes are added to it as it is being fabricated?
14. Name five materials that might be found on a piping system job site.
15. How long are most stock sections of steel pipe?
16. What are the three standard pipe ends?
17. In what order are the pipe opening sizes listed on the takeoff for a tee fitting with three different-size openings?
18. What is the purpose of a reducer on a pipe fitting?
19. What is the difference between a shutoff valve and a throttling valve?
20. What are devices that are used to connect components in a piping system so that they can be easily removed and replaced if service is needed?
21. What is the purpose of control valves?
22. Why must double-check valves be installed on hydronic systems that are connected to a public water supply?
23. List three items that might be included in a welding symbol on a drawing.
24. Name the two most commonly used types of welds in the piping industry.
25. Does the arrow side symbol appear above or below the reference line?
26. Which side of the reference line does an other side symbol appear?
27. Why are bevels cut to form v-grooves on the ends of pipe before welding?
28. What is the purpose of a consumable insert (backing) on a pipe joint?
29. A fillet weld is used to join what types of joints?

Chapter 5

Thermal-Cutting Processes

OBJECTIVES

After completing this chapter, the student should be able to:

- Describe the three major thermal-cutting processes used for pipe cutting.
- Explain methods of marking pipe prior to cutting.
- Select an appropriate oxyacetylene cutting (OFC-A) torch tip for the job.
- Light an OFC-A cutting torch and a cutting machine torch safely.
- Make freehand oxyacetylene pipe cuts in fixed and rolled positions.
- Demonstrate how to make a square cut on pipe in the horizontal rolled position, the horizontal fixed position, and the vertical position with OFC-A and plasma arc cutting (PAC) processes.
- Demonstrate an ability to set up and use an air carbon arc torch to gouge.

KEY TERMS

Air carbon arc cutting
Air carbon arc gouging (CAC-A)
Cutting tip
Dross
Electrode tip
Gouging
Heat-affected zone (HAZ)
High-frequency alternating current
Ionized gas
Kerf
Kindling temperature
Pilot arc
Plasma arc cutting (PAC)
Plasma arc gouging (PAG)
Preheat flames
Preheat holes
Slag
Standoff distance
Washing

INTRODUCTION

The three major thermal-cutting processes widely used in the pipe-welding industry are oxyacetylene cutting (OFC-A), **plasma arc cutting (PAC)**, and **air carbon arc gouging (CAC-A)**. Cutting and **gouging** are the two most commonly used thermal-cutting processes. Since OFC-A equipment can be used for both processes, it is the most commonly used in the pipe-welding industry.

(Continued)

Recent developments in **plasma arc gouging (PAG)** torch designs have made using the PAG process much easier. The PAC process has a number of advantages over the OFC-A process, so the introduction of these new gouging torches has increased the use of plasma arc in the pipe-welding industry.

CAC-A has long been used to remove welds during repair and replacement pipe work because it is not adversely affected by surface conditions such as rust or corrosion.

OXYACETYLENE CUTTING (OFC-A)

The OFC-A process, sometimes referred to as *burning*, works by burning away the steel with a stream of pure oxygen. The cutting-torch flame is used to heat the surface of the metal to its **kindling temperature**, which is the temperature at which the material will begin to burn. In the case of steel, its kindling temperature in a pure oxygen stream is when it begins to glow a dull red, which is around 1,600° to 1,800°F (870°C to 900°C).

The OFC-A method is used to cut and bevel carbon steel pipe. OFC-A cutting puts a lot of heat in the pipe, which comes from both the torch flame and the heat generated as the cutting stream burns away the steel to form a kerf. A **kerf** is the space produced during any cutting process. The combined heats cause a change in the pipe's grain structure alongside of the cut. The area where the grain structure is changed is called the **heat-affected zone (HAZ), Figure 5-1A**. In most cases, this change has little or no effect on the finished weld. However, in very cold conditions or when higher alloyed steel pipes are used, the rapid heating and cooling that occurs during an OFC-A cut can cause cracks to form in the base metal. If left alone, these cracks can spread into the surrounding metal after welding. Heating the pipe before the cut starts can help prevent cut-related cracking. A general rule of thumb is that if preheating is required for welding, then it should be used before cutting.

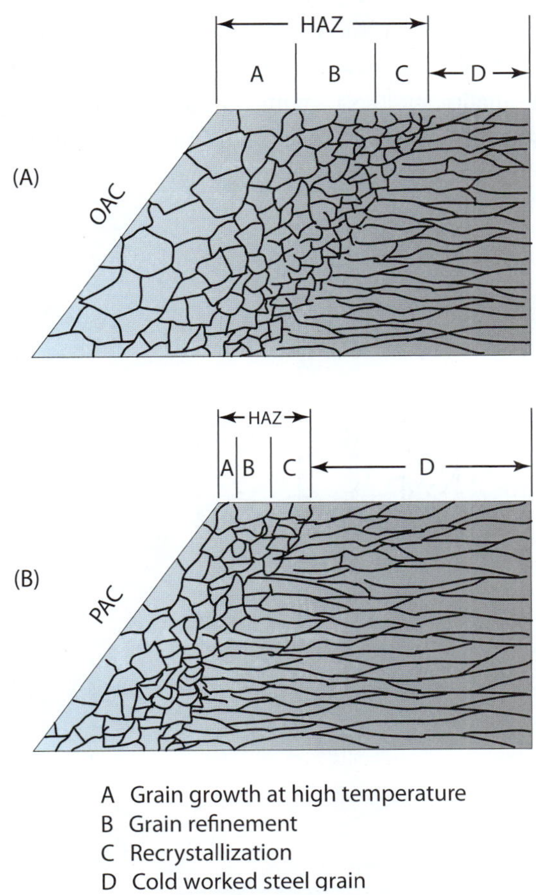

FIGURE 5-1 The HAZ in cut A, made with the OFC process, is significantly wider than the HAZ in cut B, made with the PAC process.

Oxyacetylene Gouging

The gouging process removes metal from the surface of a pipe or weld without, in most cases, cutting all the way through. Oxyacetylene gouging uses the same basic equipment as does OFC-A; the major difference is the shape of the **cutting tip** versus the gouging tip. The cutting tip is straight (**Figure 5-2A**), and the gouging tip has an angled tip (**Figure 5-2B**). The gouging tip is angled so that the flame and cutting stream can be directed parallel to the surface being gouged, **Figure 5-3**.

FIGURE 5-2 OFC cutting tip (left); (right) OFC gouging tip (right).

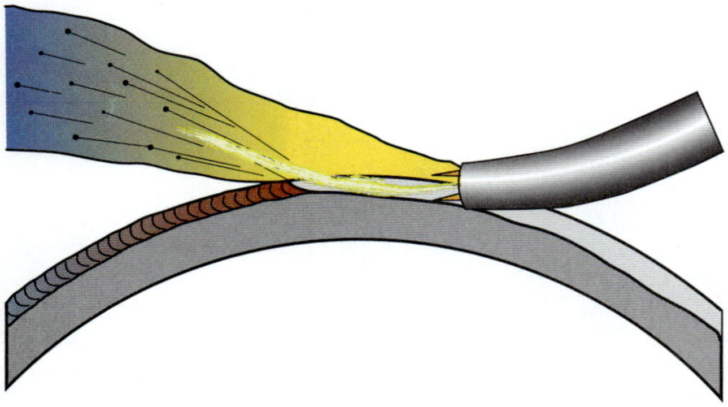

FIGURE 5-3 OFC torch gouging.

PLASMA ARC CUTTING (PAC)

PAC uses a very concentrated column of high-velocity, high-temperature **ionized gas** to rapidly melt or vaporize the metal to create a cut. The temperature of the plasma is about 43,000°F (23,900°C), **Figure 5-4**. Although the plasma arc is at a very high temperature, it does not generate a lot of heat, so it does not create as large a HAZ as does the OFC-A cutting process, Figure 5-1B. The highly concentrated plasma column can be used to cut any piping material that is electrically conductive.

Plasma Arc Gouging (PAG)

PAG, like OFC gouging, can be used to remove surface metal, cut grooves for welding, and remove old welds; however, it is a lot faster and cleaner and leaves a cut surface that needs little or no postgouging cleanup. Although most plasma cutting torches can be used for gouging, there are specially designed plasma gouging torches that do a much better job, **Figure 5-5**.

FIGURE 5-5 PAC torch gouging.

AIR CARBON ARC GOUGING (CAC-A)

In the CAC-A process, the air stream blows the molten metal away. Because the base metal does not have to react with the air stream, there is a long list of metals that can be cut, **Table 5-1**. Few cutting processes can match the speed, quality, and cost savings of this process for repair or rework. In repair or rework, the most difficult part is removing the old weld or cutting a groove so that a new

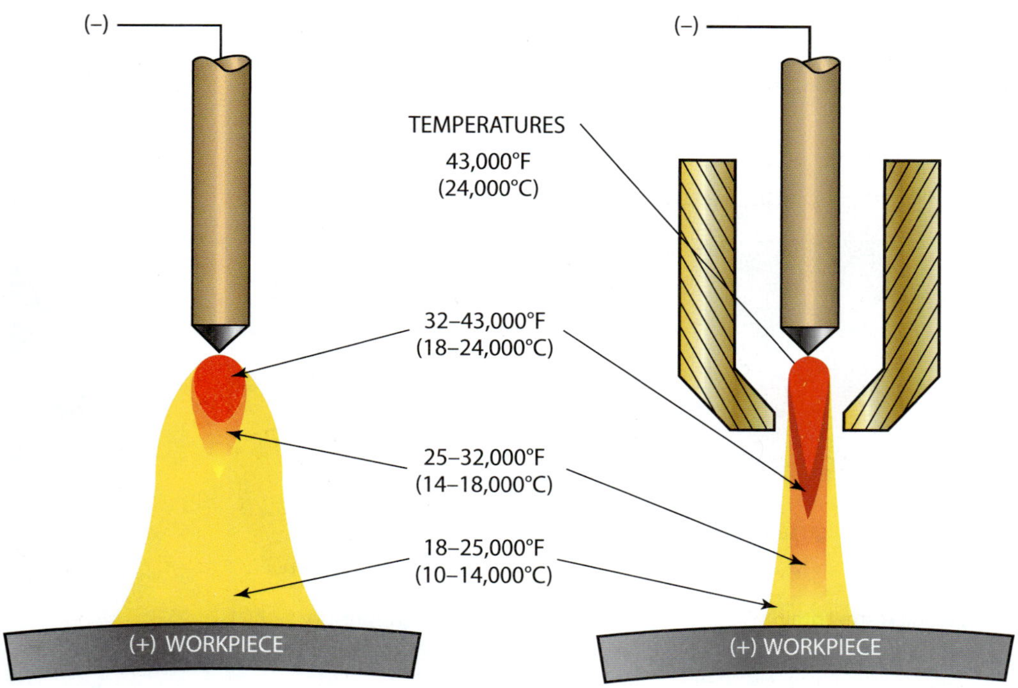

FIGURE 5-4 GTA versus PAC arc and plasma temperatures.

Base Metals	Recommendations
Carbon steel and low-alloy steel	Use DC electrodes with DCEP current. AC can be used but with a 50% loss in efficiency.
Stainless steel	Same as for carbon steel.
Cast iron, including malleable and ductile iron	Use of 13-mm or larger electrodes at the highest-rated amperage is necessary. There are also special techniques that need to be used when gouging these metals. The push angle should be at least 70°, and depth of cut should not exceed 13-mm per pass.
Copper alloys (copper content 60% and under)	Use DC electrodes with DCEN (electrode negative) at maximum amperage rating of the electrode.
Copper alloys (copper content over 60%, or size of workpiece is large)	Use DC electrodes with DCEN at maximum amperage rating of the electrode or use AC electrodes with AC.
Aluminum bronze and aluminum nickel bronze (special naval propeller alloy)	Use DC electrodes with DCEN.
Nickel alloys (nickel content is over 80%)	Use AC electrodes with AC.
Nickel alloys (nickel content less than 80%)	Use DC electrodes with DCEP.
Magnesium alloys	Use DC electrodes with DCEP. Before welding, the surface of groove should be wire brushed.
Aluminum	Use DC electrodes with DCEP. Wire brushing with stainless wire brushes is mandatory prior to welding. Electrode extension (length of electrode between electrode torch and workpiece) should not exceed 76 mm for good-quality work. DC electrodes with DCEN can also be used.
Titanium, zirconium, hafnium, and their alloys	Should not be cut or gouged in preparation for welding or remelting without subsequent mechanical removal of surface layer from cut surface.

Note: Where preheat is required for welding, similar preheat should be used for gouging.

TABLE 5-1 List of Metals That Can Be PAC-Cut

weld can be made. The air carbon arc can easily remove the worst welds even if they contain **slag** inclusions or other defects. For repairs, the arc can cut through thin layers of paint, oil, or rust and make a groove.

An air carbon arc cutting torch can be attached to most standard shielded metal arc welding (SMAW) machines for power, and to an air compressor for air. The air compressor must be capable of producing between 80 and 100 psi (550 and 690 kPa). The groove produced may take a little postcutting grinding to prepare it for welding.

PIPE CUTTING

Pipe can be cut freehand or with a pipe-cutting machine. Most freehand pipe cuts are square and made as a rough cut to get the approximate length or to create a fitting like a tee, reducer, or elbow. Although it is possible to cut a bevel on pipe or fittings freehand, these cuts vary too much to meet the requirements of most codes and standards. Pipe-cutting machines, often called *pipe-beveling machines,* can be hand-operated or motor-driven. They can produce cuts that require very little postcutting grinding or cleanup before welding begins.

Freehand Pipe Cuts

Freehand pipe cutting can be done with an OFC or PAC torch. These cuts may be done in one of two ways. On small-diameter pipe, usually under 3 in. (76 mm), the torch tip is held straight up and down and moved from the center to each side, **Figure 5-6**. This technique can also be used successfully on larger pipe.

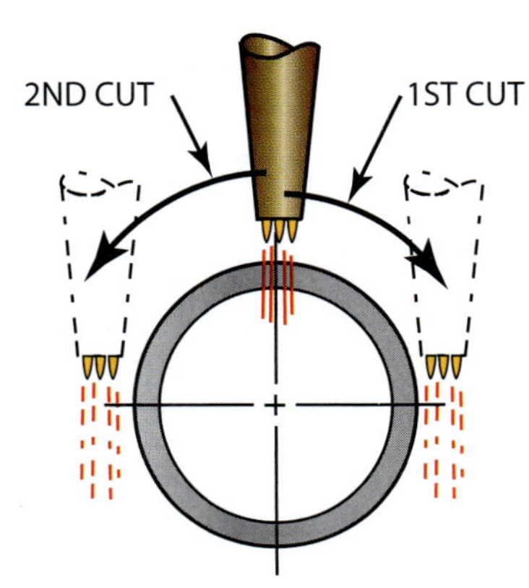

FIGURE 5-6 Technique for OF cutting of pipe.

For large-diameter pipe, 3 in. (76 mm) and larger, the torch tip is always pointed toward the center of the pipe, **Figure 5-7**. This technique is also used on all sizes of

(A)

(B)

FIGURE 5-7 (a) The welder is well braced to make a smooth, accurate cut. (b) Notice that the cutting torch is perpendicular to the pipe surface.

heavy-walled pipe and can be used on some smaller pipe sizes.

The torch body should be held so that it is parallel to the centerline of the pipe, **Figure 5-8.** Holding the torch parallel helps keep the cut square, **Figure 5-9.**

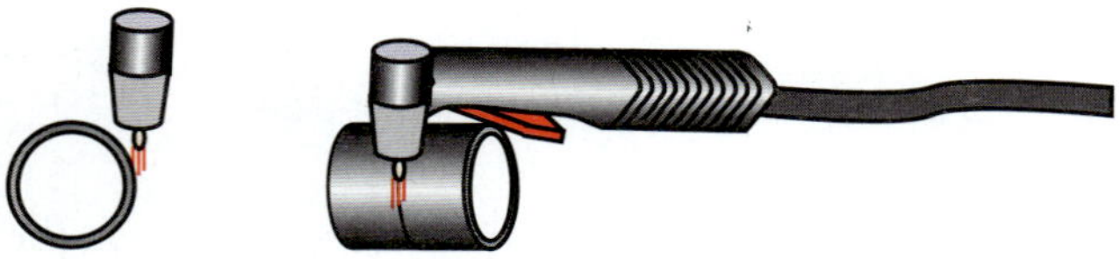

FIGURE 5-8 The plasma torch head is held parallel to the center line of the cut.

FIGURE 5-9 Notice how straight the cut can be made on this pipe joint when the torch is held properly.

CAUTION

When cutting pipe, hot sparks can come out of the end of the pipe. If you are standing at the end where the sparks are coming out, you can be burned. The other hazard that can occur is that sparks can travel some distance down the pipe before falling out of the end unnoticed, where they could start a fire.

Layout

Mark the line that you want to cut using a sharpened soapstone, scribe, or punch. Scribes and punches can both be used to lay out an accurate line. Often, a punched line is easier to see than a scribed line when cutting. To make a punched line, hold the punch, as shown in **Figure 5-10**.

FIGURE 5-10 Punching a line to be flame-cut makes it easier to see while cutting.

With the tip of the punch just above the surface of the metal, strike the punch repeatedly with a lightweight hammer. Move your hand along the line as you strike the punch; this will leave a series of marks that are easy to follow as you make the cut.

When using a piece of soapstone to mark a line, it should be sharpened properly for optimum accuracy. Holding the flat side of the soapstone against the wraparound or straight edge will ensure that the line drawn is as accurate as possible, **Figure 5-11A**. You do not have to make a heavy, wide mark to see it when you are making your cut. In fact, the sharper and more well defined the mark, the more accurately it can be cut, **Figure 5-11B**.

Larry Jeffus

(A)

Larry Jeffus

(B)

FIGURE 5-11 (A) The proper way of holding a soapstone to mark a straight line. (B) A smooth, accurate line has been drawn on the pipe.

OXYACETYLENE EQUIPMENT SETUP

When setting up any oxyacetylene equipment, you must first read all of the manufacturer's setup and safety instructions. Although there are similarities in equipment, each manufacturer may have differences in how it recommend its equipment be used. However, the general information in this section can apply to most manufacturers' oxyacetylene-cutting equipment.

Selecting the Correct Tip

Each welding torch manufacturer makes a wide range of cutting tip sizes for different material thicknesses and surface conditions. For the most part, pipe welders will be cutting relatively clean, new pipes, so the cuts being made should be clean and slag free. The major differences between welding tips are the number and size of the preheat orifices and the size of the center cutting orifice, **Figure 5-12**.

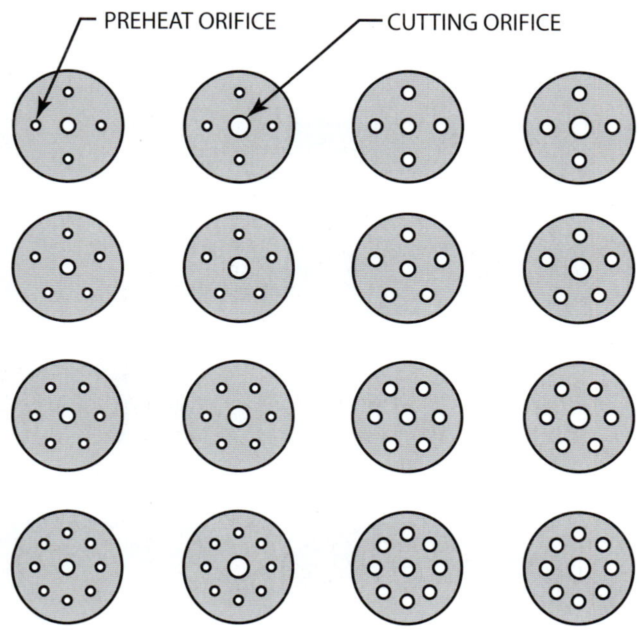

FIGURE 5-12 Examples of different cutting and preheat orifice sizes of cutting tips.

The larger the diameter of the center cutting orifice, the thicker the metal the tip will be able to cut. As the diameter and the number of preheat orifices increase, the more heat the tip generates. Higher heat can be helpful if the surface of the pipe is rusty; also, the higher heat can help the cut go faster. It also can help get the cut started quicker. **Table 5-2** lists the tip drill (orifice) size, pressure range, and metal thickness range for which the tip can be used.

Cutting tips with four preheating holes are the most commonly used to cut pipe, and the alignment of the **preheat flames** can make a difference in the quality of a square or beveled cut. On a straight-line square cut, the holes should be aligned so that one is directly on the line ahead of the cut and another is aimed down into the cut, **Figure 5-13**.

Metal Thickness in. (mm)	Center Orifice Size No. Drill	Center Orifice Size Tip Cleaner No.*	Oxygen Pressure lb/in. (kPa)	Acetylene lb/in. (kPa)
1/8 (3)	60	7	10 (70)	3 (20)
1/4 (6)	60	7	15 (100)	3 (20)
3/8 (10)	55	11	20 (140)	3 (20)
1/2 (13)	55	11	25 (170)	4 (30)
3/4 (19)	55	11	30 (200)	4 (30)
1 (25)	53	12	35 (240)	4 (30)
2 (51)	49	13	45 (310)	5 (35)
3 (76)	49	13	50 (340)	5 (35)
4 (102)	49	13	55 (380)	5 (35)
5 (127)	45	**	60 (410)	5 (35)

*The tip cleaner number when counted from the small end toward the large end in a standard tip cleaner set.

**This orifice size is larger than any tip cleaner in a standard set.

TABLE 5-2 Cutting Tip Sizes and Pressures for Different Metal Thicknesses

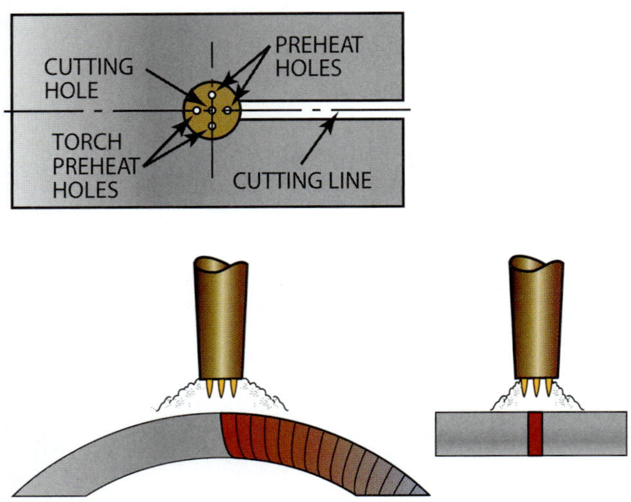

FIGURE 5-13 Proper alignment of the preheat holes and cutting tip for a square pipe cut.

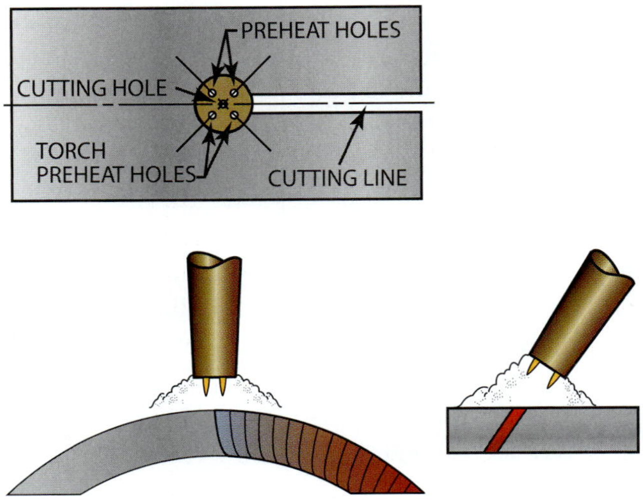

FIGURE 5-14 Proper alignment of the preheat holes and cutting tip for a bevel pipe cut.

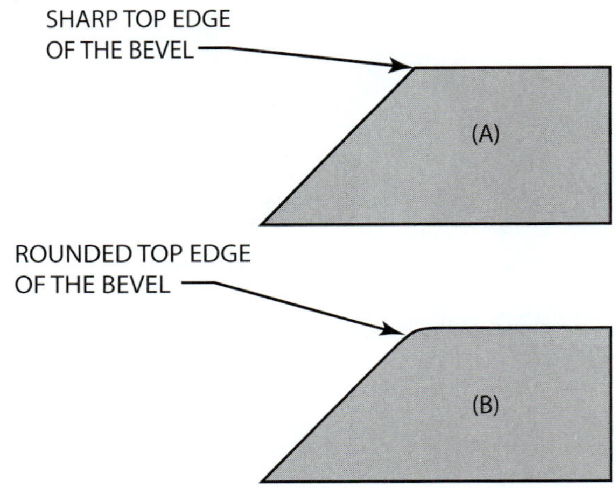

FIGURE 5-15 When properly cut, the top edge of the bevel should have a sharp edge (A), because a rounded edge (B) can contribute to underfill or undercut along the side of the cover pass.

On a beveled cut, the preheat flames should sit so that at least two of them are on the larger plate and none of them are directed on the sharp edge, **Figure 5-14**. The proper alignment of the **preheat holes** will speed up and improve the cut.

NOTE

If a cutting tip has too much preheat, the top edge of the cut can be rounded off as it is melted, **Figure 5-15**.

Cleaning a Cutting Tip

Keeping the cutting tip clean will improve the quality of cuts and reduce the postcutting cleanup. A clean cutting tip can be used to make a cut so cleanly that it might look like it was machined. A dirty tip can result in a cut not even completely separating the pipe. Here are some suggestions on how to clean a cutting tip properly:

1. Turn on a small amount of oxygen, **Figure 5-16**. This procedure is done to blow out any dirt loosened during the cleaning.
2. File the end of the tip flat, using the file provided in the tip cleaning set, **Figure 5-17**.
3. Try several sizes of tip cleaners (small, round files) in a preheat hole until the correct size cleaner is determined. It should easily go all the way into the tip, **Figure 5-18**.
4. Push the cleaner in and out of each preheat hole several times, but be careful. Excessive use will greatly increase the orifice (hole) size.

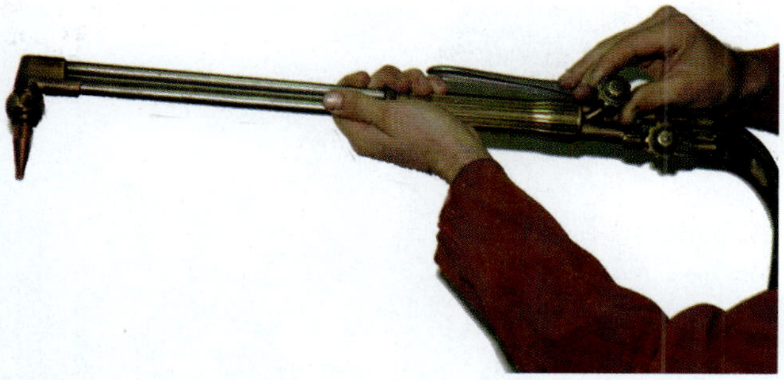

Larry Jeffus

FIGURE 5-16 A small oxygen flow will blow loosened soot and spatter out of the tip when it is being cleaned.

Larry Jeffus

FIGURE 5-17 File the end of the tip flat.

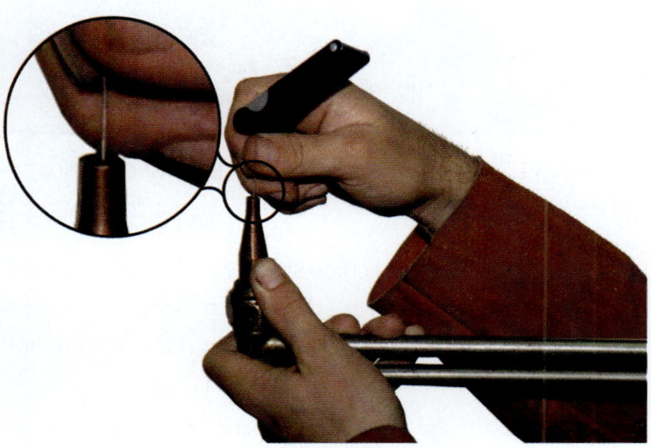

Larry Jeffus

FIGURE 5-18 Holding the tip cleaner securely will minimize the chance of its being accidentally bent while cleaning the tip.

5. Next, depress the cutting lever and, by trial and error, select the correct size tip cleaner for the center cutting orifice.

NOTE

A tip cleaner should never be forced.

Lighting the Torch

Lighting the torch when it is part of a cutting machine is different from lighting a standard cutting torch. On both machine torches and handheld cutting torches, the acetylene valve is opened slightly and lit with a spark lighter. On the handheld torch, the acetylene valve is opened until all of the smoke disappears before the oxygen valve is opened. However, on many machine torches, the cutting tip is positioned very close to the pipe surface; and no matter how much the acetylene valve is opened, it will not stop smoking, **Figure 5-19**.

Larry Jeffus

FIGURE 5-19 If the torch tip is held close to the pipe surface by a pipe-cutting machine, the acetylene, when lit, will always produce heavy black smoke.

On machine torches, therefore, it is more important that the acetylene and oxygen regulator pressures be set as close as possible to the proper working pressures. The following steps will help in setting these pressures:

1. Wear welding goggles, gloves, and any other required personal protective equipment (PPE).
2. Set the regulator working pressure for the tip size. If you do not know the correct pressure for the tip, start with the fuel set at 5 psig (35 kPag) and the oxygen set at 25 psig (170 kPag).
3. Point the torch tip away from any equipment or other workers.
4. Turn on just the acetylene valve, and only use a spark lighter to ignite the acetylene, **Figure 5-20**. If the acetylene flow is too high, the torch may not stay lit. If this happens, close the valve slightly and try to relight the torch.
5. If the flame is small, it will produce heavy black soot and smoke. In this case, turn the flame up, **Figure 5-21**. The welder need not be concerned if the flame jumps slightly away from the torch tip.

Larry Jeffus

FIGURE 5-20 Only use an appropriate tool for lighting the torch, as shown here.

Larry Jeffus

FIGURE 5-21 Increase the acetylene gas flow until the flame burns without smoke or begins to lift off or separate from the end of the torch tip.

6. Once the acetylene flame is burning smoke free, adjust the flame to a neutral setting by slowly opening the oxygen valve, **Figure 5-22**.
7. When the cutting oxygen lever is depressed, the flame may become slightly carbonizing. This may occur because of a drop in line pressure due to the high flow of oxygen through the cutting orifice.
8. With the cutting lever depressed, readjust the preheat flame to a neutral setting, **Figure 5-23**.

Larry Jeffus

FIGURE 5-22 Slowly turn on the oxygen and adjust it to a neutral flame.

Larry Jeffus

FIGURE 5-23 Press down the cutting lever and readjust the flame to a neutral flame setting.

9. The flame will become slightly oxidizing when the cutting lever is released. Since an oxidizing flame is hotter than a neutral flame, the metal being cut will be preheated faster.
10. When the cut is started by depressing the lever, the flame automatically returns to the neutral setting and does not oxidize the top of the plate.
11. When you are ready to turn off the torch, turn off the oxygen first, and then the acetylene.

NOTE

At a job site, very long hoses often are used to reach from the welding truck to the pipe being cut, **Figure 5-24**. In these cases, the pressure drop between the regulator and the torch may be so great that you must raise the regulator pressures to compensate for this pressure drop.

FIGURE 5-24 A hose reel can be used to conveniently uncoil and recoil long hose sets.

Getting Set to Cut

The slightest accidental movement of the torch when a cut is being made will result in a notch or gouge in the cut surface. Many welding codes require that any surface notches larger than 1/64 in. must be ground smooth. To keep the cut as smooth as possible, the pipe must be braced or held so that it does not move during the cut, **Figure 5-25**.

FIGURE 5-25 Secure the pipe so that it does not move during the cut.

With the torch lit and adjusted, and before starting the cut, practice moving the torch along the cut line. This does two things: first, it will make sure that nothing will obstruct the torch's free movement, **Figure 5-26**. Second, there may be loose chips of paint on the pipe surface. During the cut, these chips can cause the cutting stream to be deflected, causing a notch in the cut surface. Moving the lit torch along the cut line will blow them free, **Figure 5-27**.

FIGURE 5-26 Hold the torch tip on the line to be cut and close to the surface; next, press the cutting lever and quickly move the torch along the line to be cut to make sure that you have complete freedom of movement.

FIGURE 5-27 Moving the torch along the line will also blow off any loose paint chips or coating.

Starting a Cut

Because pipe cuts usually are started in a place other than on the edge, caution needs to be taken to prevent hot molten slag from spraying back out to the cut as it starts. This slag may damage the tip by plugging up some of the preheat orifices, or it may deflect the center cutting stream. To prevent this from happening on a handheld torch,

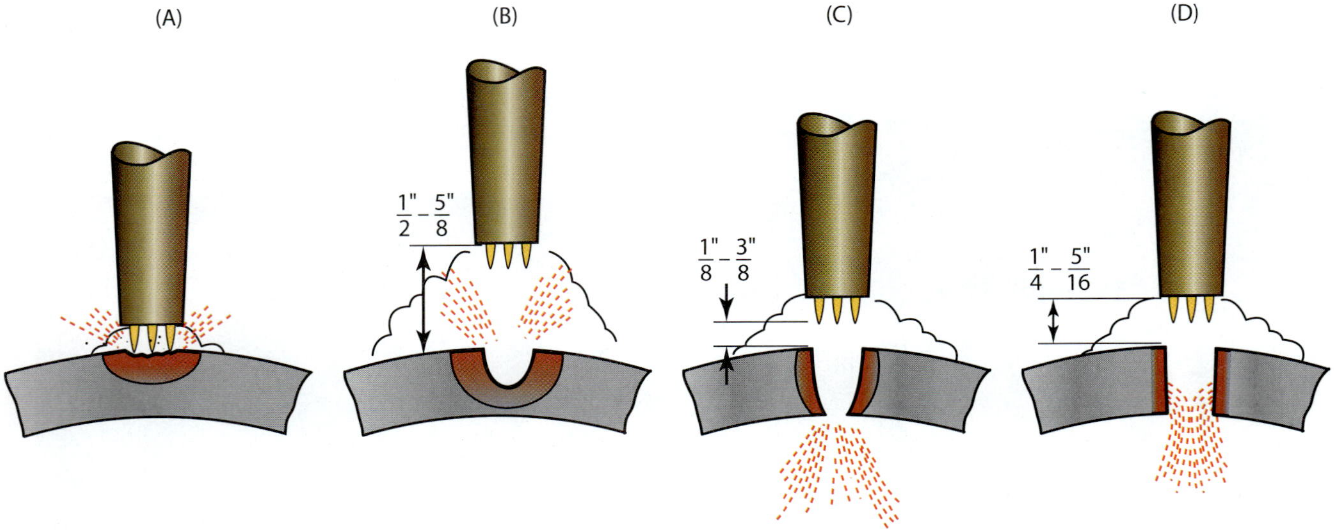

FIGURE 5-28 After the cutting torch tip has preheated the pipe (A), gradually raise the tip as the cutting lever is pressed (B). Once the pipe has been pierced, lower the tip (C). Once the piercing is completed and the tip lowered, begin the cut (D).

slowly pull the cutting stream trigger and raise the torch slightly, **Figure 5-28**. On machine torches, it is not easy (or in most cases even possible) to raise the torch tip as the cut starts, so to prevent slag from popping up onto the tip, move the torch as the cutting stream starts, **Figure 5-29**.

On both manual cutting and machine cutting, if the torch is moved one direction for a very short distance and then back the other way to complete the cut, the small slag droplet that forms as the cut starts will be blown away before it has a chance to cool. Once it cools, it cannot be cleanly blown away by the cutting stream and may cause a gouge in the cut surface just as the cut ends, **Figure 5-30**.

To start the cut faster, hold the flame so that the inner cones touch the pipe surface, **Figure 5-31**. When the metal

FIGURE 5-29 (A) The plate is heated before the cut starts, (B) the torch crank is turned, (C) and the cutting oxygen stream is started.

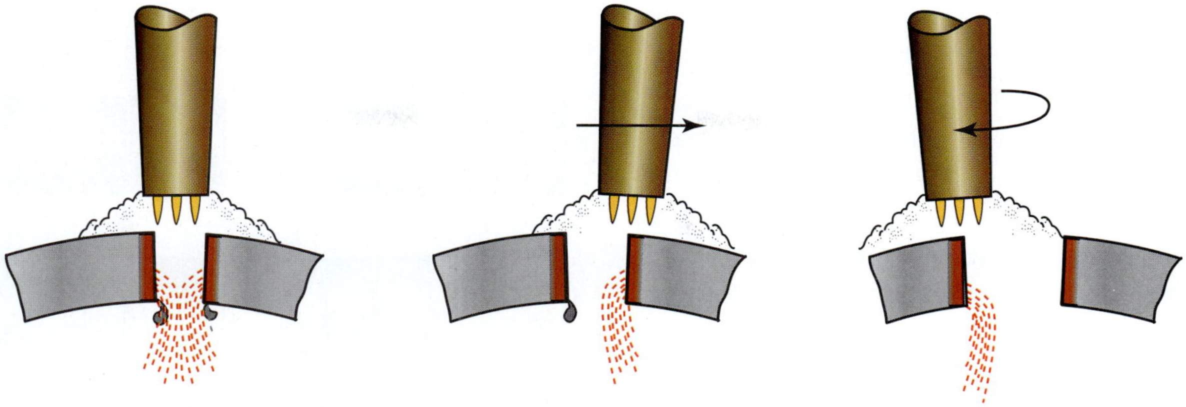

FIGURE 5-30 When starting a cut, move the torch back along the line for a short distance before continuing the cut in the opposite direction.

Larry Jeffus

FIGURE 5-31 Touching the pipe with the inner cones will disrupt the pipe surface, allowing a faster cutting start.

is hot enough to allow the cut to start (a dull red color), the torch should be raised as the cutting lever is slowly depressed, **Figure 5-32**. When the metal is pierced, the torch should be lowered again, Figure 5-28.

Larry Jeffus

FIGURE 5-32 Raising the torch prevents slag from being blown back onto the tip. If the tip gets spattered with slag, it will not make a smooth, accurate cut.

FREEHAND OXYACETYLENE PIPE CUTTING

Square Cutting Pipe

Although most freehand pipe cuts are done just to cut pipe or fittings to an approximate size so that a pipe machine can be used to make the finished bevel, freehand cuts should be made as clean and square as possible, **Figure 5-33**.

Larry Jeffus

FIGURE 5-33 Notice how little slag is remaining on the inside cut surface.

1G OFC-A Pipe Cutting

Using a properly lit and adjusted cutting torch and wearing all required PPE, you are going to make a square cut on pipe that can be rolled between cuts. When possible, rolling the pipe allows cuts to be made in the flat or near-flat position. This eliminates a lot of the problems caused when making overhead cuts on fixed pipe. Perform the following steps:

1. Use a wrap-a-round and a piece of soapstone to mark a straight line around the pipe.
2. Place the pipe horizontally on the cutting table.
3. Start the cut at the top of the pipe using the proper starting technique.

4. Move the torch backward along the line, and then forward to keep slag out of the cut.
5. Keep the tip pointed straight down.
6. When you have gone as far with the cut as you can comfortably go, quickly release the torch-cutting lever and move the flame away from the pipe.
7. Restart the cut at the top of the pipe and cut as far as possible in the other direction.
8. Stop and turn the pipe so that the end of the cut is on top and the cut can be continued around the pipe.
9. When the cut is completed, the piece that is cut off must fall free.
10. When the pipe is placed upright on a flat plate, the pipe must stand within 5 degrees of vertical and have no gaps greater than 1/8 in. (3 mm) under the cut.
11. Repeat this procedure until the cut can be made within tolerance.
12. When you have finished cutting, turn off the cylinder valves, bleed the hoses, back out the pressure regulators, and clean up your work area.

5G OF Pipe Cutting

Cutting on a horizontal pipe that is fixed and cannot be rolled requires you to transition from cutting in a flat position on top of the pipe to vertical on the sides and overhead on the bottom. It is often better to finish the cut on top so that you are not under the pipe being cut. Even though the pipe end is supported, it may not be safe to be under a pipe when it is cut free. Starting the cut on the sides and cutting downward to the bottom will let you finish the cut on top. If the cut is finished on the bottom, the cut piece will swing around, as opposed to falling free when the cut is finished on top, **Figure 5-34**.

Using a properly lit and adjusted cutting torch and wearing all required PPE, you are going to make a square cut on pipe that cannot be rolled between cuts by doing the following steps:

1. Start on the side at around the 10:30 to 11:00 position and cut down as far as you can comfortably.
2. Reposition yourself and continue the cut under the pipe.
3. Start on the other side at around the 1:00 to 2:30 position and cut downward as far as you can comfortably.
4. Reposition yourself and continue the cut under the pipe until you reach the end of the other cut.
5. Reposition yourself and restart the cut on the side, cutting upward to just before the top center.

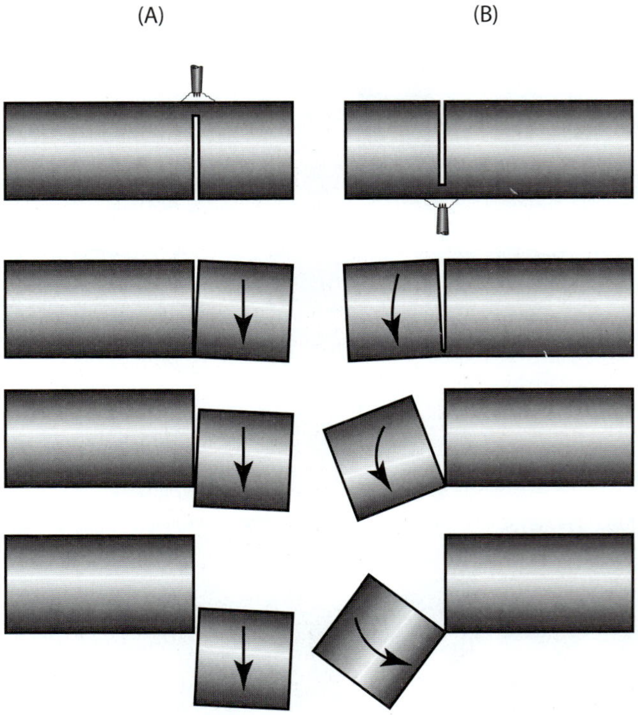

FIGURE 5-34 (A) When the cut is completed on the top, the cut piece will fall almost straight down, away from the welder. (B) However, if the cut is finished on the bottom, unless the pipe is supported, it will swing around and can hit the welder.

6. Reposition yourself and restart the cut on the other side, cutting upward until the cut is completed and the pipe falls free.
7. Stand the cut end of the pipe on a flat plate. The pipe must stand within 5 degrees of vertical and have no gaps higher than 1/8 in. (3 mm).
8. Repeat this practice until the cut can be made within tolerance.
9. When you are finished cutting, turn off the cylinder valves, bleed the hoses, back out the pressure regulators, and clean up your work area.

CAUTION

The piece of pipe being cut off must be supported so that it does not fall once it is cut free.

2G OF Pipe Cutting

Often it is easier to make a cut on vertical pipe using the technique previously illustrated in Figure 5-6, because even on pipe that is 8 in. (20 cm) or slightly larger, it may be difficult to brace your hand on the pipe. On larger pipe, bracing yourself is much easier, so the technique illustrated in **Figure 5-35** can be used.

Using a properly lit and adjusted cutting torch and wearing all required PPE, you are going to make a square

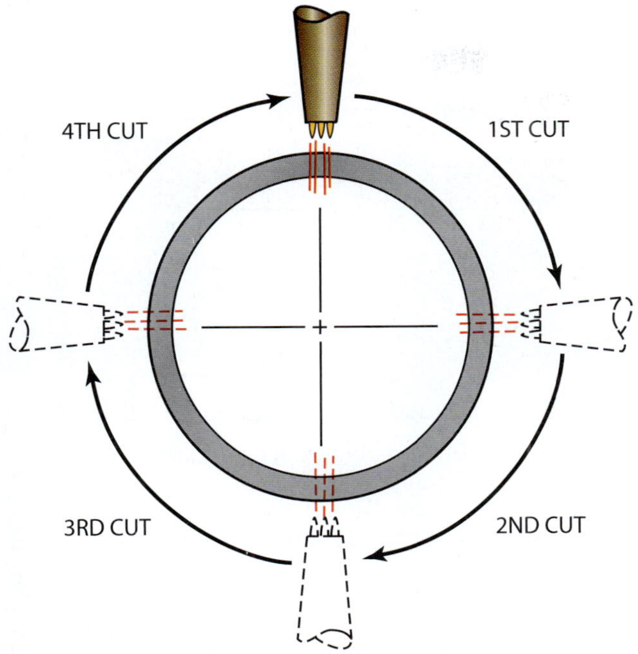

FIGURE 5-35 Sequence of cuts.

cut on pipe that is fixed in the vertical position doing the following steps:

1. Start on one side and proceed around the pipe as far as you can comfortably reach.
2. Reposition yourself and restart the cut.
3. Repeat this process until the cut is completed.
4. Because of slag, the ring may have to be tapped free.
5. Stand the cut end of the pipe on a flat plate. The pipe must stand within 5 degrees of vertical and have no gaps higher than 1/8 in. (3 mm).
6. Repeat this practice until the cut can be made within tolerance.
7. When you are finished cutting, turn off the cylinder valves, bleed the hoses, back out the pressure regulators, and clean up your work area.

Rusty Nuts and Bolts

Often the oxyacetylene torch is referred to as a *smoke wrench* because it is used to help remove rusty nuts and bolts that are so stuck they cannot be removed with other wrenches.

Nuts can be either loosened or cut off by a cutting torch. Often a nut can be removed by first heating it until it is red hot, which will cause the nut to expand. Then a wrench can be used to turn the nut off. If the nut and bolt are both heated, they expand, and the nut may not loosen. So the trick is to heat the nut as fast as possible without heating the bolt. Also, the rust in the treads between the nut and bolt will help insulate the bolt from the torch heat as the nut is heated. In addition, as the nut expands, the gap between the nut's threads and the bolt's threads further insulates the bolt from the heat. That is not to say the bolt will not get hot, but it should not get red hot. Unless the nut is being used in thin metal or sheet metal, direct the flame at the base of the nut nearest to the part because the larger mass of the part will not heat as fast as the nut, **Figure 5-36**. For a nut rusted stuck on thinner metal, start heating it at the top to keep the heat off the thinner metal as much as possible.

FIGURE 5-36 Heating a rusted nut can make it possible to unscrew it.

Cutting the nut off is almost the same as cutting the bolt head off. The only difference is that the rust that kept the bolt from heating up at the same rate as the nut can keep it from getting hot enough to cut. If the bolt does not start to cut at the same time as the nut does, stop the cutting stream and heat the bolt some more. Start the cut at the top of the nut and work the cutting stream back and forth across the nut, taking off layers of it each time. Keep this up until the entire nut and bolt are removed, **Figure 5-37**.

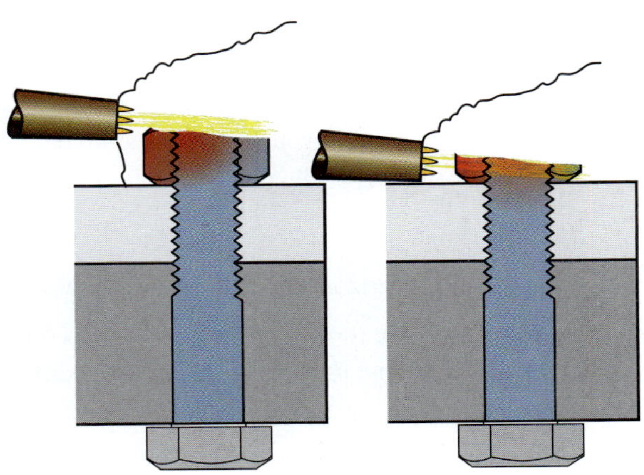

FIGURE 5-37 Cutting the nut and bolt off by starting at the top makes it easier to avoid gouging the base plate.

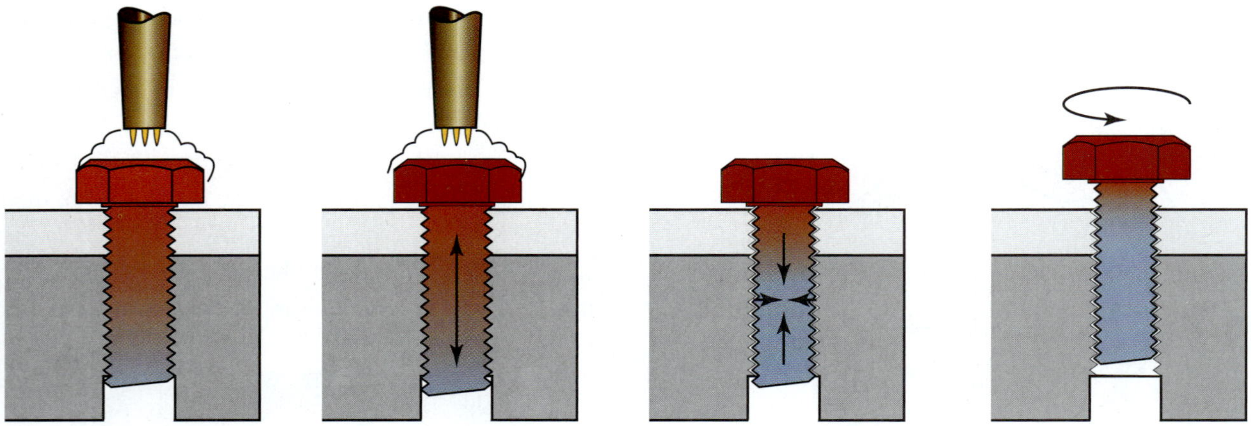

FIGURE 5-38 Heating a stuck bolt until it is red hot and then allowing it to cool can make removing it possible.

To remove a bolt that is rusted stuck in the internal threads of a part, heat the head of the bolt. If the bolt is not stuck too badly, you may be able to remove it now; but if it is really stuck, continue heating the bolt head until it is red hot. This causes the bolt to try and expand equally in all directions, but it cannot due to the part it is in. Since the hot bolt metal cannot expand outward, it will expand lengthwise. When it cools, it will contract equally in all directions. This will make the bolt just a few thousands of an inch smaller, and now it will be loose enough so it can be unthreaded, **Figure 5-38**.

OXYACETYLENE GOUGING

The standard cutting tip on an oxyacetylene cutting torch can be replaced with a gouging tip, which has a bend of around 25 degrees in the tip to allow the torch to be used at a very slight angle to the pipe surface, Figure 5-2. A gouging tip can be used to cut a groove or to remove a weld. The center cutting orifice is larger than the one found on most cutting tips, but it uses a lower-than-expected oxygen cutting pressure. The combination of a large orifice and lower pressure lets the oxygen stream blow away the metal being gouged slowly enough so it can be controlled more easily.

1G OFC-A U-Grooving of a Pipe

Using a properly lit and adjusted cutting torch and wearing all required PPE, you are going to gouge a groove around a pipe that can be rolled between cuts by doing the following steps:

1. Mark a straight line around the pipe.
2. Place the pipe horizontally on the cutting table.
3. To start the gouge the torch flame is directed on the pipe surface so that it will reach the desired dull red color required to start the cut.
4. Slowly depress the torch's cutting lever, and at the same time rotate the torch tip so it is almost parallel to the pipe's surface.
5. As the gouge begins, the torch tip can be angled down to make the groove deeper or sweep back and forth to make the groove wider.
6. The oxygen stream is actually burning the metal being removed by the oxyacetylene gouging process. The burning metal causes a lot of heat to build up in the pipe being gouged.
7. Stop and turn the pipe as needed so the cut can be continued around the pipe.
8. When the cut is completed, the piece cut off must fall free.
9. Repeat the gouging of the U-groove until you can make consistently smooth grooves that are within ±3/32 in. (2.3 mm) of a straight line and uniform in width and depth.
10. Turn off the cylinder valves, bleed the hoses, back out the pressure regulators, and clean up your work area when you are finished cutting.

1G OF Gouging Out a Pipe Weld

Using a properly lit and adjusted cutting torch and wearing all required PPE, you are going to gouge out a pipe weld in the horizontal rolled position by doing the following steps:

1. Place the pipe horizontally on the cutting table.
2. Start by pointing the torch flame at the 10:00 position.
3. As the gouge starts, rotate the torch tip to a flatter angle so that the depth of the groove is not so deep that sparks get blown back at you.
4. Move the torch along the weld toward the 2:00 position.
5. Watch the groove width to see that you are only removing the weld. Change the travel speed, the torch side-to-side movement, or both to keep the

groove the same width and depth so as to remove all the weld around the pipe.

6. Do not try to remove too much weld metal depth in one gouging pass. On thicker-walled pipe, you may need to make several gouges around the pipe to remove all of the weld metal.
7. When you have reached the 2:00 position, release the torch's cutting lever.
8. Turn the pipe so that the end of the U-groove is now at the 10:00 position so the groove can be continued around the pipe. Restart the groove at the top edge.
9. Repeat the gouging of the weld until you can make consistently smooth grooves that remove only the weld metal.
10. When you are finished cutting, turn off the cylinder valves, bleed the hoses, back out the pressure regulators, and clean up your work area.

5G OF Gouging Out a Pipe Weld

Using a properly lit and adjusted cutting torch and wearing all required PPE, you are going to gouge out a pipe weld in the fixed horizontal position by doing the following steps:

1. Fix the pipe horizontally on the cutting table.
2. Start at the 12:00 position.
3. Move the torch down along the weld as far as you can comfortably go.
4. Stop and reposition yourself and restart the cut.
5. Do not try to remove too much weld metal depth in one gouging pass.
6. When you have reached the 6:00 position, release the torch's cutting lever.
7. Reposition yourself so you can restart gouging at the 12:00 position to remove the weld on the other side.
8. Repeat the gouging of the U-groove until you can make consistently smooth grooves that remove only the weld metal.
9. When you are finished cutting, turn off the cylinder valves, bleed the hoses, back out the pressure regulators, and clean up your work area.

2G OF Gouging Out a Pipe Weld

Using a properly lit and adjusted cutting torch and wearing all required PPE, you are going to gouge out a pipe weld in the fixed vertical position by doing the following steps:

1. Fix the pipe vertically on the cutting table.
2. Start at a comfortable place on the pipe.
3. Move the torch around the weld as far as you can comfortably go.
4. Stop and reposition yourself, and then restart the cut until you have gone all the way around the pipe.
5. Do not try to remove too much weld metal depth in one gouging pass.
6. Repeat the gouging of the U-groove until you can make consistently smooth grooves that remove only the weld metal.
7. When you are finished cutting, turn off the cylinder valves, bleed the hoses, back out the pressure regulators, and clean up your work area.

PAC EQUIPMENT SETUP

PAC machines come in a variety of sizes and capacities, from small, lightweight, self-contained, and portable systems to larger, high-capacity ones. The most common plasma cutters used in the pipe-welding industry use dry, compressed air to form the plasma-cutting jet. Some units have self-contained air compressors, **Figure 5-39**, while other machines need to be connected to an external air compressor.

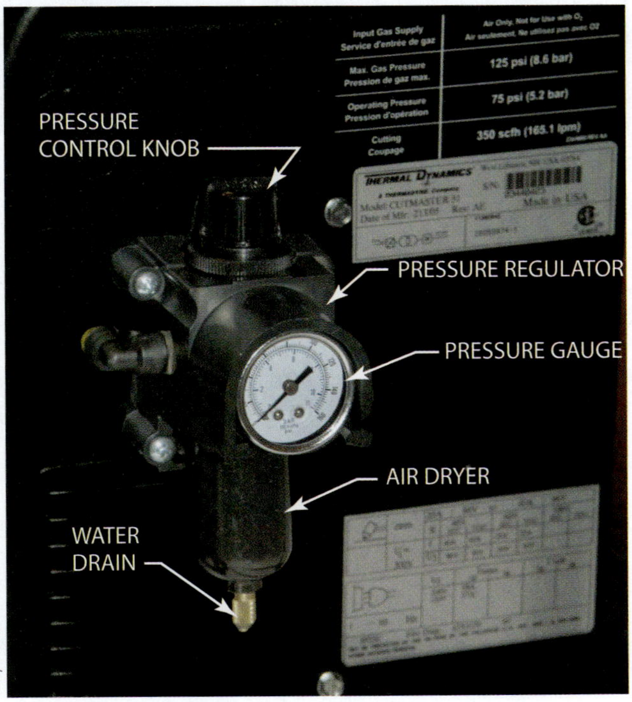

FIGURE 5-39 A pressure regulator and dryer on a self-contained plasma cutter.

When setting up any PAC equipment, you must first read all the equipment manufacturer's setup and safety instructions. However, the general information in this section can be used with most manufacturers' PAC equipment.

The **standoff distance** is the distance from the nozzle tip to the work. This distance is critical to producing quality plasma arc cuts. As the distance increases, the arc force is diminished and tends to spread out. This causes the kerf to be wider, the top edge of the plate to become rounded, and the formation of more **dross** on the bottom edge of the plate. However, if this distance becomes too small, the working life of the nozzle tip will be reduced. In some cases, an arc can form between the nozzle tip and the metal that instantly destroys the tip. Some torches' nozzle tips can be dragged along the surface of the work without shorting it out, **Figure 5-40**. This is a large help when working on pipe. This technique can cause the nozzle tip orifice to become contaminated more quickly.

FIGURE 5-40 Using a magnetic wraparound can make it easy to cut a straight line.

Most PAC torches use a **high-frequency alternating current** carried through the conductor, the electrode, and back from the nozzle tip. This high-frequency current will ionize the gas and allow it to carry the initial current to establish a pilot arc, **Figure 5-41**. After the pilot arc has been started, the high-frequency starting circuit can be stopped. A **pilot arc** is an arc between the **electrode tip** and the nozzle tip within the torch head. This is a nontransfer arc, so the workpiece is not part of the current path. If the pilot arc is left on for a long period of time, it can create enough heat to damage the torch parts. When the torch is brought close enough to the work, the primary arc will follow the pilot arc across the gap to the work, and the main plasma is started.

FREEHAND PLASMA ARC PIPE CUTTING

Often PAC is used to cut round pipe. Pipe can be a challenge to cut because the cut starts out much like a gouged groove and transitions to something like piercing a hole. It is important to keep the plasma stream straight and in line with the line that is being cut. The plasma torch will cut in the direction it is pointed, so if it is not straight, the cut may have a beveled edge.

1G PA Pipe Cutting

Using a properly set up and adjusted PAC machine and wearing all required PPE, you are going to make a square

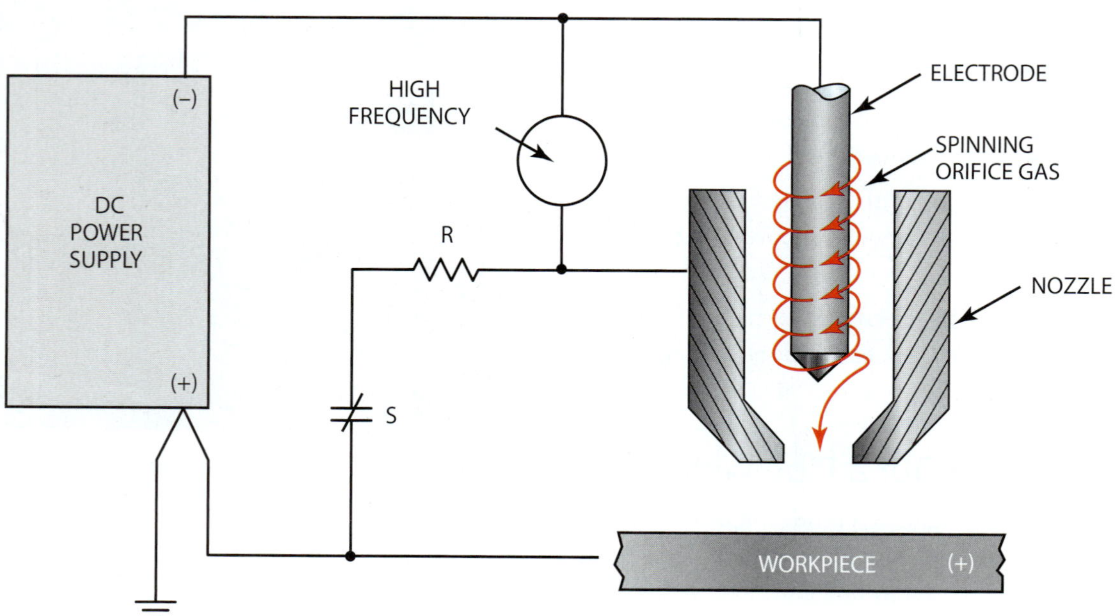

FIGURE 5-41 Running the pilot arc for long periods of time before the torch is brought close enough to the pipe to start the main plasma can damage the nozzle tip or electrode.

cut on pipe that can be rolled between cuts by doing the following steps:

1. Use a wraparound and a piece of soapstone to mark a straight line around the pipe, **Figure 5-42**.
2. Because a plasma torch does not produce a lot of heat the way that an oxyacetylene torch does, a straight edge like a magnetic strip can be used as a guide, **Figure 5-43**.
3. Place the pipe horizontally on the cutting table.
4. Start the cut at the top of the pipe by pulling the torch trigger; at the same time, start moving the torch along the straight edge to start the cut.
5. Keep the torch pointed straight toward the center of the pipe, **Figure 5-44**.
6. When you have gone as far with the cut as you can comfortably go, release the torch's cutting lever.
7. Restart the cut at the top of the pipe and cut as far as possible in the other direction.
8. Stop and turn the pipe so that the end of the cut is on top and the cut can be continued around the pipe.
9. When the cut is completed, the cut end must fall free.
10. When the pipe is placed upright on a flat plate, the pipe must stand within 5 degrees of vertical and have no gaps higher than 1/8 in. (3 mm) under the cut.
11. Repeat this procedure until the cut can be made within tolerance.
12. When you are finished cutting, turn off the plasma machine and clean up your work area.

Larry Jeffus

FIGURE 5-42 Magnetic wraparounds are a handy tool for pipe layout.

Larry Jeffus

FIGURE 5-43 OFC torches cannot be used with magnetic wraparounds because the torches produce too much heat and sparks that will damage them.

Larry Jeffus

FIGURE 5-44 Notice that the sparks from the cut are going straight across the pipe, which is a good indication that a clean cut is being produced.

5G PA Pipe Cutting

Cutting on a horizontal pipe that is fixed and cannot be rolled requires you to transition from cutting in the flat position on top of the pipe to vertical on the sides and overhead on the bottom. It is often better to finish the cut on top so that you are not under the pipe being cut. Even though the pipe end is supported, it may not be safe to be under it when it is cut free. Starting the cut on the sides and cutting downward to the bottom will let you finish the cut on top. If the cut is finished on the bottom, the cut piece will swing around, as opposed to falling free when the cut is finished on top.

Using a properly set up and adjusted PAC machine and wearing all required PPE, you are going to make a square cut on pipe that cannot be rolled between cuts by doing the following steps:

1. Start on the side around the 10:30 to 11:00 position, and cut down as far as you can comfortably go.
2. Reposition yourself and continue the cut under the pipe.
3. Start on the other side around the 1:00 to 2:30 position, and cut downward as far as you can comfortably go.
4. Reposition yourself and continue the cut under the pipe until you reach the end of the other cut, **Figure 5-45**.
5. Reposition yourself and restart the cut on the side, cutting upward to just before the top center.
6. Reposition yourself and restart the cut on the other side, cutting upward until the cut is completed and the pipe falls free.
7. Stand the cut end of the pipe on a flat plate. The pipe must stand within 5 degrees of vertical and have no gaps greater than 1/8 in. (3 mm).
8. Repeat this practice until the cut can be made within tolerance.
9. When you are finished cutting, turn off the plasma machine and clean up your work area.

FIGURE 5-45 Dragging a gloved hand along the pipe makes it easier to make a smooth cut.

2G PA Pipe Cutting

Using a properly set-up and adjusted PAC machine and wearing all required PPE, you are going to make a square cut on pipe that is fixed in the vertical position by doing the following steps:

- Start on one side and proceed around the pipe as far as you can comfortably reach.
- Reposition yourself, and restart the cut.
- Repeat this process until the cut is completed.
- Stand the cut end of the pipe on a flat plate. The pipe must stand within 5 degrees of vertical and have no gaps higher than 1/8 in. (3 mm).
- Repeat this practice until the cut can be made within tolerance.
- When you are finished cutting, turn off the plasma machine and clean up your work area.

Plasma Arc Gouging (PAG)

PAG is a recent introduction to the PAC processes. The process is similar to that of CAC-A, in that a U-groove can be cut into the metal's surface. The removal of metal along a joint before the metal is welded or the removal of a defect for repairing can easily be done using this variation of PAC, **Figure 5-46**. An advantage of using PA cutting to remove a weld is that any slag trapped in the weld will not affect the PAC gouging process.

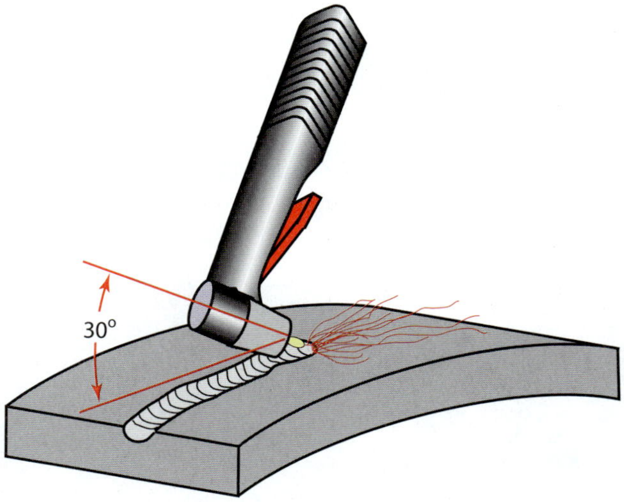

FIGURE 5-46 Plasma arc gouging.

The torch is set up with a less concentrated plasma stream. This will allow the **washing** away of the molten metal instead of thrusting it out to form a cut. The torch is held at approximately a 45-degree angle to the metal surface. Once the groove is started, it can be controlled by the rate of travel, torch angle, and side-to-side torch movement.

PAG is effective on most metals. The groove is clean, bright, and ready to be welded. PAG is especially beneficial with stainless steel and aluminum pipes because there is no reasonable alternative available. The only other process that can leave the metal ready to weld is machining, which is slow and expensive compared to PAG.

It is important to try to not remove too much metal in one pass. The process will work better if small amounts are removed at a time. If a deeper groove is required, multiple gouging passes can be used.

CAUTION

The gouging process can throw a lot more sparks than does a plasma cut. Make sure that the sparks are directed so that they will not injure anyone, damage equipment, or start a fire.

1G PA U-Grooving of a Pipe

Using a properly set up and adjusted PAC machine and wearing all required PPE, you are going to cut a U-groove around a piece of pipe in the horizontal rolled position by doing the following steps:

1. Use a wraparound and a piece of soapstone to mark a straight line around the pipe.
2. Place the pipe horizontally on the cutting table.
3. Starting at the 10:00 position, hold the torch as close as possible to a 45-degree angle, **Figure 5-47**.
4. Lower your hood and establish a plasma-cutting stream.
5. Move the torch along the line toward the 2:00 position.
6. If the width of the U-groove changes, speed up or slow down the travel rate to keep the groove the same width and depth for the entire distance around the pipe.
7. When you have reached the 2:00 position, release the torch's cutting lever.
8. Turn the pipe so that the end of the U-groove is at the 10:00 position so that the groove can be continued around the pipe.
9. When restarting the gouging, pointing the torch at the top edge of the groove will prevent sparks from being blown back at you, which would happen if you were to try and start with the torch pointed at the bottom of the groove.
10. Repeat the gouging of the U-groove until you can make consistently smooth grooves that are within ±3/32 in. (2.3 mm) of a straight line and uniform in width and depth.
11. When you are finished cutting, turn off the PAC equipment and clean up your work area.

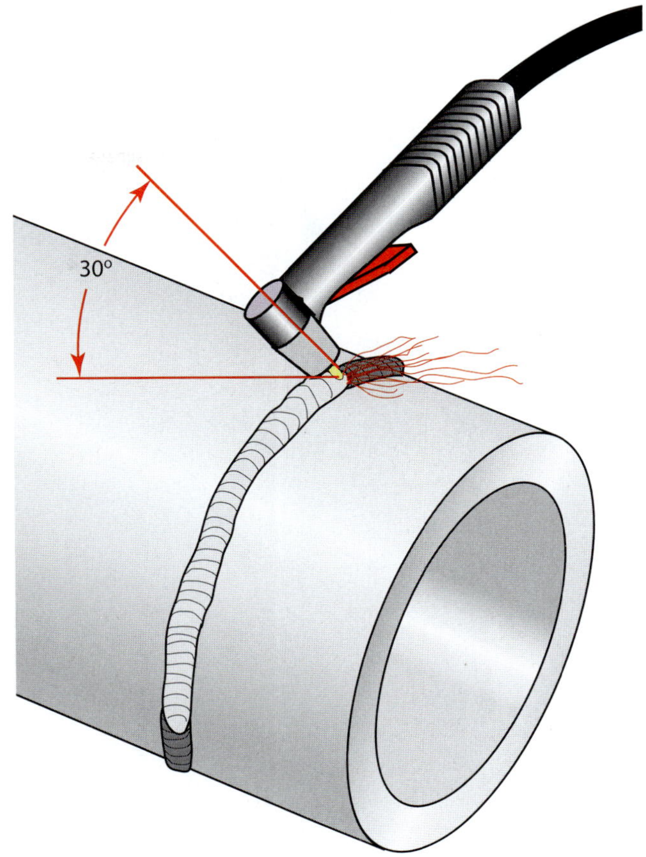

FIGURE 5-47 As the gouge progresses around the pipe, the torch must be continually rotated to keep the angle between the torch and the plate at approximately 30 degrees.

1G PA Gouging Out a Pipe Weld

Using a properly set up and adjusted PAC machine and wearing all required PPE, you are going to gouge out a pipe weld in the horizontal rolled position. As you are removing the weld metal, watch the surface of the gouge for signs of slag inclusions, lack of fusion, porosity, or other weld discontinuities. Slag inclusions will appear as black or bright spots that will be moved quickly along the gouged metal. Lack of fusion may show up as a straight line alongside of the weld or as a wavy line between weld passes. Porosity may be a cluster of spots, a line of spots, or a general area with spots by doing the following steps:

1. Place the pipe horizontally on the cutting table.
2. Starting at the 10:00 position, hold the torch as close as possible to a 45-degree angle, **Figure 5-48**.

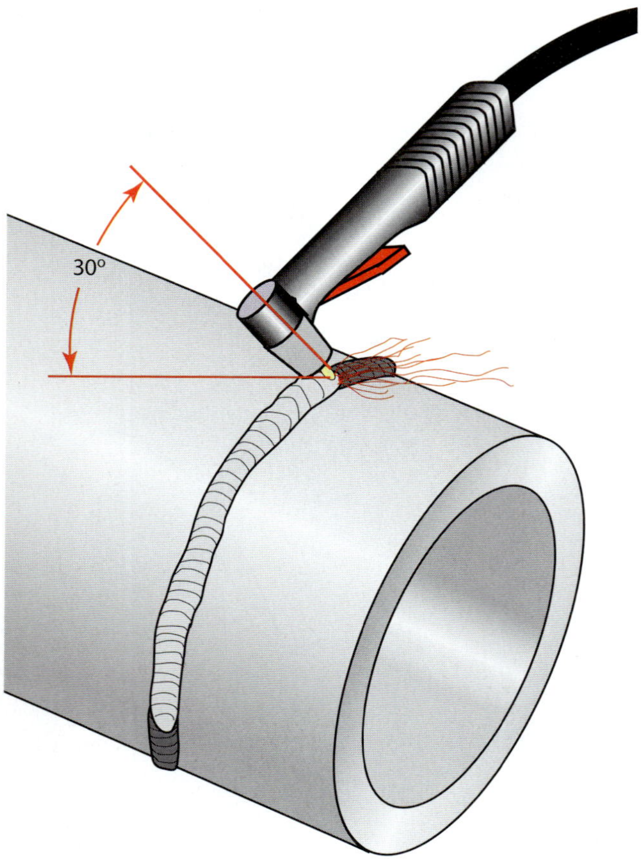

FIGURE 5-48 Rotate the pipe and continue gouging.

3. Lower your hood and establish a plasma-cutting stream.
4. Move the torch along the weld toward the 2:00 position.
5. Watch the groove width to see that you are only removing the weld. Change the travel speed, the torch side-to-side movement, or both to keep the groove the same width and depth to remove all the weld around the pipe.
6. Do not try to remove too much weld metal depth in one gouging pass. On thicker-walled pipe, you may need to make several gouges around the pipe to remove all the weld metal.
7. When you have reached the 2:00 position, release the torch's cutting lever.
8. Turn the pipe so that the end of the U-groove is at the 10:00 position so the groove can be continued around the pipe. Remember to restart the groove at the top edge.
9. Repeat the gouging of the weld until you can make consistently smooth grooves that remove only the weld metal.
10. When you are finished cutting, turn off the PAC equipment and clean up your work area.

NOTE

Trying to remove too much weld metal at one time may cause the plasma stream to be deflected into the base metal, causing a deep gouge that would have to be repaired during rewelding.

5G PA Gouging Out a Pipe Weld

Using a properly set up and adjusted PAC machine and wearing all required PPE, you are going to gouge out a pipe weld in the fixed horizontal position by doing the following steps:.

1. Fix the pipe horizontally on the cutting table.
2. Starting at the 12:00 position, hold the torch as close as possible to a 45-degree angle.
3. Lower your hood, and establish a plasma-cutting stream.
4. Move the torch down along the weld as far as you can comfortably go.
5. Stop and reposition yourself, and restart the cut.
6. Do not try to remove too much weld metal depth in one gouging pass.
7. When you have reached the 6:00 position, release the torch's cutting lever.
8. Reposition yourself so you can restart gouging at the 12:00 position to remove the weld on the other side.
9. Repeat the gouging of the U-groove until you can make consistently smooth grooves that remove only the weld metal.
10. When you are finished cutting, turn off the PAC equipment and clean up your work area.

2G PA Gouging Out a Pipe Weld

Using a properly set up and adjusted PAC machine and wearing all required PPE, you are going to gouge out a pipe weld in the fixed vertical position by doing the following steps:

1. Fix the pipe vertically on the cutting table.
2. Starting at a comfortable place on the pipe, hold the torch as close as possible to a 45-degree angle.
3. Lower your hood and establish a plasma-cutting stream.

4. Move the torch around the weld as far as you can comfortably go.
5. Stop and reposition yourself and restart the cut until you have gone all the way around the pipe.
6. Do not try to remove too much weld metal depth in one gouging pass.
7. Repeat the gouging of the U-groove until you can make consistently smooth grooves that remove only the weld metal.
8. When you are finished cutting, turn off the PAC equipment and clean up your work area.

AIR CARBON ARC EQUIPMENT SETUP

When setting up any CAC-A equipment, you must first read all the equipment manufacturer's setup and safety instructions. However, the general information in this section can be used with most manufacturers' CAC cutting equipment.

The CAC-A torch gets its electrical power from a standard welding machine. It can be permanently attached to a welding cable and air hose, or it can be attached to welding power by gripping a tab at the end of the cable with the shielded metal arc electrode holder, **Figure 5-49**. The temporary attachment can be made easier if the air hose is equipped with a quick disconnect. A quick disconnect on the air hose will allow it to be used for other air tools such as grinders or chippers. Greater flexibility for a work station can be achieved with this arrangement.

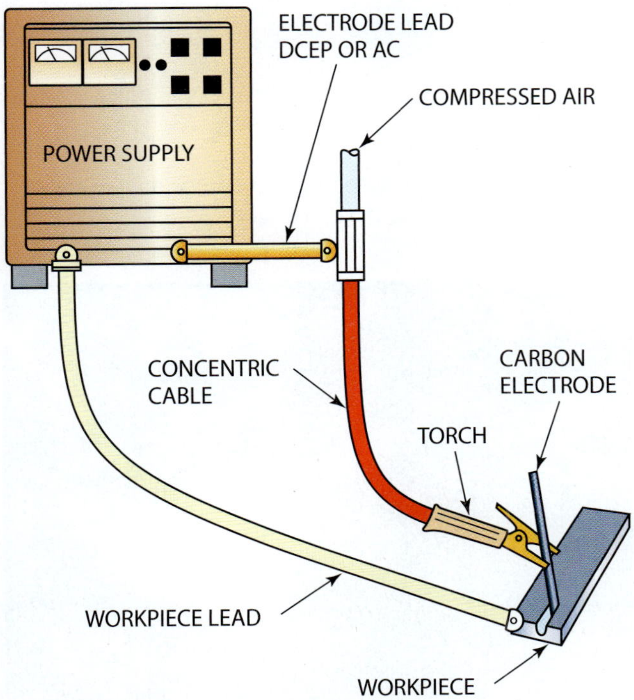

FIGURE 5-49 Air carbon arc torch setup.

The correct air pressure will result in cuts that are clean, smooth, and uniform. The air-flow rate is also important. Long air hoses or small-diameter air hoses can result in a reduced air flow at the torch. The resulting cut will be less desirable in quality.

The CAC-A highly localized heat results in only slight heating of the surrounding metal. As a result, except for work being performed in cold conditions, there is usually no need to preheat some alloyed metals to prevent hardness zones.

Air carbon arc cutting can be used to remove welds. The removal of welds can be accomplished with such success that often the part needs no postcut cleanup. On large-diameter pipes or inside pressure vessels, the root of a weld can be back-gouged so that a backing weld can be made, ensuring 100% weld penetration, **Figure 5-50**.

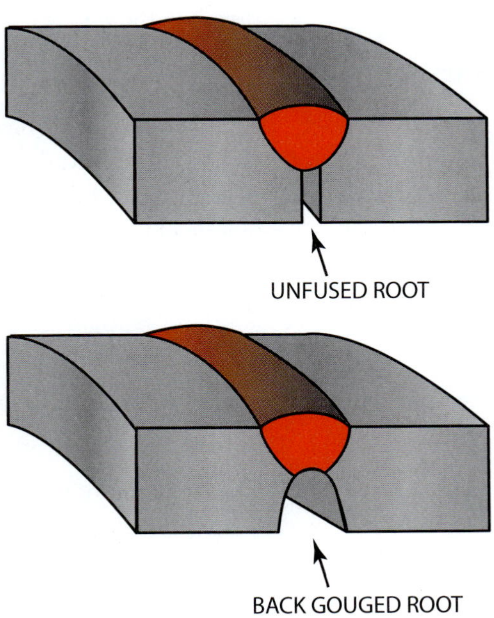

FIGURE 5-50 Back-gouging.

The electrode should extend approximately 6 in. (152 mm) from the torch when starting a cut and, as the cut progresses, the electrode is consumed. Stop the cut and readjust the electrode when its end is approximately 3 in. (76 mm) from the electrode holder. This will reduce the damage to the torch caused by the intense heat of the operation.

The groove produced along the end of a pipe is usually a J-groove. The groove produced along a joint between pipes or to remove a weld is usually a U-groove, **Figure 5-51**. Sometimes piping systems must be fitted and assembled to ensure that everything fits as designed before a groove can be cut. In these cases, a U-groove can easily be cut along the assembled edges of the joint. Both grooves are used as a means to ensure that the weld applied to the joint will have the required penetration into the metal.

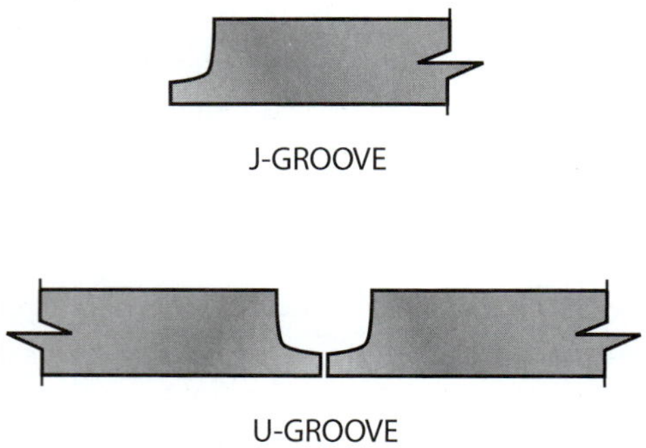

FIGURE 5-51 U-groove and J-groove.

1G Air Carbon Arc U-Grooving of a Pipe

Using an air carbon arc cutting torch and welding power supply that have been properly set up in accordance with the manufacturer's specific instructions in the owner's manual and wearing all required PPE, you are going to make a U-groove around a pipe in the horizontal rolled position by doing the following steps:

1. Mark a straight line around the pipe.
2. Place the pipe horizontally on the cutting table.
3. Adjust the air pressure to approximately 80 psi.
4. Set the amperage within the range for the diameter electrode that you are using by referring to the box the electrodes came in.
5. Check to see that the stream of sparks will not start a fire or cause any damage to anyone or anything in the area.
6. Make sure the area is safe, and then turn on the welder.
7. Using a good, dry, leather glove to avoid electrical shock, insert the electrode in the torch jaws so that about 6 in. is extending outward. Be sure not to touch the electrode to any metal parts because it may short out.
8. Turn on the air at the torch head.
9. Lower your arc welding helmet.
10. Slowly bring the electrode down at about a 30-degree angle so it will make contact with the pipe near the 10:00 position, **Figure 5-52**. Be prepared for a loud, sharp sound when the arc starts, **Figure 5-53**.
11. Once the arc is struck, move the electrode along the line toward the 2:00 position.
12. Keep the speed and angle of the torch constant.
13. When you reach the 2:00 position, lift the torch so that the arc will stop.
14. Raise your helmet and stop the air.
15. Remove the remaining electrode from the torch so it will not accidentally touch anything.
16. Turn the pipe so that the end of the U-groove is at the 10:00 position so the groove can be continued

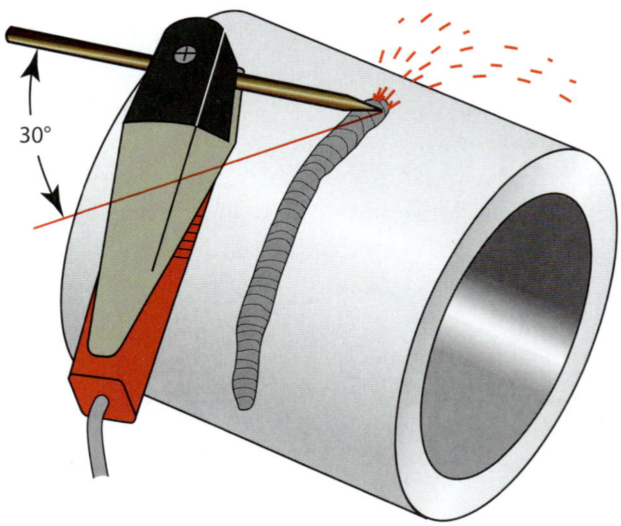

FIGURE 5-52 Hold the electrode at approximately a 30-degree angle to the pipe's surface.

FIGURE 5-53 Air carbon arc gouging produces a lot of noise, so ear protection is required.

around the pipe. Restart the groove at the top edge.

17. Repeat this process until you have made a U-groove all the way around the pipe.
18. When the metal is cool, chip or brush any slag or dross off the plate. This material should come off easily.
19. The groove must be within ±1/8 in. (3 mm) of being straight and within ±3/32 in. (2.4 mm) of uniformity in width and depth.
20. Repeat this cut until it can be made within these tolerances.
21. When you are finished cutting, turn off the CAC equipment and clean up your work area.

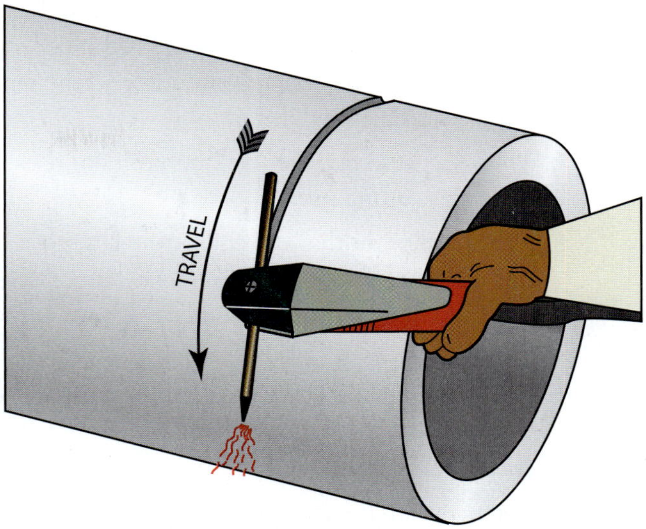

FIGURE 5-54 As the gouge progresses around the pipe, maintain a uniform 30-degree angle.

1G Air Carbon Arc J-Grooving of a Pipe

Using a properly set up air carbon arc cutting torch and welding power supply and wearing all required PPE, you are going to make a J-groove around a pipe in the horizontal rolled position by doing the following steps:

1. Place the pipe horizontally on the cutting table.
2. Adjust the air pressure and set the amperage.
3. Make sure the area is safe and turn on the welder.
4. Insert the electrode in the torch jaws so that about 6 in. is extending outward. Be sure not to touch the electrode to any metal parts because it may short out.
5. Turn on the air at the torch head.
6. Lower your arc welding helmet.
7. Slowly bring the electrode down at about a 30-degree angle so it will make contact with the pipe edge near the 10:00 position, **Figure 5-54**.
8. Once the arc is struck, move the electrode along the line toward the 2:00 position.
9. Keep the speed and angle of the torch constant.
10. When you reach the 2:00 position, lift the torch so the arc will stop.
11. Raise your helmet and stop the air.
12. Remove the remaining electrode from the torch so it will not accidentally touch anything.
13. Turn the pipe so that the end of the J-groove is at the 10:00 position so the groove can be continued around the pipe. Restart the groove at the top edge.
14. Repeat this process until you have made a J-groove all the way around the pipe.
15. When the metal is cool, chip or brush any slag or dross off the plate. This material should come off easily.
16. The groove must be within ±1/8 in. (3 mm) of being straight and within ±3/32 in. (2.4 mm) of uniformity in width and depth.
17. Repeat this cut until it can be made within these tolerances.
18. When you are finished cutting, turn off the CAC equipment and clean up your work area.

1G Air Carbon Arc Gouging Out a Pipe Weld

Using a properly set up air carbon arc cutting torch and welding power supply and wearing all required PPE, you are going to gouge out a pipe weld in the horizontal rolled position by doing the following steps:

1. Place the pipe horizontally on the cutting table.
2. Start at the 10:00 position.
3. As the gouge starts, rotate the torch so that the depth of the groove is not so deep that sparks get blown back at you.
4. Move the torch along the weld toward the 2:00 position.
5. Watch the groove width to see that you are only removing the weld. Change the travel speed, the torch side-to-side movement, or both to keep the groove the same width and depth to remove all the weld around the pipe.

6. Do not try to remove too much weld metal depth in one gouging pass. On thicker-walled pipe, you may need to make several gouges around the pipe to remove all the weld metal.
7. When you have reached the 2:00 position, release the torch's cutting lever.
8. Turn the pipe so that the end of the U-groove is at the 10:00 position so the groove can be continued around the pipe. Restart the groove at the top edge.
9. Repeat the gouging of the weld until you can make consistently smooth grooves that remove only the weld metal.
10. When you are finished cutting, turn off the CAC equipment and clean up your work area.

5G Air Carbon Arc Gouging Out a Pipe Weld

Using a properly set up air carbon arc cutting torch and welding power supply and wearing all required PPE, you are going to gouge out a pipe weld in the fixed horizontal position by doing the following steps:

1. Fix the pipe horizontally on the cutting table.
2. Start at the 12:00 position.
3. Move the torch down along the weld as far as you can comfortably go.
4. Stop and reposition yourself and restart the cut.
5. Do not try to remove too much weld metal depth in one gouging pass.
6. When you have reached the 6:00 position, release the torch's cutting lever.
7. Reposition yourself so you can restart gouging at the 12:00 position to remove the weld on the other side.
8. Repeat the gouging of the U-groove until you can make consistently smooth grooves that remove only the weld metal.
9. When you are finished cutting, turn off the CAC equipment and clean up your work area.

2G Air Carbon Arc Gouging Out a Pipe Weld

Using a properly set up air carbon arc cutting torch and welding power supply and wearing all required PPE, you are going to gouge out a pipe weld in the fixed vertical position by doing the following steps:

1. Fix the pipe vertically on the cutting table.
2. Start at a comfortable place on the pipe.
3. Move the torch around the weld as far as you can comfortably go.
4. Stop and reposition yourself and restart the cut until you have gone all the way around the pipe.
5. Do not try to remove too much weld metal depth in one gouging pass.
6. Repeat the gouging of the U-groove until you can make consistently smooth grooves that remove only the weld metal.
7. When you are finished cutting, turn off the CAC equipment and clean up your work area.

MACHINE CUTS

Pipe cutting machines can be manually operated, or they may have a drive motor. These machines operate the same whether they use a machine oxyacetylene torch or a machine plasma torch, **Figure 5-55.** Set the cutting torch up in the same way as you have for manual cutting. When the cut begins, the torch must be moved at a constant speed. If the torch moves too

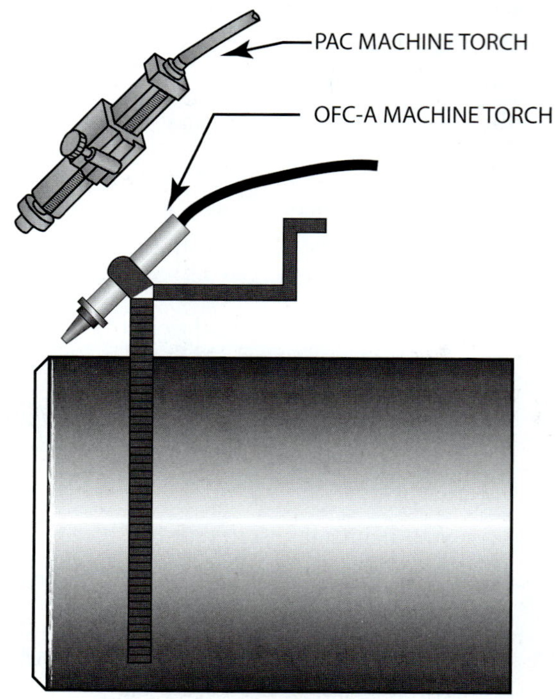

FIGURE 5-55 A pipe-cutting machine can be fitted with either an OAC or a PAC torch.

slowly, the sides of the kerf melt, making it wider. This molten metal mixes with the slag from the cut, and if that happens enough, it can become fused to the bottom edge of the cut. Removing this hard slag takes a lot of time and often requires a grinder. If the torch is moved too fast, the cut may not go all the way through the wall of a pipe.

Watching the way the slag is blown out the back of the cut is the best way of judging the cutting speed. At the correct cutting speed, the slag will be shot out in a stream, **Figure 5-56**. There will be little or no slag hanging onto the under edge of the bevel side of the cut, **Figure 5-57**, but there may be slag hanging on the scrap side of the cut, **Figure 5-58**. There is also a very distinct sound made when the cut is correct. Once you learn to recognize that sound, you do not have to watch the sparks anymore.

Larry Jeffus

FIGURE 5-58 Slag on the scrap side is common and does not pose any problem.

Larry Jeffus

FIGURE 5-56 When the torch is being used at the proper speed, very little slag is left on the inside of the cut.

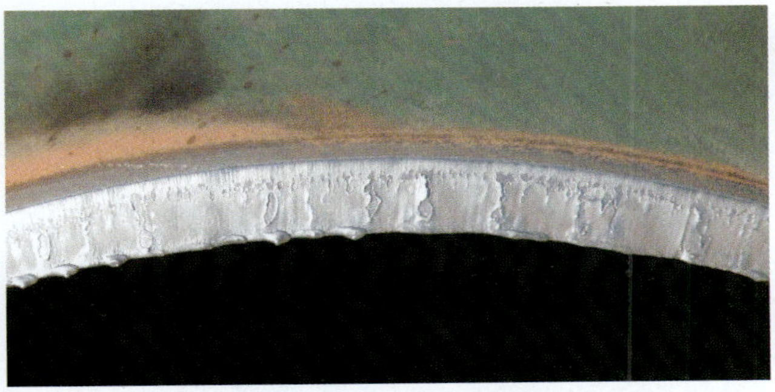
Larry Jeffus

FIGURE 5-57 A small amount of grinding will remove the oxide layer and produce the properly sized land.

On hand-cranked, manually operated, pipe cutting machines, the crank handle can be on the carriage or connected through a flexible cable, **Figure 5-59**. The advantage of flexible cable is that there is less chance that the torch might be jostled as the crank is turned.

Larry Jeffus

FIGURE 5-59 Be sure that the beveling machine cables will move freely around the pipe during the cut.

Some cutting machines use spacers to position the machine's ring so that it is centered on the pipe. Making certain that the correct spacers are used is important because if the ring is off center, the bevel cut will not be square, **Figure 5-60**.

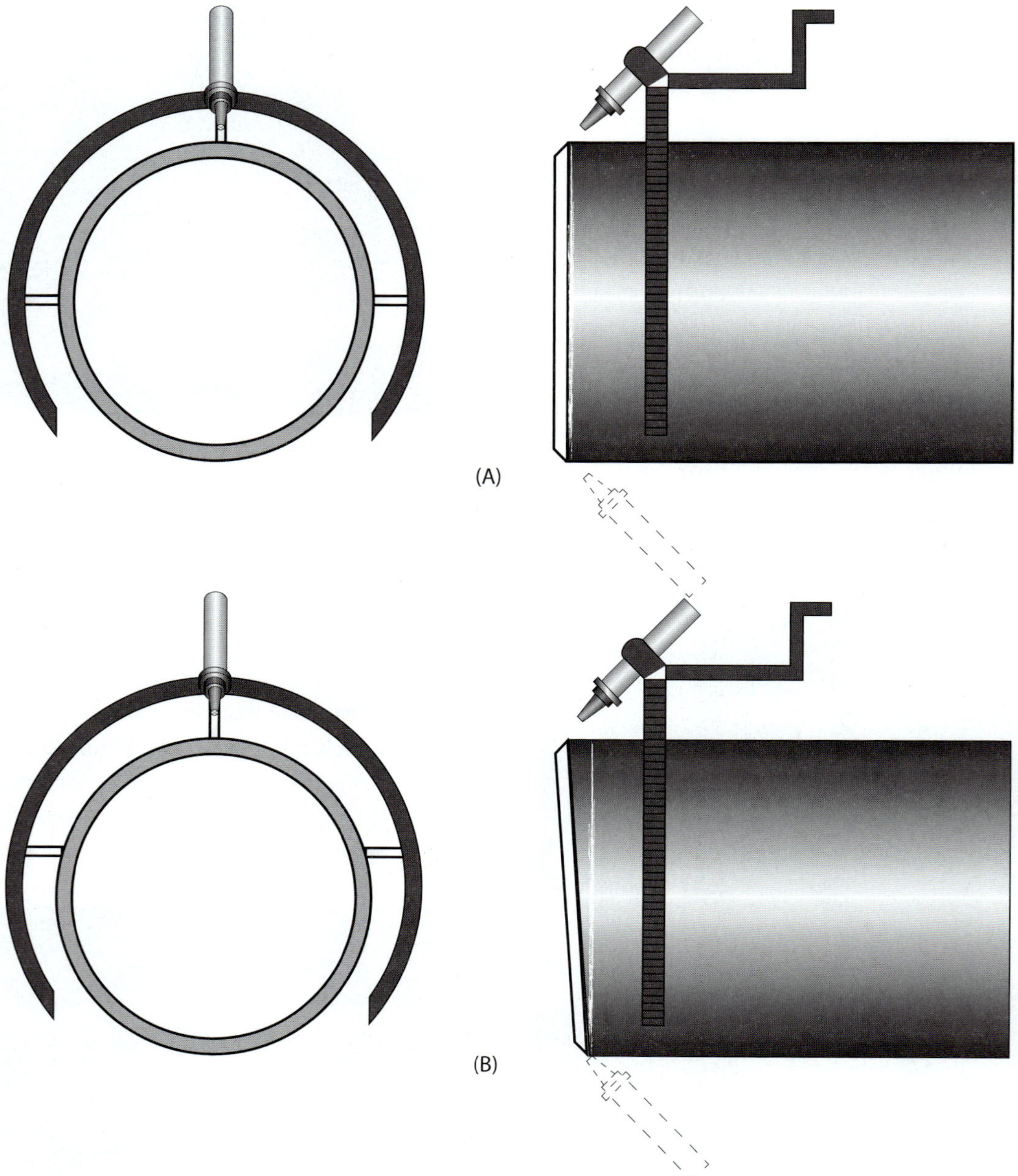

FIGURE 5-60 Make sure that the correct spacers are used so that the pipe is centered in the pipe beveling machine (A). If the incorrect spacers are used (B), the cut on the end of the pipe will not be square.

Summary

The quality of a pipe weld depends on the quality of the fit-up and the fit-up in turn depends on the quality of the bevel. So the ability to make quality pipe bevel cuts that require little or no postcutting cleanup is a skill that can be extremely valuable on the job. A poor-quality pipe cut can require grinding to make it ready to weld, which takes a lot of expensive time. In the worst-case scenario, the metal removed to fix a bad cut could make the pipe too short and therefore wasted as scrap.

One of the major factors in making a good bevel cut is the quality of the setup. Taking your time to make sure that the pressures are correct, the tips are clean, the distance between the tip and the pipe is correct, the travel speed is correct, and the other precautions described in this chapter will reduce postcut cleanup time. So do not rush to start a cut until all of the setup issues have been checked and verified. This is a case when slow and steady definitely wins the race.

Review Questions

1. What are the three major thermal-cutting processes used in the pipe-welding industry?
2. Which process works by burning away the steel with a stream of pure oxygen?
3. What is kindling temperature?
4. What is a heat-affected zone (HAZ)?
5. How is a cutting tip different from a gouging tip?
6. How does plasma arc cutting (PAC) create a cut?
7. What is plasma arc gouging (PAG) used for?
8. In the air carbon arc cutting process, what is used to blow away the molten metal?
9. Compare the position of the torch tip when freehand pipe cutting on small-diameter pipe and on large-diameter pipe.
10. List three tools used to mark the line on a pipe in preparation for cutting.
11. How does the diameter and number of preheat orifices on a torch tip affect the heat that it generates?
12. How should the holes be aligned on a torch tip to make a beveled cut?
13. If you do not know the correct pressure for a tip on a machine torch, what should you set the fuel and oxygen at when lighting the torch?
14. Why should you practice moving the lit torch along the cutting line before starting the cut?
15. What is the advantage of rolling the pipe when making a square cut?
16. If a horizontal pipe to be cut cannot be rolled, where can you finish the cut so that you are not under the pipe when it is cut free?
17. What is the tolerance for the cut end of a pipe when it stands on a flat plate?
18. How can a rusty nut be removed using a cutting torch?
19. What is a gouging tip used for?
20. What is the effect of increasing the standoff distance during PAC?
21. When gouging, how might a slag inclusion appear?
22. What can happen if too much metal is removed at one time during gouging with a plasma torch?
23. What is used to provide power to an air carbon arc torch?
24. Approximately how far should the electrode extend from the CAC-A torch?
25. If the torch on a pipe cutting machine moves too slowly, how will the cut be affected?

Chapter 6

Pipe Joint Design and Preparation

OBJECTIVES

After completing this chapter, the student should be able to:

- Discuss the elements of joint geometry and their effect on economic efficiency.
- Explain two different techniques to lay out pipe for a 90-degree saddle.
- Explain the technique to lay out pipe for a 45-degree lateral.
- Explain the technique to lay out pipe for an orange peel.

KEY TERMS

Bevel angle
Branch
Groove angle
Groove face
Header
Joint geometry
Joint root
Orange peel
Root face
Root opening
Saddle in
Saddle on
Scope

INTRODUCTION

There is a wide variety of pipe joints that a welder can be expected to fabricate for any job. **Figure 6-1** lists some examples of common types of joints. **Joint geometry** is the shape and dimensions of a joint as viewed in cross section prior to welding. In large companies, a welding engineer or mechanical engineer will specify the joint design. In smaller welding shops, the owner or supervisor will typically select the joint design based on a number of criteria, such as the type of service application, type of material, welding process, and cost. Joint designs can be selected from a prequalified list such as those provided by the American Welding Society (AWS), American Petroleum Institute (API), or American Society of Mechanical Engineers (ASME).

The pipe's application will dictate the quality and which code will apply. In some cases, a company may have their its own specific pipe welding quality manual. This manual will dictate specific requirements that will be applied to a pipe weld that is above and beyond the standard requirements presented in codes. It should be noted that each code or quality manual has a specific section titled "Scope," which describes all of the parameters regarding pipe material type, size, schedule, and other elements that this particular code or quality manual applies to. It will define the requirements for fabricating a particular type of pipe joint.

In some cases, national, state, or local laws will take precedence over a code or quality manual.

NOTE

Over time, laws and codes change, so make sure that you are using the most current and applicable ones.

(Continued)

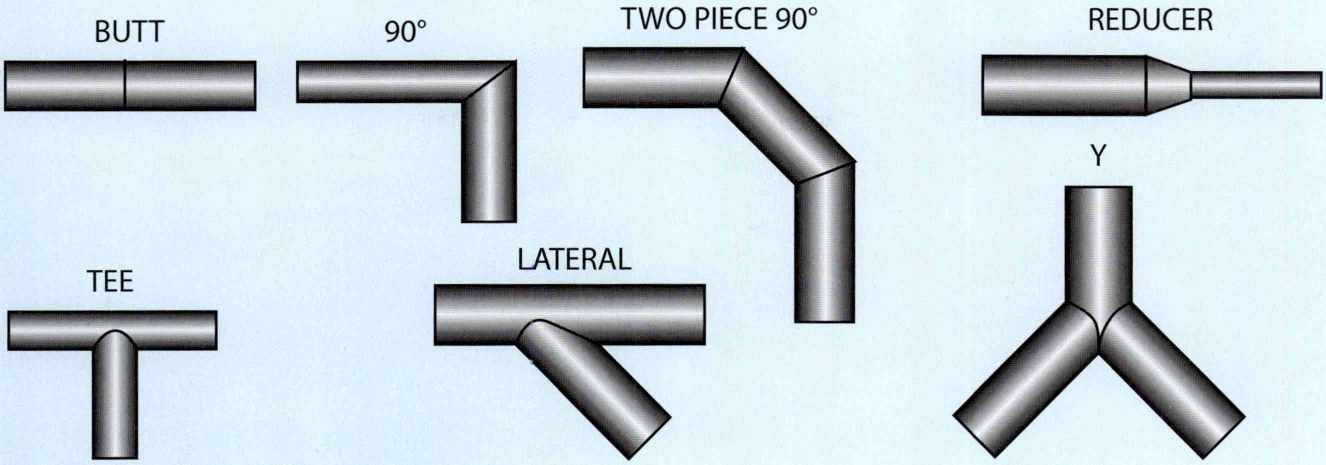

FIGURE 6-1 Typical pipe joints.

Another important consideration in joint design is accessibility. It is important that the joint should be designed in such a manner that it is possible for the welder to perform the weld that is called for. The weld on the tee in **Figure 6-2** is so close to the I-beam that the weld on the back side would have to be made using a mirror. Several other welds in this valve-pump manifold are also located so such that welding would be difficult.

The purpose of the pipe weld joint is to form a junction where two or more pipes come together. The **joint root** is the part of the pipes that are closest together where it will be welded. This joint is designed to transfer stresses uniformly throughout the pipe or pipe structure. There are four basic types of joints used to form junctions between two or more pipe members, **Figure 6-3**. They are the butt, corner, lap, and tee joints. Most often, the name of a type of joint describes the cross-sectional appearance of the joint.

There are two common types of weld joints used in most pipe-welding applications—groove welds and fillet welds.

Of the eight common groove welds, seven may be used for pipe-welding applications. These seven are square-groove, V-groove, bevel-groove, U-groove, J-groove, flare-V-groove, and flare-bevel-groove. The names of each are indicative of their appearance when seen from a cross-sectional view. The scarf groove is more common to brazing applications.

NOTE

There are a few things that should be considered when deciding which type of groove will be used. The primary consideration when preparing pipe ends is typically one of economics. The second consideration is the economics related to welding the joint. The equipment on hand or available is usually relied on for edge preparation. Sometimes this is because specialty equipment may not be available or may be too costly. The thickness of the pipe also dictates the type of edge preparation. The thinner the section, the less preparation that is required. In thin sections, a square edge may be all that is required, while thick sections may require a V-groove or U-groove.

THE 3 IN. (75-MM) SPACE BETWEEN THIS I-BEAM IS ALL THE SPACE TO MAKE THE WELD ON THE BACK SIDE OF THIS TEE FITTING.

Larry Jeffus

FIGURE 6-2 Some piping layouts leave little room for the welder.

(Continued)

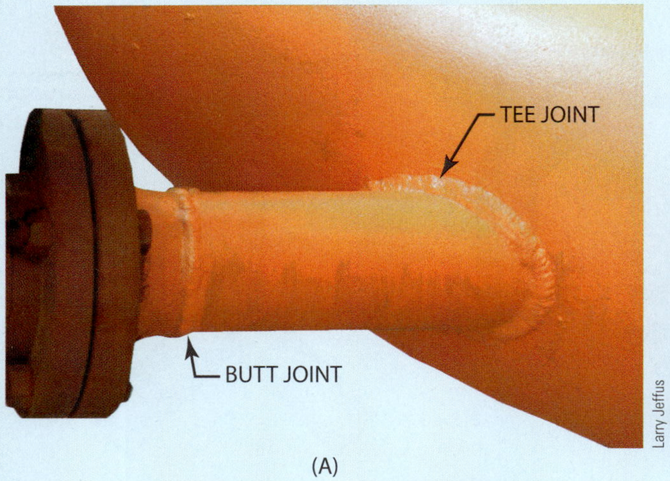

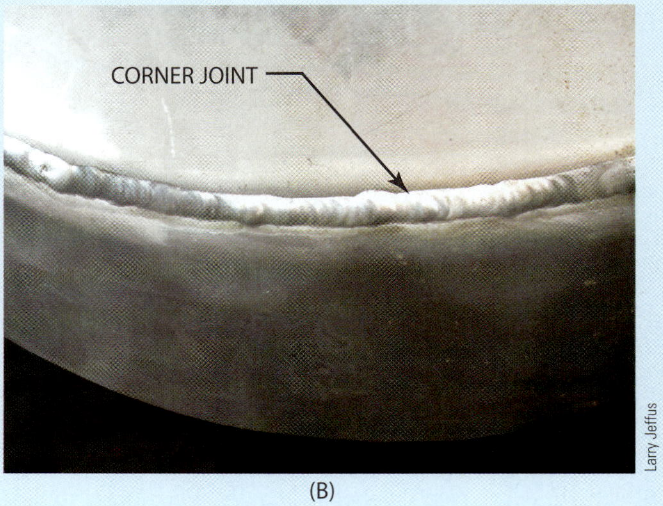

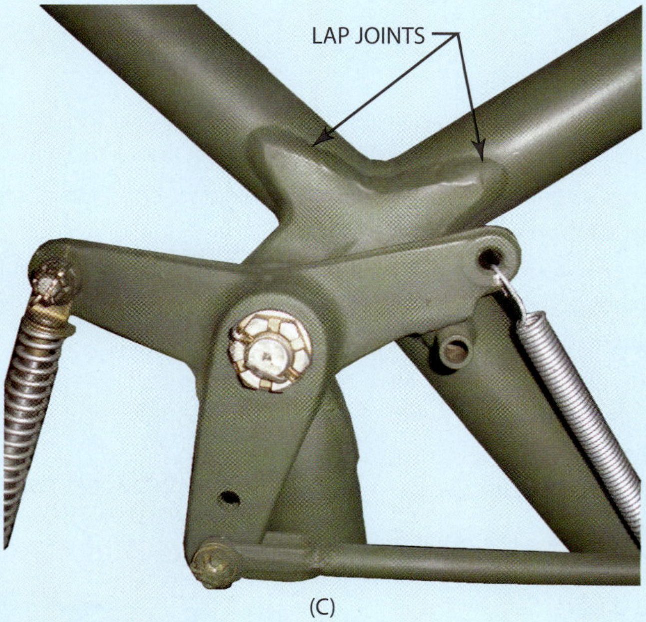

FIGURE 6-3 (A) Butt and tee joints; (B) corner joint; (C) lap joint on a Stearman Biplane throttle linkage.

JOINT GEOMETRY

Joint geometry has two elements to it—shape and dimensions, **Figure 6-4**. In order to describe the shape of the joint, it should be viewed from a profile view. This enables a person to have a good indication of the overall shape of the groove. It also enables the joint to be accurately dimensioned. In order to understand joint geometry, it is important to use standard terminology when talking about elements of the joint. The following text presents some of the common terms that would be applied to weld joint geometry.

Bevel angle is the angle formed between the prepared edge of the member and a perpendicular plane of that same member. This is normally the top surface, or the side that the preparation is being performed from.

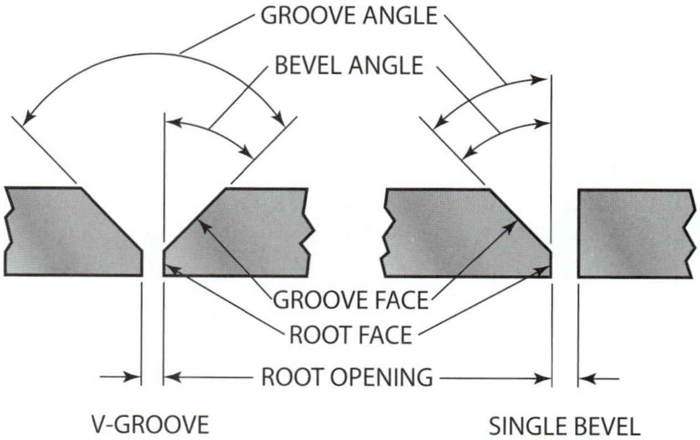

FIGURE 6-4 Groove terminology.

For example, using a typical pipe-beveling machine, this would be the angle that the torch is set at on the machine.

Groove angle is the total included angle that is formed by the two members that are to be joined. Once the edge of each member is prepared and fitted together, the total angle made by both of the members is added together to determine the groove angle.

Groove face is the surface of the members that are included in the groove. These are the surfaces that form the groove angle. When critical welds are performed, these surfaces are ground after flame cutting is performed to remove oxides and hard surfaces.

Root face is the surface that is formed by making a flat surface within the joint root. This is commonly referred to as the "land." It can be different sizes, depending on the requirements of the code or welding procedure specification that is being followed. Welders often refer to the root face or land as a "dime's" or "nickel's worth" as a reference to being the approximate thickness of either a dime or a nickel.

The **root opening** is the space between the members forming the joint root.

JOINT PREPARATION

There is a variety of equipment available for preparing pipe for welding. It ranges from machining equipment such as automated high-production machine bevellers; electrical beveling machine, **Figure 6-5**; portable drill with hole saw, **Figure 6-6**; shear, **Figure 6-7**; and hand-crank torch bevellers, **Figure 6-8**; as well as a variety of tools. In some cases,

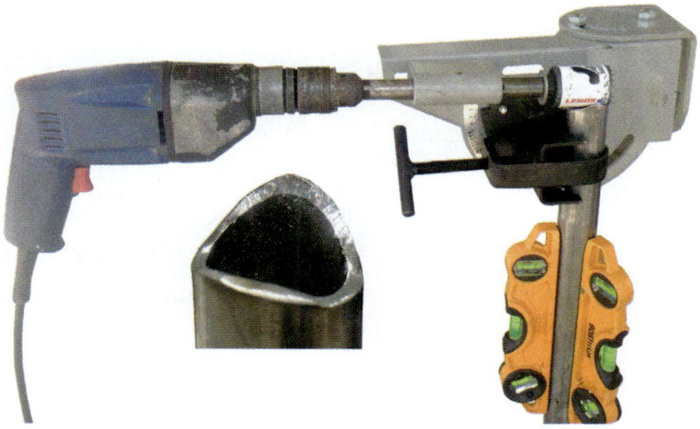

Larry Jeffus

FIGURE 6-6 Hole saw adaptor allows pipe to be beveled at almost any angle.

Larry Jeffus

FIGURE 6-7 Hand shear for notching pipe and tubing.

Larry Jeffus

FIGURE 6-5 Power-beveling machine.

Larry Jeffus

FIGURE 6-8 Hand-crank torch beveler.

a portable band saw and grinder and many other cutoff tools can be used, **Figure 6-9**, to cut off the end of the pipe squarely. Many of the smaller and more portable beveling machines can be teamed with either plasma or oxy-fuel cutting equipment.

Plasma equipment has the advantage of being a high-speed cutting process that can cut a variety of metals, but it also has a disadvantage: the cost of the equipment needed to set up a plasma beveller. Oxy-fuel equipment, on the other hand, is relatively inexpensive and can be set up with a manual hand-crank type beveller, Figure 6-8.

As mentioned previously, there are several processes and techniques to prepare pipe for welding. The most commonly used of these is the oxy-fuel torch process. One of the drawbacks of this process is that it is limited to use on iron base metals. Some of the advantages are that the process is relatively simple to perform, and the equipment that is required is simple, economical, and portable. A higher skill level is required to perform manual cutting than that required for machine cutting. Both machine and manual cutting are capable of producing high-quality cuts that result in very little postcutting cleanup. A good-quality oxy-fuel cut will have a minimum of slag on the bottom and leaves a light layer of oxide on the cut surfaces. Cleanup is simple—a hand grinder can be used to lightly grind and remove the oxide from the surfaces. Care should be taken in order to maintain the correct bevel angles. The hand grinder can also be used to face off the end of the pipe to produce the required land or root face. A half-round file can also be used to obtain the desired land and remove any slag that might be on the inside of the pipe.

In order to set up an oxy-fuel pipe beveller, you must first select the proper size of beveling machine. The ring gear and saddle assembly must be of the appropriate size for the pipe that will be beveled. A variety of equipment is available that will enable the beveling of pipe from as small as 2-in. (50-mm)–diameter pipe up to 36-in. (90-cm)–diameter pipe using a standard "saddle-type" beveling machine. Most saddle-type bevellers have blocks or spacers that allow for cutting pipe that is within a specified range. Some saddle-type bevellers have a cam knob, which allows the pipe to be preheated and then the cut to be started. At this point, the cam knob is initiated, and the torch is brought into position to start the beveled edge of the cut. This gives a lead-in to the cut that will be on the scrap side of the pipe.

OXY-FUEL PIPE CUTTING

A pipe-beveling machine torch is the best way to flame-cut a bevel end on a pipe. If a pipe-beveling machine is not available, the end can be cut at a 90-degree angle using a

Larry Jeffus

(A)

Larry Jeffus

(B)

FIGURE 6-9 (A) Carbide-tipped cutoff saw; (B) portable pipe and tubing power cutter.

hand torch, and then a hand grinder can be used to grind the required bevel angle. Some manufacturers have hand-cutting attachments that are at a 75-degree angle so that when held perpendicular to the work piece, a beveled cut is produced. Having a clean, correctly adjusted, and properly sized cutting torch will make it possible to make a smooth, clean cut that will require little or no postcutting cleanup. Torch manufacturers have charts listing the proper cutting tip for different thicknesses of materials; some include recommendations for the best torch to select for bevel cuts. The charts include the proper fuel and oxygen gas pressures.

Once the proper tip is selected and the gas pressures set, the tip should be cleaned before lighting. The time spent cleaning a torch before making a cut will save much more time than cleaning up a poor torch cut made with a dirty tip.

Often, new pipe has a thin coating applied to it, and moving the torch around the pipe before starting the cut can be used to burn it off. If this coating is left on the pipe, it can disrupt the cutting stream, causing the cut surface to be gouged. If the coating is thicker, like that used on direct bury pipe, it may be necessary to grind that coating off before starting a cut.

Larry Jeffus

FIGURE 6-10 Brace yourself when using a handheld cutting torch.

Larry Jeffus

(A)

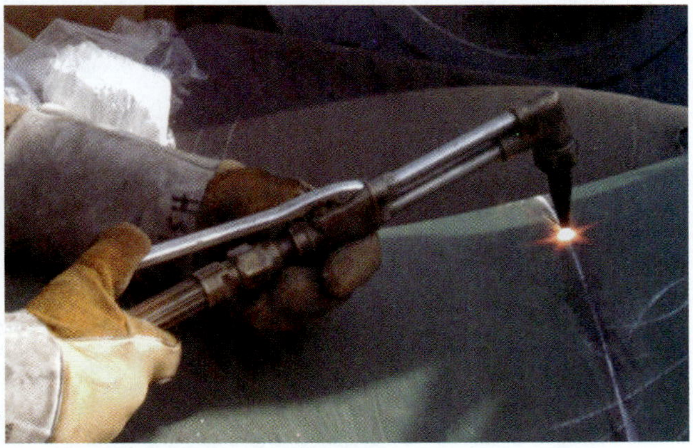

Larry Jeffus

(B)

FIGURE 6-11 (A) Raise the torch slightly as the cut begins so spatter does not blow back on the torch tip. (B) Once the cut begins, lower the torch and continue the cut.

PRACTICE 6-1

Hand-Cutting Pipe

For this practice, you will need a piece of 2–3-in. (50–75 mm)–diameter Schedule 40 pipe, a properly set up and adjusted oxy-acetylene cutting torch, a pipe wraparound, and a soapstone. You will also need personal protective equipment (PPE) for the processes being performed.

Refer to Chapter 5, "Thermal-Cutting Processes," for more information on making pipe cuts by hand.

1. Clamp the pipe horizontally in a bench or pipe vise.
2. Using a Wrap-A-Round® and soapstone, draw a chalk line around the pipe.
3. Wearing all the required PPE and using a properly set-up hand-cutting torch, light and adjust the cutting torch.
4. Make sure that there are no combustible materials that might catch fire from the cutting sparks.
5. Hold the torch in line with the pipe, **Figure 6-10**.
6. Let the pipe heat up, **Figure 6-11A**, and depress the cutting lever to start the cut, **Figure 6-11B**. To keep slag from closing in behind the cut, make a slight movement in one direction before making the cut in the opposite direction.
7. Keep the torch pointed directly at the center of the pipe as you move it along the soapstone line.
8. Once you have made the cut as far as you comfortably can while keeping the tip pointed squarely at the pipe, stop and reposition the pipe so that the cut can be continued.
9. After the cut is completed, have your instructor inspect your work for accuracy.
10. Turn off all equipment, return all tools and supplies, discard all scrap, and clean up the welding booth. ◆

PRACTICE 6-2

Machine Pipe Beveling

For this practice, you will need a piece of 6–8-in. (15–20 cm)–diameter Schedule 40 or 80 pipe, a properly set up and adjusted oxy-acetylene machine cutting torch, and a saddle-type hand-crank beveling machine matched to the size of pipe that will be cut. You will also need PPE for the processes being performed.

1. Check the saddle and make sure that the correct spacers are in place for the diameter of pipe being used. Position the beveller over the pipe in a manner that will yield a 4-in. pipe coupon.
2. Make sure that the boomer assembly is correctly in place to ensure that the beveller is secured to the pipe.
3. Light the torch and adjust the flame to a neutral setting.
4. Preheat the pipe to a dull red color to prepare it for cutting.
5. When the appropriate temperature has been reached, turn on the oxygen cutting lever and start cranking the hand crank to begin the cut.

NOTE

As with the freehand cut, it is a good idea to back up slightly to clear the slag before proceeding around the pipe. To back up, rotate the hand crank one to two revolutions in one direction, and then go back the opposite direction to make the cut. This will often result in a smoother start and stop, **Figure 6-12**. Notice how smooth the pipe-beveled cut is at the starting and stopping points, **Figure 6-13**.

6. As the cut progresses around the pipe, make sure to use a steady revolution on the hand crank. Your speed should be adjusted to make sure that a quality cut is being maintained. If your travel speed is too slow, the cut is likely to close up and

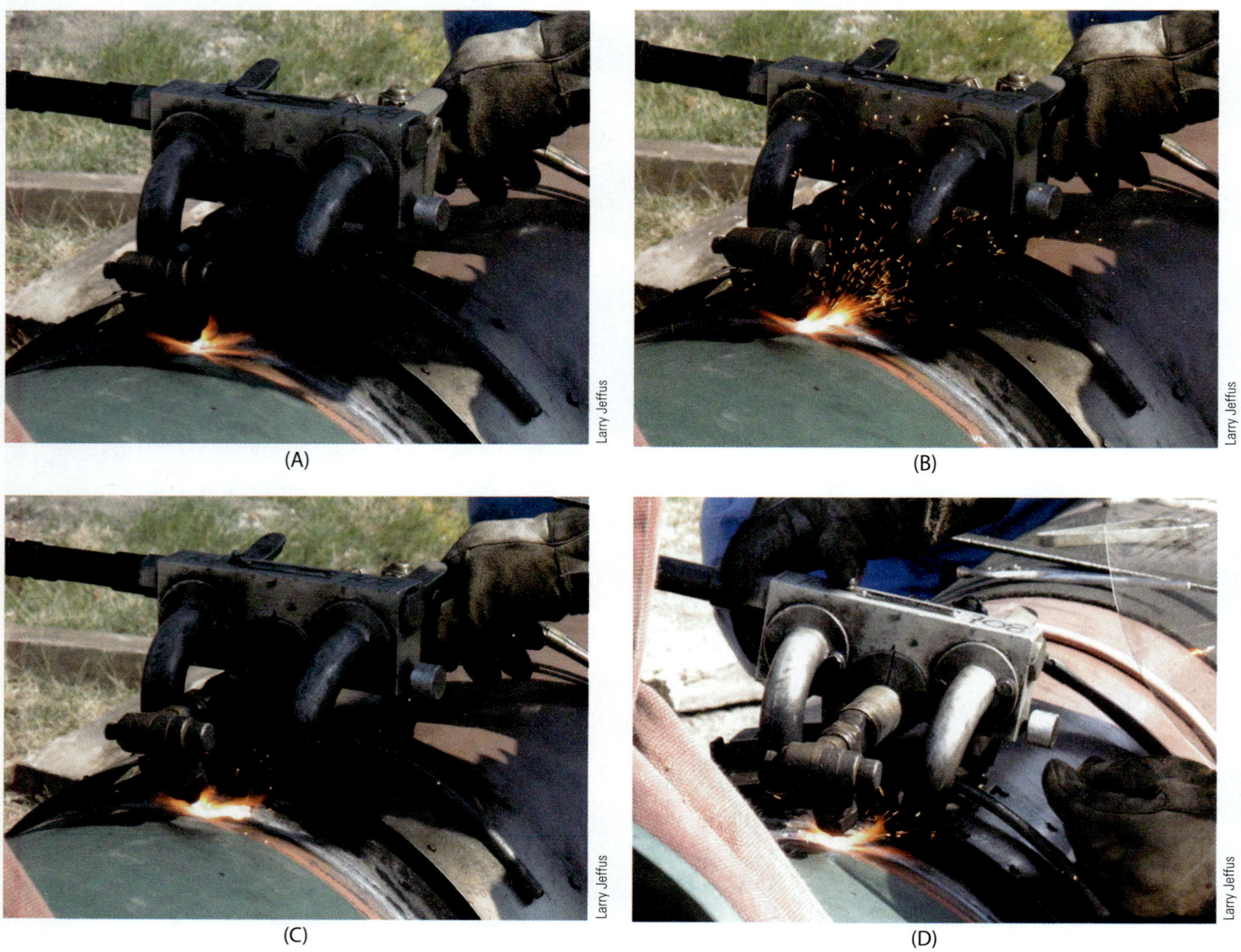

FIGURE 6-12 (A) Preheat the pipe; (B) Gradually turn on the cutting stream; (C) As the cutting stream is being turned on, slowly turn the crank and move the torch backwards for a short distance; (D) Now move the torch crank forward smoothly to make a uniform cut all the way around the pipe.

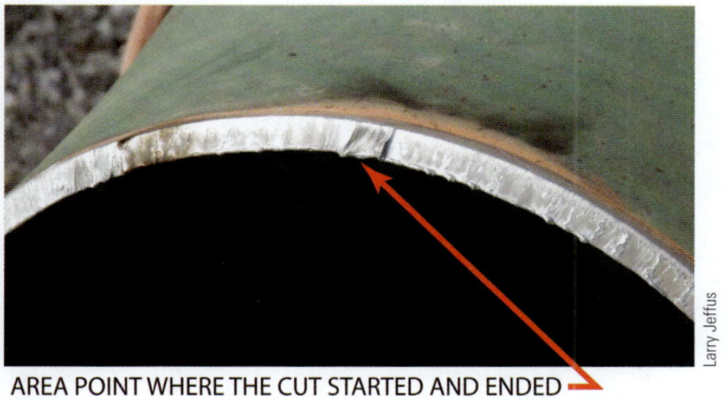

FIGURE 6-13 The procedure shown in Figure 6-12 results in a very smooth, even starting and stopping point on the beveled surface.

have excess slag left on both the scrap side and the bevel side of the pipe. Also, the coupon is not likely to drop free if slag fills the cut kerf. If your travel speed is too fast, the cut will be lost. It takes skill and practice to master producing cuts of consistent quality.

NOTE

The sound that a good cut makes is very distinctive; it is a light fluttering, and the stream of sparks can be seen flowing from the bottom of the cut.

7. Have your instructor inspect your work for discontinuities and consistently defect-free welds.
8. Turn off all equipment, return all tools and supplies, discard all scrap, and clean up the welding booth.

Once the pipe has been cut using one of the thermal-cutting processes, the groove face should be prepared for welding. Prepare the surface by grinding it with a disk, which will leave the surface clean and free of slag and surface oxides. Grinding should also leave the surface smooth without deep grooves or gouges. Position the grinder so that the bevel angle that is required by the welding procedure specification can be maintained. When grinding a piece of pipe, care should be taken to make sure that sparks are directed away from people and flammable materials. Once the surface is bright and shiny and all signs of oxide are gone from the groove surface, both the inside and outside surfaces should be wire-brushed with a grinder. A fine-grinding disk or pad may be used to remove the surface oxides on the outside of the pipe. Most codes require that surface oxides be removed a distance of 1 in. (25 mm) from the weld groove prior to welding. ◆

PIPE LAYOUT

Like many industries, the pipe trades have tools that are specific to it. Pipe layout tools enable the craftsmen to perform their work to a high standard of quality and in an economically efficient way.

Pipe Layout Tools

- **Metal rings**—Two-part, spring-loaded templates that come in different sizes to fit common pipe sizes. By using a marker in the slots or on the ends, this one tool can be used to lay out saddles, 45° and 90° elbows, and square cuts, **Figure 6-14**.

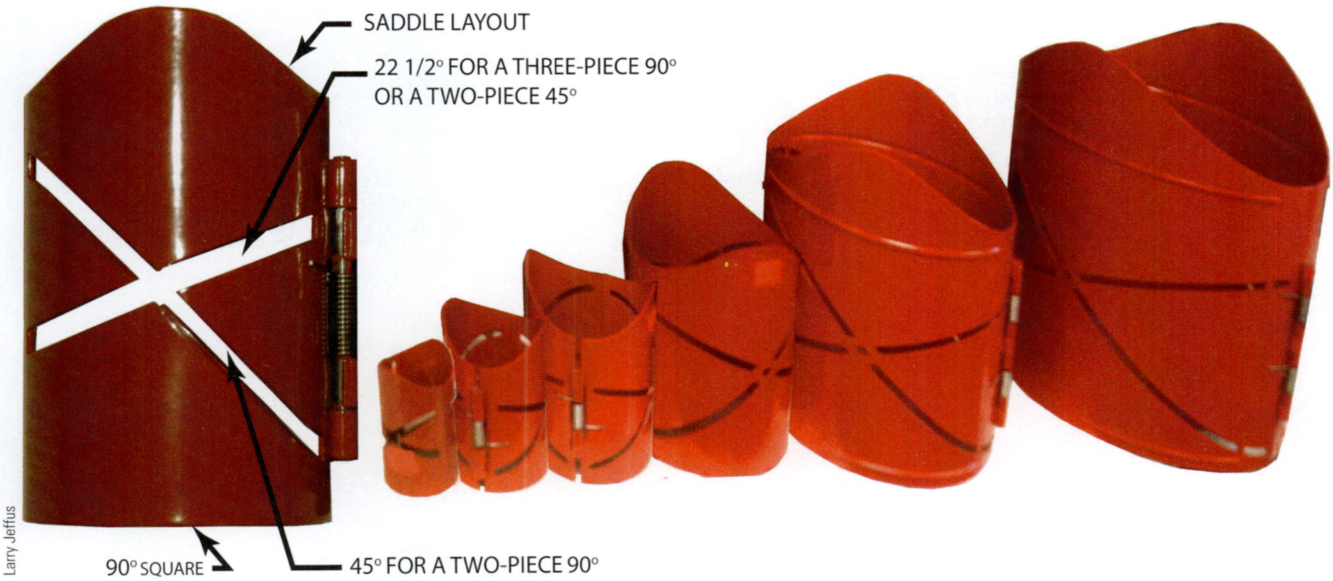

FIGURE 6-14 Metal pipe layout tools.

- **Contour radius marker**—This tool has an articulated arm that holds a piece of soapstone and allows the welder to guide the soapstone all the way around the pipe. The marker can be set at any angle needed for elbows. It can also be used to lay out saddles, laterals, and Y's, **Figure 6-15**.

Larry Jeffus

FIGURE 6-15 Contour radius marker.

- **Contour tool**—This tool has the ability to conform to any angle so that both pieces of pipe or tubing can be marked accurately. The outline of a 7/8-in. (21-mm) tube intersecting the main line at a 30-degree angle to form a lateral line can easily be traced, **Figure 6-16A, B**. Then the tool can be placed on the outside of the intersecting tubing and have that outline traced next, **Figure 6-16C.** This type of contour tool can be used to lay out any of the standard fittings, as well as intersections of multiple tubing, **Figure 6-17.**

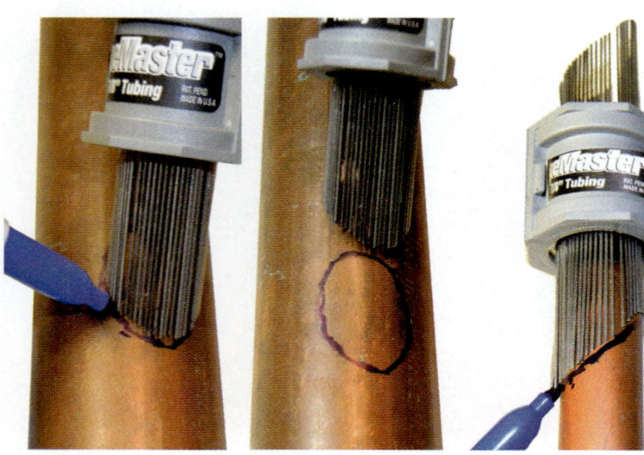

(A) (B) (C)

Larry Jeffus

FIGURE 6-16 (A) Trace around the contour tool; (B) mark the size and location of the hole to be cut in the header; and then (C) trace the contour tool on the branch.

Larry Jeffus

FIGURE 6-17 The contour tool can be used to make complex joints such as this three-tubing engine mount on a Kitfox light sport aircraft.

- **Template**—There are a number of commercially available pipe layout templates, and it is possible to make your own. The templates shown in **Figure 6-18** are designed to lay out a pipe end for a 45-degree angle for a 90° elbow and a 22½-degree angle for a 45-degree elbow on a 1-in. (24.5-mm) diameter pipe. Similar templates are available for other sizes of pipes. The templates shown in **Figure 6-19** come in sets for several different pipe sizes. One side of the template is used to draw 45-degree angles, and the other side is used for 22½-degree angles. It is possible to draw a pipe template on card stock, poster board, or similar durable paper using a protractor, compass, and ruler as shown in **Figure 6-20**. Although drawn templates have been used for over 100 years, the commercially available templates have replaced them today.

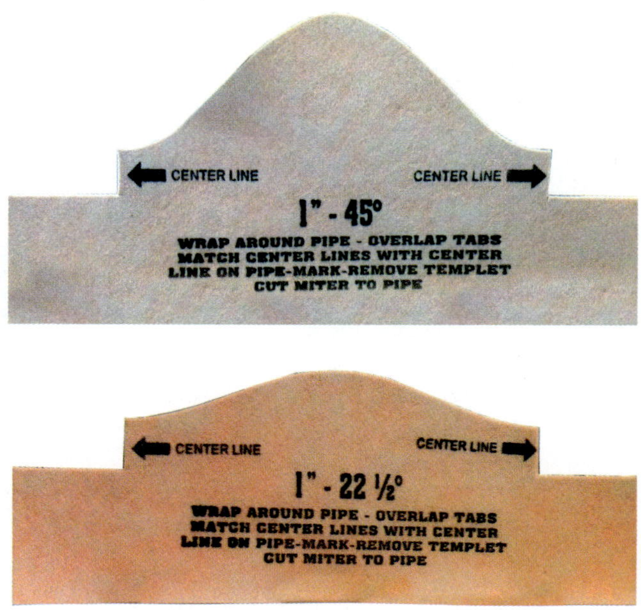

Larry Jeffus

FIGURE 6-18 Individual templates for laying out fittings.

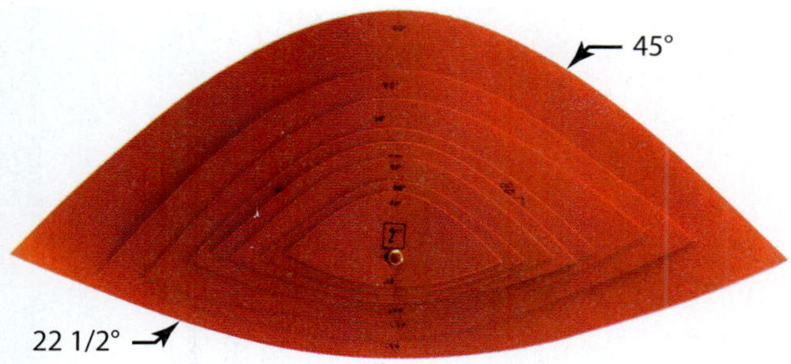

Larry Jeffus

FIGURE 6-19 A set of templates for laying out fittings.

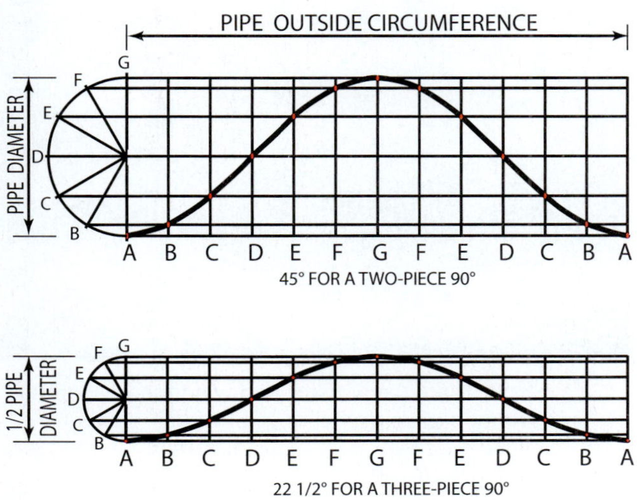

FIGURE 6-20 Examples of a welder-drawn pipe template.

Larry Jeffus

FIGURE 6-21 Curv-O-Marker® Wrap-A-Round® layout tool.

- **Wrap-A-Round®**—One of several types of heat-resistant flexible strips of material used for pipe layout work, **Figure 6-21**. They are one of the most universal layout tools because they can be used to lay out any of the different pipe joints, as well as reducers and orange peel end caps. They are available in different widths and lengths.

90-Degree Saddle on Standard Wall Pipe

When laying out pipe for 90-degree saddles, there are two techniques that can be used. One is referred to as **saddle on**, and the other is referred to as **saddle in**. This refers to how the branch pipe fits up with the header. The **branch** is often a smaller pipe that feeds into a larger pipe known as a **header**. The saddle-on technique is when the through-thickness of the branch pipe rests on the outside diameter (OD) of the header pipe. The branch pipe is beveled inward so that the bevel rests on the outside of the header pipe, **Figure 6-22.**

The other technique is the saddle-in technique. This is where the branch pipe is beveled so that it projects

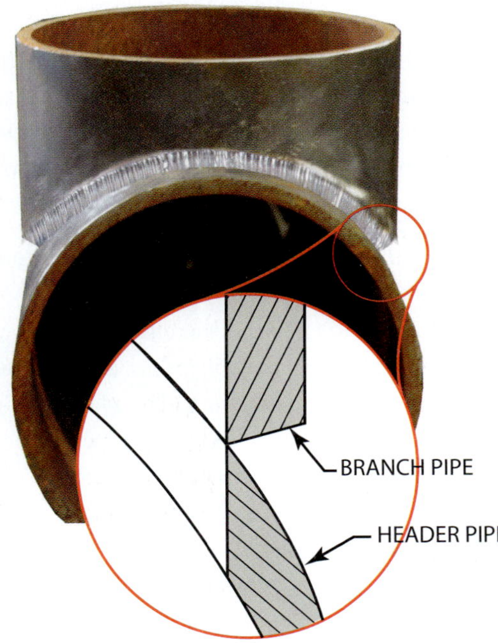

Larry Jeffus

FIGURE 6-22 Saddle-on tee fitting.

inside the header pipe. The bevel on the branch pipe rests on the bevel inside the header pipe, **Figure 6-23**. For the practices in this chapter, the saddle-on technique will be used.

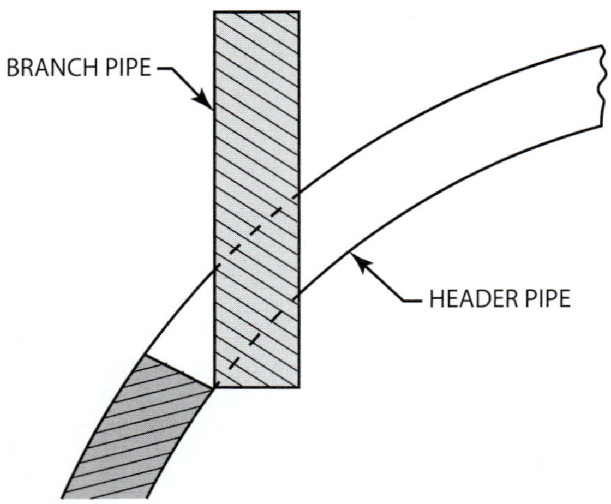

FIGURE 6-23 Saddle-in tee fitting.

Larry Jeffus

FIGURE 6-24 Mark the setback and wind the wraparound 1¼ times around the pipe.

In order to lay out a 90-degree saddle, two layouts will have to be performed: one on the header and one on the branch. These will be practiced next.

Larry Jeffus

FIGURE 6-25 Lining up the edges will ensure that a line can be marked squarely on the pipe.

PRACTICE 6-3

Branch Pipe Layout

The *header pipe* is the main pipe that several same-size or smaller pipes will feed into. The *branch* is often a smaller pipe that feeds into a larger pipe known as a header. Branches are sometimes referred to as *risers* because they rise up out of the header pipe. For this practice, we will be using a 6-in. standard wall piece of pipe for both the header and the riser.

1. Start with the branch pipe. Using a soapstone and wraparound, mark a line around the OD of the pipe. This mark around the circumference will be the reference line that all other layout lines and dimensions will come from. The placement of the reference line from the end of the pipe will be a distance equal to half the inside diameter (ID) of the pipe, **Figure 6-24**. You should overlap the wraparound on itself along approximately one-fourth the OD of the pipe to make sure that if the pipe were cut on the reference line, the pipe would stand plumb, **Figure 6-25**.
2. Make a line with a sharp, pointed soapstone so that the mark will be thin and easily seen, **Figure 6-26**.
3. Next, divide the reference line into eight equal parts, starting with 1 for the top of the pipe when it is in the flat position. Going around the reference line in a clockwise direction, the next division will be 2. At the next division line, which will be 90 degrees from the top mark, label it as 3, **Figure 6-27**. As you continue around, the next division will be 2 again.

(A)

Larry Jeffus

(B)

Larry Jeffus

FIGURE 6-26 (A) Use a sharp soapstone to draw the reference line. (B) For more accurate layouts, the reference line should be thin and easily seen.

The next division line will be 1 again; it will be 180 degrees from the first one. At this point, the pipe is divided in half. The other side of the pipe will now be a mirror of the side that was just completed. Once this side is completed, your pipe will be divided into eighths.

Larry Jeffus

FIGURE 6-27 After the reference line has been divided, number the marks.

NOTE

There are other ways of marking the eight ordinate points on the reference line. A pipe center punch can be used to mark the pipe, **Figure 6-28**. The punch has a level that can be set at the desired angles of 22½ degrees, 45 degrees, 67½ degrees, 90 degrees, etc., so that it can be located at the correct spot on the reference line. The built-in punch is then struck, leaving an indentation at the exact location needed.

Another way is to use a level and a square is shown in **Figure 6-29**. In illustration (A), the top and 90-degree side locations can be marked as shown; (B) illustrates how the top and bottom 180-degree marks can be made; (C) illustrates how the 90-degree and 270-degree marks can be made; and (D), (E), and (F) all illustrate different ways to mark the 45-degree, 130-degree, 225-degree, and 315-degree points.

4. Once you have the points marked, use a wrap-around or square to extend from the points on the reference line down to the end of the pipe. These lines, called *ordinate lines,* must be parallel to the length of the pipe and square with the reference line, **Figure 6-30**.
5. Next, you need to refer to *The Pipe Fitter's and Pipe Welder's Handbook* to find the 90-degree saddle-on chart to complete the layout, **Figure 6-31**. Typically, the larger the diameter a pipe is, the more ordinate lines there should be in order to make it more accurate.
6. When reading the chart, the first column will list the branch or riser size of the pipe and the second

FIGURE 6-28 A pipe center punch can be used to mark the pipe.

column will list the ordinate lines. The top row will list the header pipe size. Simply find the branch size pipe first, and then find the intersecting header pipe size. Within the box will be listed the correct lengths for each of the ordinate lines. In this exercise, the No. 1 ordinate line will be on the reference line. The No. 2 ordinate line will be at ¾ in. (19 mm) from the reference line, **Figure 6-32**.

7. Ordinate line No. 3 will extend 1 15/16 in. (48 mm) from the reference line.
8. Once each of the ordinate lengths are laid out, use a wraparound as a template to form a U-shaped arc that will create four other U-shaped arcs in alternating directions around the OD of the pipe. This will form what is sometimes referred to as a *fish mouth*. Start so that the U rests on point 2. Form the U shape so that it will intersect 3 at the apex and

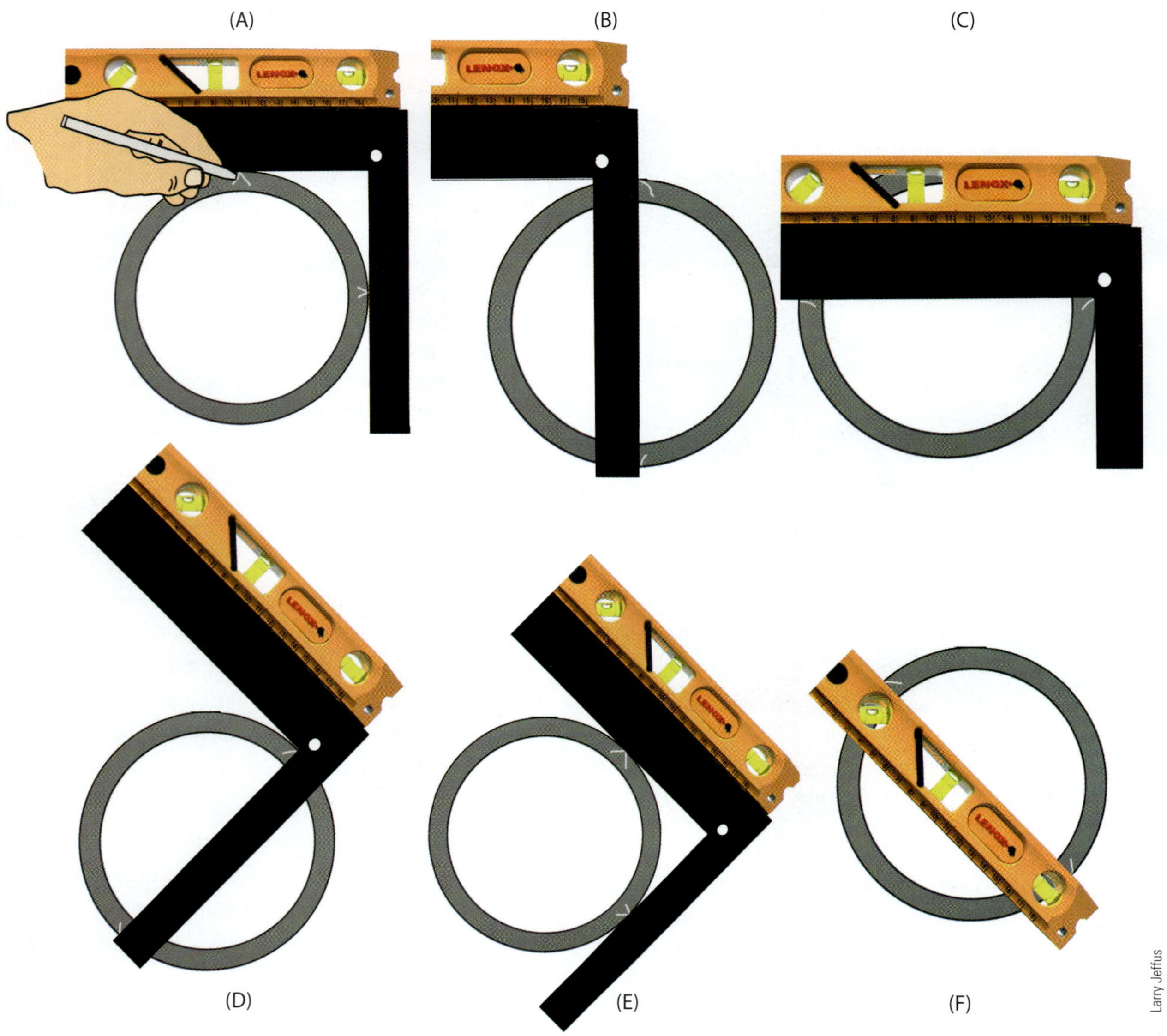

FIGURE 6-29 A square and level can be used to mark the pipe as shown here.

the 2 points that lie on either side of 3, **Figure 6-33**. Now use the soapstone to trace the U shape formed by the wraparound, **Figure 6-34**.

9. Flip the wraparound in the opposite direction and form another U shape that starts at ordinate line 2 and forms a U going through ordinate line 1 and continuing around to the 2 again, **Figure 6-35.**

Larry Jeffus

FIGURE 6-30 Draw a line through each mark on the reference line straight to the end of the pipe.

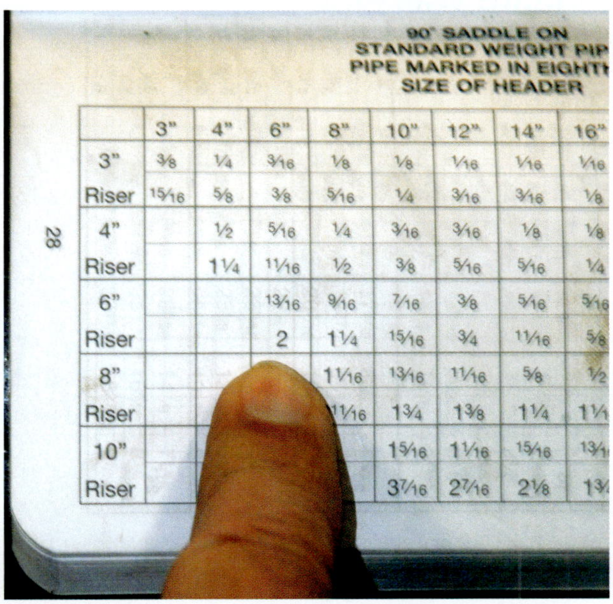

90° SADDLE ON
STANDARD WEIGHT PIP
PIPE MARKED IN EIGHT
SIZE OF HEADER

28

	3"	4"	6"	8"	10"	12"	14"	16"
3"	3/8	1/4	3/16	1/8	1/8	1/16	1/16	1/16
Riser	15/16	5/8	3/8	5/16	1/4	3/16	3/16	1/8
4"		1/2	5/16	1/4	3/16	3/16	1/8	1/8
Riser		1 1/4	1 1/16	1/2	3/8	5/16	5/16	1/4
6"			13/16	9/16	7/16	3/8	5/16	5/16
Riser			2	1 1/4	15/16	3/4	11/16	5/8
8"				1 1/16	13/16	11/16	5/8	1/2
Riser				1 1/16	1 3/4	1 3/8	1 1/4	[illegible]
10"					15/16	1 1/16	15/16	[illegible]
Riser					3 7/16	2 7/16	2 1/8	[illegible]

Larry Jeffus

FIGURE 6-31 The saddle-on chart from the *Pipe Fitter's and Pipe Welder's Handbook*.

Larry Jeffus

FIGURE 6-32 Measure and mark each ordinate line.

Larry Jeffus

FIGURE 6-33 Holding the wraparound in a U-shape, draw a chalk line through points 2, 3, and 2.

Larry Jeffus

FIGURE 6-34 The line drawn should pass through the ordinate points and be thin and easily seen.

Larry Jeffus

FIGURE 6-35 Holding the wraparound in a U-shape, draw a chalk line through points 2, 1, and 2.

10. Continue this alternating pattern until the circumference of the pipe has been laid out, **Figure 6-36.** Check your work to see that the marks made using the wraparound intersect each one of the ordinate lines where they were marked.

FIGURE 6-36 When completed, the finished line should flow smoothly from one curve to the next.

11. You are now ready to cut the branch. Using a properly adjusted oxy-fuel torch, cut along the mark that you have laid out, being sure to hold the torch perpendicular to the pipe so that it points to the center of the pipe at all times. This will result in a square cut on the pipe following the line that you laid out. At this point, you are ready to bevel the branch for proper fit on the header. Since this is a saddle-on fit, you will bevel the pipe with the torch pointing back into the pipe instead of away from it. Check it for fit on the header. If necessary, use a grinder or oxy-fuel torch to make fine adjustments to get an accurate fit-up.
12. Have your instructor inspect your work for uniformity.
13. Turn off all equipment, return all tools and supplies, discard all scrap, and clean up the welding booth. ◆

PRACTICE 6-4

Header Pipe Layout

1. The quickest and easiest way is to start by positioning the branch in the correct location on the header pipe.
2. Use a sharp soapstone or other marker and place it firmly against the edge of the branch pipe.
3. Use it as a template to trace the area that is to be cut out.
4. Once you have finished marking the location of the header opening, you are ready to make the cut.
5. Using a properly adjusted oxy-fuel cutting torch, start the cut on the inside of the cutout area that will be scrap.
6. Be sure to hold the torch perpendicular to the center of the pipe while making the cut. It is also important to make sure that you make the cut on the inside of the mark. This will make sure that the opening is not too large and that the branch does not fall through the opening.
7. Have your instructor inspect your work for uniformity.
8. Turn off all equipment, return all tools and supplies, discard all scrap, and clean up the welding booth. ◆

45-Degree Lateral On Standard Wall Pipe

The layout of a 45-degree lateral or angle less than 90 degrees will be similar to the 90-degree saddle or "T." It will consist of two layouts: one for the lateral branch and one for the header. The tools required for this are paper and pencil, a wraparound, a soapstone or similar marker, and a protractor.

PRACTICE 6-5

Lateral Layout

There are several methods to lay out a lateral. One accurate method is by making a full-scale drawing, described next.

1. Start by laying out two center lines, one for the header and the other for the branch. Make the two center lines intersect at the needed angle for the lateral. The point where the two center lines intersect will be marked *C* for the center-line intersection. At this point, draw a line perpendicular to the header center line at point *C*, called the *center circumference line,* and future dimensions will be referenced from this point, **Figure 6-37**.
2. Next, lay out the ODs of both pipes by marking a line parallel on each side of the center line. The distance that each line should be made from the center line is half the OD of each pipe.
3. Where the OD marks of the branch and header meet, mark one side as *A* and the other side as *B*.
4. Now draw a straight line connecting points *A* to *C* and *C* to *B*. Angle *ACB* will form the end of the branch layout. They will also form the opening on the header.

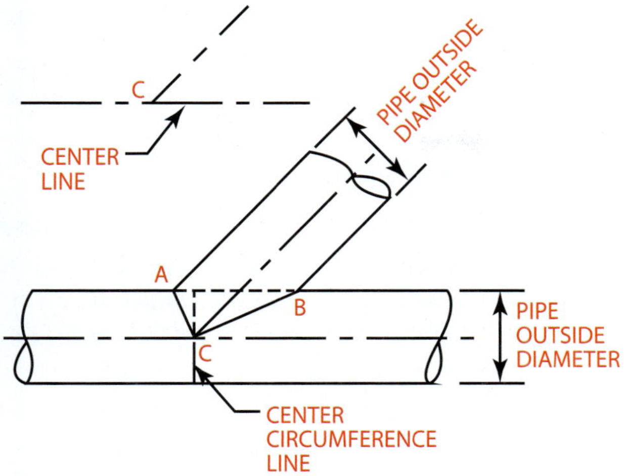

FIGURE 6-37 Laying out the header points.

> **NOTE**
>
> Your lateral layout is drawn; you are now ready to lay out the header pipe.

5. Determine the center line of the header and branch intersection point on the actual pipe. Using the wraparound, mark a circumference line located at the center-line intersection point between the header pipe and the branch pipe. This is point *C* on your drawing. Make sure that the circumference line is square with the pipe. This can be achieved by overlapping the wraparound by at least one-quarter the diameter of the pipe to make sure that it is straight.
6. Going in a clockwise direction around the pipe, number the division points as 1, 2, 3, and 4, starting at the top and proceeding clockwise around the diameter. This will result in dividing the pipe into four equal parts, **Figure 6-38**.
7. At these points, mark the lateral layout lines or ordinate lines several inches beyond the center reference line. It is important that these lines be square with the circumference line and parallel to the center line of the pipe. Measure the distance from the circumference line to points *A* and *B* on the drawing, and lay them out on the pipe.
8. Locate point *C* on ordinate lines 2 and 4 on the reference line; these will be along the sides, opposite one another.
9. Now you are ready to mark the cut lines on the header. Use a wraparound to form a template for tracing a line that runs from point *C* on one side of the pipe to point *A* at the top ordinate line and back to point *C* on the other side's ordinate line.

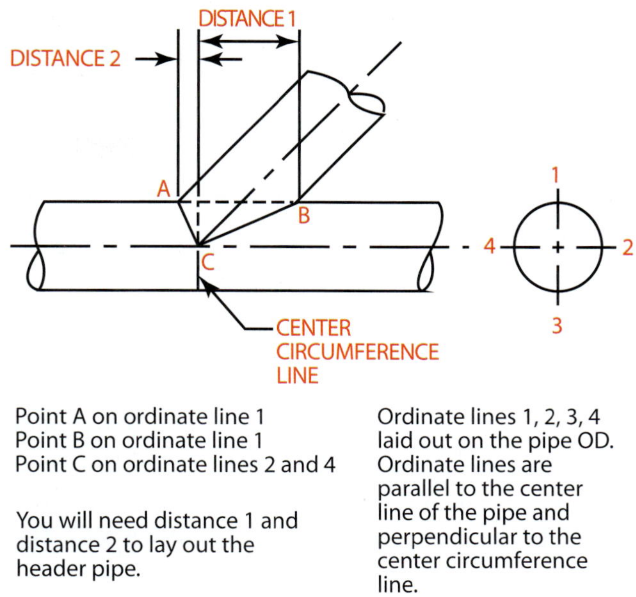

FIGURE 6-38 Connecting the header points.

Use the same technique to lay out the lines going from point *B* to point *C* on both sides of the pipe. The wraparound works well for this task because it will lie on the curved surface of the pipe to make an accurate mark for cutting the opening.

> **NOTE**
>
> You are now ready to lay out the branch pipe.

10. On the branch pipe, mark a center circumference line that will serve as a baseline that dimensions will be laid out from, **Figure 6-39**. It will be located at point *B* extending perpendicular to the other side of the branch pipe. It should be perpendicular to the center line and OD of the branch pipe.

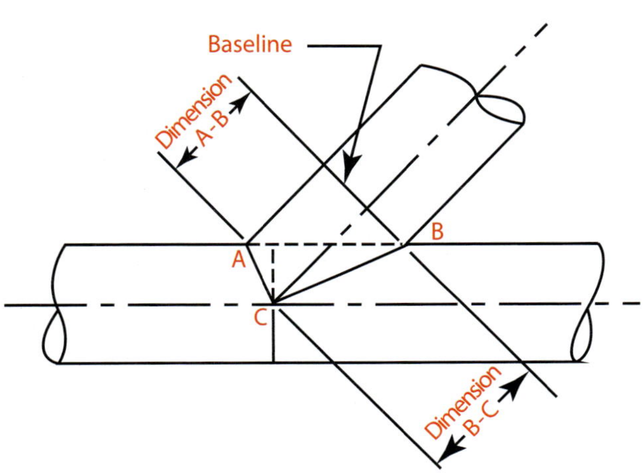

FIGURE 6-39 Laying out the branch.

11. Divide the center circumference line into quarters as you previously did with the header pipe. At the division points, mark ordinate lines at a distance of several inches along the length of the pipe. The ordinate lines will be marked 1 at the top, moving clockwise around the pipe, 2 on the side at point *C*, 3 at the bottom of the pipe, and 4 at the nine o'clock position, again at point *C*.
12. You are now ready to lay out the branch pipe. Start at the top ordinate line 1 and transfer the dimension that you measured off the drawing from the circumference reference line along ordinate line 1 to point *A*.
13. Transfer the measurement that you took from the center circumference line to both points *C* on both sides along the center line of the branch pipe. This will be ordinate lines 2 and 4.
14. Measure the distance from the center circumference line to point *A* along ordinate line 1 on the drawing. This measurement can be transferred from the drawing.
15. Measure from the center circumference line to point *C* along the center line of the branch pipe. This distance is on the drawing; mark it on ordinate lines 2 and 4 on the pipe.
16. You will now make your layout lines on the branch pipe using a wraparound and soapstone. Lay the wraparound on the branch pipe forming an arc from points *C* on both sides and passing through point *A*.
17. Rotate to the other side of the pipe and mark it in the same manner from points *C* on the sides to point *A* again.
18. Mark the cut line with your soapstone or marker.
19. Using the same technique, lay your wraparound on the pipe extending from one point *C* to point *B* and mark the pipe.
20. Rotate to the other side of the pipe and mark it in the same manner from one point *C* on the side to point *B* again.

NOTE

Your branch and header pipe are now ready to cut.

21. Using a properly adjusted oxy-fuel torch, cut the pipe, being sure to use a miter style of cut for both the header and branch. After the initial cuts are made, the pieces can be beveled for welding. Check the pieces for proper fit-up; and, if necessary, use a grinder or oxy-fuel torch to make fine adjustments.
22. Have your instructor inspect your work for uniformity.
23. Turn off all equipment, return all tools and supplies, discard all scrap, and clean up the welding booth. ◆

Orange Peel Pipe End

Sometimes it is necessary to cap off a piece of pipe, but you may not have any commercial caps available. This is when you can employ a technique known as an ***orange peel***. This is when the end of the pipe is cut, formed, and welded to cap off the end. If several orange peels are needed, it is best to make a template. There are charts available that contain the information necessary to lay out an orange peel.

The main considerations when constructing an orange peel are the diameter of the pipe that needs to be capped, the desired height of the cap, and the number of fingers (or peels) that are needed. The number of peels or fingers is typically determined by the size of pipe that is to be capped; the smaller the pipe, the fewer needed and vice versa. In most cases, the pipe should have at least five peels. Typically, if there are fewer than five peels, the cap will turn out looking square instead of rounded. Use the following formula for constructing an orange peel template:

A = Circumference of the pipe OD divided by the number of peels needed

B = Dimension A × 0.875

C = Dimension B × 0.5

H = The circumference of the pipe OD divided by 4

PRACTICE 6-6

Orange Peel Pipe End Layout

1. In order to make the template, you need to cut the material so that its length is equal to the OD of the pipe. Next, mark a baseline on the material so that it is roughly 1 in. or more from the bottom edge. All vertical measurements will be laid out from this line. Next, lay out the height of the peels, **Figure 6-40**; this will be dimension H. Next, lay out dimension A; this will be equal to the number of peels that the template will have. Lay out a center line in the middle of each peel and mark off dimensions 1/3H. Next, lay out dimensions B and C; they are the widths of the peels and will taper as they go to the top. At the top of dimension H, they will form a point at the center line. Dimensions B and C will be located at 1/3H points. Now, draw lines connecting points A, B, and C. The final step is to cut out the material between the peels. Your template is now ready to use.

FIGURE 6-40 Orange-peel pipe end cap.

FIGURE 6-41 A reducer.

2. Wrap the template around the pipe so that the two ends meet and are square. Using a piece of soapstone or other marker, trace the outline of the template onto the pipe. Now, using a properly adjusted torch, cut the pipe along the cut lines. Next, heat each of the peels to a cherry red and bend them to the desired shape. Once the peels have all been formed, tack and weld the peels together.
3. Have your instructor inspect your work for uniformity.
4. Turn off all equipment, return all tools and supplies, discard all scrap, and clean up the welding booth. ◆

PRACTICE 6-7

Pipe Reducer Layout

1. The procedure of making a reducer, **Figure 6-41**, is similar to the steps used in making an orange peel. Start by determining the OD of the pipe that you want to reduce to. Multiply the difference of the two pipe ODs by 1.3. This will give you the distance from the baseline to the final reduced pipe size. Using a wraparound, mark a baseline at the point where you want to start the reduction in pipe size. Using the dimension that you determined by multiplying the differences in the two pipe diameters, measure that distance and mark another line around the OD of the pipe. The distance from the baseline to the final reduction line will be 1.3 times the distance of the difference of the two different pipe diameters to which you are reducing.
2. Determine how many arms are needed for the layout of the reduction. The more arms that you use, the more accurate and smooth the transition of the reduction will be. Keep in mind that the more that are used, the more skill that is necessary to make the transition look good and professional. You will determine the number of arms by multiplying the difference in the pipe ODs by 1.33. This number will give you the layout for the distance of each of the arms to be laid out on the baseline. Now, mark this distance along the baseline.
3. Draw a line dividing each of the arm lines in two along the baseline. This will be the center point for each arm on the reduced end of the pipe.
4. Divide the size of the small-diameter pipe into the number of arms that were determined to be needed. This will give you the length of each of the arms on the reduced end of the pipe. Mark these

distances on the reduced end of the pipe using the center mark on the baseline as the center mark for the reduced end of the pipe.

5. Draw the layout lines by connecting the points of the arm lengths of the reduced pipe end to the size of the larger pipe arm. Once they have been laid out, use a cutting torch to trim the pieces away using a slight radial cut so that the arms will fit together smoothly on the outside. Once all of the arms are cut out, head the arms up by the base layout line and bend them in to form a nice, uniform fit-up. ◆

FABRICATED VERSUS COMMERCIAL PIPE FITTINGS

All high-strength, high-reliability, extremely critical piping systems never use welder-fabricated fittings because the structural integrity of such fittings cannot be ensured the way that the commercial fittings can be, **Figure 6-42**. However, welder-fabricated fittings are commonly used in a wide variety of projects and applications, such as handrails, sculptures, parade floats, and barbecue grills, **Figure 6-43**.

Larry Jeffus

FIGURE 6-42 Some of the many commercially available fittings.

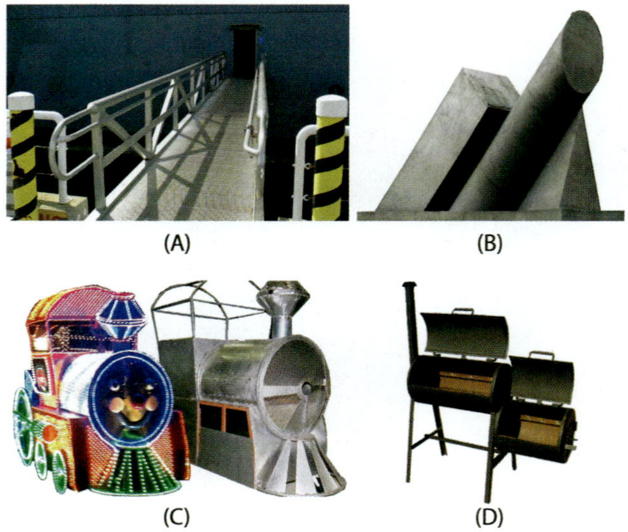

Larry Jeffus

FIGURE 6-43 (A) Gangway into the aircraft carrier *USS Yorktown* in the harbor at Charleston, South Carolina. (B) Welded sculpture at Manchester Community College in Manchester, New Hampshire. (C) Parade float at Silver Dollar City, Branson, Missouri. (D) Custom barbecue grill and smoker, Zajic's Welding in Terrell, Texas.

Summary

Understanding joint geometry is important to ensure that procedures that are required by a welding code or welding procedure are being followed. It is also necessary in order to ensure that the requirements of the weldment are being followed as called for by the designer as specified on the blueprint.

It is important to familiarize yourself with common layout tools and practice using them. Basic math concepts such as addition and subtraction of fractions and learning how to obtain common denominators are essential for a layout person to know. It is important for a fitter to be able to look at a blueprint and be able to visualize how the weldment is going to fit together. This ability comes with practice and experience, but it is helped by working with skilled colleagues who have years of experience.

Becoming a skilled craftsman is the result of a willingness to be open to new ideas and the practice of skills, with the desire to always do better. Understanding the oxy-fuel cutting process and devoting time to practicing it will help you become a true craftsman.

Modern industries, including oil refineries, tank farms, shipping, aviation, and many others, rely heavily on welders and fitters to fabricate the piping systems critical to their operation, **Figure 6-44**. Skilled welders and fitters are still needed after construction is completed to help maintain the facility, make modifications, and do repairs.

Larry Jeffus

FIGURE 6-44 There are thousands of miles of piping and hundreds of thousands of fittings in the refineries, oil depots, grain elevators, docks, and other industries along this section of the Houston Ship Channel, shown in this photo taken from the open hatch in the aft compartment of the WWII aircraft FIFI, a B-29.

Review Questions

1. What are two primary considerations for choosing weld joint design?
2. The scope of a code or quality manual contains what information?
3. What are the two most commonly used types of weld joints in pipe-welding applications?
4. What is a bevel angle?
5. Why should pipe be ground after using a thermal-cutting process?
6. What are two elements of joint geometry?
7. Using the saddle-on, technique, which piece of pipe sits on the other?
8. How many layouts must be performed in order to fit one 90-degree saddle?
9. What are ordinate lines?
10. Is it necessary to have more ordinate lines on larger-diameter pipe? If so, why?
11. What is an orange peel, and why is it used?
12. Of the eight common groove welds, how many may be used for pipe-welding applications?
13. Most codes require that surface oxides be removed a distance of ______ from the weld groove prior to welding. (Answer in either English or metric units.)
14. What is one of the most important considerations in joint design?
15. What are two advantages of plasma-cutting equipment?

Chapter 7

Pipe Fit-Up and Alignment

OBJECTIVES

After completing this chapter, the student should be able to:

- List common codes and their applications.
- List and explain some of the common sections contained in a code.
- List various fitting tools and their applications.
- List typical steps in preparing pipe for alignment and fit-up.
- Explain what the elements of offset are and why offsets are necessary.
- Calculate various lengths of pipe and angles of fittings needed.

KEY TERMS

Arc length
Code
Facing
High-low
Nominal pipe size
Offset
Rise
Run
Scope
Take out
Travel

INTRODUCTION

Welding is a tool used by many industries to join metal fabrications. Aerospace, automotive, construction, electronics, food, military, and utilities are just a few of the areas where welding is used on a regular basis. Each of these has a responsibility to the public to ensure that the products they deliver to the public and use to produce goods and services can be used without fear of catastrophic failure resulting in injury or death. In order to ensure that they meet certain standards that protect public health and safety, products are engineered and manufactured to specific specifications. These specifications or standards are often known as a **code**.

There are numerous codes, and they were often developed in response to an industry need as the result of a catastrophic failure. Some of the more commonly used codes in the United States from organizations such as the American Welding Society (AWS), the American Petroleum Institute (API), and the American Society for Mechanical Engineers (ASME) include the AWS's D1.1, a code governing structural steel; the API 1104, used in the piping industry; the ASME B31.3, used for process piping; the AWS D10.12, used for pipe-welding mild steel; the AWS D10.18, used for pipe-welding stainless steel; and the AWS D18.1, used for hygienic tube–welding stainless steel. Each code deals with a specific aspect of welding for an industry.

(Continued)

The important thing to remember about codes is that they are a set of minimum standards. In some cases, states, providences, and municipalities have adapted the codes as law; in other regions, they are a set of minimum guidelines. It is always important to also consult local and state guidelines and regulations.

CODE STANDARDS

Most codes have an introduction or foreword that states the purpose of the code—to provide engineers, contractors, and manufacturers with sound, proven engineering and operating practices. The **Scope** portion outlines what the code covers. Most codes have five or more sections that cover terms and definitions, specifications, qualification, design and workmanship, and inspection.

Within each of these sections, more information is included. For example, in the Qualification section, details will be given about the qualifications of welders as well as welding procedures. Data such as the type of tests required and information about how many samples and their locations will be specified.

Most codes give specifications on the alignment of abutting pipe ends. The goal of this information is to limit the variations in the two pipe ends that are to be joined. In most cases, the variation is limited to no more than 1/8 in. (3.175 mm). The variation around the circumference of the pipe should be as uniform as possible. When using external alignment-type clamps, the API code requires that they be used in accordance with the Welding Procedure Specification (WPS) that is being followed. In cases where it is necessary to remove the external alignment clamp prior to completing the root pass, the code requires that the root pass be completed in equal-length sections that are spaced equally around the circumference of the pipe, resulting in at least 50% of the total pipe circumference being welded prior to releasing the clamps. When internal alignment clamps are used and releasing pressure from the clamps would cause undue stress on the weld, the entire root bead must be completed prior to releasing any pressure from the clamps.

PREPARATION OF PIPE FOR FITTING

The WPS will define the type of edge preparation that will be required on the pipe. It may be necessary to perform several machining operations on the pipe ends prior to fitting in order to assure proper alignment, correct weld penetration, and prevent an excessive heat-affected zone. One of the operations is "cutting," which can be achieved mechanically or thermally. It is important to make sure that the pipe ends are cut square. This can be achieved by facing the end of the pipe after cutting.

Facing is a process where the edges of the pipe are cut mechanically so that the ends are square. If the ends of the pipe are not square, it will cause the run of the pipe to curve. In order to correct the curve, it would be necessary to fit the pipe ends in a manner that would cause a larger gap on one side of the pipe than on the other. Either of these situations will cause other problems with the weldment, such as excessive penetration, too large a heat-affected zone, and distortion.

Another operation that may be necessary is to counterbore the inside of both pieces of pipe. "Counterboring" is when a machine is used to make a light cut on the inside diameter (ID) of the pipe. It is typically a light machining that helps to clean up the inside of the pipe and ensure that the wall thickness is uniform. This helps the welder to maintain a consistent depth of penetration and helps to reduce the possibility of contaminated welds caused by oxides on the inside surface of the pipe. There are machines available that will perform mechanical cutting, facing, and counterboring. Another common method to cut pipe is with a thermal process. In this case, a flapper wheel sanding disk can be used to clean the inside of the pipe prior to welding. This will achieve the same goal as mechanical counterboring. Using a flapper wheel makes it less likely to damage or gouge the inside of the pipe; for this reason, this method is better than using an angle grinder.

Each of the steps, cutting, facing, beveling, and counterboring is important to ensure that consistent and quality welds are produced.

ALIGNMENT TOOLS

The API 1104 code requires that WPS states what type, if any, of pipe alignment clamps are to be used during welding. There are primarily two types of clamps that are used for clamping abutting pipe ends: namely, internal and external.

They perform two basic functions, which are to hold the appropriate root gap and, in some cases, to correct misshapen pipe ends. Often, material can become distorted during shipping and handling, such as becoming oval shaped. In these cases, it is necessary to correct the misshaped ends prior to welding. Use of an internal alignment clamp can apply pressure on areas of the pipe, pushing it outward so that the ID of the pipe will fit such that there is a limited amount of high-low. **High-low** is where the internal surface of the material does not line up. It results in the surfaces of the pipe not being in the same plane.

Failure to have good alignment will make putting in a quality root weld difficult. This can result in inadequate penetration due to high-low. Using what is called a Hi-Lo® gauge when fitting and tacking, as well as just prior to welding, can help determine if there is misalignment of the two pipe surfaces and how much, **Figure 7-1**. For proper alignment of the pipe, the offset of the two pipes' surfaces should be kept to a minimum. According to the API code, the offset should be no more than 1/8 in. (3.175 mm) on pipes that have the same nominal thickness.

External reforming rim clamps are designed to clamp onto the outside of the pipe and hold it in alignment. This type of clamp is also designed to reform misshapen ends. In most cases, external rim clamps are designed as standard or heavy-duty clamps. The standard type is designed to reshape Schedule 40 pipe that is X45 grade or has 45,000-pound (310,260-kPa) tensile strength that is up to 2 in. (50 mm) out of round. The heavy-duty clamps are designed to handle up to Schedule 80 pipe X45 through X80 grade that is up to 2 in. (50 mm) out of round.

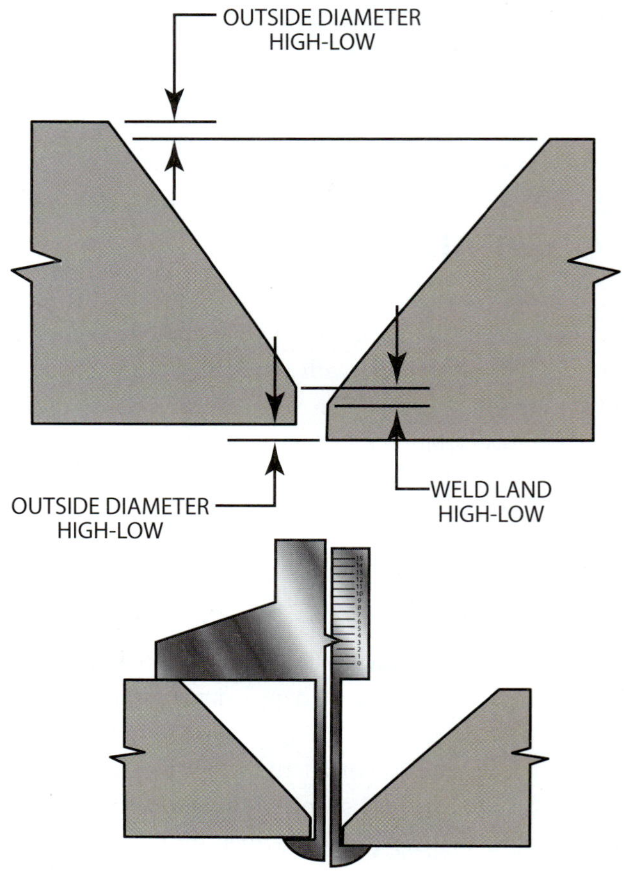

FIGURE 7-1 High-low terminology and measurement.

FITTING TOOLS

There are a variety of tools that are helpful to a pipe welder or fitter. They can be grouped into different categories; there are tools for aligning and clamping as well as those for layout. "Alignment tools" consist of tools like the ones previously mentioned that perform two functions, such as aligning pipe and clamping and holding it for welding. These can be designed for internal or external use depending on the need. In some cases, it may be too difficult to set an internal clamp in place and remove it, so an external-type system would be used instead.

Alignment pins are used to make sure that flanges are positioned so that bolt holes are in alignment for bolting. A method referred to as two hole is used in which two alignment pins are placed in the bolt holes of opposing flanges that are to be bolted together, **Figure 7-2**. The

Larry Jeffus

FIGURE 7-2 Using a level to properly align a flange on a pipe section.

purpose is to make sure that the holes of the two flanges align perfectly so that all the bolts will line up. A level is often used to make sure that the bolt holes are in horizontal alignment during the process.

Layout tools are often helpful (and in some cases necessary) to a welder or fitter. Soapstone, scribes, and paint markers are used to mark materials for cutting and fitting. Using templates can help make the layout process faster and more productive. All types of templates are available to mark pipe for cutting various angle saddles for both the branch and header, as well as templates for laying out elbows and Y-connections. Wraparounds can be used for laying out and marking pipe as well. Squares, protractors, and levels are other hand tools that are often used to lay out, fit up, and align pipe. Pipe jack stands are also very important for a welder to have, **Figure 7-3**. They serve much like a table for laying out and fitting assemblies, as well as welding them.

FIGURE 7-3 Piping sections can be supported with wooden blocks, in a process called "cribbing," or by jack stands. The advantage of cribbing is that the weight of the pipe is spread over a larger area on the ground, so it is less likely to settle under the pipe's weight.

PIPE OFFSETS

Because pipes are used to transport liquids and gases to various locations, the design and layout of piping varies greatly from job to job. In the piping industry, pipe bends, turns, and changes in direction or elevation are referred to as **offsets**. Common turns can be made using standard 90- and 45-degree fittings; however, because of design requirements and restrictions to flows, not all turns can be accomplished using these standard angles. For this reason, it is necessary to make up special angle fittings by cutting them out of the standard 90- or 45-degree fittings, **Figure 7-4.** It is also necessary to cut pipe to lengths based on these angles. The lengths are figured using trigonometry, which is the mathematical study of the sides and angles of triangles.

In order to figure offsets, you must know the related terminology. An offset is sometimes called a set, which is the

FIGURE 7-4 Before you cut, checking the accuracy of the measurement on a fitting is a good idea.

perpendicular distance between two pipe center lines that are running parallel to each other and will be joined by the offset. Offsets can be achieved through a rise or a run.

Rise is the term given to the vertical distance at the center point of bends of two pieces of pipe that are parallel to each other. This is to transition the pipe from a lower level to a higher level, **Figure 7-5**.

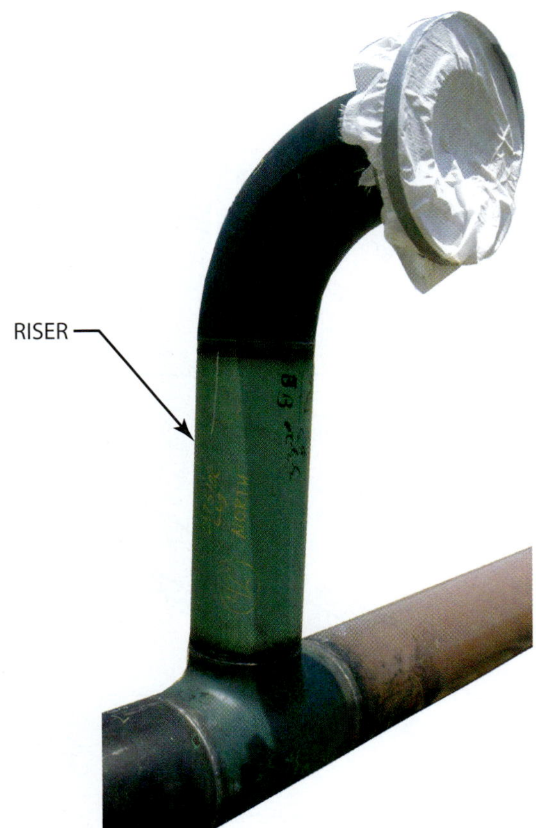

FIGURE 7-5 Any vertical pipe that intersects a horizontal pipe is called a *riser*.

Run is the term given to the horizontal distance at the center point of bends of two pipes that are parallel to each other and running in the horizontal position. **Travel** is the term given to the diagonal distance at the center point of bends of two pipes that are running parallel to each other. It can be typically thought of as the pipe that will need to be cut to join the two parallel pipes.

One way to learn the terms is to remember that some of them are related to a pipe being positioned in a specific plane. When the layout of the pipe is in the horizontal plane, the terms "offset," "travel," and "run" are used. When the pipe layout is in the vertical plane, the common terms are "offset," "travel," and "rise," **Figure 7-6**. This can be easily remembered by the fact that "rise" has to do with the vertical plane. It is also important to note that each of the terms is defined in relation to the center line of the pipe.

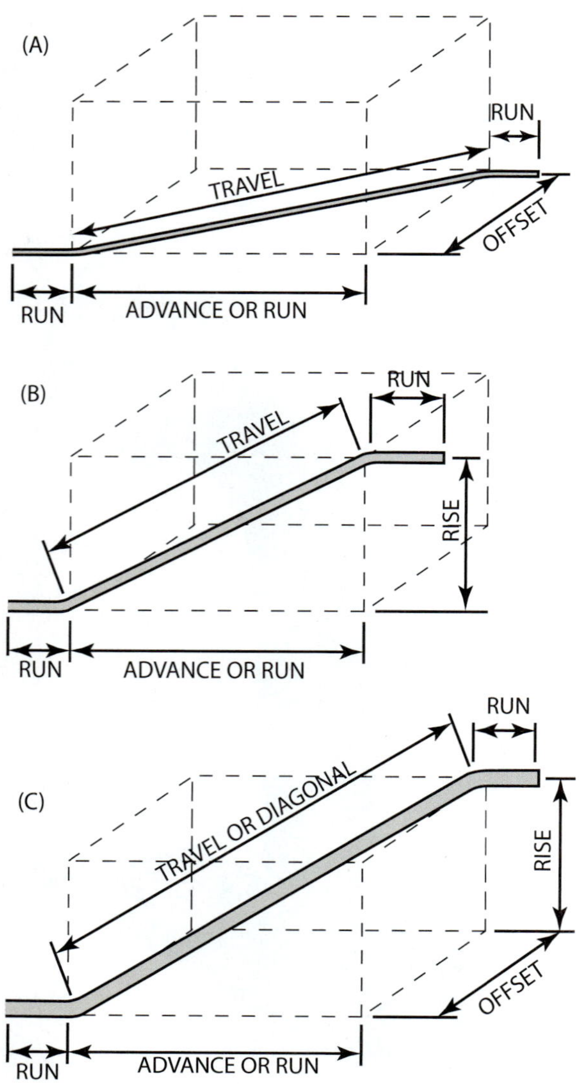

FIGURE 7-6 (A) Horizontal offset; (B) vertical offset; (C) diagonal offset terminology.

If you are familiar with right triangle trigonometry terms, each of these pipe layout terms relates to a specific term, **Figure 7-7**. "Travel" is the same thing as the **hypotenuse**, which is the leg opposite the right angle in a triangle; it is the longest of the three legs. The terms "rise," "offset," and "run" all relate to the right triangle terms "side opposite" and "side adjacent," depending on the reference angle.

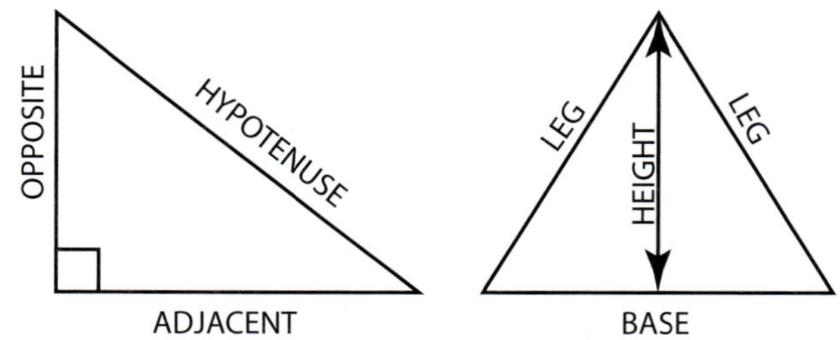

FIGURE 7-7 Triangle terminology

SOLVING UNKNOWN PIPE ANGLES AND LENGTHS USING THE PYTHAGOREAN THEOREM

There are several ways to figure lengths of pipe that are used in making offsets. One method is to use the Pythagorean theorem, which uses the formula $a^2 + b^2 = c^2$. If any two legs of a right triangle are known, the other leg can be determined by utilizing this equation. In the formula, a and b are the legs that make up the right angle of the triangle, and leg c is the hypotenuse. By plugging two leg lengths into the formula, the third length can be calculated. In our example, we will use what is commonly called a 3, 4, 5 right triangle, which has one leg length of 3, one of 4, and one of 5.

For example:

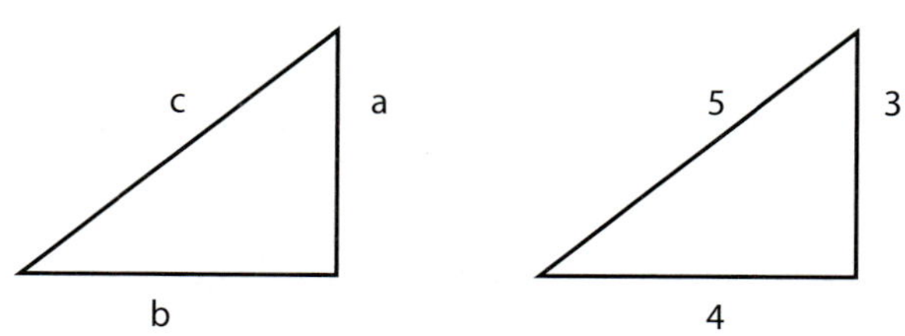

Values given: $a = 3$ $b = 4$ $c = 5$

Using the Pythagorean theorem ($a^2 + b^2 = c^2$), replace the variables a, b, and c with the values given previously:

$$3^2 + 4^2 = c^2$$
$$9 + 16 = c^2$$
$$25 = c^2$$
$$\sqrt{25} = \sqrt{c^2}$$
$$5 = c$$

Likewise, this formula can be rewritten in the following manner to solve for an unknown leg length:

$$\sqrt{a^2 + b^2} = c$$
$$\sqrt{3^2 + 4^2} = c$$
$$\sqrt{9 + 16} = c$$
$$\sqrt{25} = c$$
$$5 = c$$

Looking back at the original leg lengths, we see that c = 5, which is the longest leg (and hence the hypotenuse). Keep in mind that most pipe layouts are somewhat basic in nature, and it is the leg lengths and angles that change.

If we know the leg length of c and b but not the third leg a (rise, run, or offset), the problem could be rewritten in the following manner:

$$c^2 - b^2 = a^2$$
$$\sqrt{5^2 - 4^2} = \sqrt{3^2}$$
$$\sqrt{25 - 16} = \sqrt{9}$$
$$\sqrt{9} = \sqrt{9}$$
$$3 = 3$$

Looking back at the original equation, a = 3.

If we know the leg length of c and a but not the third leg b (rise, run, or offset), the problem could be rewritten in the following manner:

$$c^2 - a^2 = b^2$$
$$\sqrt{5^2 - 3^2} = \sqrt{4^2}$$
$$\sqrt{25 - 9} = \sqrt{16}$$
$$\sqrt{16} = \sqrt{16}$$
$$4 = 4$$

Looking back at the original equation, b = 4.

SOLVING PIPE LENGTHS USING THE ANGLES OF A RIGHT TRIANGLE

Another way to find the length of a pipe is by use of the acute angles that are present in the right triangle layout. These angles can be expressed in terms of ratios based on the length of the legs that make up the triangle. The length of the legs can vary, but it is the ratio that is important.

A *ratio* is simply the length of the legs divided by one another. There are a total of six different ratios for a right triangle and they all have names that are determined by the sides that are being compared. Ratios can be expressed as a fraction or as a decimal. In most cases, they are expressed as a decimal. These decimal numbers are constants for given angles and are based on the comparison of the two legs in relation to the specific angle that makes up the ratio. In other words, the lengths of the legs are directly related to the reference angle. If the length of one of the legs is known, the ratio or constant for a given angle can be used to determine the other lengths. In order to do this, you must learn the ratios (often called functions of the angles of the right triangle. There are three basic functions and three inverses to these basic functions. The symbol θ, called *theta,* represents the angles. These functions are:

Sine θ = opposite / hypotenuse—abbreviated Sin θ
Cosine θ = adjacent / hypotenuse—abbreviated Cos θ
Tangent θ = opposite / adjacent—abbreviated Tan θ

The inverses are directly related to the three functions as follows:

Cosecant θ = hypotenuse / opposite—abbreviated Csc θ
Secant θ = hypotenuse / adjacent—abbreviated Sec θ
Cotangent θ = adjacent / opposite—abbreviated Cot θ

The last three functions are not used as often, so we will focus on the first three, sine, cosine, and tangent. One way to remember these trigonometry functions is to use the mnemonic "SohCahToa." Each letter stands for the first letter of the functions. "Soh" is for Sine = opposite / hypotenuse, "Cah" is for Cosine = adjacent / hypotenuse, and "Toa" is for Tangent = opposite / adjacent.

Once again if we use the 3, 4, 5 right triangle, we can determine the constants for the different functions:

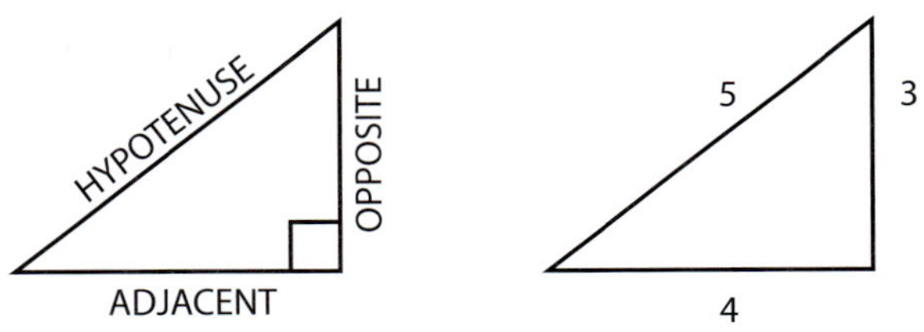

Sine θ = 3/5 = 0.6000
Cosine θ = 4/5 = 0.8000
Tangent θ = 3/4 = 0.7500

The inverse functions are:

Cosecant θ = 5/3 = 1.6666
Secant θ = 5/4 = 1.2500
Cotangent θ = 4/3 = 1.3333

Each angle will have a constant or function associated with it. These will be listed on the trigonometric function table for quick reference in case you do not have a calculator. Most tables will include the following information: degrees (and in some cases minutes) and each of the functions for each degree. This includes Sin θ, Cos θ, Tan θ, Csc θ, Sec θ, and Cot θ. Keep in mind that when working with a right triangle, you can use this information to determine the degrees of an angle with either a scientific calculator or the trigonometric function table.

If you know the length of a leg and an angle, it is possible to find another leg length. To do this, use one of the function formulas according to the information you have. In order to make this easier, and make sure that there are no mistakes, it is best to draw the function options that are possible along with the triangle. The drawing of the triangle should be labeled. Determine which angle and sides you have and are solving for.

In the following example, we have a right triangle that we know one angle θ and the side adjacent. We want to know the length of the side opposite. We first identify the sides that will be used. They are opposite and adjacent.

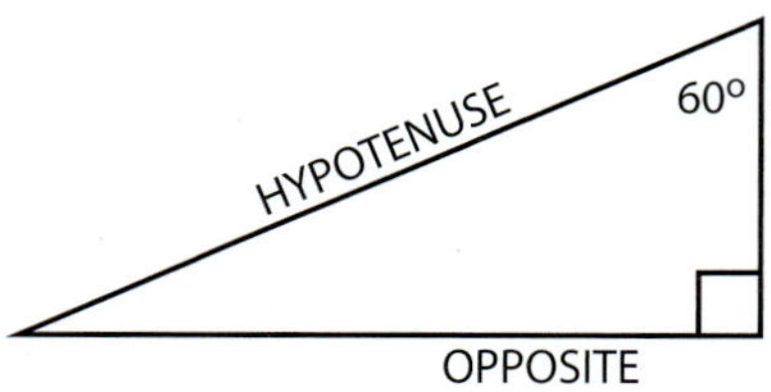

Next, write out the functions and determine which one we need to use:

A. $\text{Sin } \theta = \text{Opp/Hyp}$ B. $\text{Cos } \theta = \text{Adj/Hyp}$
C. $\text{Tan } \theta = \text{Opp/Adj}$

In this case, we see that Tan θ is the angle related to the sides opposite and adjacent.

We can rewrite the equation to make solving for the unknown easier. It is simply setting up the problem where the known values are on one side and the unknown is on the other side:

C. $\text{Adj} \times \text{Tan } \theta = \text{Opp}$

When we do this, we see that it is possible to solve for the opposite length by using the tan of θ multiplied by the side we know, which is the adjacent side. The units of measure you have are 60-degree angle and 3 in. (76.2 mm) for the side adjacent. Now plug the values into the formula:

$$\begin{aligned} \text{Adj} \times \text{Tan } \theta &= \text{Opp} \\ 3 \times \text{Tan } 60 &= \text{Opp} \\ \text{Tan } 60 &= 1.7320 \\ 3 \times 1.7320 &= 5.1961 \\ \text{Opp} &= 5.1961 \end{aligned}$$

To solve for the hypotenuse, you can use either of the two sides and either angle. Since one angle is 60 degrees, we know that the other angle has to be 30 degrees. The 60 degrees plus the 30 degrees equals 90 degrees. Then add the third angle, which is 90 degrees. The total of the three angles comes to 180 degrees.

In this case, we will use the adjacent side and the 60-degree angle, just as in the previous example. First, we identify the sides that will be used: adjacent and hypotenuse. Therefore, we know that the angle is 60 degrees, so we will plug these values into the appropriate function formula to solve for the hypotenuse.

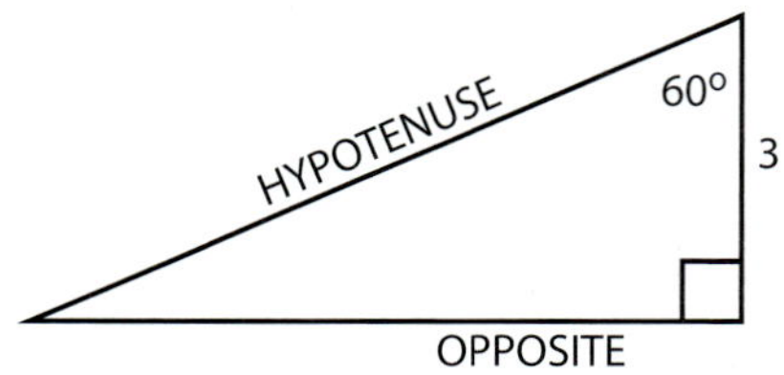

The sides that we are dealing with are the adjacent and the hypotenuse. When selecting the function, it is best to select one with the known dimension in the denominator:

$$\frac{\text{unknown}}{\text{known}}$$

For this reason, select $\text{Sec } \theta = \frac{\text{hyp}}{\text{adj}}$ as the function. Then fill in the values that are known.

$$\begin{aligned} \text{Sec } 60 &= \frac{\text{hyp}}{3} \text{ can be rewritten as} \\ \text{Adj} \times \text{Sec } \theta &= \text{Hyp} \\ 3 \times \text{Sec } 60 &= \frac{\text{hyp}}{3} \times 3 \\ 3 \times 2.0000 &= \text{hyp} \\ 6 &= \text{hyp} \end{aligned}$$

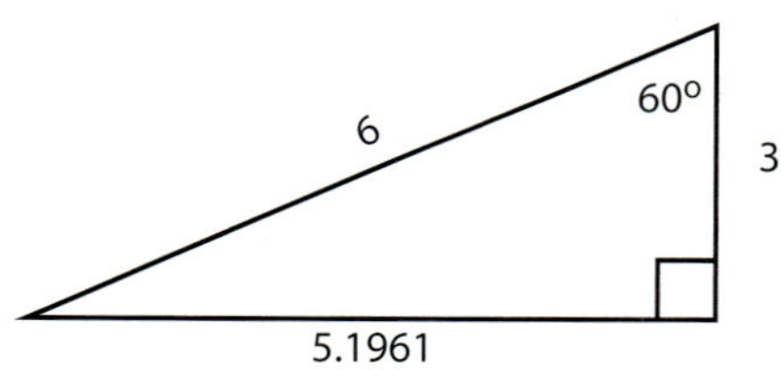

We now have values for all of the sides and all of the angles. We can go back and check them by plugging in each of the values into the appropriate function to double-check your answers. This will also give you practice and training to remember the six different functions.

MARKING BUTT WELD ELBOWS FOR NONSTANDARD ANGLES

Sometimes there will be a need to use an elbow that is not the standard 45 or 90 degrees. For these situations, it will be necessary to mark and cut a specific degree of elbow. If the elbow is not marked correctly or cut with great accuracy, there will be a great likelihood that the ends will be oval instead of round. This can leave a very large gap for the welder to fill and cause distortion in the weld joint. We will use three steps to determine and mark the length of an arc that will give the correct degree of angle that is needed. **Arc length** is the distance measured on a curved line. A specific arc length will give a specific degree of angle. **Figure 7-8**

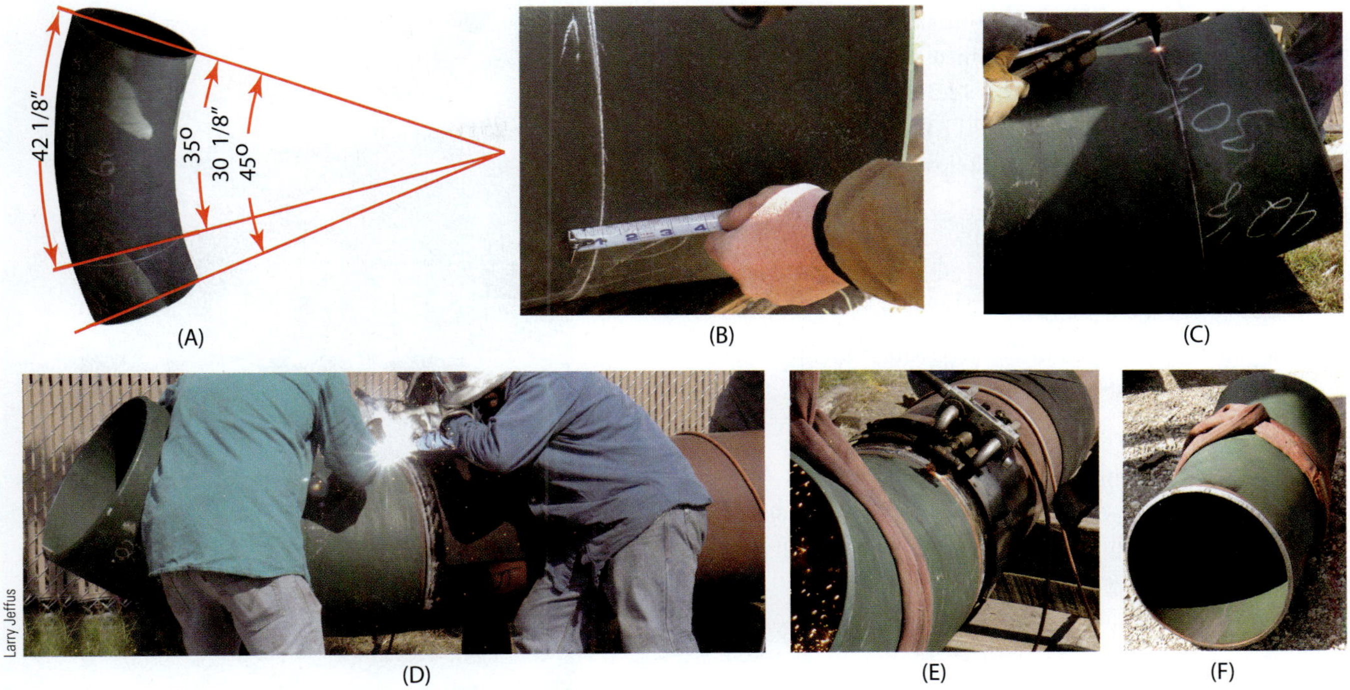

Larry Jeffus

FIGURE 7-8 (A) Determine the angle needed and calculate the length of the inside and outside radii. (B) Lay out a line around the pipe that is set back 1 in. (25 mm) from the determined length to allow the elbow to be machine beveled later. (B) Make a square freehand cut along the cut line. (C) Tack weld the elbow onto a straight section of pipe. (D) Use a cutting machine to cut a bevel at the determined length needed for the 35-degree elbow. (E) The finished cut leaves the end of the elbow round.

goes through the following steps to modify a manufactured 45-degree elbow. The following paragraphs describe the step-by-step process of modifying an elbow.

In order to make a proper fit-up, the angle must be accurately marked. There are three steps to obtain the information necessary to mark the correct arc length and resulting angle. Step 1 is to determine the radians for the given angle that is to be cut. Step 2 is to determine the inside and outside radii for the size pipe that is to be cut. Step 3 is to determine the inside and outside arc lengths to be marked and cut.

For Step 1, the formula for determining the radians for a given degree is

$$\text{Angle } \theta = \frac{\theta\pi}{180} \text{ rad}$$

You can also remember that $\frac{\pi}{180} = .017453$, which is a constant. Then all you have to do is multiply that by the degree of angle for which you need the radians. For example, if you need to cut an elbow to 42 degrees, you can just multiply it by 0.017453:

$$42° \times .017453 = 0.733026 \text{ rad}$$

In Step 2, we need to know the inside and outside radii for the butt weld elbow. Let's take, for example, the 6-in. standard pipe. We multiply the nominal size 6 in. (152.4 mm) by 1 1/2. This gives us 9 in. (228.6 mm), which is the center radius measured from the point where the planes of the two ends of the butt weld elbow faces intersect, also known as the "vertex." Using the center radius as a reference point, we can determine the inside and outside radii. We use the outside diameter (OD) of 6-in. (152.4-mm) pipe, which is 7.625 in. (193.675 mm), so we divide this in half. This gives 3.3125 in. (84.14 mm). Now all we have to do is add the 3.3125 in. (84.14 mm) to the center radius, which is 9 in. (228.6 mm), and we get 12.3125 in. (312.74 mm). If the butt weld elbow is lying with one face sitting on a table or other flat surface and we measure to the top height of the other face, that height will be 12.3125 in. (312.74 mm). This result will give the outside radius of the elbow. We can subtract 3.3125 in. (312.74 mm) from the inside radius, which is 9 in. (228.6 mm), and we get 5.6875 in. (144.46 mm). This gives the inside radius. With the face of the butt weld elbow sitting on a table or other flat surface, we could measure to the underside portion of the other weld face at the outside edge of the pipe at its lowest point; it should be 5.6875 in. (144.46 mm).

Step 3 is to determine the arc length. This will be done by using the information that we have already obtained and plugging it into the formula for arc length. That formula is:

$$\text{Radius} \times \text{rad} = \text{arc length}$$

So, to take one example, for a 6-in. pipe that needs to be cut to 42 degrees, we have determined the inside radius to be 5.6875 in. The outside radius is 12.3125 in. The number of radians based on 42 degrees is 0.733026. So, given the formula for arc length, we get the following:

Radius × rad = arc length

Inside radius 5.6875 in. × 0.733026 = 4.169 in., or just less than $4\frac{3}{16}$ in.

Outside radius 12.3125 in. × 0.733026 = 9.025 in., or just over 9 in.

We can now mark the arc length on both the inside radius and the outside radius of the pipe. These areas are often referred to as the "throat" and "outside" of the elbow. They can be marked with a tape measure or pipe wraparound. It will be easiest if the pipe is quartered, and then the distances can be marked on the inside radius and the outside radius. After these points are marked, use a wraparound or framing square and a soapstone or other appropriate marker to mark around the OD of the pipe. It should now be ready for cutting. **Figure 7-9** shows the modified elbow, installed with its protective coating applied, being tested to ensure that the coating is thick enough to protect the pipeline for many years of safe service.

FIGURE 7-9 Checking the thickness of the protective coating on a pipe elbow.

TAKE OUTS AND ROOT OPENING

When figuring the length that pieces of pipe need to be cut to in order to fit up properly, there are two things that need to be considered. The first is the root opening, which is the distance between two pieces of pipe at the joint root. This figure is often specified by the code or welding procedure, or it can be left to the welder's discretion. It has to be considered for pipe and elbow lengths when laying out and fitting pipe. In many cases, the root opening is going to be between 1/16 in. and 1/8 in. for each weld joint. The overall length of the pipe assembly in **Figure 7-10** is 250 in. To determine the total length of pipe needed, you must subtract the width of the 10 flanges and the 10 boot openings.

250 in.	Total length of pipe assembly
−10 in.	Minus total width of all flanges
−1 1/4 in.	Minus total width of root openings
238 3/4 in.	Equals actual length of pipe required

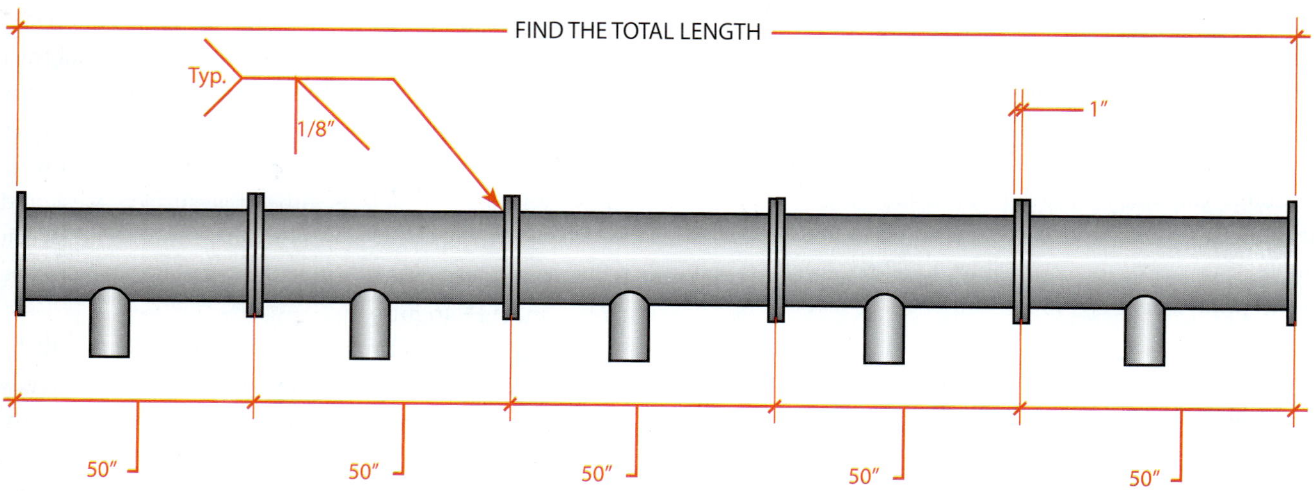

FIGURE 7-10 Calculating the total length of pipe needed to fabricate an assembly.

Take outs have to be figured for each pipe layout that has fittings or elbows, and they are needed to maintain the location of center lines of pipes when turns and transitions are used. A **take out** is the distance that an elbow extends the center line past the end of a piece of pipe. It is always important to subtract the root openings and take outs from the overall run of a piece of pipe. Failure to do so will result in the pipe being too long.

To figure the take out of a particular butt weld elbow, you can use the following formula:

$$\text{Take out} = \tan\frac{\theta}{2} \times \text{radius of elbow}$$

It is important to remember that the radius is 1.5 × the nominal pipe size. So with a 6-in. pipe, for example, the radius would be 9 in.

Given we have two 6-in. butt weld elbows that we need to cut to a 30-degree angle, we use the previous formula to figure the take out:

$$\text{Take out} = \tan\frac{30^\circ}{2} \times 9$$
$$\text{Take out} = \tan 15^\circ \times 9$$
$$\text{Take out} = 0.2679 \times 9$$
$$\text{Take out} = 2.4111 \text{ in.}$$

Be sure to remember to figure the take out twice because there are two of them. We also need to add in the root opening for two welds; in this case, we will make each 1/8 in. So the total will be two take outs, at 2.4111, and two root openings, at 1/8 in. each.

$$\text{Take out } 2 \times 2.4111 = 4.8222 \text{ or approximately } 4\frac{13}{16} \text{ in.}$$

$$\text{Root opening } 2 \times \frac{1}{8} \text{ in.} = \frac{1}{4} \text{ in.}$$

$$\text{Total amount of take out} = 5.072 \text{ in., or approximately } 5\frac{1}{16} \text{ in.}$$

FIGURE 7-11 Tubular steel racecar frames must be strong to protect the driver in the case of a high-speed accident on the race track.

Summary

Pipe fit-up and alignment are critical in ensuring the quality of a weldment, **Figure 7-11**. In most cases, there are codes that serve as specifications governing the minimum requirements that the weldment must meet. They serve as engineering and fabricating guidelines for the protection and well-being of the public. There are a lot of formulas to remember as a pipe fitter or welder. Pocket reference books and charts are available to help with this job. Many calculators are specifically designed to help with fitting pipe applications.

Review Questions

1. Describe what codes are and how they came about.
2. List five sections that might be contained in a code.
3. Name two types of clamps that are used in the piping industry.
4. Explain what hi-low means.
5. Explain what the term "two hole method" refers to and why it is important.
6. Write the formula for the Pythagorean theorem.
7. How many different trigonometry functions are there?
8. What is the mnemonic that can be used to help remember the primary trigonometric functions?
9. What is trigonometry?
10. The perpendicular distance between two pipe center lines that are running parallel to each other and will be joined by the offset is known as ________ or ________.
11. In the piping industry pipe bends, turns, and changes in direction or elevation are referred to as ________.
12. The pipe-fitting term "travel" is the same thing as the trigonometry term ________.
13. The terms "offset," "travel," and "run" are used when the layout of the pipe is in the ________ plane.
14. When the pipe layout is in the ________ plane, the common terms are "offset," "travel," and "rise."
15. The term "take out" refers to the distance that an elbow extends the ________ past the end of a piece of pipe.

Chapter 8

Shielded Metal Arc Welding of Pipe

OBJECTIVES

After completing this chapter, the student should be able to:

- Describe what a tack weld is, its importance, and the role that it plays in the weld.
- Explain the effect of welding rod angle on penetration and buildup.
- Explain the role and importance of the root and hot pass.
- Describe the various positions for pipe welding.

KEY TERMS

Arc length

Cold lap

Cover pass

Drag travel angle

Fast fill

Fast freeze

Hot pass

Keyhole

Push travel angle

Root pass

Slag inclusions

Stringer bead

Tack weld

Wagon tracks

INTRODUCTION

Shielded metal arc welding has several advantages that makes it a good choice for pipe welding. One is its ease of portability. The use of motor generators and inverter-type machines makes it possible to get to jobs that are located in remote or otherwise difficult-to-reach areas, **Figure 8-1**. Because of the simplicity of the process, the equipment required is relatively inexpensive. The wide variety of rod types available allows many different types of metals to be welded. Depending on the rod selected, welding can take place in all positions and can result in high-quality, low-hydrogen welds.

This chapter will focus on the shielded metal arc welding (SMAW) of pipe. Several skills are necessary to produce high-quality welds in pipe. Some of the skills that will be covered in this chapter are tack-welding the pipe and making the various passes to complete the weld. It is important that these skills be practiced and mastered.

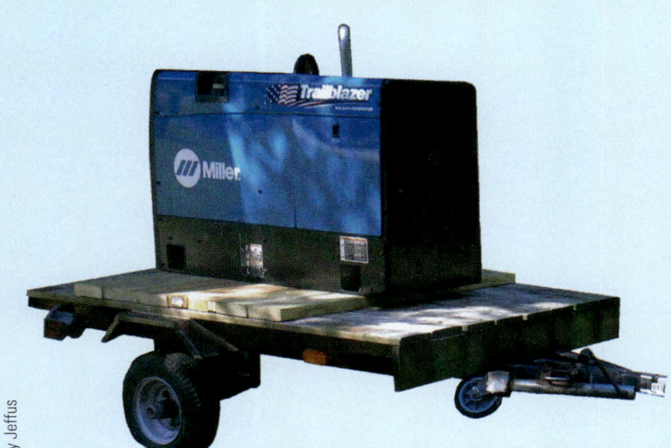

Larry Jeffus

FIGURE 8-1 Portable engine-driven welder.

ARC STRIKES

When an arc is struck and the metal melts and cools quickly (which is known as being "quenched"), it becomes very hard and brittle. That is exactly what happens to a pipe's surface when an arc is struck. If the arc strike occurs inside the area to be welded, its effect will be completely erased by the heat of the weld. When the hard brittle arc strike spot is remelted, it becomes part of the weld bead and no longer can cause a problem. That is why it is important that you only strike an arc in the groove to be welded. An arc strike that occurs outside the groove will leave hard brittle spots that will cause problems even if they are ground off following the weld, **Figure 8-2.** The crack on this arc strike is not large enough to fail this bend test, but the arc strike itself would cause the bend test specimen to fail. Because of the brittleness of these arc strikes outside of the weld groove, cracks tend to start, so most welding codes consider them defects that must be repaired.

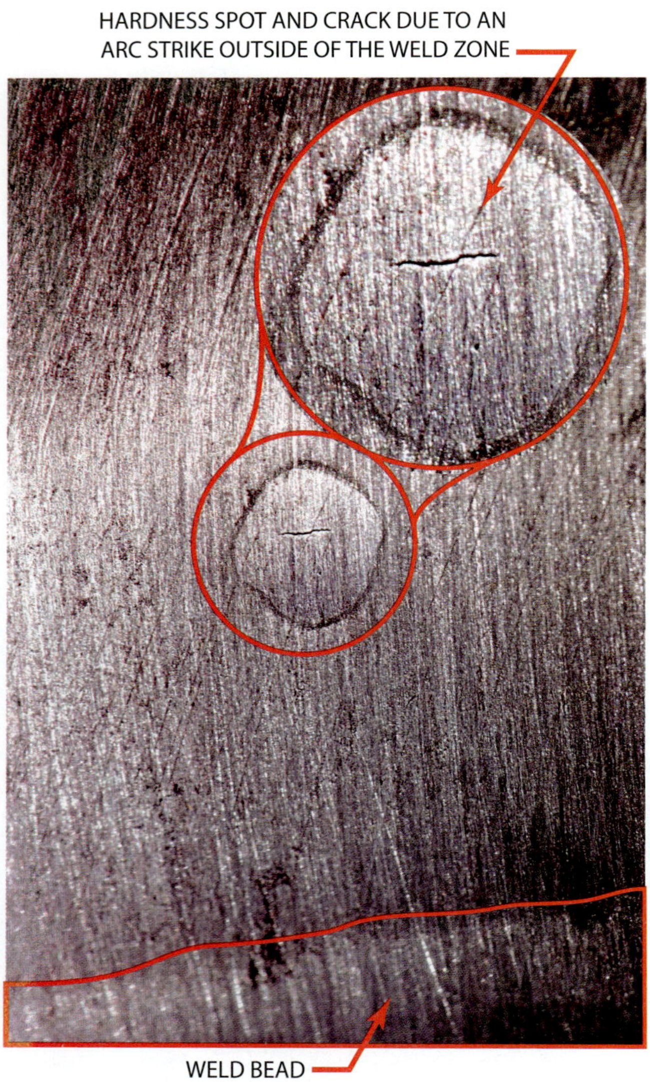

FIGURE 8-2 A small crack in a pipe caused by an arc strike is much like a chip in a car windshield. Over time, both cracks spread. The crack in the windshield can be an inconvenience, but a crack in a pipe can cause a catastrophic failure.

TECHNIQUE FOR STRIKING AN ARC

Although the process of striking an arc on a pipe takes only a fraction of a second, there are a number of important things that need to happen in that instant of time. First, the arc should be struck ahead of the point that the weld will begin so that the weld can be made back over the spot that the arc was struck, **Figure 8-3.** Second, it takes a short time for the end of the electrode to heat up so that the protective gaseous cloud can form as the flux begins to vaporize. Also, as the end of the electrode heats up, the arc will begin to stabilize. Any weld metal deposited before this gaseous cloud forms can have a number of defects, including porosity and hydrogen embrittlement. So it is important that the end of the electrode be held high enough so that filler metal is not deposited until the gaseous cloud is completely formed. Once the arc is stable and the gaseous cloud has formed, lower the electrode over the spot where the weld is to begin. Hold it at this location as the weld pool begins to build, and it increases to the desired size. Once the molten weld pool is properly formed in shape and size, begin moving the electrode slowly along the joint back over the area of the arc strike so that the arc strike is remelted and becomes part of the weld.

FIGURE 8-3 Using both hands to guide the electrode can help prevent arc strikes.

ELECTRODE MANIPULATION

SMAW pipe welds can be made using stringer beads, weave beads, or both. Often, the root pass will be made as a stringer bead with little or no electrode manipulation. Some of the small movements used on stringer beads include the following:

- Rocking the electrode holder back and forth to tilt the end of the rod to point at one side of the groove and then the other. This concentrates the arc and arc force on the sides of the weld to help sidewall fusion and reduce penetration of the center of the weld. This technique can be helpful when there may be excessive root face buildup or inadequate sidewall tie-in, **Figure 8-4**.

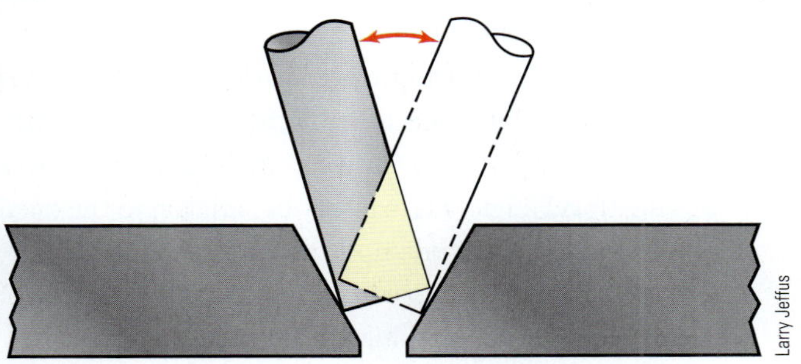

FIGURE 8-4 Rocking the electrode back and forth can help with sidewall penetration.

- A very slight circular pattern or side-to-side pattern can be used to ensure that the root face is melted and the keyhole remains open.
- A slight whipping technique can be used if the keyhole begins to get too large.
- A T-motion with a whip-and-pause technique can be used if the root opening is too wide and the root weld tries to separate, leaving an opening in the center of the root weld.
- A zigzag or Z-weave is made by slowly moving the electrode back and forth across the weld pool and moving the electrode slightly forward with each zigzag movement, **Figure 8-5**. Remember not to weave low-hydrogen electrodes wider than 2.5 times the electrode wire diameter; that is, a 1/8-in. (3-mm) E7018 electrode can be woven no more than 5/16 in. (8 mm). Naturally, the weld pool itself will be wider than the weave since it extends beyond the sides of the electrode.

Some welders like to use a drag technique. Start with the electrode pointed between 10 degrees off of perpendicular to perpendicular of the pipe's surface in the direction of travel. When done properly, this technique will result in a keyhole developing along the leading edge of the weld pool and closing up along the trailing edge as the weld progresses.

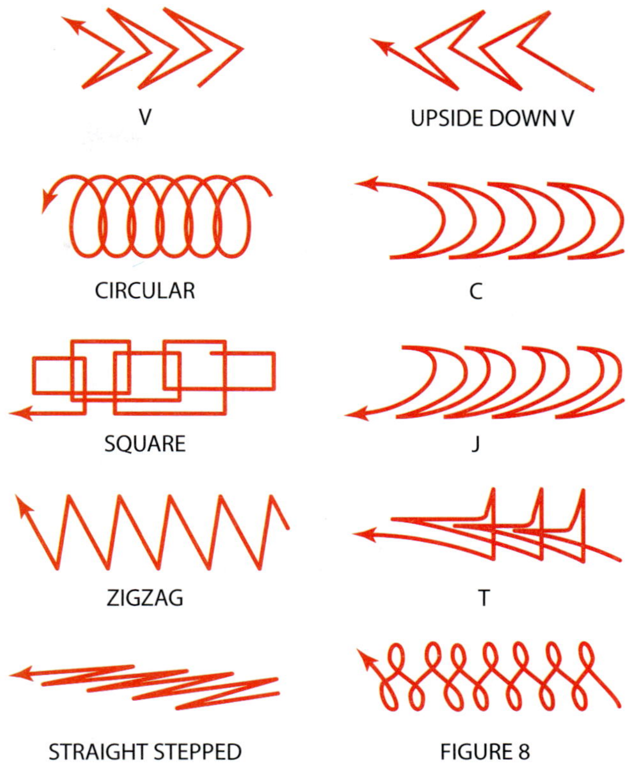

FIGURE 8-5 Examples of weave patterns.

USE BOTH HANDS FOR WELDING

It is a good idea to practice welding all the passes with both hands; do half the pipe using your dominant hand and the other half using your nondominant hand. There will be times that it may be difficult or nearly impossible to make some welds completely with your dominant hand, so it is a good idea to practice for that situation. Always keep in mind that welding is an art form, and there are many ways to achieve the same result. Regardless of which technique you adopt, practicing it is the key to making consistent, high-quality welds.

WELD PASS

There are several variables that affect how a weld pass is made:

- **Amperage setting**—Use the electrode manufacturer's recommendation for the amperage setting as a guide.

Many codes require that welding machines be calibrated; but not all school welding machines are calibrated, so the machine settings may not be exact. For that reason, you should make test welds before starting work on a practice. Because the welding position changes as a weld progresses around a pipe in the 5G horizontal fixed position, it is often necessary to make changes in the welding current. This is especially true when making the root pass and hot pass. On many pipeline jobs, a welder has a helper whose job is to make slight changes in the current when the welder asks. As little as a 1- or 2-amp change can make a difference in a weld's quality; so if you are going to make an amperage adjustment, it should be a small one.

- **Direction of travel**—In general, it is more common to use a **drag travel angle**, sometimes referred to as a "trailing angle." The term means that the electrode will be pointing back at the weld pool. **Push travel angle**, sometimes called "leading travel angle," means that the electrode will be pointed ahead of the weld pool in the direction of travel.

  As an electrode burns hot gases, molten metal and slag are propelled from the end of the rod. Using a push travel angle can cause molten metal and slag to be pushed ahead of the weld pool. This can result in slag entrapment as well as **cold lap**, which is when molten weld metal is deposited on top of the base metal without fusing the two together. Using the push travel angle will reduce weld penetration; it can be used in a vertical position to help hold molten metal in place.

  Using the drag travel angle helps to increase weld penetration. As the rod angle changes from the vertical position to more of a drag travel angle, penetration increases. Along with the change in weld penetration, weld reinforcement increases as well.

- **Electrode angle**—The electrode angle, sometimes called the "rod angle," relates to the angle between the welding electrode and center of the pipe. It is the same angle that can be measured between the electrode and the pipe's or weld's surface. At any given point as a weld progresses around a pipe, the relative position of the welding electrode to vertical or horizontal may change; but the angle between the center of the pipe's or weld's surface may need to remain relatively consistent.

  The electrode angle relates to the direction of travel, as well as to the right, left, downward, or upward alignment of the electrode to the weld bead's orientation. For example, a weld may be made with a 10-degree drag travel angle or a 10-degree push travel angle. Also, for a horizontal weld, the electrode may have both a 30-degree upward angle and a 10-degree drag travel angle, **Figure 8-6**. In addition to the need to point the electrode upward for a horizontal weld, it may need to be pointed to the right or left side of the groove when making a filler pass.

- **Arc length**—The term **arc length** refers to the distance from the end of the electrode to the molten weld pool. The rod needs to be positioned so that it is close to the root of the joint; in some cases, the flux on the electrode may actually be touching the sides of the weld groove. Maintaining a short arc length also will help reduce the chance of undercut and possible **slag inclusions**, which are pieces of slag

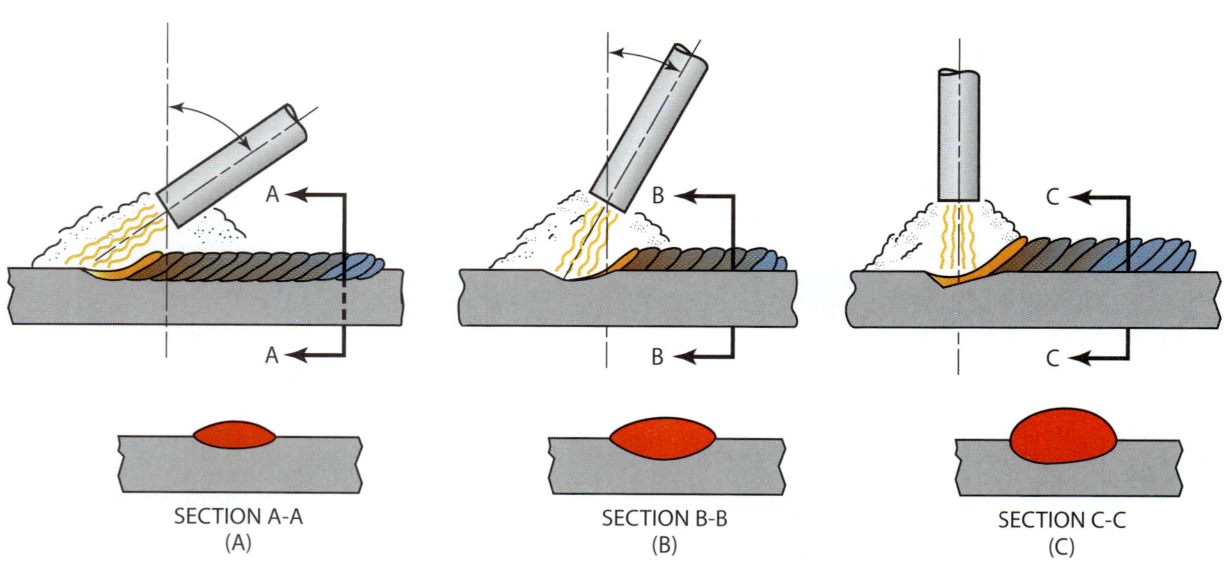

FIGURE 8-6 Electrode angle and weld penetration.

that get trapped under weld metal, usually along the sides where undercut or deep V-grooves are formed between weld passes, **Figure 8-7**.

FIGURE 8-7 Slag entrapment alongside a weld bead.

Slag inclusions along both sides of pipe weld beads that show up on an x-ray are often referred to as "wagon tracks." This term is used by welders and inspectors to describe marks that look on an x-ray like ruts made by wagon wheels. Maintaining a short arc length can help control the weld bead shape and undercut, thus reducing slag inclusion.

- **Type of electrode**—For many years, E6010 and E6011 electrodes were used for almost all of the root and hot pass welds on pipe. However, that is changing today, and many pipe-welding codes and standards allow or require only low-hydrogen electrodes such as E7018.

E6010 and E6011 electrodes are referred to as having fast fill and fast freeze characteristics. The term **fast fill** refers to the rod's ability to deposit metal quickly. The term **fast freeze** refers to the electrode's ability to solidify very quickly. These qualities help when making open root welds. In addition, the rods produce a large quantity of shielding gases, welding fumes, or smoke, which can protect the inside surface of the molten weld metal when doing an open root weld. As a result of the quantity of welding fumes, there is very little slag left on the surface of welds with E6010 and E6011 electrodes.

The flux on most low-hydrogen welds produces heavy layers of slag and a much smaller quantity of welding fumes. In the past, there were concerns about the ability of the gaseous cloud of shielding gases to protect the open root weld face from atmospheric contamination. Tests have shown that a weld with 100% low-hydrogen filler metal is superior to welds made with E6010 and E6011 root and hot passes.

- **Preheating**—The primary purpose of preheating is to slow the rate of thermal cycling of the metal around the weld; that is, it helps to control the heat-affected zone (HAZ). If the base metal is thermally cycled too fast, it will become hard and brittle. Preheating is specified by welding codes and standards for each type, thickness, and application of pipe.
- **Interpass temperature**—The **interpass temperature** is the temperature of a weldment as measured between weld passes. Some codes limit the interpass temperature to 500°F for steel pipe. As the pipe heats up from multiple weld passes, it will become harder to control the weld bead shape. Too high of a temperature caused by the welding heat can weaken the surrounding pipe. So from time to time, it may be necessary to stop welding and let the pipe cool down.

TACK WELDS

A **tack weld** is a weld for the purpose of holding parts in proper alignment until final welding can take place, **Figure 8-8**. In most cases, tack welds will become a permanent part of the weldment. Because of this, they must meet the same requirements as the final welds. In some cases, the tack welds are remelted and fused into the root pass, eliminating all evidence of the original tack weld. The requirements for tack welds will be the same as those designated in the welding procedure specification (WPS). This would include elements such as cleaning, preheat, and postweld heat treatment. It is important to keep in mind that all discontinuities or defects need to

FIGURE 8-8 Tack weld.

be prevented and that some, such as cracking and porosity, can be propagated into the rest of the weld. In effect, the tack welds should be treated the same as the root weld passes.

PRACTICE 8-1

Tack Welds

For this practice, you will need two pieces of 6-in. (152.4-mm). Schedule 80 pipe approximately 4 in. (100 mm) in length, an angle grinder, a half-round bastard file, a wire brush, and an 1/8-in. (3-mm) E6010 electrode for making the tack welds. You will also need personal protective equipment (PPE) for the processes being performed.

1. Mild steel pipe is often prepared by beveling with the oxy-fuel cutting process.

NOTE

Read and follow all the pipe-cutting equipment manufacturer's instructions, and see Chapter 7, "Thermal-Cutting Processes," for details on cutting pipe bevels.

NOTE

There are several common bevel angles for pipe preparation, such as 30 degrees, 37.5 degrees, and 45 degrees. For this practice, we will use a 30-degree bevel angle; this will result in a 60-degree groove angle.

2. Once the beveller is positioned to 30 degrees, the torch should be positioned on the pipe where the cut is to be made.
3. Adjust the torch to a neutral flame and make the cut.
4. Remove any scale and oxide from the cut surfaces and adjacent areas by grinding and wire brushing, **Figure 8-9**.

Larry Jeffus

FIGURE 8-9 Cleaning the groove and pipe face will prevent weld contamination.

5. Light slag can be removed from the edge of the cut surface by using a half-round file and running it from the inside of the pipe, holding it at a low angle, and knocking lightly attached slag from the inside of the pipe.
6. Check to see that the end of the pipe is lying flat and square with the outside diameter (OD) of the pipe. You can check this by placing it on a flat surface and checking for light coming from under the surface. You can also use a framing square, lay it on the OD of the pipe, and lay the other end over the cut edge of the pipe.

NOTE

In most cases, if the beveling equipment is working well, your cut should be square, and you should see little to no light coming from under the root face.

7. Before starting to grind, you need to wear all the required PPE for grinding and follow the grinder's and grinding wheel manufacturer's safety and operating instructions.
8. If necessary, use an 8-in. (200-mm) grinder to square the end of the pipe by holding the pipe securely, placing the grinder flat on the surface of the pipe root face, and lightly applying pressure on the pipe so that a small chamfer is formed on the root face surface. Grind a little at a time and keep checking until the end of the pipe is square.

NOTE

Grind as little as possible so the chamfer is 1/8 in. (3 mm) or less in size.

9. Before grinding the end of the pipe, make sure to secure the pipe by clamping it to a table or by using a vise or other appropriate clamping tool.
10. Grind the groove face down to bright, shiny metal, which ensures that surface oxides have been removed. When you start to grind the groove face, be sure to position the grinder so that it touches the surface and maintains the correct 30-degree bevel angle.
11. Apply light pressure to the metal surface with the grinder, and follow the contour of the pipe surface. You should move the grinder back and forth at a quick speed for a distance of about a quarter of the circumference of the pipe.
12. Rotate the pipe and work the grinder around the entire diameter of the pipe. As you change directions while moving the grinder back and forth

around the contour of the pipe, be sure to release pressure from the surface to prevent flat spots and gouges on the groove face.

13. Grind oxides from the outside of the pipe adjacent to the joint for a distance of about 1 in. (25 mm).
14. Check the inside of the pipe for scale and oxides, and use a rotary file, a sanding flap, or both to clean the inside of the pipe until the surface is bright and shiny.
15. Once you have cleaned the oxide off the inside and outside of the pipe, you are ready to fit the two pieces together.
16. Check to make sure that the inside surfaces line up and there is little to no mismatch. "Mismatch" is where the inside surfaces of the pipe do not line up, causing high-low on the alignment of the pipe. Excessive mismatch and high-low make it difficult to get a uniform and defect-free root weld.

NOTE

It is important not to feel rushed. Take your time when fitting the pipe. Proper fit-up is important because it is the foundation of a successful pipe weld.

17. After preparing the pipe sections as described in the previous steps, you are going to tack-weld them together.
18. Welding on the 6-in. (150-mm) pipe will require it to be tacked in four different places, 90 degrees apart.
19. Take a bare welding rod that is the thickness of the desired root opening and bend it into a V-shape. Take one of the pipe coupons and lay it bevel side up, and place the remaining pipe coupon bevel side down on top of the rod and other coupon. A spacer wedge is often used when fitting joints in the field, **Figure 8-10**.

Larry Jeffus

FIGURE 8-10 A root gap chisel can be used to adjust the root space.

20. Line up the two pieces so that the inside of the pipes match up and give a smooth uniform fit-up with an approximately 1/8-in. (8-mm) root gap and about a 3/32-in. (2-mm) root face. On seamed pipe, lining up the pipe seams can help eliminate some mismatch, **Figure 8-11**.

Larry Jeffus

FIGURE 8-11 Lining up the seams on a pipe can help avoid mismatching.

21. Keep in mind that as the tack welds are made, the welds will shrink as they cool, causing the gap to close slightly.
22. Test the machine settings by making a short weld on a piece of scrap metal.

NOTE

The process described in the section entitled "Technique for Striking an Arc," earlier in this chapter, must be followed when making tack welds.

23. Make the first tack weld approximately 1 in. (25 mm) long and ensure that the two coupons are fused together at the weld.
24. Check to make sure that the alignment and root opening are staying consistent because weld shrinkage can cause the pipe root gap to increase on the opposite side. If necessary, readjust it before the second tack weld is made.
25. Place the second tack 180 degrees from the first tack.
26. Once the first two tack welds are made, you can remove the rod and place the third and fourth tack welds halfway between the first two welds. This will give you four tack welds spaced approximately 90 degrees apart.
27. Grind a taper on both ends of the tack weld down to a featheredge.

NOTE

Use a thin grinding disk to taper the tack welds on both ends. This will serve as a point to start welds from, as well as a point to run up to on end welds. By tapering the tacks, it will help provide a better weld tie-in and a bed to level out the tack welds.

28. Have your instructor inspect your work for discontinuities and consistent defect-free welds.
29. Additional tack welds can be made between the first four tack welds if additional practice is needed.
30. Turn off all equipment, return all tools and supplies, discard all scrap, and clean up the welding booth. ◆

ROOT PASS

The **root pass** is the first weld pass to be made. It serves as the foundation that all other passes will be laid on. It serves to establish penetration and seal the joint.

The root weld should bridge the gap across the root opening. This is when the rod angle will play an important role in making sure that there is complete penetration and sufficient weld reinforcement.

The techniques used to make a root pass vary from welder to welder and region to region. One technique is to have the root opening spaced so that the welding rod will not pass through the joint. The welding rod will just reach the root face. If the rod were able to pass completely through the joint, the root gap would be much too large, making it very difficult to make a quality weld with uniform reinforcement. Doing this will ensure that the root face is melted by the arc and complete fusion is obtained between the root faces of the pipe.

ROOT-PASS 1G POSITION

In the 1G position, the pipe is placed in the horizontal position, but it can be rolled so that most of the welding takes place near the flat position. This exercise will help you to improve your welding skills going around a curved surface. Fit and tack the pipe together as described in the section entitled "Tack Welds" in Practice 8-1.

The root pass will be made with an E6010 electrode. The E6010 is a good choice for doing this pass because of its fast-fill/fast-freeze characteristics.

PRACTICE 8-2

Root-Pass 1G Position Rolled

For this practice, you will need 6-in. (150-mm) Schedule 40 or 80 pipe, 1/8-in. (3-mm) E6010 and E7018 electrodes, a grinder, a half-round bastard file, and a wire brush. You will also need PPE for the processes being performed. Prepare your short pipe sections as described in the section entitled "Tack Welds" in Practice 8-1.

1. Start the root weld on one of the tack welds.
2. Position the pipe so that you can start welding between the ten o'clock and eleven o'clock positions and weld to between the twelve o'clock to one o'clock positions.

NOTE

It is important to hold a very short arc when making the root pass, so pipe welders often find it beneficial to grasp the electrode holder at the end so that one gloved finger can be rested lightly against the electrode, **Figure 8-12**. This lets you feel when the electrode flux is gently touching the sides of the pipe groove when you have a very short arc.

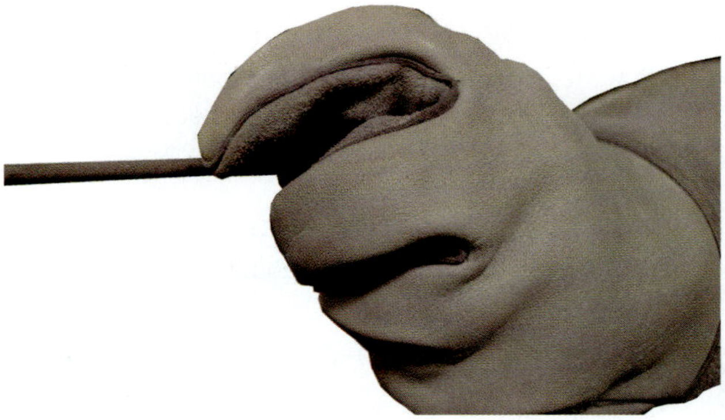

Larry Jeffus

FIGURE 8-12 Resting your finger lightly on the electrode will allow you to feel when the flux touches the groove face.

NOTE

Arc strikes outside the weld groove are considered a defect by most welding codes, so it is important that the arc be struck only in the groove.

3. Use your index and middle finger to guide the rod into the joint to make sure that you do not strike the arc outside the weld groove. Arc strikes cause hard spots on the base material that can cause very small cracks to form.
4. Once you start the arc, allow enough time for it to establish itself and become stable, and then start your travel. Be sure to place the tip of the rod down into the root of the joint so the rod flux gently touches the groove face. This will help ensure complete penetration.
5. Keeping a short arc helps to control the weld heat and weld pool, as it concentrates the arc on the root face to ensure complete root face fusion.

> **NOTE**
>
> A **keyhole** is the small opening at the leading edge of a weld pool caused by the arc melting away the base metal, **Figure 8-13**. As the weld progresses, the leading edge of the molten weld pool continues to open into a keyhole as the filler metal continues to fill it along the back edge of the keyhole. Using the keyhole welding technique is a good way of ensuring complete root fusion and penetration.

Larry Jeffus

FIGURE 8-13 Keyhole.

6. You will be able to see a keyhole by looking on the trailing side of the electrode between the end where the rod is burning off and the start of the weld pool.
7. There are two things to look for in this area:
   a. First, make sure that the keyhole is not getting much larger than the diameter of the electrode.
   b. Second, make sure that the distance between the electrode and weld pool is half the diameter of the rod being used or less. Note the following:
      i. If the distance gets too great, the travel speed is too fast.
      ii. If there is no gap between the electrode and the weld pool, the travel speed is too slow, and the keyhole is likely to be lost resulting in lack of fusion.
8. When you reach the twelve o'clock position, stop and rotate the pipe so that the weld end is between the two o'clock and three o'clock positions.
9. Repeat this process until the root pass is complete.
10. Have your instructor inspect your work for discontinuites and consistent defect-free welds.
11. The weld should have uniform buildup and penetration, as well as fusion to the root and along the toe of the weld.
12. Repeat this practice until you can make a smooth, consistent weld every time.
13. Turn off all equipment, return all tools and supplies, discard all scrap, and clean up the welding booth. ◆

> **NOTE**
>
> Once you have the joint completed, you can practice beading on the outside of the pipe, **Figure 8-14**. This would be similar to building a pad on plate.

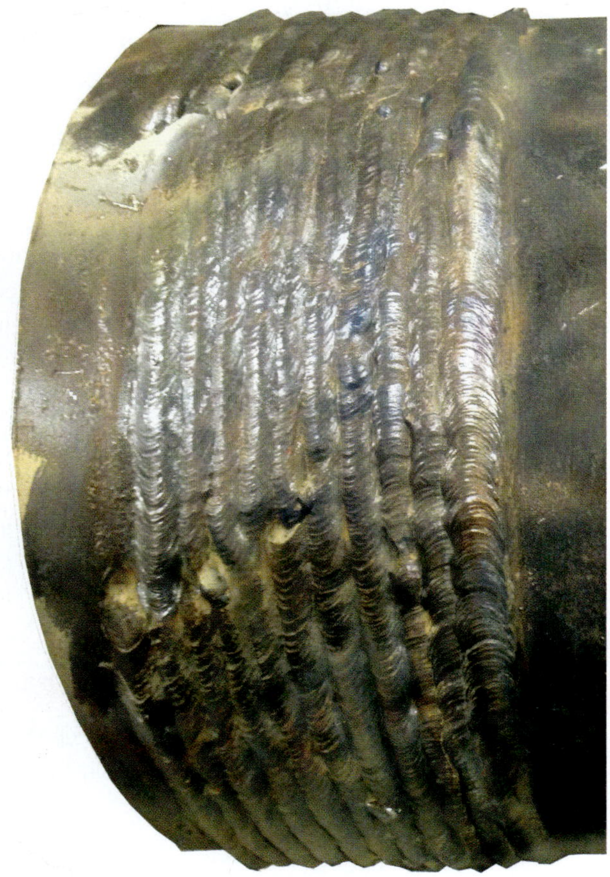

Larry Jeffus

FIGURE 8-14 There are several advantages to practice beading on scrap pipe. It is a good way to build your welding skills; warm up, much as a baseball pitcher does; and use up scrap pipe.

HOT PASS

The term "hot pass" has been used for years to identify the weld pass that is used to clean up slag inclusions and fix a lack of penetration. This is because in the early years of pipe welding, the standard practice was to crank up the amperage and "burn out" slag. Today, slag alongside the root weld is typically ground out using a narrow grinding disk. Higher amperage is no longer used to burn out slag, but the hot pass is mostly used to smooth out the weld buildup and to ensure good side-wall fusion.

In some cases, slag can be floated free, but it is not necessary to use high-amperage settings to get the job done. Often, root passes have a convex weld face that can trap small pieces of slag along one side or both sides. This trapped slag may be difficult to impossible to remove by chipping. The hot pass is used to wash away any of this trapped slag that may be along the sides of the root pass by melting the sides of the groove so the trapped slag can float free. It can be used to help level out the root-weld face and also to remove and fill any undercut. The hot pass will be made using an uphill progression, which also promotes deeper penetration and fill. Care must be taken to make sure that the weld pool does not get too large and out of control.

PRACTICE 8-3

Hot-Pass 1G Position

Using the weld coupon from Practice 8-2, 1/8-in.(3-mm)-diameter E6011 electrodes, a properly set-up and adjusted welding machine, and all the required PPE, you are going to make a hot pass.

1. Strike an arc between the ten o'clock and eleven o'clock positions in the groove directly on top of the root pass. Maintain a longer-than-normal arc length—that is, 1/8 in. to 3/16 in. (3.175 to 4.76 mm) for a 1/8-in.-diameter electrode—until the arc stabilizes. Bring the electrode closer to the weld surface to begin the hot pass.

NOTE

The primary purpose of the hot pass is to clean up the face of the root pass without adding substantially to the filling of the weld groove.

2. Keep the weld face flat or slightly concave by traveling fast enough and using a slight weave so that it does not trap slag along its sides, **Figure 8-15**.
3. Keep an eye on the leading edge of the weld pool to ensure that any trapped slag along the root pass is being melted free.
4. When the weld has progressed to the twelve o'clock to one o'clock position, stop and roll the pipe so the end of the weld is at the ten o'clock to eleven o'clock positions so the next weld can begin there.
5. Repeat this process until the hot pass is completed around the pipe.
6. Once the hot pass is complete, clean the weld thoroughly with a chipping hammer and wire wheel.

Larry Jeffus

FIGURE 8-15 A very short arc length can help control the weld face and prevent slag entrapment.

7. Have your instructor inspect your work for discontinuities and consistent defect-free welds.
8. The weld should have uniform buildup, as well as fusion to the groove sides.
9. Repeat this practice until you can make a uniform consistent weld each time.
10. Turn off all equipment, return all tools and supplies, discard all scrap, and clean up the welding booth. ◆

FILLER PASSES

Filler passes come next. They are used to fill up the weld groove to within 1/16 to 1/8 in. (1.5 to 3 mm) of the top of the groove, **Figure 8-16.** The number of filler passes required will vary depending on the weld groove depth. In order to maximize the amount of buildup with each filler pass, they are made with complete fusion to the previous weld beads and sides of the groove, but without deep penetration. Each filler weld bead must be chipped and cleaned of all slag before the next filler weld is made.

Larry Jeffus

FIGURE 8-16 Leave enough space in the groove for the cover pass.

The size and profile of the groove will determine which size electrode will be appropriate. Smaller-diameter electrodes will allow easier control of the weld pool when out of position and be less likely to overfill small-diameter or thin-walled pipe grooves. However, on thicker walled pipe sections, it would be better to use a larger-diameter electrode so the joint can be filled more efficiently with fewer weld passes. In addition, the thicker pipe can withstand the higher heat input from larger-diameter electrodes and is less likely to be overheated. Overheating makes it difficult to control the weld bead.

You should stagger the weld starts and stops so they do not all end up in the same location, **Figure 8-17**.

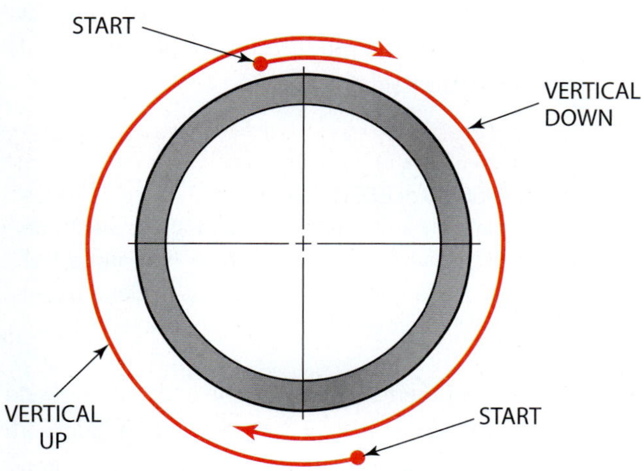

FIGURE 8-17 Most weld imperfections occur when starting or stopping the weld bead, so staggering the weld beads reduces the impact that they would have if they were all in the same area.

PRACTICE 8-4

Filler-Pass 1G Position

Using the weld coupon from Practice 8-3, 3/32-in.- or 1/8-in. (2-mm or 3-mm) diameter E7018 electrodes, a properly set-up and adjusted welding machine, and all the required PPE, you are going to make a filler pass in this practice.

The filler passes will be made with uphill progression because for beginning welders, the likelihood of trapping slag is very high if the electrode is run downhill. Most experienced pipe welders run downhill when codes allow it. For uphill welds, a slight Z-weave can be used for these passes. It is important to pay close attention to the electrode angle and travel speed because both will affect the ability to control the weld pool.

1. Strike an arc between the ten o'clock and eleven o'clock positions in the groove directly on top of the hot pass. As before, maintain a longer-than-normal arc length until the arc stabilizes. Bring the electrode closer to the weld surface to begin the filler pass.
2. Keep the weld face flat or slightly concave by traveling fast enough and using a slight weave so that it does not trap slag along its sides.
3. Keep an eye on the weld pool size (particularly the trailing edge of the weld pool) to ensure that it does not begin to sag as a result of an excessively large weld pool.
4. When the weld has progressed to the twelve o'clock to one o'clock position, stop and roll the pipe so that the end of the weld is at the ten o'clock to eleven o'clock positions. The next weld can begin at that point.
5. Repeat this process until the filler pass is completed around the pipe.
6. Once the filler pass is completed around the pipe, clean the weld thoroughly with a chipping hammer and wire wheel.
7. Have your instructor inspect your work for discontinuities and consistent defect-free welds.
8. Continue making filler passes in this same manner until the weld groove has been filled to within 1/16 to 1/8 in. (1.5 to 3 mm).
9. The weld should have uniform buildup.
10. Repeat this practice until you can make a uniform consistent weld each time.
11. Turn off all equipment, return all tools and supplies, discard all scrap, and clean up the welding booth. ◆

COVER PASS

The **cover pass** is the final weld pass used to complete a weld. It may be made with one single bead or multiple beads, depending on the width of the groove, **Figure 8-18**. They are sometimes referred to as the "cap pass," or "cap." To begin the cover pass, the last filler pass should be about 1/16 to 1/8 in. (1.5 to 3 mm) below the surface of the pipe. The outside edge of the pipe surface and the groove face serve as a guide for you to follow as you make the cover pass. As you weave the pass, you should bring the electrode to the junction of the groove face and surface of the pipe, allowing the arc to melt the base metal to ensure that there will be good fusion along this edge.

FIGURE 8-18 Single-weld-bead cover pass.

In order to have good tie-in of one weld to the next and not have a visible arc strike remaining after the weld is completed, strike the arc in the groove approximately 1 in. (25 mm) ahead of the weld where you stopped. Once the arc is established, bring the electrode back into the weld crater that was left at the end of the bead when you stopped welding. This is where you want to start depositing weld metal and continuing the weave pattern.

PRACTICE 8-5

Cover-Pass 1G Position

Using the weld coupon from Practice 8-4 1G Filler Pass, 3/32-in.- or 1/8-in.-(2 or 3 mm) diameter E7018 electrodes, a properly set-up and adjusted welding machine, and all the required PPE, you are going to make a cover pass in this practice.

1. Strike an arc between the ten o'clock and eleven o'clock positions in the groove directly on top of the filler pass. As before, maintain a longer-than-normal arc length until the arc stabilizes. Bring the electrode closer to the weld surface to begin the cover pass.

> **NOTE**
>
> The cover pass should not have more than 1/8 in. (3 mm) of buildup and be no more than 1/8 in. (3 mm) wider than the groove opening, **Figure 8-19**.

2. The finished cover pass weld face should be uniformly built up so that it curves from one edge of the groove to the other. A weld that abruptly rises along the edge can produce a stress point that could cause future weld cracking.

> **NOTE**
>
> Compared to E6010 or E6011 electrodes, E7018 electrodes produce a relatively small gaseous cloud. If the electrode is weaved excessively (i.e., greater than two and a half times the electrode diameter), the gaseous cloud might not be adequate to prevent weld contamination.

3. Because of the restriction on the width of the weld bead, the cover pass may need to be made with more than one pass. If more than one pass is required, each weld pass should cover the preceding weld pass by approximately one half, **Figure 8-20**.
4. Keep an eye on the weld pool size (particularly the trailing edge of the weld pool) to ensure that it is consistent in size and shape.

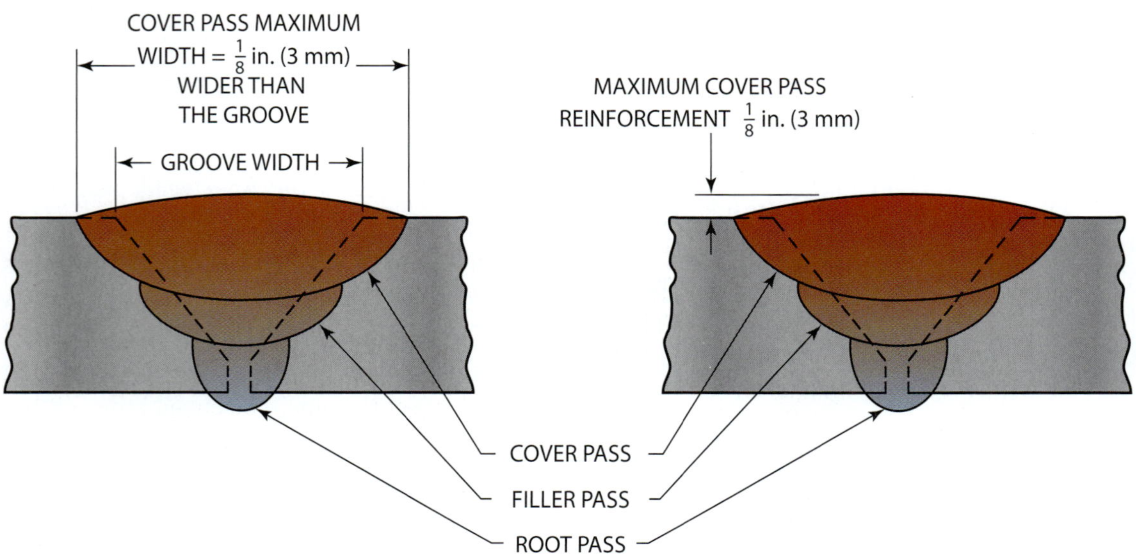

FIGURE 8-19 Cover pass buildup and width specifications.

FIGURE 8-20 Multiple-weld-bead cover pass.

5. When the weld has progressed to the twelve o'clock to one o'clock position, stop and roll the pipe so the end of the weld is at the ten o'clock to eleven o'clock positions. The next weld can begin there.
6. Repeat this process until the cover pass is completed around the pipe.
7. Once the cover pass is completed around the pipe, clean the weld thoroughly with a chipping hammer and wire wheel.
8. The weld should have uniform buildup.
9. Have your instructor inspect your work for discontinuities and consistent defect-free welds.
10. Repeat this practice until you can make a uniform, consistent weld each time.
11. Turn off all equipment, return all tools and supplies, discard all scrap, and clean up the welding booth. ◆

WELD PHYSICS—SURFACE TENSION

"Surface tension" is the force that holds a molten weld pool in place when making an out-of-position weld. An example of surface tension is a small drop of water on the bottom of a glass plate. When the droplet is small, it has a smooth blister shape that tightly hugs the bottom of the plate. However, as the drop increases in size, it will begin to sag as gravity pulls it down. If the droplet gets too large, gravity will actually cause most of it to drip off the plate. This is the same thing that happens with a molten weld pool. If it is kept small enough, it will have a smooth, uniform shape regardless of the weld position; however, if the weld pool gets too large, gravity will cause it to sag. An excessively large weld pool may actually run down the face of a vertical weld or drip off of an overhead weld bead.

2G PIPE-WELDING POSITION

When welding in the 2G position, the pipe will be fixed vertical so that the weld will be horizontal around it. The major challenge with the 2G position is keeping the weld uniformly shaped since gravity tries to pull the molten weld metal down. If the metal is pulled down, the finished weld will not be uniform in shape.

PRACTICE 8-6

2G Position Vertical Fixed

For this practice, you will need 6-in. (150-mm) Schedule 40 or 80 pipe, 1/8-in. (3-mm) E6010 and E7018 electrodes, a grinder, a half-round bastard file, and a wire brush. You will also need PPE for the processes being performed. Prepare your pipe coupons as described previously.

1. Assemble the pipe section as previously described, and grind the ends of the tack welds to a featheredge.

You can now proceed to the root pass.

Root Pass

2. Using an E6010 electrode, start the root pass on one of the tack welds. You can use your index and middle finger to guide the electrode into the joint to make sure that you do not strike the arc outside of the weld groove.
3. Once you start the arc, allow enough time for the arc to establish itself and become stable, before starting your travel.
4. Keep a very short arc by placing the electrode tip down into the root of the joint.
5. It is important to maintain the keyhole at the front edge of the weld pool to insure that the weld will have good root penetration, fusion, and to help build up the weld's root face.
6. The electrode should be held directly in line with the pipe groove and at an approximately 5-degree to 15-degree drag angle.

NOTE

The angle can be varied depending on the amount of penetration and weld reinforcement that is needed. The greater the angle the greater the buildup and the less penetration the weld will have, and a more perpendicular angle will produce less buildup.

7. If the keyhole begins to close up, decrease the drag angle and/or increase the weld amperage. If the

keyhole begins to increase in size increase the drag angle, use a slight whipping motion, and/or decrease the amperage.

8. Ideally you should stop the root weld on top of one of the tack welds by breaking the arc by quickly moving the electrode tip back over the weld bead. It is important not to fill the weld crater by stopping your forward motion for a brief moment as you normally would do when ending a SMA weld. Filling the weld crater would close the keyhole and make restarting harder.
9. Chip and wire brush the root pass. If you were not able to stop the weld on a tack weld, use a thin grinding disk to feather the end of the weld. Feathering the end of the weld bead makes it possible to immediately establish 100% root penetration and reinforcement.
10. When the root pass is completed, inspect the root face for uniform penetration and reinforcement. If an area does not have 100% penetration, a thin grinding disk can be used to remove the root pass in that area so it can be rewelded.
11. The weld should have uniform buildup.
12. Totally clean the weld surface and have your instructor inspect your work for discontinuities and consistent defect-free welds.

You can now proceed to the hot pass.

Hot Pass

13. Grind any excessively large areas of buildup on the root weld and focus on removing the excess convexity and any slag at the toes.

NOTE

If the welds that are being performed are to be used in critical or code weld applications, it is best to use a grinder and remove any discontinuities of large size. The grinder can be used to contour the shape of the root pass and prepare it for the hot pass.

14. Starting by setting the welding amperage slightly higher than the setting you used for the root pass, you are going to use 1/8 in. (3 mm) diameter E6011 electrodes to make the hot pass.
15. To help remove any trapped slag along the toe of the root pass use a T motion with a whip-and-pause technique. The manipulation of the electrode for the T should be only wide enough to ensure that the toes of the welds are broken down by the arc.
16. As with the root pass, the electrode should be held directly in line with the pipe groove and at an approximately 5 to 15-degree drag angle. The arc force will help hold the molten weld pool in place until it solidifies.
17. The stops and starts of the hot pass must not be at the same locations of the stops and starts of the root pass.
18. When it is time to end the weld bead, hold the arc on the weld pool for a fraction of a second to fill the weld crater before quickly breaking the arc back over the weld bead, **Figure 8-21**.

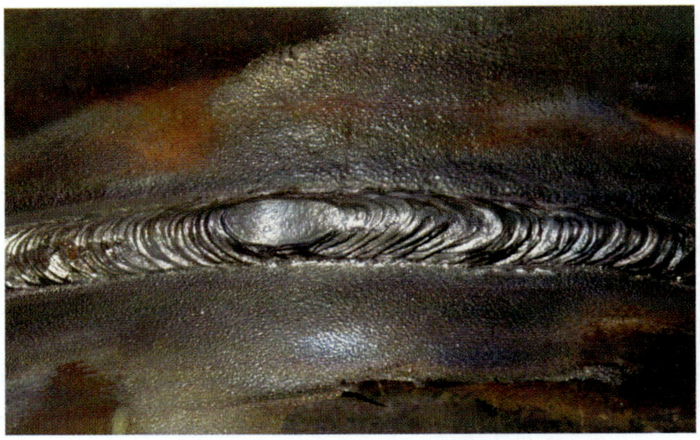

Larry Jeffus

FIGURE 8-21 A properly filled weld crater.

19. The weld should have uniform buildup.
20. Totally clean the weld surface, and have your instructor inspect your work for discontinuities and consistent defect-free welds.

You can now proceed to the filler pass.

Filler Passes

NOTE

Horizontal welds are best made as stringer beads. **Stringer beads** are individual beads that are not weaved and are not much wider than the welding electrode diameter. They are easier to make in this position.

21. The hot pass should be ground to level it out and remove any discontinuities or defects that might remain.
22. The filler passes will be made using a 1/8 in. (3 mm) diameter E7018 electrode.

NOTE

It is better to make a larger number of small weld beads than to make a few larger weld beads in the 2G position because smaller weld beads are easier to control and are less affected by the pull of gravity.

23. The filler pass is made up of a number of weld passes. The first weld pass is made around the back and lower side of the weld groove. Most of the weld metal will rest on the bottom pipe groove face, but the top edge of the weld must not form a deep V groove that can trap slag, **Figure 8-22**.

Larry Jeffus

FIGURE 8-22 A properly shaped filler pass.

24. The electrode will be angled in an upward direction of approximately 30 degrees. The electrode trailing angle should be between 75 degrees and 90 degrees.
25. The second filler pass is made so that most of it is on the upper groove face so that it joins that surface to approximately the center of the first weld pass. By making this weld along the top and back it is almost overhead so that it is not as likely to sag downward from gravity.
26. Do not overheat the pipe, take your time, and allow it to cool if necessary between weld beads and weld passes.
27. Complete the filler pass placing one bead along the lower groove and the next ones above it, **Figure 8-23**.

Larry Jeffus

FIGURE 8-23 The next weld bead will rest on the previous weld bead in the 2G position.

28. Stop the filler passes when the groove is filled within 1/16 in. to 1/8 in. (1.5 to 3 mm) of the groove.

NOTE

The goal is to make sure that the filler passes end up parallel to the outside surface of the pipe. It is important the bottom pass not roll or have metal sagging out beyond the outside surface of the pipe. This will cause excess buildup on the lower welding passes and make the cover pass difficult to level out.

You can now proceed to the cover pass.

Cover Pass

29. The cover pass caps the weld and must have a smooth uniform curved shape arcing from the bottom edge of the groove to the top edge. To accomplish this the cover pass(es) are made just like the filler passes.
30. Grind any high or uneven areas of the filler pass so that it will be easier to make a uniform cover pass.
31. Keeping the cover pass shape and reinforcement uniform is important; for a large part, its appearance is used to judge your whole weld. Use the bottom edge of the groove as a guide, but do not overlap it more than 1/16 in. (1.5 mm). Keeping the weld beads small makes it easier to control them.
32. Remember not to overheat the weld.
33. Place each additional weld pass so that it overlaps the surface of the previous weld by 1/3 to 1/2. Judge the size of each weld pass so that you do not wind up with only a small space along the top edge to fit the last pass.
34. The weld should have uniform buildup.
35. Have your instructor inspect your work for discontinuities and consistent defect-free welds.
36. Repeat this practice until you can make a uniform consistent weld each time.
37. Turn off all equipment, return all tools and supplies, discard all scrap, and clean up the welding booth. ◆

5G PIPE WELDING POSITION

In the 5G position, the pipe is parallel to the horizon and fixed, not rolled as it was in the 1G position. In the 5G position parts of the weld are made in the flat, vertical, and overhead positions. The challenge is that the welds are constantly transitioning between positions, which requires constant changing of the welding techniques and in some cases changes in welding current settings.

PRACTICE 8-7

5G Position Horizontal Fixed

For this practice, you will need 6-in. (150-mm) Schedule 40 or 80 pipe, 1/8-in. (3-mm) E6010 and E7018 electrodes, grinder, half-round bastard file, and wire brush. You will also need PPE for the processes being performed. Prepare your pipe coupons as described previously.

1. Assemble the pipe section as previously described and grind the ends of the tack welds to a featheredge.
2. Once the pipe is tacked together and positioned, you should have your instructor check the fit-up and positioning just as a welding inspector would. Once the instructor has checked the coupon, you can start your weld.

You can now proceed to the root pass.

Root Pass

3. Perform the root pass with the weld progression down. This is referred to as "downhill welding." This is a common pipe-welding technique, especially for pipelines and thinner-walled pipe. One of the main advantages to running downhill passes is that it is faster than welding uphill. Also note that it will be necessary to increase the amperage setting above the setting that would be used for running an uphill bead. The amperage settings will vary from machine to machine and depend on the joint fit-up. In general, for a 1/8-in. (3-mm) E6010 electrode, the amperage will be in the range of 90–100 amps.
4. Using an E6010 electrode, start the root pass on one of the tack welds around the eleven o'clock position and weld through the twelve o'clock position. You can use your index and middle finger to guide the electrode into the joint to make sure that you do not strike the arc outside the weld groove.

NOTE

Breaking the weld into quarters gives you an opportunity to stop and monitor your progress. It also makes it easier to reposition yourself without having to make such a large transition. On large-diameter pipe, breaking the weld into quarters can help control distortion. It can also be controlled by having two welders welding at the same time on opposite sides of the pipe.

5. Once you start the arc, allow enough time for the arc to establish itself and become stable, and then start your travel.
6. Be sure to place the tip of the electrode down into the root of the joint. You should be able to feel the flux on the electrode resting on the root face.
7. The electrode angle that you will use should be slightly trailing to almost a 90-degree angle. The force of the arc will help to hold the weld pool in place. The travel speed should be quick.
8. In order to get complete root face fusion, these surfaces must be melted by the heat of the welding arc.
9. You should be able to see a very small gap between the weld pool and the end of the rod. If the distance is too great, the travel speed is too fast, resulting in the weld pool not bridging the root gap.
10. You should be able to hear the jetting sound of the electrode coming through the root of the weld joint. This is an indication that the keyhole is open and that you are getting 100% penetration. If you cannot hear the jetting sound of the arc, there is a good chance that the keyhole has closed and you are not getting the required penetration.

Once the root pass has been completed, inspect it for defects and discontinuities, as well as any high or low spots. Often with a downhill bead that is not manipulated, the bead will end up convex. Convex weld beads can contribute to wagon tracks. For this reason, it is important that the center be ground to obtain a uniform bead that is flat to a slightly U-shaped groove. Care must be taken not to grind the root pass too thin, or the hot pass may burn through it. In general, high areas should be ground to obtain a uniform bead. All slag should be removed as well. Have your instructor look at the weld and provide feedback.

You can now proceed to the hot pass.

Hot Pass

11. A hot pass will help to ensure penetration and remove small amounts of slag that may be trapped along the toe of the weld. It can be used to help level out the root and will also help to remove and fill any undercut.
12. The hot pass will be made using an uphill progression.
13. Uphill welds travel slower and deposit more weld metal than do downhill welds; therefore, they put more heat into the weld than do downhill welds, so a lower amperage setting should be used for uphill welds.
14. Care must be taken to make sure that the weld pool does not get too large and out of control.

15. Start the hot pass at the five o'clock position and weld through the six o'clock position and continue to the eleven o'clock position.
16. When it is time to stop a weld bead to change electrodes, fill the weld crater and break the arc back over the weld bead.
17. Clean the weld and grind the weld crater to a featheredge to make it possible to produce a seamless tie-in between the two welds. Always follow the recommended procedure for starting a weld.
18. Once the hot pass is complete, clean the weld thoroughly with a chipping hammer and wire wheel mounted on your grinder. Inspect the pass checking for defects and discontinuities that may be present.
19. Have your instructor look at the weld and provide feedback.

You can now proceed to the fill pass.

Fill Passes

20. The fill passes will be made with 3/32-in. or 1/8-in. (2-mm or 3-mm) E7018 electrodes.

NOTE

The size and profile of the groove will determine which size electrode will be appropriate. The smaller, 3/32-in. (2-mm) electrode will allow easier control of the weld pool when out of position. If the groove is larger because of the pipe wall thickness (such as when welding on Schedule 80 pipe as opposed to Schedule 40), it would be better to use a larger electrode. When deciding on the diameter of filler electrode, consider that a large filler electrode in a smaller-profile joint will likely overfill the joint as you are performing the weld. A smaller filler electrode in a larger-profile joint will underfill the joint, resulting in less efficiency in the fabrication process, causing more passes to be made to fill the joint.

21. The fill passes will need to be made with uphill progression. Because of the fluid puddle and flux, the likelihood of trapping slag is very high if the electrode is run downhill.
22. As you begin the fill passes, pay close attention to your start and stop points. You should stagger them so that they do not all end up in the same location.
23. A slight zigzag weave can be used for these passes. It is important to pay close attention to the electrode angle and travel speed. This is especially important at the four o'clock and seven o'clock positions on the pipe. These are the areas where, if you are not careful, the weld pool will become too large and sag. This is why it is important to watch the weld puddle, not the arc.
24. Clean the weld and grind the weld crater to a featheredge to make it possible to produce a seamless tie-in between the two welds. Always follow the recommended procedure for starting a weld.

NOTE

The last filler pass should be about 1/16 in. (1.5 mm) below the surface of the pipe. The number of weld passes needed to fill the joint to that point will be determined by the schedule pipe being welded, the groove profile, the travel speed, and the diameter and type of welding electrode being used. Normally, it will take between two to three fill passes to fill the joint on Schedule 80 pipe.

25. Once the filler passes are complete, clean the weld thoroughly with a chipping hammer and wire wheel mounted on your grinder. Inspect the pass checking for any defects and discontinuities that may be present.
26. Have your instructor look at the weld and provide feedback.

You can now proceed to the cover pass.

Cover Pass

27. The cover pass is the final pass used to complete a weld.
28. Because the effects of gravity are different on different parts of the 5G positions, weave beads are often used for the fill and cover passes since it is easier to control the weld with a weave.
29. The edge of the pipe outside surface and the groove face will serve as a guide for you to follow as you are welding the cover pass.
30. As you weave the pass, you should bring the electrode to the junction of the groove face and surface of the pipe, allowing the arc to melt the base metal to ensure that there will be good fusion along this edge.
31. As you travel around the diameter of the pipe, you will have to change electrodes and restart the weld. Use the recommended procedure for stopping and starting a weld bead so that the finished cover pass is uniform.
32. In order to have a good tie-in of one weld to the next, you want to start approximately 1 in. (25 mm) ahead of the weld where you stopped.
33. The 5G position is a combination of several plate positions that transition from one to the other.

34. Continue to practice, and once you are finished, have your instructor inspect the weld and provide you with feedback.
35. Once the cover passes are complete, clean the weld thoroughly with a chipping hammer and wire wheel mounted on your grinder.
36. Have your instructor inspect your work for discontinuities and consistent defect-free welds.
37. Repeat this practice until you can make a uniform, consistent weld each time.
38. Turn off all equipment, return all tools and supplies, discard all scrap, and clean up the welding booth. ◆

PRACTICE 8-8

6G Position 45-Degree Fixed

In the 6G position, the pipe is placed so that the center line of the pipe is at a 45-degree angle and fixed, **Figure 8-24**. This position is a combination of the 2G and 5G positions, and for this reason, it is considered to be one of the most difficult tests to pass. In most cases, a welder that qualifies in the 6G position will be qualified for all of the other pipe positions. Because access to some welds will be limited, this is a good position to practice using both your dominant and nondominant hand to make welds.

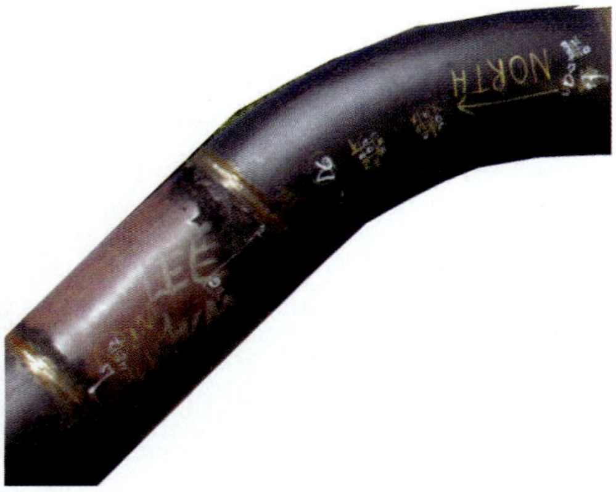

FIGURE 8-24 6G pipe fittings on a cross-country natural gas line.

For this practice, you will need 6-in. (150-mm) Schedule 40 or 80 pipe, 1/8-in. (3-mm) E6010 and E7018 electrodes, a grinder, a half-round bastard file, and a wire brush. You will also need PPE for the processes being performed. Prepare your pipe coupons as described previously. Once the pipe is tacked together and positioned, you should have your instructor check the fit-up and positioning just as a welding inspector would. Once the instructor has checked the coupon, you can start your weld.

Root Pass

1. Perform the root pass with the E6010 electrode.
2. Position the pipe so that a tack weld is near the top of the pipe at about the eleven o'clock or one o'clock position. Strike the arc on that tack weld, and weld over the top of the pipe through the twelve o'clock position, moving either left or right depending on where you begin. Continue the weld down the other side with a downhill progression.
3. Making the weld in quarters will give you an opportunity to stop and monitor your progress. If the pipe has a large diameter, making the weld in quarters will help to control distortion as well.
4. A very slight circular pattern can be used to ensure that the root face is fused and the keyhole remains open. If the keyhole gets too large, some welders will use a slight whipping technique to close it, while others prefer a drag technique.
5. If you use the drag technique, the electrode angle should range from being directed at about a 10-degree angle in the direction of travel to straight in at the center of the pipe. Using this technique, a keyhole will develop; as you travel, the gap will open around the electrode and close on the backside as the weld pool forms and solidifies.
6. Your arc length should be short, and you should be able to just feel the flux on the electrode resting on the root face. The travel speed should be fairly fast.
7. You should be able to see a very small gap between the weld pool and the end of the rod. If the distance is too great, the travel speed is too fast, which will result in the weld pool not bridging the root gap.
8. Once you come to the end of a weld, it is important to chip and clean it, and then feather the end of the bead with a grinder so that a smooth transition can be made from bead to bead.
9. Use the area that you ground to a feather as a sort of on-ramp to get the weld started. This helps to make sure that you maintain uniform weld reinforcement and gives consistent weld penetration and tie-in.
10. Once you have completed the first quarter of the pipe, you can move to the second quarter, which will be opposite the first quarter.
11. Complete the root weld by alternating opposite quarters of the pipe.

> **NOTE**
>
> As you develop your welding skills, try to weld about half the pipe at a time. If you started at the eleven o'clock position and welded through the twelve o'clock position, you will weld to about the five o'clock position. If you started at the one o'clock position and welded through the twelve o'clock position, you will weld to the seven o'clock position.

12. Once the root pass is complete, clean it thoroughly with a chipping hammer and wire wheel mounted on your grinder. Inspect the pass checking for any defects and discontinuities that may be present.
13. Have your instructor look at the weld and provide feedback.

You can now proceed to the hot pass.

Hot Pass

14. The hot passes will be made with 3/32-in. or 1/8-in. (2-mm or 3-mm) E7018 electrodes.
15. The hot passes will need to be made with uphill progression.
16. Stagger the start and stop points so they do not all end up in the same location.
17. A slight zigzag weave can be used for these passes.
18. Clean the weld and grind the weld crater to a featheredge to make it possible to produce a seamless tie-in between the two welds. Always follow the recommended procedure for starting a weld.
19. Once the filler passes are complete, clean it thoroughly with a chipping hammer and wire wheel mounted on your grinder. Inspect the pass checking for any defects and discontinuities that may be present.
20. Have your instructor look at the weld and provide feedback.

You can now proceed to the filler pass.

Filler Passes

21. The fill passes will be made with 3/32-in. or 1/8-in. (2-mm or 3-mm) E7018 electrodes.
22. The fill passes will need to be made with uphill progression using a zigzag weave.
23. The first weld pass is made around the back and lower side of the weld groove. Most of the weld metal will rest on the bottom pipe's groove face, but the top edge of the weld must not form a deep V-groove that can trap slag.
24. The electrode will be angled in an upward direction of approximately 30 degrees. The electrode trailing angle should be between 75 degrees and 90 degrees.
25. The second filler pass is made so that most of it is on the upper groove face so that it joins that surface to approximately the center of the first weld pass.
26. Complete the filler pass placing one bead along the lower groove and the next ones above it. Clean the weld and grind the weld crater to a featheredge before starting the next electrode.
27. Stop the filler passes when the groove is filled within 1/16 in. (1.5 mm) of the groove.
28. Once the filler passes are complete, clean it thoroughly with a chipping hammer and wire wheel mounted on your grinder. Inspect the pass checking for defects and discontinuities that may be present.
29. Have your instructor look at the weld and provide feedback.

You can now proceed to the cover pass.

Cover Pass

30. The cover pass will be made with E7018 electrodes.
31. Grind any high or uneven areas of the filler pass so that it will be easier to make a uniform cover pass.
32. Keep the cover pass shape and reinforcement uniform. Use the bottom edge of the groove as a guide, but do not overlap it more than 1/16 in. (1.5 mm). Keeping the weld beads small makes it easier to control them.
33. Remember to not overheat the weld.
34. Place each additional weld pass so that it overlaps the surface of the previous weld by 1/3 to ½. Judge the size of each weld pass so that you do not wind up with only a small space along the top edge to try to fit the last pass.
35. The weld should have uniform buildup.
36. Have your instructor inspect your work for discontinuities and consistent defect-free welds.
37. Repeat this practice until you can make a uniform, consistent weld each time.
38. Turn off all equipment, return all tools and supplies, discard all scrap, and clean up the welding booth. ◆

Summary

Pipe welding requires a great amount of skill in order to master it. The curve of the pipe requires constant readjustment to make high-quality, consistent welds. Welding is both an art and a science. There are several different styles and techniques that can be utilized and achieve the same or similar results. It is helpful to consult with other welders who have been working in the field for many years. You can often pick up tips that many have tried and have proven successful. Above all, it is important to practice on a consistent basis to improve your skills.

Review Questions

1. What is the purpose of a tack weld?
2. Why is the root pass so critical?
3. What effect does drag travel angle have on a weld, as opposed to push travel angle?
4. Describe the position of the pipe when welding in the 1G position.
5. Describe the position of the pipe when welding in the 2G position.
6. What is the function of the hot pass?
7. Describe the position of the pipe when welding in the 5G position.
8. Which diameter rod allows the most control of the weld pool when welding out of position, the 3/32 in. or 1/8 in. (2 mm or 3 mm)?
9. When welding in the 5G position, what two areas on the pipe are most likely to experience sagging of the weld pool?
10. When making fill passes with E7018 electrodes in the 5G position, why is it suggested that welds be made with vertical progression uphill?
11. Why is a weave bead preferred for welds made in the 5G position?
12. When making the hot pass, the amperage does not always need to be increased, so why would it be considered a hot pass?
13. Why would a larger-diameter electrode be used for welding on heavier-walled pipe?
14. Describe the welding position of the pipe when it is in the 6G position.
15. Why would it be important to stagger the starts and stops in a weld, as opposed to having them all in the same location?

Chapter 9

Gas Metal Arc Welding of Pipe

OBJECTIVES

After completing this chapter, the student should be able to:

- List the four main types of metal transfer.
- Explain how to adjust the welding machine variables to obtain the different types of metal transfer.
- List the steps necessary to prepare pipe prior to welding it.
- Explain the importance of tack welds, including the number and quality required.
- Explain the importance of weld nozzle angle.

KEY TERMS

Axial spray transfer

Electrode extension

Gun angle

Inductance

Mismatch

Pinch effect

Pulsed-arc metal transfer

Short circuit metal transfer

Work angle

INTRODUCTION

Gas metal arc welding (GMAW) was originally developed as a process for welding aluminum. Through the years, technology has changed the process into one that can perform a variety of tasks. Some of the elements that have shaped this process are the wider variety of consumables available today. The sizes and types of these filler metals have opened the door to new applications. Changes in shielding gases as well as changes in power sources have led to improved operator appeal and reduced discontinuities for a wider variety of applications.

The integration of digital electronics with GMAW machines has led to a new metal transfer process called Surface Tension Transfer (SST)®, which was first introduced by Lincoln Electric. Other equipment manufacturers have now introduced similar processes. All of these new processes are known generically as "modulated current welding processes."

TYPES OF METAL TRANSFER

The GMAW process is unique in that there are six modes of transferring the filler metal from the wire to the weld. Each mode of metal transfer has its own unique characteristics. The mode of metal transfer is the mechanism by which the molten filler metal is transferred across the arc to the base metal. The modes of metal transfer are:

- Short circuiting transfer (GMAW-S)
- Axial spray transfer
- Globular transfer
- Buried-arc transfer
- Pulsed-arc transfer (GMAW-P)
- Modulated current

Essential variables such as filler wire size, type, amperage, voltage, wire feed speed, and shielding gases all affect the type of metal transfer that is obtained. Each type has specific benefits that make it useful to certain applications. This flexibility is one of the reasons that the gas metal arc welding process is so versatile.

Pulsed-arc and modulated current transfer methods require welding machines and wire feeders that are specifically designed to provide the specialized current and wire feeder control needed to perform these methods of metal transfer.

Short Circuit Metal Transfer (GMAW-S). This method of metal transfer has the advantage of low heat input, which allows for welding of thin materials. Given the correct welding parameters, it allows for welding materials as thin as 0.024 in. (0.6 mm) and up to about 3/16 in. (4.7 mm) without edge preparation. The process allows for all-position welding, as well as progression downhill.

The first mode of metal transfer for the gas metal arc process was axial spray transfer using 100% argon shielding gas. As development of the process continued, the use of small amounts of oxygen was introduced. The oxygen provided a more stable arc and enabled the use of the process on ferrous metals. During the 1950s, improvements in the welding power sources and the use of smaller-diameter electrodes made possible the development of the short circuiting mode of metal transfer. This mode of transfer allowed for the welding of thinner sections of metal and reduced the amount of heat input. It also allowed welding to take place in all positions.

There are four basic stages in the short circuit transfer—the short circuit, the pinch, the arc, and droplet formation, **Figure 9-1**. **Short circuit metal transfer** occurs when the electrode comes into physical contact with the molten weld pool or solid base metal upon initiation of the weld. Once the contact is made, the welding voltage rapidly reduces because there is a dead short and the resistance is lower. Without the arc resistance, the amperage starts to increase. **Inductance** is the property in an electrical circuit that directly affects the rate of change in current flow. The rate of change of the current flowing through the arc decreases as the inductance increases. It has the effect of dampening the harsh arc restarting, which also reduces weld spatter. Since the GMAW-S process can occur 20 to 200 times a second, inductance is only slowing down the change by milliseconds.

Some welding power sources have the inductance preset, while other power sources can be adjusted. On some power sources, this adjustment control is referred to as the "pinch." The **pinch effect,** the second stage of the short circuit transfer process, is the force that squeezes the molten end of the electrode off so that the droplet can separate to restart the arc. The pinch effect is caused when the electromagnetic field around the molten end of the filler becomes strong enough to cause it to neck down and separate. The third stage of the short circuit transfer

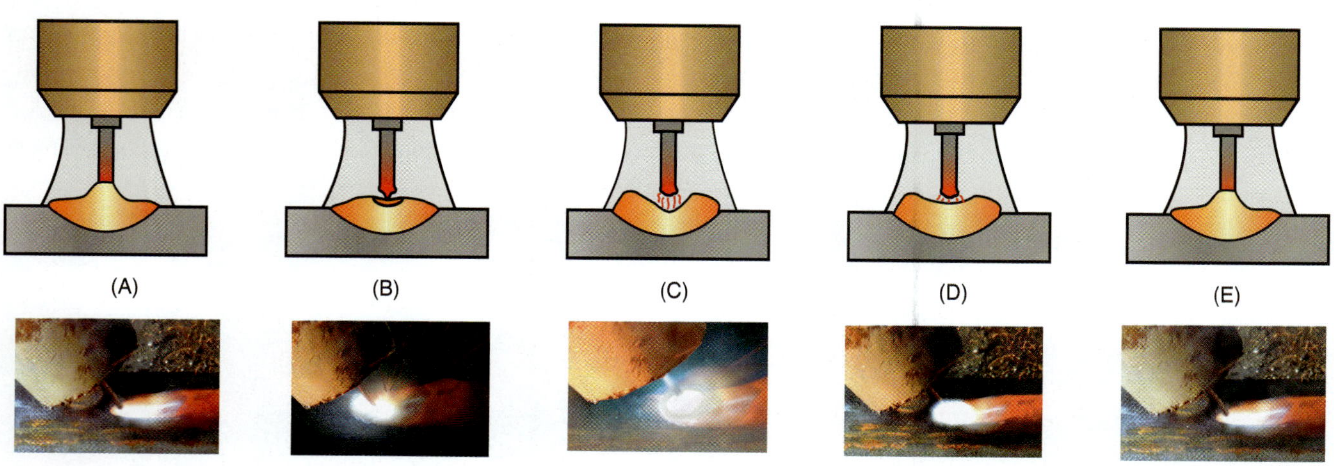

FIGURE 9-1 Short circuiting transfer.

process occurs when the small droplet of molten filler metal is separated, allowing it to be absorbed into the molten weld pool. The fourth and final stage is when the arc is reestablished and its resistance begins to cause the current flow to reduce. The end of the filler wire begins to melt, forming a small ball that quickly grows in size until it is slightly larger than the wire diameter. It is at this point that the end of the electrode contacts the molten weld pool and the process begins again.

GMAW-S has the lowest current (heat) of all the GMAW transfer methods. The low heat input to the weld can result in lack of fusion along the sides. For that reason, many pipe fabricators have allowed short-arc transfer to be used only on lower-strength, less critical applications.

Globular Metal Transfer The globular metal transfer process is identified by a large molten glob (ball) of metal that forms on the end of the electrode. When it gets large enough, gravity pulls it off the end of the electrode, where it drops into the weld pool, **Figure 9-2**. This metal transfer method was popular in the 1960s and 1970s. It uses a solid core wire and carbon dioxide (CO_2), or a mixture of carbon dioxide and argon for shielding gas.

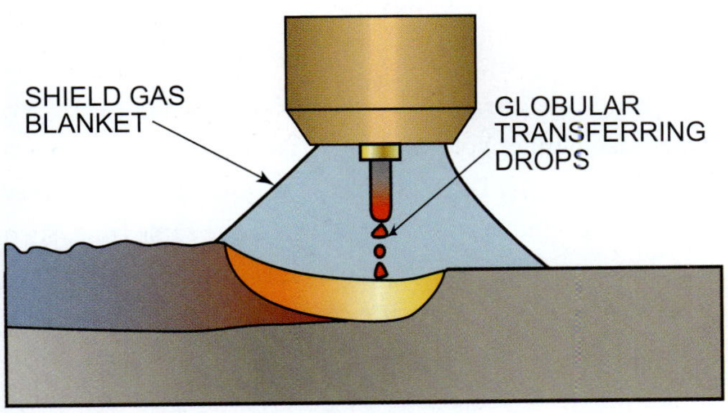

FIGURE 9-2 Globular transfer.

Axial Spray Transfer The **axial spray transfer** process is identified by the pointing of the wire tip from which hundreds of very small drops are projected per second axially (in the center of the arc) across the arc gap to the molten weld pool, **Figure 9-3**. Because of the sound and appearance of the metal transfer, this process is sometimes called "spray arc" or "spray transfer." It takes place at higher current and energy levels than short circuiting and globular transfer. It uses a solid or metal core wire, along with a mixture of shielding gases such as argon and oxygen or argon and carbon dioxide with CO_2 percentages of less than 20%. The amount of current necessary is just above the transition level required for globular transfer.

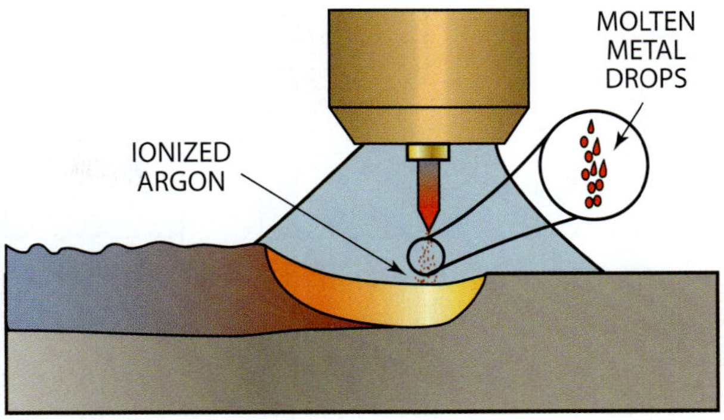

FIGURE 9-3 Axial spray transfer.

This current will result in a steady stream of drops being propelled by arc forces at high velocity in the direction that the wire is pointing, so it can be used in vertical and overhead positions without losing control of the transfer. Because the drops are very small and directed at the molten weld pool, the process is spatter free.

The high current levels give it deep penetrating capacity with only a slight buildup, so it tends to have fewer fusion issue defects and requires less cleanup.

Pulsed-Arc Metal Transfer (GMAW-P) The **pulsed-arc metal transfer** method uses a specially designed welding machine that can provide a rapidly pulsing current. The power supply cycles between a peak current (the high current) and a background current (the lower current), **Figure 9-4**. The average current for a pulse spray weld is figured using the peak and background currents, along with the duration of each. It is possible to control the metal transfer to one metal droplet per pulse using this process.

Some of the advantages of the GMAW-P transfer include the fact that by pulsing the current, it is possible to reduce the overall heat input to the weld and still achieve complete joint fusion. Pulsed arc also has little or no spatter. It is more adaptable to robotic and automated welding applications.

Buried-Arc Transfer This process has a very forceful arc that is driven below the surface of the molten weld pool, **Figure 9-5**. With the shorter arcs, the drop size is small, and any spatter produced as the result of short circuits is trapped in the cavity produced by the arc. This process is used for high-speed mechanized welding of thin sections.

Modulated Current Metal Transfer This is the newest GMAW process. It uses an extremely modified version of GMAW-S and has a very sophisticated computer-controlled welding machine. The controller and welding machine together are capable of providing the voltage and amperage required to produce the seven different parts of the complex waveform of power needed to smoothly

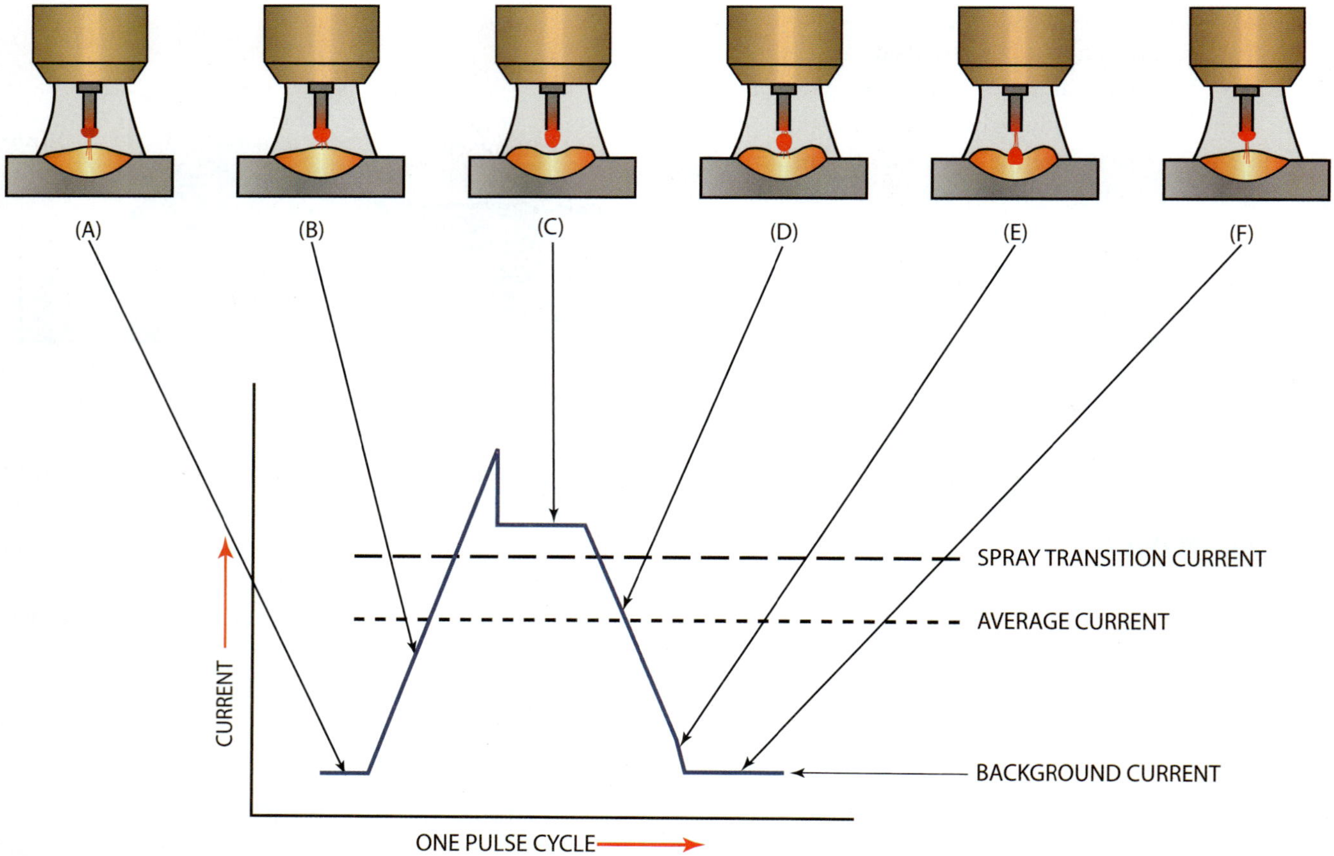

FIGURE 9-4 Pulsed-arc transfer.

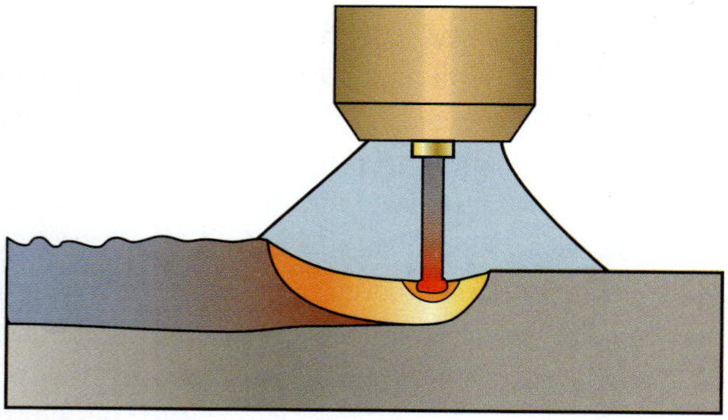

FIGURE 9-5 Buried-arc transfer.

transfer the molten metal when making a weld, **Figure 9-6**. This method of metal transfer provides the welder an amazing amount of control and efficiency when making pipe welds.

WELDING, VOLTS (POTENTIAL), AND AMPS (CURRENT)

The terms "voltage," "volts," "potential," and "electromotive force" all relate to the same aspect of electricity, which is the electrical pressure. The terms "amperage," "amps," "ampere," and "current" also refer to an aspect of electricity, which is volume. Volts and amps are independent properties of electricity, For example, it is possible to have very high voltage and very low amperage at the same time, such as when a static electricity spark jumps from your finger to a doorknob. It is the pressure of the electrons (voltage) that forces them across the inch or so gap between your finger and the doorknob. Also, it is possible to have very high amperage and at the same time only have a low voltage. For example, a stick welder can produce 350 amps or more, but it only has an electrical pressure of around 20 volts. The combination of high voltage and high amperage can be deadly (e.g., lightning).

The combination of volts and amps produces heat. The basic formula for heat is volts $\times$ amps $=$ joules (heat). If either the volts or amps increase, so does the heat; and if either decreases, the heat does likewise.

CC Versus CP Welding Machines

On shielded metal arc welding (SMAW) machines, a welder can set the desired amperage (current), and it will remain constant during the weld. SMAW machines are referred to as "constant current (CC) machines." There are no voltage controls on CC machines. However, on GMAW

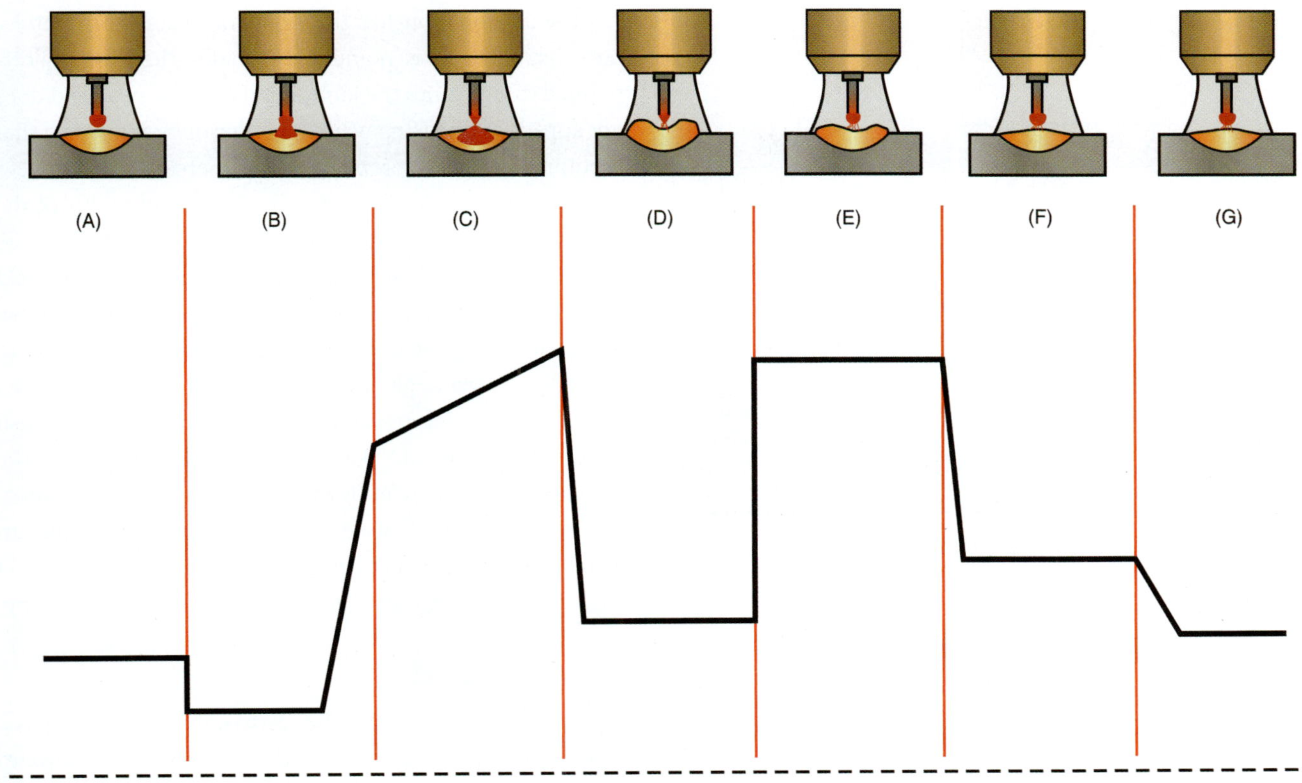

FIGURE 9-6 Modulated current transfer.

machines, a welder can adjust the voltage and it will remain constant during the weld. GMAW machines are referred to as "constant voltage (CV)" or "constant potential (CP)" machines.

If the welding leads of a CC machine were electrically shorted together, the amperage would remain about the same, and the voltage would decrease a little. But if a CP welder's leads were shorted together, the voltage would remain constant, and the amperage would uncontrollably soar until it reached the machine's maximum output.

Understanding CP and GMA Welding

The voltage on a GMAW machine is set directly by turning a knob, but changing the wire feed speed indirectly sets the amperage. The faster the wire is fed, the higher the amperage has to rise to melt the larger quantity of filler metal entering the arc. As the wire feed speed increases, the arc length and weld bead penetration decrease and weld buildup increases. As the voltage increases, the arc length increases and weld bead penetration and weld buildup decrease. In other words, there must be a balance between the voltage, amperage, and wire feed to produce an acceptable GMA weld. It is important to stay within the range of settings or else your welds will most likely fail.

GMAW VARIABLES

In the GMAW process, it is essential to understand other variables such as electrode extension, work and gun angles, travel speed and direction, electrode positioning, and the impact that they have on the weld.

Electrode Extension

Electrode extension is the distance from the end of the contact tube to the arc measured along the welding electrode, **Figure 9-7**. Electrode extension is often referred to as **stickout.** As the electrode extension increases, some of the welding heat is absorbed in the filler wire. The loss of this heat results in a weld that has less penetration and more buildup. If the electrode extension is extreme, the weld may have no fusion with the base plate. A shorter electrode extension will result in an increase in penetration and a reduction in buildup.

By carefully watching the molten weld pool during a root pass, it is possible to prevent burnthrough by raising the gun to lengthen the electrode extension, which will reduce penetration. Or if it appears that the root weld is building up too much and may not be getting complete fusion, the gun can be lowered to reduce the electrode extension and increase the penetration.

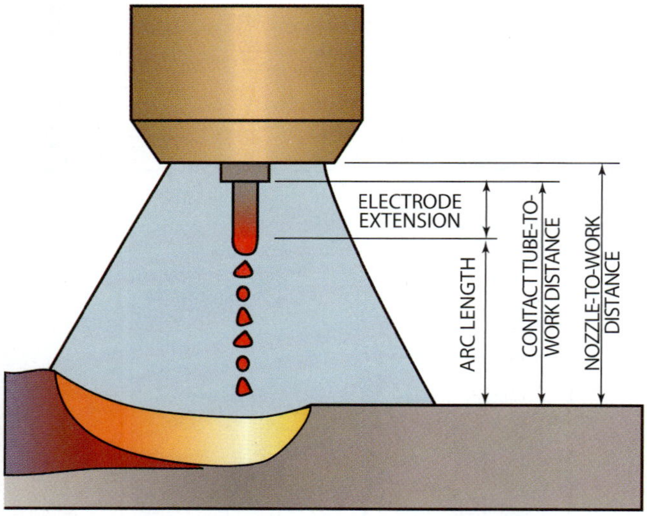

FIGURE 9-7 Electrode-to-work distance diagram.

Work Angle and Gun Angle

Changing both the work angle and the gun angle can control the weld bead penetration and buildup without having to change the machine settings.

The **gun angle** is the angle between the center line of the gun and the weld as it relates to the direction of travel, **Figure 9-8A**. The **work angle** is the angle between the center line of the welding gun and the work, **Figure 9-8B**.

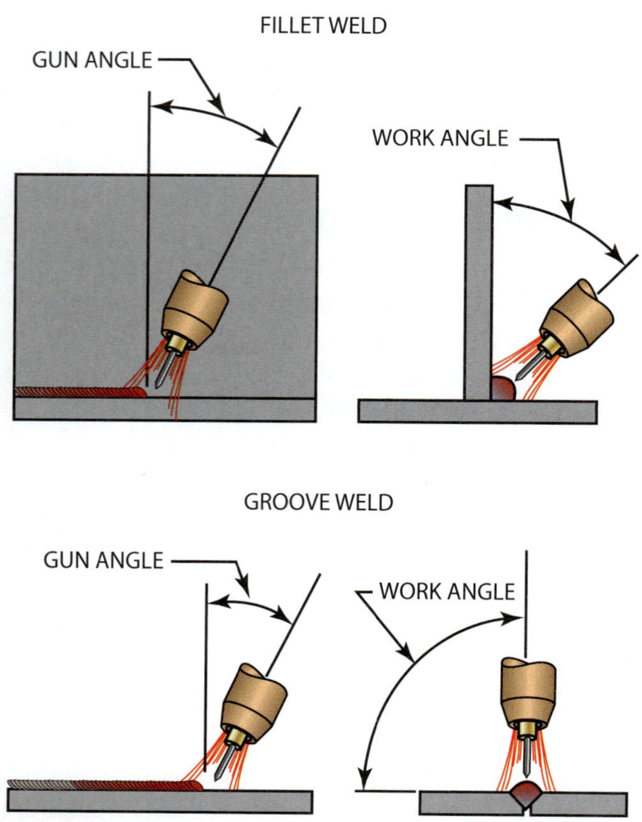

FIGURE 9-8 Travel and work angles.

The arc force pushes the depth of fusion in the direction that the gun is pointed. Normally, the gun angle is pointed directly into the joint; but if one side of a weld does not appear to be fusing well, the gun can be rotated so that it points toward the problem area to increase the fusion.

The same arc force affects the depth and width of the weld being produced when the gun angle is changed. Knowing this is significant so that the welder can control the weld pool and penetration as a weld progresses around a pipe. It is common to use a trailing angle, sometimes referred to as a "drag travel angle," for vertical down-welds in the 5G position. A leading travel angle, sometimes called a "push travel angle," refers to pointing the welding nozzle in the direction of travel. Leading angles can be used in the vertical up-position of 5G welds to help ensure that the molten metal stays in place. Both push travel angle and drag travel angle penetration are affected by the size of the root opening.

Travel Speed

Travel speed is very important, especially for the root pass. If the travel speed is too slow, then the excessive buildup it causes can prevent the welding heat from penetrating to the root face, resulting in a lack of root fusion. An excessive travel speed can produce a very thin root weld that is incapable of withstanding the postweld shrinkage forces, and it will crack along its center line.

Electrode Positioning

The electrode must be positioned so that it and the arc are directed onto the leading edge of the molten weld pool to keep a smooth even metal transfer in GMAW-S. It is essential to maintain a steady weld nozzle placement and angle and focus on keeping the welding wire close to the leading edge of the weld pool. This will help to ensure that uniform penetration is maintained, **Figure 9-9**. If the welding

Larry Jeffus

FIGURE 9-9 Keep the electrode close to the leading edge of the weld pool.

wire is placed too far ahead in the weld pool, it is likely to blow through the joint, resulting in pieces of welding wire sticking through the root of the joint. These bits of wire are referred to as "whiskers." But if the wire is positioned too far back on the trailing edge of the molten pool, there may not be adequate heat along the leading edge to fuse the weld completely.

TACK WELDS

As noted already in previous chapters, the importance of tack welds cannot be understated. The diameter of the pipe to be welded will determine the number of tack welds that will be necessary to maintain joint alignment during welding. For smaller-diameter pipe, such as less than 4 in. (100 mm), as few as three tacks may be used. For pipe diameters larger than 4 in. (100 mm), four or more tacks may be necessary. For complete root fusion and root reinforcement, the tack weld must be no less than flush and have no more than 1/8 in. (3 mm) of buildup to meet many codes. The tacks should be free of defects and ground down to a gradual feather taper on both ends.

ROOT PASS

As discussed in previous chapters, the **root pass** is the first weld pass that is made, and the foundation that all other passes will be laid on. It serves to establish penetration, root fusion, and seal the joint. It is important to take the time to properly prepare the pipe by grinding the weld face and root face smooth and free from gouges and oxides. Also, make sure that the fit-up is proper, with a uniform root opening all the way around the pipe before starting to weld. This will help to produce a consistent weld more easily. The root pass is made using the short circuiting mode of metal transfer. It is essential to make sure that the root edge of the groove is melted and fused with the root weld bead. Failure to achieve complete penetration and fusion results in failed welds and weld tests.

Backing Gas

In some cases, depending on the type of pipe being welded, it can be desirable and necessary to have a backshielding or purge gas. Metals such as stainless steel, titanium, and other reactive metals should have the back purge gas to protect the molten weld pool from damaging gases such as nitrogen and oxygen. This can be accomplished by using a purge dam on both sides of the weld joint. Several different types of dams are available. They can be sophisticated, such as a bladder dam type system that is inflated with the same argon that is used to purge the pipe. Simpler systems may be used as well, such as cutting a piece of plywood to fit the inside diameter (ID) of the pipe. When welding smaller-diameter pipe, water-soluble or thermally disposable barriers are useful. It is essential when using such devices that you place them an appropriate distance from the weld heat to prevent damaging the devices while welding. It is important to tape off the outside of the weld joint and use an oxygen sensor during purging to ensure that an acceptable purge is achieved.

WELDING PRACTICES

The GMAW practices in this chapter are designed around the basic GMAW machine technology. Once you have mastered the technique required to make welds with the basic GMAW machines, only a slight change in your techniques will be required to make welds using the advanced GMAW process with pulsed-arc transfer and modulated current machines. The setup for some of these computer-controlled advanced machines only requires that you enter the information for the filler metal, base metal, shielding gas, and joint geometry, and then the welder sets itself.

PRACTICE 9-1

Fillet Weld

Using a constant voltage power supply properly set up according to the manufacturer's specifications for GMAW, .035-in. (0.9-mm) ER70-S filler wire, wire cutters or mig pliers, channel lock-type pliers, a grinder, wire brush, and all the required personal protective equipment (PPE), you are going to weld pipe to plate in this practice.

Fillet welds are used to join a pipe to a socket weld flange, blank-off plate, or base plate, or in other similar architectural or design applications. Blank-off plates are used to seal the end of a pipe when the pipe is being used for a handrail or other similar application and where the pipe will not be carrying a fluid. Base plates are used to attach a pipe structure to another structure, such as a construction column of a concrete floor, a railing on a ferry, or a roller coaster support column, **Figure 9-10**.

In most of the cases discussed here, the fillet weld used to join the plate to the pipe end must be strong enough to withstand the physical forces applied to the structure, **Figure 9-11**. Often it is impossible to make a weld on the inside of the pipe, so the outside fillet weld must have complete root fusion and be large enough to hold up under service forces.

The term "2F position" refers to a fillet weld made in what most welders would consider a flat position. A flat fillet weld is one where the face of the weld is horizontal, which

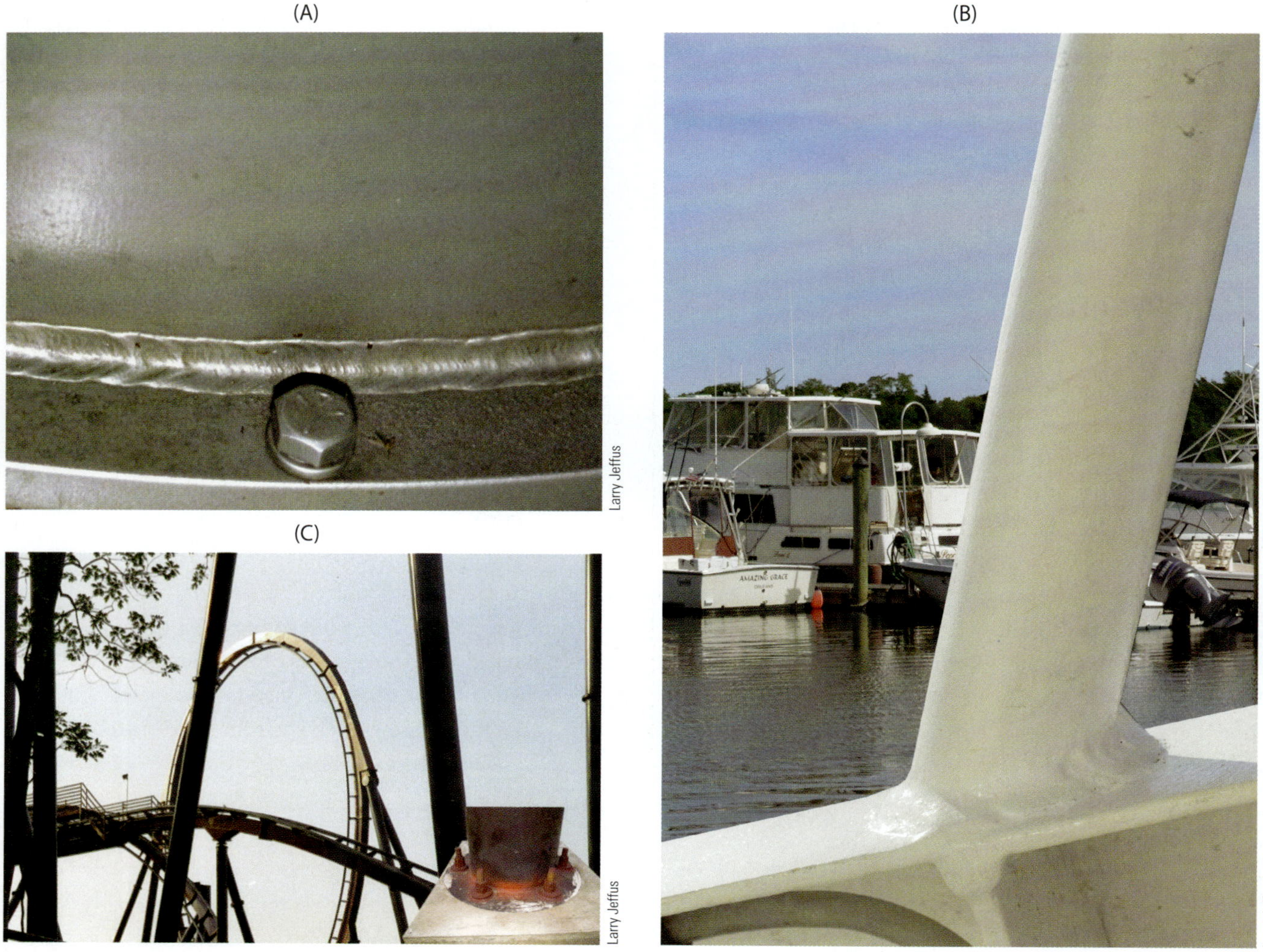

FIGURE 9-10 Fillet welds on pipe. (A) Water tank leg; (B) railing on ferry boat; (C) roller coaster support column.

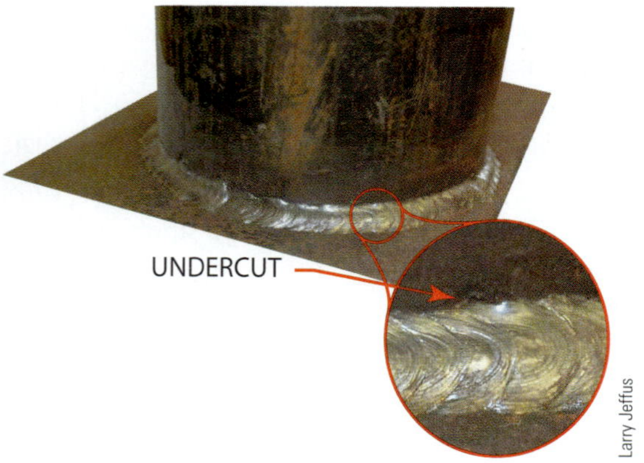

FIGURE 9-11 Undercut caused by momentarily allowing the arc to wander out of the joint; it instantly melted the edge of the pipe.

would require that the pipe be held at a 45-degree angle, **Figure 9-12A.** However, most often a fillet weld would be made in the 90-degree angle shown in **Figure 9-12B.** In Practice 9-1, the pipe may be welded in either a 45-degree or 90-degree angle.

1. The end of the pipe may or may not be beveled for welding. Beveling is often used on thick-walled pipe that is to be used in severe service. For this practice, however, a bevel is not required.
2. Tack-weld the pipe to the plate at four locations with 1-in. (25 mm)–long tack welds. Make sure that the roots of the tack welds are completely fused.
3. Grind the ends of the tack welds to a featheredge. Use a thin grinding stone so that the side of the pipe is not ground away or the plate gouged.
4. Start the weld on top of one of the tack welds so that the weld is fully formed and hot enough to fuse the plate and pipe surfaces all the way to the root. Watch the leading edge of the weld to make sure that there is not a notch forming, which would indicate a lack of root fusion.
5. As the weld progresses, make sure the electrode is making contact with the molten weld pool in the correct spot and at the correct gun angle. Keeping these

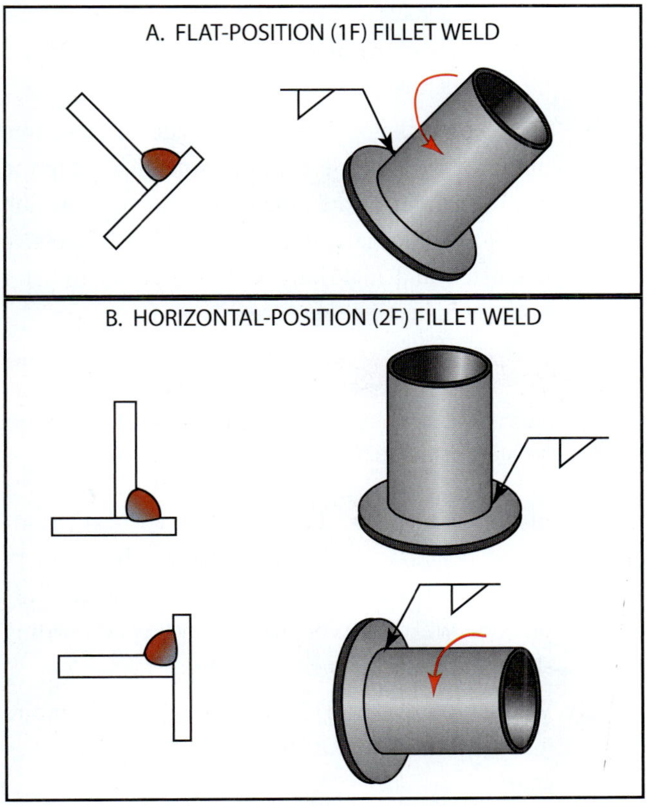

FIGURE 9-12 Flat versus horizontal pipe fillet welds.

items consistent as the weld progresses around the pipe can be challenging, but it is very important to do.

6. Continue welding until you have reached the middle of the next tack weld.
7. Reposition the pipe and start the next weld at the end of the last one, on top of the tack weld.
8. Continue welding as before until the root weld around the pipe is completed.
9. Grind the starting and stopping points if they are more than 1/16 in. (1.5 mm) above the rest of the root weld.
10. Shift the pipe so that the next weld pass is not being started at the same location as the starts and stops of the root pass.
11. To taper the fillet weld's starting point gradually increase the size of the weld as you begin moving along the joint by slowly increasing the side-to-side weave until the weld bead has reached the desired size. This tapered start will make it possible to have a smooth/flush ending to the weld when it is completed around the pipe.
12. Continue the weld pass until you have traveled approximately one-quarter of the way around the pipe. At that point, start traveling a little faster so that the weld size tapers down. This will make restarting the weld easier without having excessive buildup.
13. Repeat this process until the weld pass is completed around the pipe. Make three more weld passes around the pipe and plate.
14. Repeat this practice until a uniform weld with complete root fusion can be consistently made. ◆

PRACTICE 9-2

Pipe 1G Position, Rolled

For this practice, you will need 6-in. (150 mm) Schedule 40 or 80 pipe, .035-in. (0.9-mm) ER 70-S wire, a grinder, and a wire brush. The use of a tapered welding nozzle will reach farther into the joint, which helps to give the proper stickout for the root pass. You will also need PPE for the processes being performed.

Prepare pipe coupons using the following technique:

1. Mild steel pipe is often prepared by beveling with the oxy-fuel process.

NOTE

Read and follow all the pipe-cutting equipment manufacturer's instructions, and also see Chapter 7, "Thermal-Cutting Processes," for details on cutting pipe bevels.

NOTE

There are several common bevel angles for pipe preparation, such as 30 degrees, 37.5 degrees, and 45 degrees. This practice will use a 30-degree bevel angle; this will result in a 60-degree groove angle.

2. Once the beveller is positioned to 30 degrees, position the torch on the pipe where the cut is to be made.
3. Adjust the torch to a neutral flame and make your cut.
4. Remove any scale and oxide from the cut surfaces and adjacent areas by grinding and wire brushing.
5. Light slag can be removed from the edge of the cut surface by using a half-round file and running it from the inside of the pipe, holding it at a low angle and knocking lightly attached slag from the inside of the pipe.
6. Check to see that the end of the pipe is flat and square with the outside diameter (OD) of the pipe. You can check this by laying the pipe down on a flat surface and checking for light coming from under the surface. You can also use a framing square and

place it on the OD of the pipe, and put the other end over the cut edge of the pipe.

NOTE

In most cases, if the beveling equipment is in good working order, your cut should be square, and you should see little to no light coming from under the root face.

7. Before starting to grind, you need to be wearing all the required PPE for grinding and following the grinder and grinding wheel manufacturer's safety and operating instructions.
8. Use an 8-in. (200-mm) grinder to square the end of the pipe by holding the pipe securely, placing the grinder flat on the surface of the pipe root face, and lightly applying pressure on the pipe so that a small chamfer is formed on the root face surface. Grind a little at a time, and keep checking until the end of the pipe is square.

NOTE

Grind as little as possible so the chamfer is 1/8 in. (3 mm) or less in size.

9. Before grinding the end of the pipe, make sure to secure it by clamping it to a table using a vise or other appropriate clamping tool.
10. Grind the groove face to bright, shiny metal. This ensures that surface oxides have been removed. When you start to grind the groove face, be sure to position the grinder so that it touches the surface and maintains the correct 30-degree bevel angle.
11. Apply a light pressure on the metal surface with the grinder and follow the contour of the pipe surface. You should move the grinder back and forth at a quick speed a distance of about a quarter of the circumference of the pipe.
12. Rotate the pipe and work the grinder around the entire diameter of the pipe. As you change directions while moving the grinder back and forth around the contour of the pipe, be sure to release pressure from the surface to prevent flat spots and gouges on the groove face.
13. Grind oxides from the outside of the pipe adjacent to the joint for a distance of about 1 in. (25 mm).
14. Check the inside of the pipe for scale and oxides, and use a rotary file, a sanding flap, or both to clean the inside of the pipe to a bright, shiny surface.
15. Once you have cleaned the oxide off the inside and outside of the pipe, you are ready to fit the two pieces together.
16. Check to make sure that the inside surfaces line up and there is little to no **mismatch** (where the inside surfaces of the pipe do not line up, causing high-low on the alignment of the pipe). Excessive mismatch and high-low make it difficult to get a uniform, defect-free root weld.

NOTE

If the pipe is not seamless, lining up the seams will often help eliminate mismatch.

17. In the 1G position, the pipe is placed in the horizontal position but can be rolled so that most of the welding takes place near the flat position. This practice will help you to improve your welding skills going around a curved surface.
18. Welding on the 6-in. (150-mm) pipe will require it to be tacked in four different places, 90 degrees apart.
19. Take a bare welding rod the thickness of the desired root opening and bend it into a V-shape. Take one of the pipe coupons and lay it down bevel side up.
20. Place the welding rod on top of the pipe, and then place the remaining pipe coupon on top of the rod and other coupon, bevel side down.
21. Line up the two pieces so that the inside of the pipes match up and give a smooth, uniform fit-up with approximately a 1/8-in. (3 mm) root gap and about a 3/32-in. (2 mm) root face.
22. Keep in mind that as the tack welds are made, the welds will shrink as they cool, causing the gap to close slightly.
23. The machine should be adjusted for short circuit metal transfer since it is assumed that you will be using a conventional constant voltage machine.
24. The voltage should be set between 17 and 18.5 volts and a wire speed of approximately 200–300 in. per minute (500–750 cm per minute).
25. Make the first tack weld, ensuring that the two coupons are fused together at the weld. Place the second tack 180 degrees from the first tack.
26. Check to make sure that the alignment and root opening are staying consistent.
27. Once the first two tack welds are made, you can remove the rod and place the third and fourth tack

welds halfway between the first two welds. This will give you four tack welds, spaced approximately 90 degrees apart.

28. The tacks should be 1–1.5 in. (25–38 mm) long.
29. Grind a taper on both ends of the tack weld down to a featheredge.
30. Fit and tack the pipe together as previously described in the TACK WELD paragraphs above.
31. The root weld is started on one of the tack welds.
32. Position the pipe so that you will weld starting between the two and three o'clock positions and weld to between the ten and eleven o'clock positions.
33. Use both hands while performing the weld: one hand to hold the GMAW gun and the other hand to steady yourself so you can have a smooth, constant travel speed and welding gun–to-work distance.
34. It is important to maintain a consistent nozzle angle as you travel around the diameter of the pipe. Since this position is primarily flat, it is recommended that a drag travel angle of approximately 10–15 degrees in the direction of travel be used.
35. The closer to the two or three o'clock position (where the travel around the pipe is closer to the vertical position), the more need there is to use a slight push angle to offset the effects of gravity.
36. Continue working, rotating the pipe to the same position each time.
37. Repeat this process until the root pass is complete.
38. Repeat this practice until you can make a smooth, consistent weld each time.
39. The weld should have uniform buildup and penetration, as well as fusion to the root and along the toe of the weld.
40. Have your instructor inspect your work for discontinuities and consistent defect-free welds.
41. Turn off all equipment, return all tools and supplies, discard all scrap, and clean up the welding booth. ◆

PRACTICE 9-3

2G Position, Vertical Fixed

For this practice, you will need 6-in. (150-mm) Schedule 40 or 80 pipe, .035-in. (0.9-mm) ER 70-S wire, a grinder, and a wire brush. The use of a tapered welding nozzle will reach farther into the joint, helping to give the proper stickout for the root pass. You will also need PPE for the processes being performed. Prepare your pipe coupons as described in Practice 9-2.

When welding in the 2G position, the pipe will be vertical to the horizon and fixed so that the weld will be horizontal.

1. Fit the pipe with approximately a 1/8-in. (3-mm) root opening and approximately a 3/32-in. (2-mm) root face, and make four equally spaced tack welds, as described in Practice 9-2.
2. Position the assembled pipe fitting, and have your instructor check the fit-up and positioning as a welding inspector would before you start your welds.
3. Start the root weld on one of the tack welds. Once the molten weld pool has formed, start your travel.
4. Keep the wire electrode focused on the front third of the weld pool.
5. It is essential to make sure that the root of the weld is being completely fused on both sides of the joint, so a slight up-and-down motion may be used to bridge the gap between the two weld coupons.
6. Do not use excess manipulation of the weld gun, as this can cause the root bead to become too large and may cause undercut on the top of the weld or excess roll on the bottom half of the bead.
7. The welding gun should be positioned so that wire feeds into the front third of the weld pool. The gun should be slightly at a drag travel angle. Note the following:
    a. If there is too much weld buildup on the inside of the pipe, you could be using too much drag travel angle and/or too slow travel speed.
    b. If too little weld reinforcement is on the inside of the pipe, you may be using too much push travel angle or too high travel speed.
    c. Likewise, if the wire is being fed too far away from the leading edge of the weld pool, there tends to be a lack of weld reinforcement on the inside of the pipe.
8. Once the root pass is in place, inspect it for convexity and tie in at the toe of the weld.
9. The appearance of the root bead should be relatively flat across the face of the weld and should be fused into the groove face.
10. There should not be any lack of fusion of the weld onto the groove face. If you suspect that there may not be fusion along the toe, that area should be ground out.

11. If the weld is excessively convex, the bead should be ground down to ensure that the following welds can tie in and have proper fusion.
12. All starts and stops of the weld bead should be ground to ensure proper tie-in and fusion. This will also help to ensure uniform buildup on following weld passes.
13. Have your instructor look at the weld and provide feedback.

Hot Pass

The next pass is most often referred to as the "hot pass." However, this does not necessarily mean that the welding machine settings are going to be considerably hotter than when running the root pass. The hot pass is typically placed over the root pass. It is essential that the toe of the root pass and the toe of the hot pass not lie on top of one another. The toe of GMA welds can have discontinuities such as lack of fusion. This is the reason why the welds should not have the toes overlapping. It is important not to carry too large a weld pool when putting in the fill passes. This can result in cold lap and lack of fusion.

14. Set the welding machine voltage and wire feed speed a little higher than was used for the root pass.
15. The hot pass will help to ensure fusion and penetration of the root pass.
16. Care must be taken to make sure that the weld pool does not get too large and out of control.
17. Using a slight elliptical pattern and not carrying too large a bead are important to prevent the weld from sagging or resulting in a lack of fusion caused by the weld bead rolling on top of the base metal.
18. As with other GMA welds, it is essential to keep the wire feeding into the leading edge of the weld pool.
19. Inspect the pass checking for defects and discontinuities that may be present.
20. Have your instructor look at the weld and provide feedback.

Filler Pass

The fill passes should be made as stringer-type beads. When putting in the fill passes, it is essential that the beads not be placed too far out on the bottom surface of the groove face too soon. This causes excess buildup on the bottom toe of the weld and increases the likelihood of possible excess buildup or excess roll on the final pass. It also leads to the likelihood of undercut or lack of fill on the final top stringer bead. For this reason, it is a good idea to add slightly more fill on the topside of the groove than on the bottom side during the first layers.

21. Start by placing the first fill pass above the hot pass, tying it in by overlapping the hot pass by about half a bead width.
22. The preceding passes should then be stacked from bottom to top.
23. The fill passes should tie in at both toes.
24. If there is excess roll or lack of fusion, the weld should be ground down. Only grind enough to ensure that the lack of fusion has been removed and the weld contour is consistent.
25. It is important to keep the toe of the second weld from lying directly on top of the toe of the previous weld, **Figure 9-13**. If this cannot be avoided, then that area needs to be ground down to eliminate any lack of fusion.

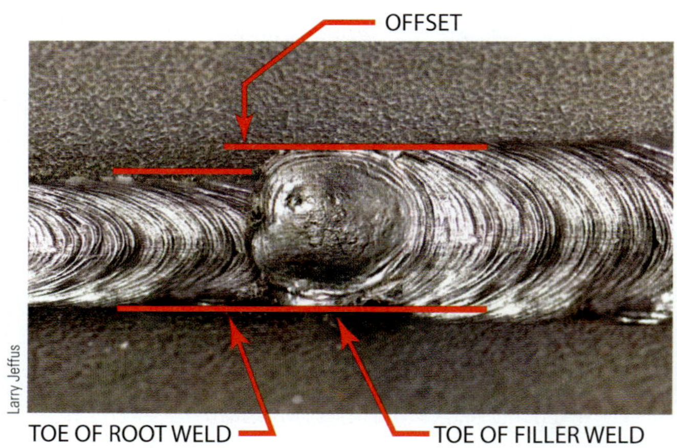

FIGURE 9-13 The bottom toe of the second weld is incorrectly lined up on top of the toe of the first weld, which can trap impurities.

26. It is best to overlap one bead on another by approximately one-half.
27. Continue to place fill stringer beads in the groove until the joint is filled to the desired level of approximately 1/16 to 1/8 in. (1.5 to 3 mm) below flush.
28. It is essential to make sure that if there is an interpass temperature requirement, that it is not exceeded during the fill and prior to the cap pass. It is a good idea to allow the pipe to cool. If the temperature of the pipe is too high when the final weld pass is made, it will often make the weld pool hard to control and develop a tendency to sag.

29. The cover pass or cap is the finishing weld that completes the joint, and it is used to finish filling the groove and produce a uniform weld width and reinforcement. This pass is normally made as a series of narrow weld beads placed side by side, but sometimes it can be made as a single, wide bead.
30. Inspect the pass checking for defects and discontinuities that may be present.
31. Have your instructor look at the weld and provide feedback.

Cover Pass

The cover pass is the final weld pass. It will be capped off with filler pass stringer beads. Although in some welding positions, it can be made as a single weld in the 2G position, it is almost always made with multiple stringer beads. The number of stringer beads required depends on the size of the groove that is to be welded.

It is essential to allow some time for the pipe to cool down between layers. If the pipe overheats, the stringer beads will sag, similar to when the welding amperage is too high. The weld gun angle is important with both the fill and cap passes. If the gun angle has too much trailing angle, the bead will tend to have excess roll. This can be corrected, and the roll reduced, by changing the angle to direct it straight at the center of the pipe.

32. Just as you would build a wall by laying a foundation at the bottom and building upward, do the same with the cover pass. Make the first weld along the bottom edge, and then each following weld just above it until the cover pass is completed.
33. You should use the edge formed by the groove face as a guide to maintain a straight weld bead around the pipe.
34. It is also essential to make sure that the surfaces are melted by the heat of the arc. This can be achieved by directing the electrode on the edge of the molten weld pool closest to these surfaces. There are many different oscillation patterns that can be used. One is a cursive lowercase "e" pattern; no matter what pattern you choose, though, be sure not to work it too wide.
35. Travel speed and welding gun angle are very important to control the size of the weld and ensure that it does not have excessive roll or overlap.
36. It is vital to stagger the starts and stops so that they do not end on top of the starts and stops of the previous passes, and to make sure that each bead ties into the previous bead.
37. Each bead should overlap the previous one by approximately one-third to one-half. By starting at the bottom of the weld groove and using that edge as a guide, you now use each previous bead as a guide as you work around the diameter of the pipe. This will help to keep the beads parallel with the weld groove.
38. Occasionally check the interpass temperature to see that the weld is not being overheated. Allow it to cool if necessary.
39. Once you are finished with the weld, allow the piece time to cool down. It should have uniform buildup and penetration, as well as fusion to the root and along the toe of the weld.
40. Have your instructor inspect your work for discontinuities and consistent defect-free welds.
41. Repeat this practice until you can make a smooth, consistent weld each time. Pipe welding takes a lot of time and practice to master.
42. Turn off all equipment, return all tools and supplies, discard all scrap, and clean up the welding booth. ◆

PRACTICE 9-4

5G Position, Horizontal Fixed

In the 5G position, the pipe will be parallel to the horizon and fixed. It will be the same position as the 1G, except that the pipe will be stationary. This position gives the challenges of welding in the flat, vertical, and overhead positions.

For this practice, you will need 6-in. (150-mm) Schedule 40 or 80 pipe, .035-in. (0.9-mm) ER 70-S wire, a grinder, and a wire brush. The use of a tapered welding nozzle will allow the welder to reach farther into the joint, helping to give the proper stickout for the root pass. You will also need PPE for the processes being performed. Prepare your pipe coupons as described previously in Practice 9-2.

Tack Weld

1. Fit the pipe with approximately a 1/8-in. (3-mm) root opening and approximately a 3/32-in. (2-mm) root face, and make four equally spaced tack welds, as described in Practice 9-2.
2. Once the pipe is correctly fit and the root gap is the desired uniform amount, place the pipe in the horizontal fixed position. Position it so that the tack

on the top side is in approximately the one o'clock position. Have your instructor check the fit-up and positioning of the pipe before continuing.

Root Pass

The application of the pipe will typically determine the progression of welding. Cross-country pipelines are typically welded with downhill progression. Powerhouse and similar applications are usually welded with the progression up. For this practice, the root pass will be performed with travel progression down.

3. Start welding on the tack weld that is in the one o'clock position. Check to see that you can move freely from the one o'clock position through the six o'clock position, stopping at the seven o'clock position.
4. Start the weld on the tack weld, using a drag travel angle; your weld gun angle should be between 15 to 20 degrees in the direction of travel.
5. As you transition from the top of the tack weld to the open root, and as you progress more in a downhill angle, you will want to increase your weld gun angle to between 45 and 55 degrees. This will help to hold the weld pool in place as you travel down.
6. As you travel down, you will need to make sure that both sides of the root face are melted and fusing during welding. It may be necessary to use a small oscillation of the gun to ensure that the edges are being fused.
7. Stop when you have reached the seven o'clock position tack weld.
8. Prior to starting welding on the other side, you will want to grind both the start and stop points of the weld. They should be ground to a tapered edge, which will allow you to weld on top of the tack welds.
9. Start welding again on the tack at the one o'clock position and weld through the twelve o'clock position until you reach the tack that is at the seven o'clock position.
10. It is recommended to weld through the twelve and six o'clock positions to help provide a more uniform weld.
11. Have your instructor inspect your work for discontinuities and consistent defect-free welds.
12. Repeat this practice until you can make smooth, consistent, and defect-free root welds each time.

Fill Passes

The remaining fill and cap passes will be performed with the weld progression in the uphill direction. It may be necessary to slightly increase the welding settings because after the root pass has been completed, there is more thermal mass to absorb the welding heat. If the settings are not increased, there may be a lack of fusion between the filler passes.

13. On material with thickness of 1/4 in. (6 mm) or more, an electrode extension of approximately 1/2 in. (13 mm) will yield a weld with better fusion.
14. Weld passes should not be much wider than 1/2–5/8 in. (13–16 mm)
15. Continue making weld passes until the beads are approximately 1/16 in. (1.5 mm) below flush with the surface of the outside of the pipe.
16. As you start the fill passes, it is important to control your travel speed to prevent the weld pool from becoming too big and starting to sag. It also helps to control the thickness of the passes. Passes that become too thick can be plagued with fusion problems.
17. Start welding in an area that is not directly on previous starts or stops. By not having all the starts and stops in one area, it decreases the likelihood of a weld failure.
18. Stagger the starts and stops by welding through the five o'clock position, on to the seven o'clock position, and finally the eleven o'clock position.
19. The fill passes can be welded using an inverted-V technique or a triangle technique.
20. It is important to stay close to the leading edge of the weld pool to ensure that fusion is taking place, regardless of which technique you use. Keeping the heat of the arc as close to the leading edge as possible is also necessary to ensure weld fusion.

Cover Pass

21. Use the edge of the groove face to act as a guide for applying the cap pass. The final bead should be uniform in width, and reinforcement should have a height of no more than 1/8 in. (3 mm).
22. It is essential that the edge of the groove face at the top surface of the pipe be melted and fused with the cap bead. Weld bead overlap can occur if the weld pool is allowed to flow out onto the top surface of the pipe without fusing. In a bend test, this lack of fusion can cause the cap pass to fail.

23. Continue working around the pipe until the joint is filled.
24. Once you are finished with the weld, allow the piece time to cool.
25. Repeat this practice until you can make a smooth, consistent weld each time.
26. The weld should have uniform buildup and penetration, as well as fusion to the root and along the toe of the weld.
27. Have your instructor inspect your work for discontinuities and consistent defect-free welds.
28. Turn off all equipment, return all tools and supplies, discard all scrap, and clean up the welding booth. ◆

PRACTICE 9-5

Pipe 6G Position, 45-Degree Fixed

The 6G position is approximately halfway between the 2G and 5G positions. The pipe's center line is fixed at a 45-degree angle. This simulates several different positions on one piece of material. The 6G position is most likely the most difficult position to weld in.

For this practice, you will need 6-in. (150-mm) Schedule 40 or 80 pipe, .035-in. (0.9-mm) ER 70-S wire, a grinder, and a wire brush. The use of a tapered welding nozzle will reach farther into the joint, helping to give the proper stick-out for the root pass. You will also need PPE for the processes being performed. Prepare your pipe coupons, as described in Practice 9-2.

1. Fit the pipe with approximately a 1/8-in. (3-mm) root opening and approximately a 3/32-in. (2-mm) root face, and make four equally spaced tack welds, as described in Practice 9-2.
2. The root gap must be consistent all the way around the groove. Root gaps that vary in width can cause lack of fusion or burnthrough.
3. Have your instructor inspect your work for discontinuities and consistent defect-free welds.

Root Pass

The root pass will be performed with the weld progression down. This is a common technique for welding of cross-country pipelines. One of the main advantages to running downhill passes is that it is faster than welding uphill.

4. The welding machine should be set for short circuit metal transfer. This will require 17–19 volts and approximately 200–250 in./min. (500–640 cm/min.) wire feed speed.
5. The root pass can be made by starting at about the eleven o'clock position or the one o'clock position and welding through the twelve o'clock position, progressing downhill.

NOTE

In the field, welds like this are done with only two sections, but for large-diameter pipes and for new welders, you can make the weld in quarters.

6. Making welds in quarters gives you opportunities to stop and monitor your progress. Also, making welds in quarters will help to control distortion.

NOTE

As your skill increases, you will be able to weld half of a 6- or 8-in. (150- to 200-mm) pipe without stopping.

7. Always grind your start and stop points on your welds. Once you have completed the first quarter of the pipe, you can move to the second quarter.
8. This weld is made in the opposite direction to the first quarter. Remember that all root passes will be with a downhill progression.
9. Complete the root weld by alternating opposite quarters of the pipe. Work toward the goal of welding about half of the pipe at a time. If you started at the eleven o'clock position and welded through the twelve o'clock position, you will weld to about the five o'clock position. If you started at the one o'clock position and welded through the twelve o'clock position, you will weld to the seven o'clock position. Work toward the goal of never stopping or starting directly at the twelve o'clock or six o'clock position.

Once the root pass has been completed, inspect it for defects and discontinuities, as well as any high or low spots. The bead should be flat to slightly concave. Prior to running the hot pass, it is important to clean and grind the root pass. Any areas that have cold lap or lack of fusion should be ground, removing the lack of fused weld metal. It is essential to make sure that the root bead has a smooth contour that allows the next pass to fuse in and be flat to slightly concave. If the bead is too convex, it will increase the likelihood of lack of fusion and that wagon tracks would be found on an x-ray. If your bead ends up convex, grind it to obtain a uniform bead that is flat to slightly u-groove-shaped. Care must be taken to not grind the root pass too thin, or else the hot pass may burn through it.

Hot Pass

10. The hot pass should be made at close to or somewhat higher settings than the root pass. You should slightly increase your voltage settings and wire feed speed.
11. The hot pass will help to ensure fusion and penetration of the root pass. The hot pass will be made using an uphill progression. The uphill progression will promote deeper penetration and fill.
12. Care must be taken to make sure that the weld pool does not get too large and out of control. The technique will be similar to the one used on the 2G position.
13. Using a slight elliptical pattern and not carrying too large a bead is important to prevent the weld from sagging or getting lack of fusion caused by the weld bead rolling on top of the base metal.
14. As with other GMA welds, it is important to keep the wire feeding into the leading edge of the weld pool.
15. The weld should have uniform buildup and penetration, as well as fusion to the root and along the toe of the weld.
16. Have your instructor inspect your work for discontinuities and consistent defect-free welds.

Fill Passes

Upon completion of the hot pass, it is important to check and make sure that the beads are fused in well at the toe of the welds. A technique that is similar to the one used in the 2G position will be helpful.

17. It is essential to start your fill passes high enough on the hot pass face so that excessive buildup does not occur along the bottom edge of the groove.
18. Using an elliptical pattern and slightly slowing down at the top of the pattern will add a little more weld metal to the top of the weld to help counteract gravity's tendency to pull the weld bead downward, which can overfill the bottom edge. If the bottom edge of the weld is overfilled, it will make it more difficult to achieve a smooth, uniform cover pass.
19. The beads should be placed in the joint as stringer beads and should overlap each other between a third and a half of the previous bead.
20. You should fill the joint to where it is within about 1/16 in. (1.5 mm) of being filled. This will give you a guide to follow for making the final pass.
21. The weld should have uniform buildup and penetration, as well as fusion to the root and along the toe of the weld.
22. Have your instructor inspect your work for discontinuities and consistent defect-free welds.

Cover Pass

The cover pass will be the final weld pass and will be capped off with stringer beads. The number required will depend on the size of the groove that is to be welded. Once again, it is important to stagger the starts and stops so that they do not end on top of the starts and stops of the previous passes.

23. It is critical to make sure that you allow some time for the pipe to cool in between layers. If the pipe overheats, the stringer beads will sag, similar to when the welding amperage is too high.
24. Weld gun angle will be important with both the fill and cap passes.
25. If the gun angle has too much trailing angle, the bead will tend to have excess roll. This can be corrected and the roll reduced by changing the angle to direct it straight at the center of the pipe.
26. The weld should have uniform buildup and penetration, as well as fusion to the root and along the toe of the weld.
27. Have your instructor inspect your work for discontinuities and consistent defect-free welds.
28. Turn off all equipment, return all tools and supplies, discard all scrap, and clean up the welding booth. ◆

MODULATED CURRENT TRANSFER

The modulated current transfer method of GMAW has made a major improvement in the efficiency and quality of GMA pipe welds. In addition, it can be used to make x-ray quality welds in other metals, such as stainless steel and nickel, all while reducing the spatter and fumes associated with conventional GMAW.

Modulated Current Process

Both Lincoln Electric and Miller Electric offer welding equipment designed to produce the highly specialized welding current profile needed to perform modulated current transfer welding. There are some differences in how each manufacturer's systems operate, but the theory used to reduce spatter and increase fusion while controlling heat input is similar. In both systems, the wire feed speed is constant and the welding current is modulated, controlled up and down, at the same rate that metal is being transferred. The control of the modulated current process is far more sophisticated than with the traditional GMAW-P process.

In the modulated current process, a computer senses when the droplet of molten metal is almost ready to separate from the filler wire and instantly reduces the welding current, Figure 9-6D. Once the arc has been reestablished, the computer increases the current to the higher level for a softer restarting of the arc. The higher current level of the reestablished arc allows it to melt back the filler metal to form a molten ball of metal on the end. After the molten ball is created, the current level is reduced so the end of the wire and molten ball are pushed downward into the molten weld pool.

Effect of Modulated Current on Welds

The three major effects that modulated current provides are reduced spatter, lower heat input, and better fusion.

Reduced Spatter In modulated current transfer, because the current is lowered as the molten ball of metal is transferred into the weld pool, the explosive reestablishment of the arc is softer, occurring over a very short period of time. This stops the explosive effect that throws tiny bits of molten metal (i.e., spatter) out of the weld zone of the traditional GMAW-S. The current is lowered a second time just before the molten ball makes contact with the surface of the weld pool. Since the welding current was lowered before the short between the filler and base metal, this traditional GMAW-S explosive event is also eliminated.

Lower Heat Input Although the molten weld pool can be larger with modulated current, there is actually less heat input to the base metal as a result of the periods of lower current flow during two stages of the metal transfer. The smoother transfer of metal means that the molten pool is not as violently pushed around by the explosive shorting and reestablishment of the arc, which allows a larger molten weld pool to be easily controlled.

Better Fusion Even with the lower heat input, modulated current obtains better root penetration and bead fusion than does traditional GMAW-S because the molten weld pool fluidity can be accurately controlled. This is most evident when pipe root welds are being made because complete fusion will occur even on the inside edge of the root face and inside pipe wall.

PRACTICE 9-6

Modulated Current Pipe Weld 6G Position

For this practice, you will need 6-in. (150-mm) Schedule 40 or 80 pipe, .035-in. (0.9-mm) ER 70-S wire, a grinder, and a wire brush. According to manufacturer specifications, a properly set-up GMAW machine capable of modulated current welding. You will also need PPE for the processes being performed.

Root Pass

1. Prepare the pipe coupons according to the equipment manufacturer's recommendations for bevel angle, root face, and root opening.
2. After tack welding is completed, grind the ends of the tack welds to a featheredge.
3. Place the pipe in the 6G position.
4. Start the root pass on a tack weld around the twelve o'clock position.
5. Use a 5-degree to 10-degree trailing angle with a 3/8–5/8-in. (10–16 mm) electrode extension.
6. The electrode can be pointed in the center of the weld and does not have to be pointed at the leading edge.
7. Use a side-to-side motion at the beginning of the weld. This will prevent the molten metal's surface tension from pulling it back toward the larger mass of metal. It will also allow the welding heat to become established so that complete fusion will occur.
8. Once the weld has been established and the weld pool is flowing evenly down the groove, the side-to-side weave can be stopped.
9. As the weld passes the nine o'clock position, gravity may begin to pull down on the center of the molten weld pool. A slight side-to-side motion can be used to help keep the molten metal in place.
10. Stop the weld when you have reached the center of the bottom tack weld.

NOTE

The end of any weld that does not end on a tack weld must be ground down to a featheredge to prevent lack of fusion or other discontinuity in the weld from occurring as the weld is restarted.

11. Repeat this process to make the root weld on the other side.
12. The face of the root weld should be flat or slightly concave, with a uniform buildup.
13. Grind starts and stops and any areas that have higher reinforcement, such as those that might occur over tack welds.

Filler Passes

14. Make the filler passes in the same way as you did with the GMAW-S process.

Cover Passes

15. The cover pass is made using the same downhill technique used in the filler passes. Be sure not to build up the cover pass more than 1/8 in. (3 mm).
16. Have your instructor inspect your work for discontinuities and consistent defect-free welds.
17. Repeat this practice until you can make a smooth, consistent weld each time. Pipe welding takes a lot of time and practice to master.
18. Turn off all equipment, return all tools and supplies, discard all scrap, and clean up the welding booth. ◆

Larry Jeffus

FIGURE 9-14 Brittle fracture of pipe to plate fillet weld.

Summary

Settings and the technique for welding in the 6G position will be similar to the 2G position. Just as in the 2G position, stringer beads should be used to help control the weld pool size and bead shape. Welding variables such as stickout, gun angle, voltage, and wire feed speed will affect the final outcome. Keep practicing, and do not get discouraged. Pipe welding requires a great amount of skill to master. The curve of the pipe requires constant readjustment to make high-quality, consistent welds. Welding is both an art and a science. There are several different styles and techniques that can be utilized and achieve the same or similar results. It is often helpful to consult with other welders who have been working in the field for many years. You can often pick up tips that many have tried and have proved successful. Above all, it is important to practice consistently to improve your skills.

It is essential to make pipe welds that are defect free because even small defects can result in catastrophic failure. This pipe support on an office chair, **Figure 9-14**, suffered a brittle fracture—the part shows no bending. The pipe-to-plate weld broke so cleanly that the two parts fit back together like pieces of a puzzle. This failure was the result of a small undercut all the way around the toe of the weld.

Pipe welds in a piping system, like the chair support, are subjected to cyclic loading. Cyclic loading is much like repeatedly bending a wire; bend it enough times and it will break. If the wire has a small nick, it will break even more quickly. The chair pipe support breaking was an inconvenience, but a pipe weld in a refinery failing can be disastrous.

Review Questions

1. Name the four common types of metal transfer discussed in this chapter.
2. List three advantages of GMAW-S.
3. What effect does increasing inductance have on the welding current?
4. Which form of metal transfer depends on gravity for its operation?
5. What is the main disadvantage of the axial spray process?
6. Name four advantages of the pulsed spray mode of metal transfer.
7. What is electrode extension, and by what other name is it sometimes called?
8. As electrode extension increases, what happens to the weld?
9. With decreased electrode extension, what happens to the weld?
10. What are whiskers?

Chapter 10

Flux Cored Arc Welding of Pipe

OBJECTIVES

After completing this chapter, the student should be able to:

- List the types of fluxes and their applications.
- Explain the American Welding Society (AWS) classification of carbon steel flux cored wire.
- Explain why flux cored wire has higher current densities for the same machine settings and how this affects the deposition rate.
- Explain why the flux core arc welded (FCAW) process can be a good choice for pipe-welding applications.

KEY TERMS

Backhand technique
Deposition rate
Dual shield
Electrode extension
Machine welding
Self-shielded

INTRODUCTION

The flux cored arc welding (FCAW) process was developed in the mid- to late-1950s. It is very similar to the gas metal arc welded (GMAW) process. The main difference is the use of a tubular wire electrode that contains either a rutile-based flux that is acidic and designed for welding in all positions or a lime-based flux that is alkaline and used for welding in flat and horizontal positions.

The American Welding Society (AWS) classification for flux cored electrodes is similar to that of the gas metal arc welding classification. It uses an "E" to signify "electrode," then a set of two digits followed by a "T," and then another number that is preceded by a hyphen. The first number following the "E" gives the minimum tensile strength in ten thousands of pounds per square inch. The next number signifies the position that the weld can be performed in. A "0" signifies flat and horizontal fillets and a "1" signifies all positions. The "T" indicates that the wire is tubular in shape, indicating that it is designed to contain flux. The last digit gives information related to the chemical composition and the operating characteristics.

Some wires are designed to be welded using an external shielding gas known as **dual shield**, while flux cored wires are designed to be welded using no external shielding gas, known as **self-shielded**. New manufacturing processes and technologies have given the industry a greater number of wire sizes. Flux cored wire is available in diameters ranging from 0.028–0.062 in. (0.3–1.6 mm)

When the correct welding settings are employed, fluxed cored wires have the advantage of having higher deposition rates than either gas metal arc welding (GMAW) or the shielded metal arc welding (SMAW) process. At lower current settings, the wire melts and forms a globule on the end just before melting off. When the current density is higher, the wire will pinch down to a needlepoint.

APPLICATION OF FCAW TO PIPE

Some of the characteristics of the FCAW process are:

- It is a low hydrogen welding process, and today, stronger pipe such as X70 grade and higher are being used. Pipe of this grade is more sensitive to hydrogen cracking.
- The flux cored process has one of the most efficient deposition rates of all welding processes. The **deposition rate** is the amount of metal that can be deposited in the weld joint in a given unit of time. For this reason, it makes the process attractive to companies involved in pipe construction.
- It is easily adaptable to machine process applications. Machine welding applications require less operator skill than manual welding applications. **Machine welding** is when welding is done by equipment that is under constant supervision, monitoring, and control by a welding operator.
- At one time, the process would have been limited to in-shop, 1G welds only. Improvements in welding equipment and manufacturing techniques for producing the flux cored wire and a wider array of filler metals and flux alloys have increased the use of FCAW for pipe.
- A variety of elements can be added to the flux to change the chemistry of the flux so that specific metallurgical and mechanical properties can be achieved. Some alloys can be added to the flux core to help give the weld strength and impact resistance. Deoxidizers can help to reduce the amount of preweld cleaning that is necessary. Fluxes can be designed to produce a fast-freezing slag that helps to hold the molten weld pool in place while welding out of position.
- Some fluxes are made for single pass, and others can be used on multipass welds. In many cases when a single pass filler wire is used to make a multipass weld, the weld can crack, become porous, or have other defects because of excessive buildup of alloys in the filler wire or flux.
- The flux cored arc welding process does not lend itself to making root welds without some type of backing or reinforcement. This is typically in the form of a root weld made with some other welding process or the use of some type of backing bar or strip. In some industries, such as shipbuilding, ceramic backing is often used, and once the weld is made, the backing strip is removed. For most pipe applications, the root weld is made using some other low-hydrogen process such as GMAW or gas tungsten arc welding (GTAW).
- The gas metal arc process is preferred because of the speed at which the weld can be performed and the ability to control the heat input.
- Because the process is fast, it is more difficult for a welder to move at the speeds that would be required on smaller-diameter pipe. For this reason, the process is better applied to larger-diameter pipe where time savings would be realized more easily.

Root Pass Welds

In all the practices in this chapter, the root welds are to be performed with the GMAW process. The root weld should be applied using short circuit metal transfer or with one of the newer modulated short circuit current types of metal transfer such as Lincoln Electric's Surface Tension Transfer (STT)® or Miller Electric's Regulated Metal Deposition (RMD)®. Both of these systems require specific equipment that, in operation, will provide improvements over regular short circuit metal transfer. One of the advantages is that both will produce a thicker root pass weld. This will be an advantage when welding with the FCAW process by providing a pass that will reduce the chance of burnthrough. If necessary, the root pass can be made using E6010 electrodes and cleaned up using a narrow grinding wheel.

It is important to remember to preheat the pipe if the ambient temperature is below 70°F (21°C), especially if the pipe is greater than Schedule 80. Once the root pass is complete, it is important to grind the weld to make sure that there is not cold roll or lack of fusion. Refer to Chapter 9, "Gas Metal Arc Welding of Pipe," as a review.

PRACTICE 10-1

1G, Position Rolled

For this practice, you will need 6 in. (150-mm) Schedule 80 pipe, .035-in. (0.9-mm) E70T or E71T wire, a grinder, and a wire brush. You will also need personal protective equipment (PPE) appropriate for the processes being performed and a welding shade lens of a minimum number 10 and suggested number 11 or 12, depending on the amperage at which welding is taking place. Refer to ANSI Z49.1 for specific recommendations.

Hot Pass

1. Grind the root pass to make it uniform, but care must be taken to make sure that the root pass is not ground

too thin. This can result in excessive melt-through on the first pass made with the FCAW process.

2. Place the majority of welding on the top quarter of the pipe by positioning the pipe so that you will weld starting between the two o'clock and three o'clock positions and weld to between the ten o'clock and eleven o'clock positions.
3. Use both hands while performing the weld. Use one hand to hold the GMAW welding gun, and use the other hand to steady your travel and contact tip-to-work distance. It will be important to maintain a consistent nozzle angle as you travel around the diameter of the pipe.
4. This weld is made primarily in the flat position and uses a drag travel angle of approximately 10 degrees to 15 degrees into the direction of travel. As the two o'clock or three o'clock position is approached, the travel around the pipe is closer to the vertical position, where a slight push angle will be needed to offset the effects of gravity.
5. Because of the nature of the flux cored wire, it is not designed to weld with downhill progression. For this reason, do not weld too far past the ten o'clock or two o'clock position.

NOTE

The **electrode extension** is the distance from the end of the contact tip to where the wire electrode is melted off. It is important to follow the manufacturer's recommendations for electrode extension. Some FCAW electrodes require long electrode extensions that preheat the tip of the electrode before it melts off and enters the arc. This preheating is required so that the flux reacts as designed once it enters the arc. Other FCAW electrodes may use a very short electrode extension.

6. As electrode extension increases, resistance increases, along with voltage, in the welding circuit. With this increase in resistance, heating of the wire and flux core takes place.
7. Repeat the procedure, rotating the pipe to the same position each time.

Filler Passes

8. Start the weld between the ten o'clock and eleven o'clock position in the groove directly on top of the hot pass.
9. Keep the weld face flat or slightly concave by traveling fast enough to prevent excess buildup. The travel speed will be determined by the weld bead buildup and width.
10. Watch the weld pool size, and particularly the trailing edge of the weld pool, to ensure that it does not begin to sag as a result of an excessively large weld pool, **Figure 10-1**.
11. When the weld has progressed to the twelve o'clock to one o'clock position, stop and roll the pipe so that the end of the weld is at the ten o'clock to eleven o'clock position. The next weld can begin there.
12. Repeat this process until the filler pass is completed around the pipe.
13. Once the filler pass is completed around the pipe, clean it thoroughly with a chipping hammer and wire wheel.
14. Have your instructor inspect your work for discontinuities and consistent defect-free welds.
15. Continue making filler passes in this same manner until the weld groove has been filled to within 1/16 to 1/8 in. (1.5 to 3 mm).
16. The weld should have uniform buildup and be made with stringer beads, with little to no weaving of the puddle.
17. Repeat this practice until you can make a uniform, consistent weld each time.
18. Turn off all equipment, return all tools and supplies, discard all scrap, and clean up the welding booth.

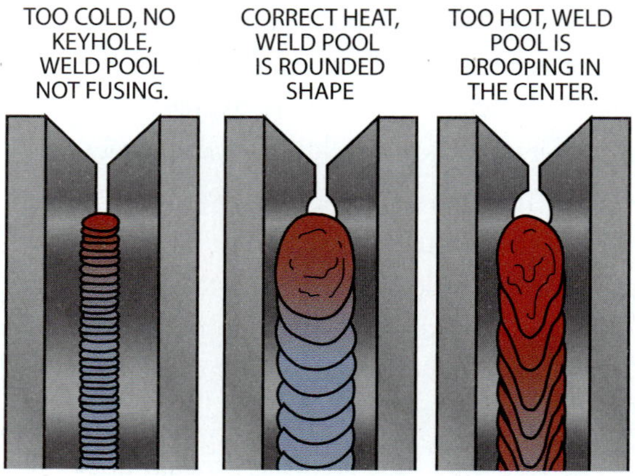

FIGURE 10-1 Watch the leading edge of the weld for a keyhole and the shape of the trailing edge of the weld as a way to tell if the heat is too low or too high.

Cover Pass

19. Start the weld between the ten o'clock and eleven o'clock position in the groove directly on top of the filler pass.

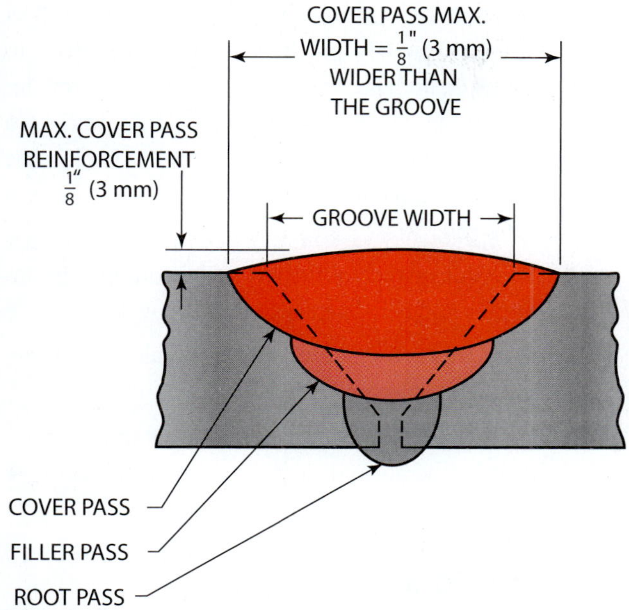

FIGURE 10-2 The weld bead reinforcement and width must be controlled carefully.

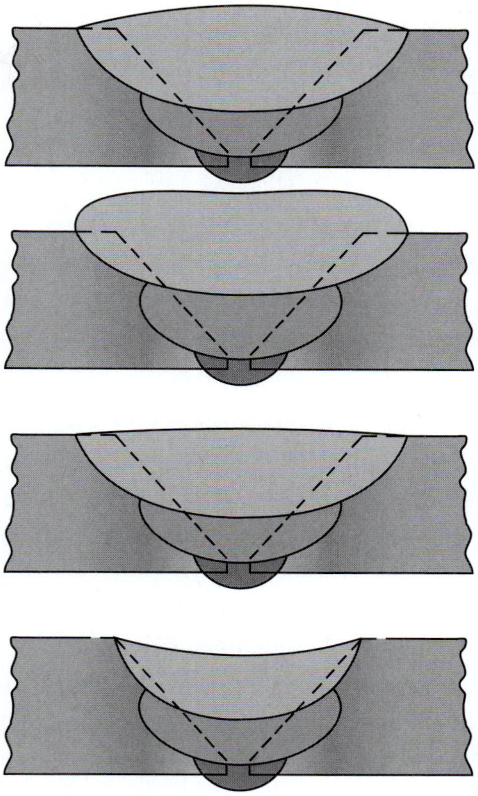

FIGURE 10-3 The weld reinforcement must have a smooth transition from the toe of the weld across the face of the weld.

20. The cover pass should not have more than 1/8 in. (3 mm) of buildup and be no more than 1/8 in. (3 mm) wider than the groove opening, **Figure 10-2**.
21. The finished cover pass weld face should be uniformly built up so that it curves from one edge of the groove to the other, **Figure 10-3**. A weld that abruptly rises along the edge can produce a stress point that could cause future weld cracking.
22. Because of the restriction on the width of the weld bead, the cover pass may need to be made with more than one pass. If more than one pass is required, each weld pass should cover the preceding weld pass by approximately one-half, **Figure 10-4**.
23. Watch the weld pool size (and particularly the trailing edge of the weld pool) to ensure that it is consistent in size and shape.
24. When the weld has progressed to the twelve o'clock to one o'clock position, stop and roll the pipe so that the end of the weld is at the ten o'clock to eleven o'clock position. The next weld can begin there.
25. Repeat this process until the cover pass is completed around the pipe.
26. Once the cover pass is completed around the pipe, clean it thoroughly with a chipping hammer and wire wheel.
27. The weld should have uniform buildup.
28. Have your instructor inspect your work for discontinuities and consistently defect-free welds.
29. Repeat this practice until you can make a uniform, consistent weld each time.
30. Turn off all equipment, return all tools and supplies, discard all scrap, and clean up the welding booth. ◆

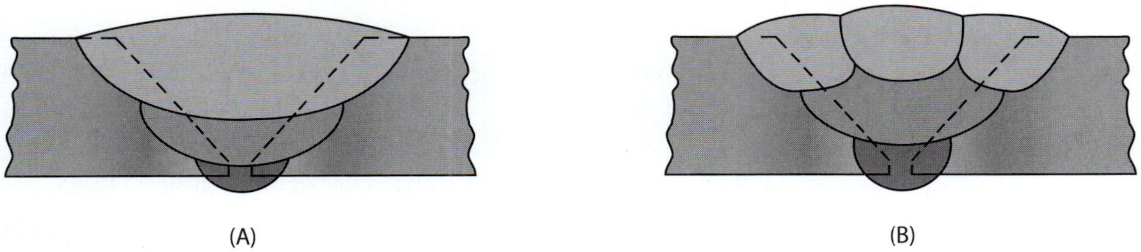

FIGURE 10-4 The cover pass can be made as one or more weld beads as long as the weld face is smooth and uniform.

PRACTICE 10-2

2G Position, Vertical Fixed

For this practice, you will need 6-in. (150-mm) Schedule 80 pipe, .045-in. (1.2-mm) E71T wire, a chipping hammer, a grinder, and a wire brush. You will also need PPE appropriate for the processes being performed, and a welding shade lens of a minimum number 10 and suggested number 11 or 12, depending on the amperage at which welding is taking place. Refer to ANSI Z49.1 for specific recommendations.

1. The root weld should already be made using the GMAW process. It should be ground down to make sure that there are no areas that might have lack of fusion or excess roll. Care should also be taken to make sure that the root pass is not ground too thin, as this can result in excessive melt-through on the first pass made with the FCAW process.
2. Use a stringer bead technique. It will result in lower heat input that can help improve weld toughness in cold weather applications.
3. The first pass after the root will be placed directly over the root or hot pass. A working of the torch in a slight zigzag pattern will help to ensure good tie-in at the toe of the weld.
4. The torch angle should be a 70-degree to 80-degree drag travel angle. This means that the torch will be angled 20 to 30 degrees away from the direction of travel. Remember that using the drag travel angle or **backhand technique** will provide deeper penetration and a more convex weld bead. Knowing this will allow you to adjust your technique to obtain the necessary weld shape.
5. Be sure to chip and wire-brush each pass to make sure that no slag is trapped between beads. The next bead will be placed just below the first filler pass. It should overlap the previous bead by approximately one-half. It should also blend in smoothly; if there is any roll, it should be removed by grinding, taking care not to gouge into the previous bead or base metal.
6. Having a weld bead with a roll or improperly grinding a bead can result in lack of fusion caused by an inability to access the weld joint properly. Continue stacking the beads, working from the bottom of the joint to the top.
7. The number of filler pass beads that will be necessary will depend on the wall thickness of the pipe being welded, as well as the travel speed.
8. Once you are within 1/8 in. (3 mm) of the top of the joint, you will put in the cap passes. Prior to making a cap pass, you should allow some time for the workpiece to cool. Heat can build up quickly in a weld coupon, and if not given time to cool, some of the heat can cause the beads to sag, resulting in excess roll.
9. Have your instructor inspect your work for discontinuities and consistently defect-free welds.
10. Repeat this practice until you can make a uniform, consistent weld each time.
11. Turn off all equipment, return all tools and supplies, discard all scrap, and clean up the welding booth. ◆

PRACTICE 10-3

5G Position, Horizontal Fixed

For this practice, you will need 6-in. (150-mm) Schedule 80 pipe, .045-in. (1.2-mm) E71T wire, a chipping hammer, a grinder, and a wire brush. You will also need PPE appropriate for the processes being performed, and a welding shade lens of a minimum number 10 and suggested number 11 or 12, depending on the amperage at which welding is taking place. Refer to ANSI Z49.1 for specific recommendations.

1. As with the previous FCAW welds on pipe, complete the root pass with the GMAW process. The pass should be ground down to make sure that there are no areas that might have lack of fusion or excess roll. Care should also be taken to make sure that the root pass is not ground too thin, as this can result in excessive melt-through on the first pass made with the FCAW process.
2. The pipe is to be set up in the horizontal fixed position. Weld progression with the flux cored process will be uphill. It is best to start at or near the six o'clock position; however, some welders like to start at the five o'clock or seven o'clock position, welding through the six o'clock position so that the start and stop do not end on the very bottom of the pipe.
3. The torch positioning will be a drag travel angle of 10 to 20 degrees in the direction of travel. This will prevent slag and any spatter from being thrown ahead in the joint and being welded over, resulting in trapped slag between the weld layers.
4. As the weld progresses from underneath the pipe, the torch angle should change slightly to point

toward the center of the pipe. As progression reaches the ten o'clock to eleven o'clock or two o'clock to one o'clock position, the same angle needs to be maintained and move back to more of a slight drag travel angle. As travel reaches either of these two locations, the tendency is to move to a push travel angle. You will have to resist this tendency, as it will change the weld bead shape and increase the chance of trapping slag.

5. Torch oscillation should be a slight zigzag or slight weave. Care should be taken to maintain a quick travel speed to prevent excess weld buildup.
6. Different techniques are used in completing a 5G flux cored weld on pipe. Depending on the welding procedure specifications, sometimes a weave bead technique is used; in other cases, stringer beads are used. Weave beads tend to result in slower travel speeds, resulting in more heat input into the piece. Stringer beads tend to result in faster travel speeds and a little less heat input. With some materials, greater heat input can result in lower impact strength, which may be an issue with pipe that is used in cold service applications.
7. The weave technique is a side-to-side zigzag straight across the joint. Movement should be relatively quick. If you are using the stringer technique, it will be similar to the first passes that were used. Simply do a slight side-to-side motion and move quickly in the uphill direction. Regardless of the technique that you use, the slag systems for flux cored wires will freeze quickly, forming a dam to hold the molten weld pool in place. It is still important to allow the workpiece to cool between passes, especially the cover pass. Failure to do so can cause excess penetration, as well as cause the cover pass to sag.
8. Have your instructor inspect your work for discontinuities and consistently defect-free welds.
9. Repeat this practice until you can make a uniform, consistent weld each time.
10. Turn off all equipment, return all tools and supplies, discard all scrap, and clean up the welding booth. ◆

PRACTICE 10-4

Pipe 6G Position, 45-Degree Fixed

For this practice, you will need 6-in. (15-mm) Schedule 80 pipe, .045-in. (1.2-mm) E71T wire, a chipping hammer, a grinder, and a wire brush. You will also need PPE appropriate for the processes being performed, and a welding shade lens of a minimum number 10 and suggested number 11 or 12, depending on the amperage at which welding is taking place. Refer to ANSI Z49.1 for specific recommendations.

As with the previous flux cored welds on pipe, complete the root pass with the GMAW process. The pass should be ground down to make sure that there are no areas that might have lack of fusion or excess roll. Care should also be taken to make sure that the root pass is not ground too thin, as this can result in excessive melt-through on the first pass made with the FCAW process.

In the 6G position, the pipe is at a 45-degree angle. For this reason, weld passes will be done using the stringer technique. This allows greater control of the weld bead and heat input.

1. Start welding at approximately the six o'clock position, and work up toward the nine o'clock or three o'clock position. Direct the wire in at the center of the groove. You should work it with a slight zigzag or wiggle motion up and down in the joint. This will help to give good tie-in at the toe of the weld. This pass should be just slightly larger than the first root pass.
2. Run a similar pass on the opposite side of the pipe in the three o'clock to twelve o'clock or nine o'clock to twelve o'clock positions. Be sure to grind down your start and stops. Make the next sequence of welds opposite the first two.
3. The remaining welds should be stringer welds starting at the bottom side of the joint and working up the joint. Continue welding filler passes, making sure to grind down the starts and stops. Be sure to stagger the start and stop points, offsetting them just slightly. This can help to prevent an accumulation of discontinuities in one area of the pipe.
4. Once the filler passes are between 1/8 and 1/16 in. (3 and 1.5 mm) below the outer surface of the pipe, you are ready to make the cover passes. Just as with the previous welds, be sure to allow some time for the pipe to cool. The cover pass should also be performed using stringer beads. Be sure to overlap the beads between one-quarter and one-half.
5. Have your instructor inspect your work for discontinuities and consistently defect-free welds.
6. Repeat this practice until you can make a uniform, consistent weld each time.
7. Turn off all equipment, return all tools and supplies, discard all scrap, and clean up the welding booth. ◆

Summary

Advances in flux core filler wire, welding machines, and techniques has resulted in many more codes and standards, allowing pipe welds to be made with the FCAW. As a result, there is an increased demand for FCA welders capable of filling these emerging jobs in the piping industry.

Review Questions

1. ________ -based fluxes are acidic and can be used in ________ positions.
2. ________ -based fluxes are alkaline and can be used in ________ positions.
3. Fluxes can be designed to produce a fast-freezing slag that can be used for what purpose?
4. In the AWS classification system for carbon steel flux cored electrodes, what does the "T" stand for?
5. What is one of the main advantages that makes the FCAW process attractive to the piping industry?
6. What problem is encountered when welding pipe of a higher grade or strength?
7. When welding pipe with the FCAW process, what welding process would be a good choice for welding the root?
8. When welding the root pass, if specialized equipment is unavailable, what mode of metal transfer should be used?
9. Why is it important to allow the workpiece to cool a bit prior to making the cover pass?
10. Why is it important to stagger starts and stops on the pipe as you weld?
11. Which welding technique will provide deeper penetration and a more convex weld bead?
12. The slag systems for flux cored wires will freeze quickly, forming a dam to hold the molten weld pool in place. True or False?
13. What would be an advantage of self-shielded wires over dual-shield wires?

Chapter 11

Gas Tungsten Arc Welding of Pipe

OBJECTIVES

After completing this chapter, the student should be able to:

- Discuss the advantages and disadvantages of the freehand and walking the cup techniques.
- Discuss how root opening relates to different techniques for adding the filler rod.
- Explain the causes of suck-back and how to control it.
- Explain the importance of tack welds.

KEY TERMS

Freehand

Hot pass

Suck-back

Walking the cup

INTRODUCTION

The concept of using an eternal shielding gas for welding was experimented with by H. M. Hobart and P. K. Devers in the late 1920s. They used helium and argon as a gas to shield the welding arc and weld pool. During the 1930s, the gas tungsten arc welding (GTAW) process was still in its infant stage. By the 1940s, it was imperative that a process be developed to weld magnesium and aluminum to meet the aircraft industry's needs as a result of the U.S. entry into World War II.

A man named Russell Meredith, who worked for Northrup Aircraft, was given the charge to develop a method to produce high-quality welds on these materials. He was given a patent for a GTAW torch that brought the process to a level that enabled welding to be made on a production-level basis. The process was later licensed to Linde Air Products Company and given the trade name Heliarc®. These developments resulted in a welding process that could produce high quality welds in a large variety of materials, **Figure 11-1**.

Larry Jeffus

FIGURE 11-1 GTAW is ideal for critical welds, such as this one on an aircraft engine mount, where a weld failure can result in a catastrophic event.

(Continued)

The skill necessary to perform pipe welds in position is greater than that of welding on plate. As a welding student, you should already have welding experience on plate in the 1G, 2G, 3G, and 4G positions and any metals that you might pipe-weld. If you have yet to master-plate, you should perfect those skills before moving to pipe.

PREPARATION OF PIPE FOR WELDING

The first step is to prepare the ends of the pipe to meet the requirements of the welding procedure. Mild steel pipe is often prepared by beveling with the oxy-fuel process. There are several common bevel angles for pipe preparation, such as 30 degrees, 37.5 degrees, and 45 degrees. For this practice, we will use a 30-degree bevel angle, which will result in a 60-degree groove angle. Once the beveller is positioned to 30 degrees, the torch should be positioned on the pipe where the cut is to be made. Then adjust the torch to a neutral flame and make your cut.

Next, you will need to remove any scale and oxide from the cut surfaces and adjacent areas by grinding and wire-brushing. Light slag can be removed from the edge of the cut surface by using a half-round file and running it from the inside of the pipe, holding it at a low angle and knocking lightly attached slag from the inside of the pipe. Then check to see that the end of the pipe is flat and square with the outside diameter (OD) of the pipe. You can check this by laying it on a flat surface and checking for light coming from under the surface. You can also use a framing square and lay it on the OD of the pipe and lay the other end over the cut edge of the pipe, **Figure 11-2**. In most cases, if the beveling equipment is in good working order, your cut should be square and you should see little to no light coming from under the root face. If there is some, you can use an 8-in. (200-mm) grinder to face the end of the pipe. This can be done by holding the pipe secure, lay the grinder flat on the surface of the pipe root face, and lightly apply pressure on the pipe so that a small chamfer is formed on the root face surface, **Figure 11-3**. This should be done only a slight amount, and you should end up with a chamfer with less than 1/8 in. (3 mm) of width. Do this a little at a time until the end of the pipe is square.

FIGURE 11-2 Use a square to check the end of the beveled pipe.

FIGURE 11-3 Use a grinder to square the end and make the root face.

The next step is to grind the groove face to bright shiny metal, which ensures that surface oxides have been removed. Before grinding, make sure to secure the pipe by clamping to a table or by using a vise or other appropriate clamping tool. Be sure to use the following personal protective equipment (PPE): hearing protection, safety glasses, and face shield. When you start to grind the groove face, be sure to position the grinder so that it contacts the surface and maintains the correct 30-degree bevel angle. Apply a light pressure to the metal surface with the grinder and follow the contour of the pipe surface. You should move the grinder back and forth at a quick speed a distance of about a quarter of the circumference of the pipe. Rotate the pipe and work the grinder around the entire diameter of the pipe.

As you change directions while moving the grinder back and forth around the contour of the pipe, be sure to release pressure from the surface to prevent flat spots and gouges on the groove face.

Next, remove oxides from the outside of the pipe adjacent to the joint for a distance of about 1 in. (25 mm). Check the inside of the pipe for scale and oxides. With a rotary file and a sanding flap, you can clean the inside of the pipe to bright, shiny metal as well, **Figure 11-4**. Once you have cleaned the oxide off the inside and outside of the pipe, you are ready to fit the two pieces together. Check to make sure that the inside surfaces line up and there is little to no mismatch, which is where the inside surfaces of the pipe do not line up, causing high-low on the alignment of the pipe. This will make it difficult to get a uniform, and defect-free root weld. If the pipe is not seamless, lining up the seams will often help eliminate the mismatch.

FIGURE 11-4 These boiler tubes have been cleaned and shaped so that they are ready for welding.

GTAW PIPE TECHNIQUES

Welding is not only a science, but an art as well. Therefore, there may be many different ways of doing a job that will produce the same high-code-quality results.

There are two different methods of controlling the arc length when making a GTAW. The most commonly used technique is called **freehand**. In the freehand technique, the GTAW torch is held so that the tungsten is positioned just above the weld, **Figure 11-5**. Often, a welder will brace the arm or elbow for better control; however, it is still considered freehand welding.

FIGURE 11-5 Freehand welding.

The other technique, **walking the cup**, is done by resting the cup on the workpiece, **Figure 11-6**. The torch is then manipulated in one of several different ways to move it down the weld joint. The arc length is changed by varying the tungsten stickout, cup size, and the angle at which the torch is held. This technique is used on carbon steels and stainless steels that are typically greater than gage thickness. It is probably most commonly used for putting in weld passes on pipe when high-quality, low-hydrogen welds are required.

FIGURE 11-6 Cup walking.

One method for walking the cup is to start by resting the edges of the cup lightly on the groove face. This gives the cup two points of contact—one on each groove face.

NOTE

It is important not to apply too much pressure to the torch; this can cause damage to the torch neck and increase operator fatigue.

Next, sway your arm to one side, **Figure 11-7**. The side that you are rolling toward should stay stationary while the opposite side gently slides on the opposite groove face. As you continue to swing your arm around, you will plant the torch on the opposite groove face, release pressure on the previously stationary side, and then gently let it slide along the opposite groove face. This will create a zigzag pattern and will propel the torch down the groove. By varying the pattern, the travel speed can be faster or slower. It will also change the appearance of the bead by keeping the ripple pattern closer or farther apart. As you practice the technique, you will become smoother with the movements, resulting in a more uniform-looking weld bead.

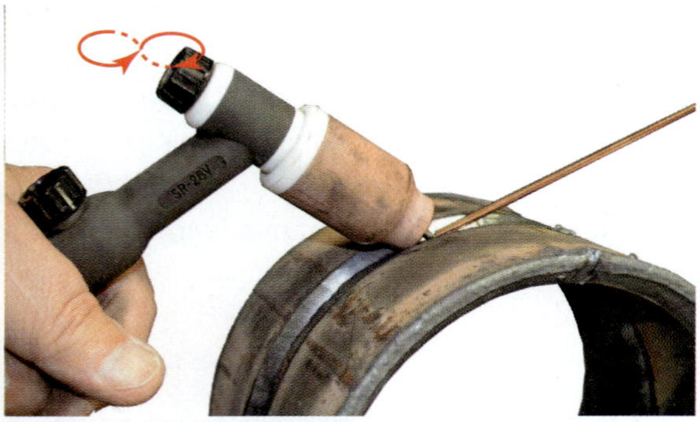

Larry Jeffus

FIGURE 11-7 The top of the torch makes a figure eight when walking the cup along the groove. This technique is referred to as "cup walking."

Another technique is to simply rest the cup on the groove face as before and wiggle the torch side to side, allowing the cup to maintain contact with the groove faces and at the same time moving the torch forward. It is best to use a smaller size cup when using this technique, which can give excellent results on the root pass of a groove weld.

TECHNIQUES FOR ADDING FILLER METAL

When walking the cup, there are two techniques for adding the filler wire. Depending on the joint fit-up, the welder might choose one over the other. One of these is referred to as the lay wire technique, in which the filler rod is laid down in the root of the joint, **Figure 11-8**. The root opening should be such that the filler wire will lie in the joint but catch on the root face, not allowing it to pass all the way through.

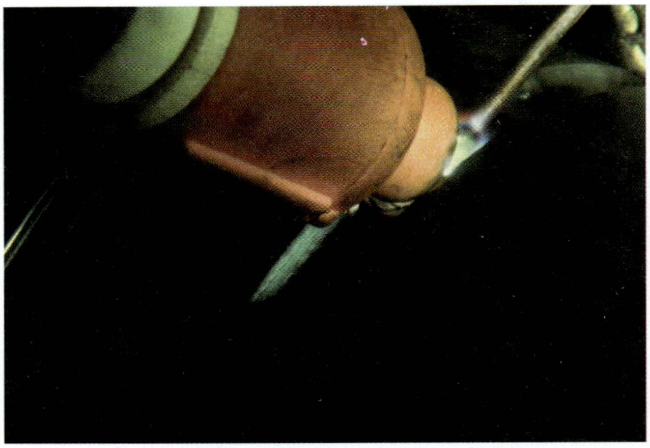

Larry Jeffus

FIGURE 11-8 Laying the filler metal in the groove as the weld is made.

The torch is then walked over the filler as the weld progresses. It is important to have the correct amperage setting and travel speed in order to obtain complete joint penetration and tie in at the root of the weld.

The other technique is to feed the filler rod through the root opening to the back of the joint. The joint fit-up and preparation is critical with this technique. The tack welds must be good and solid and evenly spaced around the pipe. They should be spaced and of a size so that the root opening will not close up as welding progresses around the pipe. For 6-in. (150-mm) pipe, as used for your practice, four evenly spaced tacks approximately 1 in. (25 mm) in length and tapered at each end should hold the root gap open and keep it consistent. The root opening should be just large enough that the filler rod can be fed through the gap, not flop side to side, and not get stuck by the root gap closing on it.

With this fit-up, you can now feed the filler rod into the weld pool from the back of the joint, **Figure 11-9**. Start by placing the rod through the joint and resting it on a tack weld that is opposite the side of the joint that is to be welded. Rest the filler rod on the side of the tack weld that the torch travel will be coming to first. By doing this, you will be able to steady the filler rod and have more control feeding it. As the weld progresses around the pipe, you will change the angle that the filler rod is feeding into the backside of the weld. At the same time, you will be feeding the filler rod through the joint and into the back of the weld. Weld reinforcement can be varied by the amperage setting, as well as where the filler rod is being placed in the weld pool.

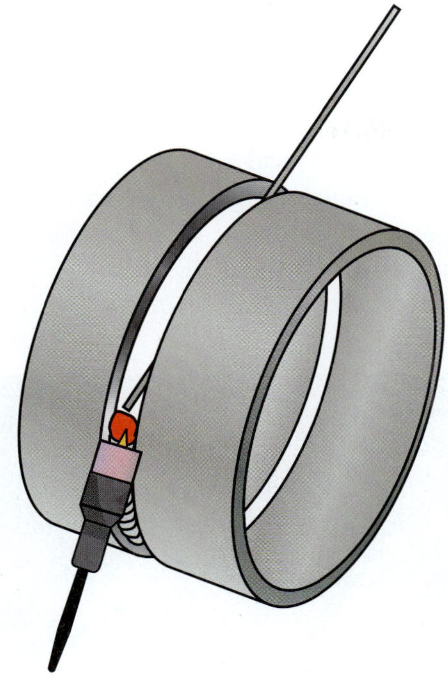

FIGURE 11-9 Feeding the filler through the root opening as the weld is made.

It is important that when using any technique, the root face is broken down so that complete fusion takes place. Failure to fuse the root face can show up as lack of fusion on an x-ray or break out on a bend test.

PRACTICE 11-1

Mild Steel Pipe, 1G Position

Welding pipe in the 1G position will be similar to welding plate in the 1G position. Just as when welding plate in 1G, the filler metal is added from above the weld joint and the pipe is rolled to maintain the deposition of weld metal from above the joint. This can be achieved by using a weld positioner or by simply welding and stopping to reposition the pipe. Greatest weld depositions can be achieved by welding in the 1G position; in addition, it requires less skill to master. For this reason, the first exercise will be to perform welds in this position.

For this practice, you will need 6-in. (150 mm) Schedule 40 pipe, a 3/32-in. E70-S3 mild steel filler rod, a grinder, a half-round bastard file and wire brush, various shielding gas cups ranging in sizes from 6 to 10, a 3/32-in. (2.4 mm) 2% thoriated or 1.5% lanthanated tungsten electrode, a collet, and a gas lens matching the tungsten size being used. You will also need PPE for the processes being performed.

After preparing the material as described earlier, you will tack-weld two pipe nipples together. For 6–8-in. (150–200 mm) pipe, four 1-in. (25-mm) tacks equally spaced around the pipe will be adequate.

If the pipe is smaller, three tack-welds at approximately 120 degrees apart will be suitable. This can be done by taking a filler rod that is about 1/32 in. (0.8 mm) larger than the root gap that is needed. For example, if a 3/32-in. (2.4-mm) root gap is called for, use a 1/8-in. (3-mm) welding wire.

1. Take and bend the rod into a V-shape.
2. Place the one pipe nipple on a flat surface, weld groove side up.
3. Place the gapping rod on top of the pipe and then place the other pipe nipple on top of it, weld groove side down.
4. Place a tack-weld on the side of the pipe that the open ends of the gapping rod come out. Once that tack-weld is completed, remove the rod and check the opening around the pipe. Make adjustments as necessary by spreading or closing the root gap to keep it uniform. You can use the rod to make sure that the root opening is consistent.
5. Place the next tack-weld 180 degrees opposite the first tack weld. Continue checking the root opening to make sure that it is consistent.
6. Place the next tack-weld 90 degrees from the previous weld.
7. The fourth and final tack weld will be 180 degrees from the last tack weld. This will give you four evenly spaced tack welds approximately 90 degrees apart, **Figure 11-10.**
8. Now grind both ends of the tack weld so that they are tapered to help form a ramp. This will enable you to start the weld on the tack and blend the weld down into the root. It also allows the weld to be tapered back onto the tack at the end of the weld.

Larry Jeffus

FIGURE 11-10 The tack welds should be approximately 1 in. (25mm) long.

A GTA welding pipe alignment tool can be used to both clamp the pipe ends in alignment and allow access to make the tack welds, **Figure 11-11**. Specialty tools like this make it easier and faster to accurately fit up pipe joints.

Larry Jeffus

FIGURE 11-11 A GTAW pipe alignment tool for tack welding.

You are now ready to start welding:

1. Place the pipe in a piece of angle iron that is V side up and has a piece of flat stock tacked to the bottom side so that the pipe will be able to be steady and hold the pipe in position. A small notch can be cut in the angle iron to allow for clearance as the cover pass is applied.
2. For this weld position, the lay wire technique or traditional freehand technique will work best. Check your argon shielding gas flow; it should be between 15 and 20 cubic feet per hour (cfh) (7 and 9.5 L/min.). Strike the arc down in the weld joint on one of the tack welds. Hesitate until you see that the weld pool is the size desired. Move forward and watch to make sure that the root faces on both pipes are melted and that the tack weld blends into them as well.
3. Once both sides of the root face are melting, add the filler metal. The filler metal should feed through the root face but be a very close fit, **Figure 11-12**.
4. Using the lay wire technique, begin walking the cup as described earlier, making sure to tie in the side walls of the groove face. Using this technique, there should be no break in the filler rod and the weld pool. The filler rod should be feeding in at the leading edge of the weld pool. The travel speed should be rapid to make sure that there is no excessive buildup and that the weld pool does not get too big. If the weld pool gets too large, the excess heat and oversized weld pool will shrink as the weld cools, causing **suck-back**, which is when the weld pool shrinks as it cools and the bead becomes concave, resulting in lack of reinforcement on the root side of the weld.

Larry Jeffus

FIGURE 11-12 The root surface will have a heavy oxide layer since no backing gas is used for these practice welds.

You should weld from the two o'clock to the eleven o'clock position, welding from tack to tack. Once you cover that distance, stop, rotate the pipe, and start again, **Figure 11-13**.

Larry Jeffus

FIGURE 11-13 The weld must be uniform and fused into the groove face, as shown in the enlarged view.

Use a small-diameter cup so that the arc can be more easily focused at the root of the joint. Do not use too wide a root opening. For thinner-walled pipe such as Schedule 40 and smaller, use the appropriate size tungsten. When selecting tungsten size, it is important to make sure the welder considers the thickness of the pipe. The thinner the material being welded, the less amperage that is required, and the less amperage, the smaller the tungsten needed to do a good job. If you are unsure about the tungsten size and amperage range, consult a tungsten manufacturer's guide.

The second pass is often referred to as a **hot pass**. The reason for this is that in some cases, the amperage is increased to burn out any slag or areas that may have cold lap or lack of fusion. Using the GTAW process should not require that the amperage be increased for the second pass. Slag inclusions will not be an issue with this process. Cold lap and lack of fusion should be seen and corrected before getting to the second pass. With the second pass, the welder can again use the cup walking technique.

1. For the second pass, increase the cup size to a number 7 or larger. The technique will be similar to the root pass, but your torch will have three points of contact this time—one on the groove edge of each side and one additional point of contact with the first weld pass.
2. Continue in the same manner as with the root pass, welding a distance of about one-fourth the diameter around the pipe. Be sure to stagger your start and stop points so that as you make your passes, they do not all end up at the same location. Leave approximately 1/16 in. (1.5 mm) above the weld so that it can be finished out.
3. Next will be the final or cap pass for the weld. Depending on the size of cup that was chosen for the second pass, you may select a larger cup for the final weld pass. Using a larger cup will make it easier to cover a greater distance faster, speeding up the travel speed as well as providing better shielding gas coverage.
4. On the final pass, you will again use the cup walking technique as described previously. It is important to remember to not grip the torch too tightly. This can cause the welder to fatigue more quickly. Another thing to remember is when resting the torch on the workpiece, make sure not to apply too much pressure to the pipe. Applying too much pressure can cause the torch to slip on the work surface. As with the previous pass, you will have three points of contact with the pipe. The cup will ride on the weld bead and on each side of the weld groove.
5. Continue to practice until you can produce a weld that is uniform in penetration, buildup, and bead spacing. ◆

PRACTICE 11-2

Mild Steel Pipe, 2G Position

For this practice, you will need 6-in. (150-mm) Schedule 40 pipe, a 3/32-in. (2.4-mm) E70-S3 mild steel filler rod, a grinder, a half-round bastard file and a wire brush, various shielding gas cups ranging in sizes from 6 to 10, a 3/32-in. (2.4-mm) 2% thoriated or 1.5% lanthanated tungsten electrode, a collet, and a gas lens matching the tungsten size being used. You will also need PPE for the processes being performed.

1. Start by fitting two prepared pipe nipples together as described previously. They should be clean and free of slag, scale, rust, and grease, Figure 11-13. Use a 1/8-in. (3-mm) filler rod to set the root gap between the two pieces of pipe. Check to make sure that the inside surfaces line up and there is little to no mismatch. Mismatch will make it difficult to get a uniform, defect-free root weld. If the pipe is not seamless, lining up the seams will often help eliminate the mismatch.
2. Once the pipe is tack welded together, it should be placed in the 2G position. This is where the pipe is fixed vertical so that the weld groove will run horizontal.
3. Start the arc in the groove of the joint on one of the tack welds.
4. Heat the tack so that a weld pool is formed.
5. Start the travel by making a zigzag motion with the torch. The zigzag pattern should be just wide enough so that the point of the tungsten is directed at the edge of the root filler junction. The filler rod should be placed in the weld groove at the root. The lay wire technique will work well with this position. The filler wire should be placed in the groove tangent to the root.
6. Continue to walk the cup around the pipe, keeping a steady speed and uniform weave pattern until reaching the next tack. Remember to continue on to the tack weld and end the weld after the weld pool has reached about the middle of the tack.
7. Move around the pipe, continuing to weld from tack to tack until the root pass is complete.
8. The second pass, which will be a fill pass, will be performed in a similar manner as the root pass.

The amperage will need to be increased to between 100 and 110, depending on the individual welding machine and the thickness of the root pass. The torch should have three points of contact on this pass—one on each side of the groove face, and one on the root pass. Stagger the starts and stops so that each weld pass does not have the same start and stop point. The filler rod once again should be held tangent to the pipe, and the weave pattern will be a zigzag pattern, just slightly larger than the first pass. Keep a quick and steady travel speed so that the weld pool does not get too large, causing the weld pool to sag or overheat the workpiece. Also, allow the workpiece to cool off between weld passes.

9. The cover pass can be made using a two- or three-bead technique, depending on the total weld groove angle and root opening. The next-to-last pass should be approximately 1/16 in. (1.5 mm) below the surface. Try not to carry too large a weld pool, because it can cause the bead to sag, resulting in excessive roll. Carrying an excessively large weld pool is often caused by too slow a travel speed or too much heat.
10. Maintain a smooth and steady travel speed. Weld bead placement is another important factor in making a high-quality weld. They should be placed so that the toes of the welds do not overlap on top of one another. There should be an overlap of one-third to one-half of the weld beads on each other. The welder should start at the bottom and stack the beads on top of each other. Again, using the cup walking technique will result in a zigzag motion, rocking the cup up and down to form a consistent weave pattern. By properly practicing these techniques, your skill will increase, and soon you will be making consistent welds. ◆

PRACTICE 11-3

Mild Steel Pipe, 5G Position

For this practice, you will need 6-in. (150-mm) Schedule 40 pipe, a 3/32-in. (2.4-mm) E70-S3 mild steel filler rod, a grinder, a half-round bastard file and a wire brush, various shielding gas cups ranging in sizes from 6 to 10, a 3/32-in. (2.4-mm) 2% thoriated or 1.5% lanthanated tungsten electrode, a collet, and a gas lens matching the tungsten size being used. You will also need PPE for the processes being performed.

1. Start by fitting two prepared pipe nipples together as described previously. They should be clean and free of slag, scale, rust, and grease. Use a 1/8-in. (3-mm) filler rod to set the root gap between the two pieces of pipe. Check to make sure that the inside surfaces line up and there is little to no mismatch. This will make it difficult to get a uniform, and defect-free root weld. If the pipe is not seamless, lining up the seams will often help eliminate the mismatch.
2. Once the pipe is tack-welded together, it should be placed in the 5G position. This position is such that the pipe will be placed in the horizontal fixed position. This results in part of the weld being made in the flat position, part in the vertical, and part in the overhead position.
3. Start the arc in the groove of the joint on one of the tack welds. Heat the tack so that a weld pool is formed.
4. Start the travel by making a zigzag motion with the torch. The zigzag pattern should be just wide enough so that the point of the tungsten is directed at the edge of the root filler junction. The filler rod should be placed in the weld groove at the root.
5. Another acceptable technique is to freehand the root instead of walking the cup. With this technique, you should use a slight forward movement and then a slight backstep movement. Using this technique, it is important to watch the root edge. A slight keyhole should form at the edge of the root. If the keyhole gets too large, you should reduce the amperage or slightly increase the amount of filler rod that you are adding.
6. Continue to work your way around the pipe, keeping a steady speed and uniform weave pattern until reaching the next tack. Remember to continue up on to the tack weld, and end the weld after the weld pool has reached about the middle of the tack.
7. Move around the pipe, continuing to weld from tack to tack until the root pass is complete.
8. Once you have completed the root pass, check to make sure that there are no high spots on the pipe. You can remove them with a grinder.
9. The second pass, which will be a fill pass, will be performed in a similar manner by walking the cup on the root pass. The amperage will need to be increased to between 100 and 110, depending on the individual welding machine and the

thickness of the root pass. The torch should have three points of contact on this pass—one on each side of the groove face, and one on the root pass. Stagger the starts and stops so that each weld pass does not have the same start and stop point. The filler rod once again should be held lightly against and tangent to the pipe, and the weave pattern will be a zigzag pattern just slightly larger than the first pass. Keep a quick and steady travel speed so that the weld pool does not get too large, causing the weld pool to sag or overheat the workpiece. Also, allow the workpiece to cool off between weld passes.

10. The cover pass can be made by walking the cup and making a weave bead. Prior to the last pass, the fill should be approximately 1/16 in. (1.5 mm) below the surface. Try not to carry too large a weld pool because it can cause the bead to sag, resulting in excess buildup. Carrying an excessively large weld pool is often caused by too slow a travel speed or too much heat. It is important to maintain a smooth and steady travel speed. Again, using the cup walking technique will result in a zigzag motion. Think of walking the cup as being like rolling a large barrel from one side to the other to form a consistent weave pattern. By properly practicing these techniques, your skill will increase, and soon you will be making consistent welds. ◆

PRACTICE 11-4

Mild Steel Pipe, 6G Position

For this practice, you will need 6-in. (150-mm) Schedule 40 pipe, a 3/32-in. (2.4 mm) E70-S3 mild steel filler rod, a grinder, a half-round bastard file and a wire brush, various shielding gas cups ranging in sizes from 6 to 10, a 3/32-in. (2.4-mm) 2% thoriated or 1.5% lanthanated tungsten electrode, a collet, and a gas lens matching the tungsten size being used. You will also need PPE for the processes being performed.

1. Start by fitting two prepared pipe nipples together as described previously. They should be clean and free of slag, scale, rust, and grease. Use a 1/8-in. (3-mm) filler rod to set the root gap between the two pieces of pipe. Check to make sure that the inside surfaces line up and there is little to no mismatch. This will make it difficult to get a uniform, and defect-free root weld. If the pipe is not seamless, lining up the seams will often help eliminate the mismatch.
2. Once the pipe is tack-welded together, it should be placed in the 6G position, such that the pipe will be placed at a 45-degree angle in a fixed position. This will be similar to the 2G position, except that it will be tipped to the side at a 45-degree angle. For the root pass, you can use either the cup walking technique, the lay wire technique, or freehand the root with the forward and back-step technique.
3. Start the arc in the groove of the joint on one of the tack welds. Heat the tack so that a weld pool is formed.
4. Start the travel by making a zigzag motion with the torch. The zigzag pattern should be just wide enough so that the point of the tungsten is directed at the edge of the root filler junction. The filler rod should be placed in the weld groove at the root. The lay wire technique will work well with this position. The filler wire should be placed in the groove tangent to the root.
5. Continue to walk the cup around the pipe, keeping a steady speed and uniform weave pattern until reaching the next tack. Remember to continue on to the tack weld, and end the weld after the weld pool has reached about the middle of the tack.
6. Move around the pipe, continuing to weld from tack to tack until the root pass is complete.
7. The second pass, which will be a fill pass, will be performed in a similar manner as the root pass. The amperage will need to be increased to between 100 and 110, depending on the individual welding machine and the thickness of the root pass. The torch should have three points of contact on this pass—one on each side of the groove face, and one on the root pass. Stagger the starts and stops so that each weld pass does not have the same start and stop point. The filler rod once again should be held tangent to the pipe, and the weave pattern will be a zigzag pattern just slightly larger than the first pass. Keep a quick and steady travel speed so that the weld pool does not get too large, causing the weld pool to sag or overheat the workpiece. Also, allow the workpiece to cool off between weld passes.
8. The cover pass can be made using a two- or three-bead technique, depending on the total weld groove angle and root opening. The next-to-last pass should be approximately 1/16 in. (1.5 mm) below the surface. Try not to carry too large a weld pool because this can cause the bead to sag, resulting

in excessive roll. Carrying an excessively large weld pool is often caused by too slow a travel speed or too much heat.

9. It is important to maintain a smooth and steady travel speed. Weld bead placement is another important factor in making a high-quality weld. They should be placed so that the toes of the welds do not overlap one another. There should be an overlap of one-third to one-half of the weld beads on each other. The welder should start at the bottom and stack the beads on top of each other. Again, using the cup walking technique will result in a zigzag motion, rocking the cup up and down to form a consistent weave pattern. By properly practicing these techniques, your skill will increase, and soon you will be making consistent welds. ◆

Summary

It is important to remember that welding is an art and a science. There are several different styles and techniques that can be utilized and achieve the same or similar results. It is often helpful to consult with other welders who have been working in the field for many years. You can often pick up tips that many have tried and have proved successful. Above all, it is important to practice on a consistent basis in order to improve your skills.

Review Questions

1. The GTAW process was originally developed for what purpose?
2. Which GTAW technique is most commonly used?
3. Using the cup walking technique on the root pass gives how many points of contact between the cup and the groove face?
4. Using the lay wire technique, what size should the root opening be?
5. What root opening should be used to feed the filler rod from the back of the weld joint?
6. How many tacks should be required for pipe between 6 and 8 in. (150 and 200 mm) in diameter, and what should the spacing be between the tacks?
7. What are some techniques that you can use to help prevent suck-back?
8. A weld pool that is too large is often caused by what?

Chapter 12

Pipe Welding with Multiple Processes

OBJECTIVES

After completing this chapter, the student should be able to:

- Explain the relationship between a welding procedure specification (WPS) and effects that it can have on the efficiency of producing a weld fabrication.
- Explain which combination of welding processes can be more efficient and why.
- Describe the steps of preparing and welding pipe using two different processes.

KEY TERMS

Chamfer

Inert gas

Mismatch

Welding procedure specification (WPS)

INTRODUCTION

When doing any welding, there are several factors that must be taken into consideration when deciding how to proceed. Two main considerations are the application of the end product and the efficiency of the fabrication processes being used. In fabricating a weldment, the application will dictate the necessary specifications that are required. In most cases, this will involve a code or quality manual that gives detailed specifications that a weldment must meet. For the manufacturer to be able to produce the fabrication and make a profit, it must do so in as efficient a manner as possible. In the case of welded fabrications, efficiency is most often controlled by the welding process or processes to be used. For this reason, it is extremely important to have a good **welding procedure specification**, which is a document that provides in detail the required variables for a particular application to ensure repeatability by properly trained welders and welding operators. The welding procedure and processes employed play a major role in the quality and efficiency of the weld produced.

This is why it is important to know the advantages and disadvantages of welding processes. In many cases, technology has improved the welding processes.

For example, when dealing with high-quality welds with little to no postweld cleaning required, the gas tungsten arc welding (GTAW) process might be selected. However, if speed, control of heat input, and welder skill are factors, the gas metal arc welding (GMAW) process might be preferred. Cost of the equipment and the amount of work that the fabricator can expect to do with the required equipment are often considerations in the selection of the process as well. Some of the new equipment for performing GMAW may have great benefits regarding speed and ease of use, but the overall cost of the equipment may be a deterrent that would keep many small welding shops from making that investment.

(*Continued*)

In many cases, decisions like this will be decided by engineers and management rather than the welder. However, those decisions will ultimately affect the welders, as they will be the ones actually performing the work. For this reason, the skill of the welder should be broad in nature. You should not limit yourself to the knowledge and practice of only one or two processes. For training purposes, we will work with two procedures—one utilizing a root weld with the GTAW process and fill and cap passes with the shielded metal arc welding (SMAW) process. The second procedure will use the GMAW process for the root pass, and the fill and cap pass will be done with SMAW.

PREPARATION OF PIPE FOR WELDING

As with all welding, one of the first and most important steps is the preparation of the materials to be welded. You will start by preparing the ends of the pipe so that it meets the requirements of the welding procedure. Mild steel pipe is often prepared by beveling with the oxy-fuel process. There are several common bevel angles for pipe preparation, such as 30 degrees, 37.5 degrees, and 45 degrees. For this practice, we will use a 37.5-degree bevel angle; this will result in a 75-degree groove angle. Once the beveller is positioned to 37.5 degrees, the torch should be positioned on the pipe where the cut is to be made. Next, adjust the torch to a neutral flame and make your cut.

Next, you will need to remove any scale and oxide from the cut surfaces and adjacent areas by grinding and wire-brushing. Light slag can be removed from the edge of the cut surface by using a half-round file and running it from the inside of the pipe, holding it at a low angle and knocking lightly attached slag from the inside of the pipe. Next, check to see that the end of the pipe is flat and square with the outside diameter (OD) of the pipe. You can check this by laying it on a flat surface and checking for light coming from under the surface. You can also use a square by laying it on the OD of the pipe and laying the other end over the cut edge of the pipe, **Figure 12-1**. In most cases, if the beveling equipment is in good working order, your cut should be square and you should see little to no light coming from under the root face. If there is some, you can use an 8-in. (200-mm) grinder to face the end of the pipe. Light grinding can be done by holding the pipe securely and laying the grinder flat on the surface of the pipe root face, and then lightly applying pressure on the pipe so that a small chamfer is formed on the root face surface, **Figure 12-2**. After grinding, the chamfer should be 1/8-in. (3-mm) or less in width. A **chamfer** is a flat-edge preparation that is placed on the end of a bevel joint. Do this a little at a time until the end of the pipe is square.

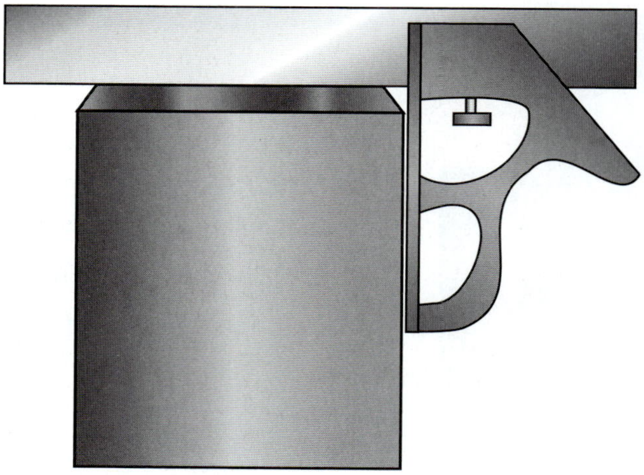

FIGURE 12-1 Check the squareness of the bevel.

FIGURE 12-2 Use a grinder to prepare the root face.

The next step is to grind the groove face to bright, shiny metal; this ensures that surface oxides have been removed. It is very important to remove the oxides left behind by the thermal cutting processes. The reason for this is that the melting temperature of the oxide is higher than the melting temperature of the base metal. If it is not removed, it may cause inclusions, porosity, or both.

Before grinding, make sure to secure the pipe by clamping to a table or by using a vise or other appropriate clamping tool. You should be sure to use the following personal protective equipment (PPE): hearing protection, safety glasses, and face shield. When you start to grind the groove face, be sure to position the grinder so that it makes contact with the surface and maintains the correct 37.5-degree bevel angle. Apply a light pressure to the metal surface with the grinder, and follow the contour of the pipe surface. You should move the grinder back and forth at a quick speed for a distance of about a quarter of the circumference of the pipe. Rotate the pipe and work the grinder around the entire diameter of the pipe. As you change directions while moving the grinder back and forth around the contour of the pipe, be sure to release pressure from the surface to prevent flat spots and gouges on the groove face.

Next, remove oxides from the outside of the pipe adjacent to the joint for a distance of about 1 in. (25 mm). Check the inside of the pipe for scale and oxides. With a rotary file and a sanding flap, you can clean the inside of the pipe to bright, shiny metal as well. Once you have cleaned the oxide off the inside and outside of the pipe, you are ready to fit the two pieces together. Check to make sure that the inside surfaces line up and there is little to no mismatch. **Mismatch** is where the inside surfaces of the pipe do not line up, causing high-low on the alignment of the pipe. Any mismatch in the pipes will make it difficult to get a uniform, and defect-free root weld. If the pipe is not seamless, lining up the seams will often help eliminate the mismatch.

GTAW PIPE TECHNIQUES

Welding is not only a science, but an art as well. For this reason, it should be understood that there are several different techniques that will yield acceptable results—namely, high-quality welds that will meet the requirements of a specific code.

There are two different techniques that are commonly used with the GTAW process. One of the most commonly used is the "freehand" technique. This type of technique is used when welding metals like aluminum and magnesium. It is also used on thin-gauge material that is often welded in the sheet metal industry. The freehand technique is when the welder rests a hand on the workpiece or other support and controls the arc length by the way the welder holds the torch. By making slight changes, the arc length can be increased or decreased.

The other technique, called "cup walking," is done by resting the cup on the workpiece. The torch is then manipulated in one of several different ways to move it down the weld joint. The arc length is changed by varying the tungsten stickout, the cup size, and the angle that the welder holds the torch. This technique is used on carbon steels and stainless steels that are typically greater than gauge thickness. It is probably most commonly used for putting in weld passes on pipe when high-quality, low-hydrogen welds are required.

One method for cup walking is to start by resting the edges of the cup lightly on the groove face. This gives the cup two points of contact—one on each groove face. It is important not to apply too much pressure to the torch; this can cause damage to the torch neck and increase operator fatigue. Next, sway your arm to one side. The side that you are rolling toward should stay stationary while the opposite side gently slides on the opposite groove face. As you continue to swing your arm around, you will plant the torch on the opposite-side groove face, release pressure on the previously stationary side, and then gently let it slide along the opposite groove face. It is somewhat similar to a ratcheting movement with a wrench. This technique will give a zigzag pattern and propel the torch down the groove. By varying the pattern, the travel speed can be faster or slower. It will also change the appearance of the bead by keeping the ripple pattern closer or farther apart. As you practice the technique, you will become smoother with the movements, resulting in a more uniform-looking weld bead.

Another technique is to simply rest the cup on the groove face as before and wiggle the torch from side to side, allowing the cup to maintain contact with the groove faces and at the same time moving the torch forward. It is best to use a smaller cup size when using this technique. This type of technique can give excellent results on the root pass of a groove weld.

GTAW ROOT WELD

For this multiple-process pipe weld, the root pass will be done with the GTAW process. This process is often selected when high-quality welds on special alloy materials are required and a minimal amount of postweld cleaning is desired.

When cup walking, there are two techniques for adding the filler wire. Depending on the joint fit-up, the welder might choose one over the other. One of these is referred to as the "lay wire technique." With this technique, the filler rod is laid down in the root of the joint. The root opening should be such that the filler wire will lie in the joint but catch on the root face, not allowing it to pass all the way through. The torch is then walked over the filler as the weld progresses. It is important to have the correct amperage setting and travel speed in order to obtain complete joint penetration and tie-in at the root of the weld.

The other technique is to feed the filler rod through the root opening to the back of the joint. The joint fit-up and preparation is critical with this technique. The tack welds must be good and solid and evenly spaced around the pipe. They should be spaced and of a size so that the root opening will not close up as welding progresses around the pipe. For 6-in. (150-mm) pipe (as used for your practice), four evenly spaced tacks approximately 1 in. (25 mm) in length and tapered at each end should hold the root gap open and keep it consistent. Smaller-diameter pipes may require fewer tack welds, and larger-diameter pipes may require more tack welds. The root opening should be just large enough that the filler rod can be fed through the root gap, not flop side to side and get stuck in the root gap closing on it. With this fit-up, you can now feed the filler rod into the weld pool from the back of the joint. Start by placing the rod through the joint and resting it on a tack weld that is opposite the side of the joint that is to be welded. Rest the filler rod on the side of the tack weld that the torch travel will be coming to first. By doing this, you will be able to steady the filler rod and have more control of feeding it. As the weld progresses around the pipe, you will change the angle that the filler rod is feeding into the backside of the weld. At the same time, you will be feeding the filler rod through the joint and into the back of the weld. Weld reinforcement can be varied by the amperage setting, as well as where the filler rod is being placed in the weld pool.

It is important that when using any technique, the root face is broken down so that complete fusion takes place. It is also important to practice using both hands to walk the cup. Depending on the position of the pipe, it is very likely that you will have to walk the cup using both hands. Failure to fuse the root face can show up as lack of fusion on an x-ray or break out on a bend test.

GMAW ROOT WELD

Running a root pass with the GMAW process has several advantages over the GTAW process, the main one being GMAW's travel speed, which is considerably faster than the GTAW process. GMAW also has less heat input and therefore less distortion. It also requires less operator skill than using the GTAW process.

When welding pipe with the GMAW process, it will be important to use both a tapered tip and a tapered nozzle. This will allow you to reach farther into the joint while welding, which allows you to utilize a shorter stick-out length. Many times, this will result in a smoother arc. You will also need to use an ER 70S series wire that is .035 in. (0.9 mm) in diameter. For the GMAW process, the use of an inert gas shielding mixture of 75% argon and 25% CO_2 prevents any reaction with the weld pool or arc.

It is important that you spend time to properly prepare the pipe and fit it up prior to the start of welding. This will help to make producing a consistent weld easier. The root pass will be made using the short-circuiting mode of metal transfer. It is important to make sure that the root edge of the groove is melted and fused with the root weld bead. Failure to have complete penetration and fusion will result in failed welds and weld tests.

The welding nozzle angle is an important variable. In general, it is more common to use a drag travel angle, sometimes referred to as a "trailing angle." "Drag travel angle" means that the welding nozzle will be tipped forward in the direction of travel, with the welding nozzle pointing back at the weld pool. "Push travel angle," sometimes called "leading travel angle," refers to pointing the welding nozzle in the direction of travel.

There are many important elements to consider for each technique. Remember that using a drag travel angle tends to increase weld penetration. It also results in a higher and typically more convex weld bead. Using the push travel angle will tend to reduce weld penetration; it will also cause the weld bead to be more flat-to-concave in appearance. A slight push travel angle can be used in the vertical position to help ensure that the molten metal stays in place. Both push travel angle and drag travel angle penetration will be affected by the size of the root opening.

It is important to keep a steady weld nozzle placement and angle. The welding wire needs to be focused close to the leading edge of the weld pool. This will help to ensure that uniform penetration is maintained. If the welding wire is placed too far ahead in the weld pool, it is likely to blow through the joint, resulting in pieces of welding wire sticking through the root of the joint. These bits of wire are referred to as "whiskers."

PRACTICE 12-1

Mild Steel Pipe, 1G Position

Welding pipe in the 1G position will be similar to welding plate in the 1G position. Just as when welding plate in 1G, the filler metal is added from above the weld joint, and the pipe is rolled to maintain the deposition of weld metal from above the joint. This can be achieved by using a weld positioner or by simply welding and stopping to reposition the pipe. Greatest weld depositions can be achieved by welding in the 1G position; in addition, it requires less skill to master. For this reason, the first practice will be to perform welds in this position. It is important that with any welds, you are to make sure that time is taken to carefully prepare the materials before welding.

For this practice, you will need 6-in. (150-mm) Schedule 80 pipe and a 3/32-in. (2.4-mm) E70-S3 mild steel filler rod, a grinder, a half-round bastard file and wire brush, various shielding gas cups ranging in sizes from 6 to 10, a 3/32-in. (2.4-mm) 2% thoriated or 1.5% lanthanated tungsten electrode, a collet, and a gas lens matching the tungsten size being used. You will also need PPE for the processes being performed. For the GTAW process, the use of the inert gas argon will protect the welding pool, arc, and electrode from the atmosphere.

After preparing the material as described earlier, you will tack-weld two pipe coupons together. For 6- to 8-in. (150- to 200-mm) pipe, four 1-in. (25-mm) tacks equally spaced around the pipe will be adequate. If the pipe is smaller, three tacks at approximately 120 degrees apart will work. This can be done as follows:

1. Take a filler rod that is about 1/32 in. (0.8 mm) larger than the root gap that is needed. For example, if a 3/32-in. (2.4-mm) root gap is called for, use a 1/8-in. (3-mm) welding wire.
2. Bend the rod into a V-shape to be used as a spacer to set the root opening.
3. Place the one pipe coupon on a flat surface with the weld groove side up.
4. Place the spacer rod on top of the pipe, and then place the other pipe coupon on top of it and weld groove side down.
5. Start by placing a tack weld on the side of the pipe that the open ends of the spacer rod come out.
6. Once that tack weld is completed, remove the rod and check the opening around the pipe. Make adjustments as necessary by spreading or closing the root gap to keep it uniform. You can use the rod to make sure that the root opening is consistent.
7. Place the next tack weld 180 degrees opposite the first tack weld. Continue checking the root opening to make sure that it is consistent.
8. Place the next tack weld 90 degrees from the previous weld.
9. The fourth and final tack weld will be 180 degrees from the last tack weld. This will give you four evenly spaced tack welds approximately 90 degrees apart, **Figure 12-3.**
10. Now grind both ends of the tack weld so that they are tapered to help form a ramp. This will enable you to start the weld on the tack and blend the weld down into the root. It also allows the weld to be tapered back onto the tack at the end of the weld.

Larry Jeffus

FIGURE 12-3 Grinding the tack welds to a featheredge will make it easier to maintain 100% root penetration.

You are now ready to start welding, as follows:

1. Place the pipe in a piece of angle iron that has the v-side up and has a piece of flat stock tacked to the bottom so that the pipe will be able to be held steady in position. A small notch can be cut in the angle iron to allow for clearance as the cover pass is applied. With this weld position, the lay wire technique or traditional freehand technique will work best.
2. Check your argon shielding gas flow; it should be between 15 to 20 cubic feet per hour (cfh) (7 to 9.5 L/min.).
3. Strike the arc down in the weld joint on one of the tack welds. Wait until you see that the weld pool is the desired size. Move forward and watch to

make sure that the root faces on both pipes are melted and that the tack weld blends into them as well.

4. Once both sides of the root face are melting, add the filler metal. The filler metal should feed through the root face but be a very close fit.
5. Using the lay wire technique, begin cup walking as described earlier, making sure to tie in the side walls of the groove face. With this technique, there should be no break in the filler rod and the weld pool. The filler rod should be feeding in at the leading edge of the weld pool. The travel speed should be rapid to make sure that there is no excessive buildup and that the weld pool does not get too big. If the weld pool gets too large, the excess heat and oversized weld pool will shrink as the weld cools, causing "suck back," which is when the weld pool shrinks as it cools and the bead becomes concave, resulting in lack of reinforcement on the root side of the weld.
6. You should weld from the two o'clock to the eleven o'clock position, welding from tack to tack.
7. Once you cover that distance, stop, rotate the pipe, and start again.

Tips for making the root pass weld include using a small-diameter cup so that the arc can be more easily focused at the root of the joint. Do not use too wide a root opening. For thinner-walled pipe, such as Schedule 40 and smaller, use the appropriate size of tungsten. The thickness of the pipe determines the size of tungsten to select. The thinner the material being welded, the less amperage that is required; and the less amperage, the smaller the tungsten needed to do a good job. If you are unsure about the tungsten size and amperage range, consult a tungsten manufacturer's guide.

After the root weld, all remaining passes will be performed using the SMAW process with E7018 electrodes. Starting out, using a 3/32-in. (2.4-mm) diameter rod is best. Once your skill has increased, you can work toward using a 1/8-in. (3-mm) electrode. If you have a weld positioner, use it, and welding should take place in either the eleven o'clock or two o'clock position. You should run weave beads, being careful not to carry too much metal. Your wave pattern should not be greater than two-and-a-half times the diameter of the rod that is being used. Practice until you get a weld that has uniform root reinforcement, buildup, and width. Have your instructor look at the welds and provide you with feedback. ◆

PRACTICE 12-2

Mild Steel Pipe, 2G Position

For this practice, you will need 6-in. (150-mm) Schedule 80 pipe and a 3/32-in. or 1/8-in. (2.4-mm or 3-mm) E70-S3 mild steel filler rod, along with 3/32-in. or 1/8-in. (2.4-mm or 3-mm) E7018 electrodes, a grinder, a half-round bastard file, a wire brush, various shielding gas cups ranging in size from 6 to 10, a 3/32-in. (2.4-mm) 2% thoriated or 1.5% lanthanated tungsten electrode, a collet, and a gas lens matching the tungsten size being used. You will also need PPE for the processes being performed.

1. Start by fitting two prepared pipe coupons together as described previously. They should be clean and free of slag, scale, rust, and grease.
2. Use a 1/8-in. (3-mm) filler rod to set the root opening.
3. Check to make sure that the inside surfaces line up and there is little to no mismatch. Eliminating as much mismatch as possible will help you achieve a uniform and defect-free root weld. If the pipe is not seamless, lining up the seams will often help eliminate the mismatch.
4. Once the pipe is tack-welded together, place it in the 2G position. The pipe should be fixed in a vertical orientation so that the weld groove will be running horizontal. Start the arc in the groove of the joint on one of the tack welds.
5. Heat the tack so that the desired size of weld pool is formed.
6. Start the travel by making a zigzag motion with the torch. The pattern should be just wide enough so that the point of the tungsten is directed at the edge of the root.
7. The filler rod should be placed in the weld groove at the root. Slight pressure should be placed on the filler rod as it is being placed in the weld groove. The lay wire technique works well with this position.
8. The filler wire should be placed in the groove tangent to the root. Continue to walk the cup around the pipe, keeping a steady speed and uniform weave pattern to the next tack. Remember to continue onto the tack weld and end the weld after the weld pool has reached about the middle of the tack.
9. Move around the pipe, continuing to weld from tack to tack until the root pass is completely around the pipe.

As an alternative technique, you can feed the filler rod in from the backside of the weld joint. To do this, perform the following steps:

1. Feed the filler rod through the pipe and rest it on a tack weld that is opposite the side where the weld is taking place.
2. As the weld progresses around the joint, follow the weld with the filler rod while resting it on the tack weld. It is best to have the rod resting on the side of the tack weld that you are coming up on first.
3. As you travel around, feed the rod through the opposite side weld groove. In order to do this, you must have a piece of rod long enough so that you will not be constantly starting and stopping. It will take some practice to get your hand steady to be able to direct the filler rod into the back of the weld and feed the weld pool at the same time. ◆

FILLER AND COVER PASS—SMAW PROCESS

After completing the root pass, you will switch over and use the SMAW process. The root pass should be tied in at the toe and be just slightly convex to flat across the face. If you have placed the root pass correctly, there will be no reason to grind it. The filler passes are made next, using an E7018 electrode with either a 3/32-in. or 1/8-in.-(2.4-mm or 3-mm)–diameter rod. Since this is a horizontal weld, the beads should be welded using a stringer bead format. Stringer beads are individual beads that are not weaved and are not much wider than the welding electrode diameter.

They are easier to make in this position and look better than a bead that is weaved. The first pass made with the SMAW process should be placed directly over the root pass. This is the pass that is typically referred to as the "hot pass." Since the root weld was made with the GTAW process and should be correctly tied in, there should be no real reason to run the first SMAW pass much hotter than normal. This pass should completely cover the root pass. The rod will be angled in an upward direction of approximately 30 degrees. The rod trailing angle should be between 75 degrees and 90 degrees.

Once the pass is complete, chip and clean the weld using a chipping hammer or file and a wire wheel on the grinder. Inspect it to make sure that it does not have too much roll. It should be flat to slightly convex. If there is excessive roll, use the grinder to remove it, and contour the bead surface to be flat to or slightly higher than the top of the weld. The toes of the weld should be tied in so that there is good fusion and no chance of trapping slag on future passes. The next filler pass should be placed so that it overlaps the previous pass by about half. It should be placed so that it is on the upper side of the weld groove. A rod angle similar to the first one will be used. Any variation will be used in this process to ensure that the bead is placed in the correct location. Be sure to chip and clean each pass as previously described. It is important that each bead that is laid down tie in to the surrounding weld beads and base metal.

Continue making filler passes until you are within approximately 1/16 in. (1.5 mm) below the surface of the pipe OD. Doing so will allow you to use the edge junction of the groove face and the outside of the pipe as a reference guide for making your final weld passes. In general, it is better to have a little more fill at the top of the groove than at the bottom of the groove. This helps to ensure that you do not have excessive roll on the bottom of the weld groove and that you do not end up with underfill on the top of the weld joint.

When you start to run the cover passes, the number of stringer beads required and the width across the top of the groove opening will determine the size of welding rod that you will use. Keep in mind that the schedule of the pipe also plays a role in this decision; however, the bevel angle and the distance across the top of the weld groove is primarily determined by the wall thickness of the pipe being welded. As you complete each stringer bead, be sure to clean it by chipping as well as using a grinder mounted with a wire wheel. Each and every pass should be thoroughly cleaned. It is also important to make sure that the piece does not become overheated, so the interpass temperature should be monitored during welding.

The cover pass can be made using a two- or three-bead technique, depending on the total weld groove angle and root opening. The next-to-last pass should be approximately 1/16 in. (1.5 mm) below the surface. It is important not to carry too large a weld pool because this can cause the bead to sag, resulting in excessive roll. Carrying an excessively large weld pool is often caused by a too-slow travel speed or too much heat. It is important to maintain a smooth and steady travel speed. Weld bead placement is another important factor in making a high-quality weld. Beads should be placed so that the toes of the welds do not end up on top of one another. There should be an overlap of one-third to one-half on each other. The welder should start by placing the first stringer bead at the bottom side of the weld groove and stack the beads one on top of the other. By properly practicing these techniques, your skill will increase, and soon you will be making consistent welds.

PRACTICE 12-3

Mild Steel Pipe, 5G Position

For this practice, you will need 6-in. (150-mm) Schedule 80 pipe and a 3/32-in. (2.4-mm) E70-S3 mild steel filler rod, as well as a 3/32-in. or 1/8-in. (2.4-mm or 3-mm) E7018, a grinder, a half-round bastard file, a wire brush, various shielding gas cups ranging in size from 6 to 10, a 3/32-in. (2.4-mm) 2% thoriated or 1.5% lanthanated tungsten electrode, a collet, and a gas lens matching the tungsten size being used. You will also need PPE for the processes being performed.

1. Start by fitting two prepared pipe coupons together as described previously. They should be clean and free of slag, scale, rust, and grease.
2. Use a 1/8-in. (3-mm) filler rod to set the root gap between the two pieces of pipe. Check to make sure that the inside surfaces line up and there is little to no mismatch. Mismatch will make it difficult to achieve a uniform and defect-free root weld. If the pipe is not seamless, lining up the seams will often help eliminate the mismatch.
3. Once the pipe is tack-welded together, place it in the 5G position, such that the pipe will be in a horizontal, fixed position. This results in part from the weld being made in the flat position, part in the vertical, and part in the overhead position.
4. Perform the root pass using the GTAW process. You will use 100% argon for the shielding gas. The flow rate should be set between 15 and 20 cfh (7 and 9.5 L/min.).
5. Start the arc in the groove of the joint on one of the tack welds.
6. Heat the tack so that a weld pool is formed.
7. Start the travel by making a zigzag motion with the torch. The zigzag pattern should be just wide enough so that the point of the tungsten is directed at the edges of the root. The filler rod should be placed in the weld groove at the root. Another acceptable technique is to freehand the root instead of cup walking. With this technique, you should use a slight forward movement and then a slight back-step movement. Using this technique, it is important to watch the root edge. A slight keyhole should form at the edge of the root. If the keyhole gets too large, you should reduce the amperage or slightly increase the amount of filler rod that you are adding.
8. Continue to work your way around the pipe, keeping a steady speed and uniform weave pattern to the next tack. Remember to continue onto the tack weld and end the weld after the weld pool has reached about the middle of the tack.
9. Move around the pipe, continuing to weld from tack to tack until the root pass is complete.
10. Once you have completed the root pass, check to make sure that there are no high spots on the pipe. If there are any, you can remove them with a grinder.

The second and remaining pass will be performed with the SMAW process. If the root pass was performed correctly, there is no real reason to run a hot pass. The weld will be made with 3/32-in. or 1/8-in. (2.4-mm or 3-mm) E7018 electrodes. The size and profile of the groove will determine which size rod will be appropriate. The smaller 3/32-in. (2.4-mm) rod will allow easier control of the weld pool when out of position. If the groove is larger because of the pipe wall thickness, such as when welding on Schedule 80 pipe as opposed to Schedule 40, it would be better to use a larger electrode.

When deciding on the diameter of filler rod, consider that a large filler rod in a smaller-profile joint is more likely to overfill the joint as you are performing the weld. A smaller filler rod in a larger-profile joint is more likely to underfill the joint, resulting in less efficiency in the fabrication process and causing more passes to be made to fill the joint.

The filler passes made with the 7018 electrodes will need to be made with an uphill progression. Because of the fluid weld pool and flux, the likelihood of trapping slag is very high when running the rod downhill. As you begin the filler passes, pay close attention to your start and stop points. You should stagger them so they do not all end up in the same location. A slight Z-weave can be used for these passes. It is important to pay close attention to the rod angle and travel speed. This is especially important at the four o'clock and seven o'clock positions on the pipe. These are the areas that if you are not careful, the weld pool will become too large and sag. This is why it is important to watch the weld pool and not the arc.

The number of weld passes needed to fill the joint will be determined by the schedule of pipe being welded, the groove profile, the travel speed, and the diameter and type of welding rod being used. Normally, it will take between two to three fill passes to fill the joint on Schedule 80 pipe.

It is important to remember that the 5G pipe weld is very similar to the 3G plate weld. Because the effects of gravity are going to react similarly on the two weld positions, weave beads are often used for the filler and cover passes in the 5G position. The cover pass is the final weld pass used to complete a weld. They are sometimes referred to as the "cap pass." To begin the cover pass, your next-to-last pass should be about 1/16-in. (1.5 mm) below the surface of the pipe. The edge of the pipe outside surface and the groove face will serve as a guide for you to follow as you apply the cover pass. As you weave the pass, you should bring the rod to the junction of the groove face and surface of the pipe, allowing the arc to break down the base metal to ensure that there will be good fusion along this edge. As you travel around the diameter of the pipe, you will have to change rods and restart the weld.

In order to have good tie in of one weld to the next, you want to start approximately 1 in. (25 mm) ahead of the weld where you stopped. Strike the arc, and bring the rod back into the shelf that was left when you stopped welding. This is where you want to start depositing weld metal and continuing the weave pattern. By starting the arc ahead of the weld, it affords an opportunity for the rod to warm up and the arc to stabilize. As you continue your weave pattern and weld over the area where the arc was started, the arc marks will be remelted and welded over. As you weld around the pipe, keep in mind that this position is a combination of several plate positions that transition from one to the other. Continue to practice, and once you are finished, have your instructor inspect the weld and provide you with feedback.

It is a good idea to practice welding all of the passes with both hands; do half of the pipe using your dominant hand and the other half using your nondominant hand. There will be times that it may be difficult or nearly impossible to make some welds with your dominant hand, so it is a good idea to practice for that situation. Always keep in mind that welding is an art, and there are many ways to achieve the same result. Regardless of which technique you adopt, practicing it is the key to making consistent, high-quality welds. ◆

PRACTICE 12-4

Mild Steel Pipe, 6G Position

For this practice, you will need 6-in. (150-mm) Schedule 80 pipe and a 3/32-in. (2.4-mm) E70-S3 mild steel filler rod, 3/32-in. or 1/8-in. (2.4-mm or 3-mm) E7018 electrodes, a grinder, a half-round bastard file, a wire brush, various shielding gas cups ranging in size from 6 to 10, a 3/32-in. (2.4-mm) 2% thoriated or 1.5% lanthanated tungsten electrode, a collet, and a gas lens matching the tungsten size being used. For the root pass, you will need argon shielding gas with a flow rate between 15 and 20 cfh (7 and 9.5 L/min.). You will also need PPE for the processes being performed.

1. Start by fitting two prepared pipe coupons together as described previously. They should be clean and free of slag, scale, rust, and grease.
2. Use a 1/8-in. (3-mm) filler rod to set the root gap between the two pieces of pipe. Check to make sure that the inside surfaces line up and there is little to no mismatch.
3. Once the pipe is tack-welded together, it should be placed in the 6G position, such that the pipe will be placed at a 45-degree angle and fixed position. This position will be similar to the 2G position, except that it will be tipped to the side at a 45-degree angle.
4. For the root pass, you can select one of the following three techniques. First, you can use the cup walking and lay wire techniques together. Second, you can feed the filler rod from the back of the joint and walk the cup. Or third, you can freehand the root with the forward and back-step technique. The latter technique is more commonly used on 2-in. (50-mm) diameter and smaller pipe. Whichever technique you choose, be sure to start the arc in the groove of the joint on one of the tack welds.
5. Heat the tack so that a weld pool is formed.
6. Start the travel by making a zigzag motion with the torch. The zigzag pattern should be just wide enough so that the point of the tungsten is directed at the edges of the root. The filler rod should be placed in the weld groove at the root. The lay wire technique will work well with this position. The filler wire should be placed in the groove tangent to the root.
7. Continue to walk the cup around the pipe, keeping a steady speed and uniform weave pattern to the next tack. Remember to continue onto the tack weld and end the weld after the weld pool has reached about the middle of the tack.
8. Move around the pipe, continuing to weld from tack to tack until the root pass is complete.

The second pass will be a filler pass performed with an E7018 electrode. These electrodes are designed to be operated in the vertical position, with weld progression in the upward direction. When making these welds, stagger all of the start and stops so that they do not end up in the same location. If all the starts and stops are in the same area, that can lead to potential discontinuities that result in the failure of the weld.

The 6G weld will be very similar to the 2G, except that it will be tipped on its side. For this reason, it is best to make the filler passes using the stringer technique, as follows:

1. Start by placing the first filler pass directly over the root pass.
2. Place the next pass just above the previous pass. You should overlap it by about half of the bead.
3. The third filler pass will go below the second filler pass and overlap it by about one-third to one-half. As a rule of thumb, you can direct the center of the electrode at the toe of the previous weld. It is best if you can add a little more weld on the top bead than the bottom. This will help ensure that there is enough filler on the top weld bead when it is time to put the final weld pass on.
4. Continue to place the filler passes in the groove until the surface of the stringer beads are about 1/16 in. (1.5 mm) below the surface of the pipe. The edge of the groove will serve as a guide for applying the cap pass. The profile of the weld should be approximately flush to slightly higher at the top of the groove. By doing this, it helps to prevent undercut.

The cover pass will be the final weld pass and will be capped off with stringer beads. The number required will depend on the size of the groove that is to be welded. Once again, it is important to stagger the starts and stops so that they do not end on top of the starts and stops of the previous passes. The size of the stringer beads should be close to the size of the welding rod that is used to make the passes.

1. It is important to make sure that you allow some time for the pipe to cool down between layers. If the pipe overheats, the stringer beads will sag similar to what happens if the welding amperage is too high.
2. Rod angle will be important with both the filler and cap passes. If the rod is held at too much of a drag travel angle, the bead will tend to have excess roll. This can be corrected by changing the angle to direct it more straight at the center of the pipe. Continue welding, remembering to clean in between each pass. Once you have completed the weld, have your instructor inspect the weld and provide you with feedback.
3. The cover pass can be made using a two- or three-bead technique, depending on the total weld groove angle and root opening. It is important to not try and carry too large of a weld pool. This can cause the bead to sag resulting in excessive roll. Carrying an excessively large weld pool is often caused by too slow of travel speed or too much heat.
4. It is important to maintain a smooth steady travel speed. Weld bead placement is another important factor in making a high-quality weld. They should be placed so that the toes of the welds do not overlap on top of one another. There should be an overlap of one-third to one-half on each other. The welder should start at the bottom and stack the beads on top of each other. Again, using the cup walking technique will result in a zigzag motion rocking the cup up and down to form a consistent weave pattern. By properly practicing these techniques your skill will increase, and soon you will be making consistent welds. ◆

Summary

Pipe welding requires a great amount of skill to master it. The curve of the pipe requires constant readjustment to make high-quality, consistent welds. As a student of pipe welding, you might find it helpful to practice making dry runs around the pipe. This can help you get a feel for following the contour of the pipe and actually see the angle at which you are directing the rod or welding nozzle. There are several different styles and techniques that can be utilized to achieve the same or similar results. It is often helpful to consult with other welders who have been working in the field for many years. You can often pick up tips that many have tried successfully. Above all, it is important to practice regularly to improve your skills.

Review Questions

1. What is a welding procedure specification (WPS)?
2. What are two things that have a major impact on the quality and efficiency of a weldment?
3. Why might the GTAW process be chosen over the SMAW process?
4. What type of weld beads should be used for welding with the SMAW process in the 2G position?
5. Why is it important to monitor welding interpass temperatures?
6. During welding, too large a weld bead is often caused by what?
7. When making a 2G SMAW cover pass, what will typically determine the number of stringer beads used?
8. What is the typical shielding gas flow rate for a root pass using the GTAW process?
9. List the three techniques that can be used for making the root pass with the GTAW process.
10. When welding in the 2G and 6G positions, if the rod is held at too much of a drag travel angle, the bead will tend to have ______________________________.
11. The cover pass should be made when the fill passes are how far from the outside surface of the pipe?
12. How should weld beads be overlapped in order to make high-quality welds?
13. When learning to weld pipe, it is a good idea to practice all weld passes using both ______________________________.
14. When welding with the GTAW process, how should you determine the diameter of tungsten that should be used?

Chapter 13

Machine and Automatic Pipe Welding

OBJECTIVES

After completing this chapter, the student should be able to:

- Describe automatic welding.
- Describe the equipment necessary to automate pipe welding.
- List the welding processes that are best suited for automated pipe welding.
- List the goals of automating pipe welding.
- List the advantages and disadvantages of stationary pipe welding equipment.
- List the advantages and disadvantages of orbital pipe welding equipment.

KEY TERMS

Automatic welding

Dual-torch

Dual-torch, tandem welding

Lateral walking

Orbiting pipe welding

Single-wire torch

Tandem-torch

INTRODUCTION

Automatic welding is when the arc, filler metal, joint travel, and joint guidance are all controlled by a machine. Some type of pipe welding automation has been around for nearly 100 years. The earliest automation process used rollers to roll the pipe under the welding head so that the weld could be made on the top of the pipe. In the early 1960s, orbital welding heads capable of traveling on a track around a pipe were introduced. The first ones were used by the aerospace and nuclear power industries to meet their need for a highly reliable and cost-effective way of producing pipe welds. The only welding process available to early automation used bare wire electrodes.

The type of tube/pipe GTA welder shown in **Figure 13-1** is available for a number of small-diameter tubes and pipes. The

Larry Jeffus

FIGURE 13-1 Automated tubing welder.

(Continued)

system allows backing gas to protect the root surface of the weld, **Figure 13-2**, and shielding gas to protect the weld face, **Figure 13-3**. The joint may use a consumable insert as filler.

FIGURE 13-2 Root surface.

FIGURE 13-3 Face surface.

Robots are available that can machine the fittings, assemble the system, and then make the welds, **Figure 31-4**.

Today, there are four processes that are well suited to automation and are the most commonly used. They are the gas metal arc welding (GMAW) process, flux-cored arc welding (FCAW) process, submerged arc welding (SAW) process, and gas tungsten arc welding (GTAW) process. Other processes that are used include hybrid laser arc welding, plasma arc welding (PAW), modulated current GMAW, and metal cored FCAW.

- Hybrid laser arc welding—Uses a laser with a GMAW or GTAW arc to produce high-speed, deep-penetrating welds that have narrow, heat-affected zones with excellent metallurgical and mechanical properties, **Figure 13-5**.
- PAW—Uses a concentrated, superheated plasma gas column to rapidly melt the pipe or tubing much as GTAW does, but much faster. Filler metal may or may not be added to the weld.

FIGURE 13-4 Welding robot.

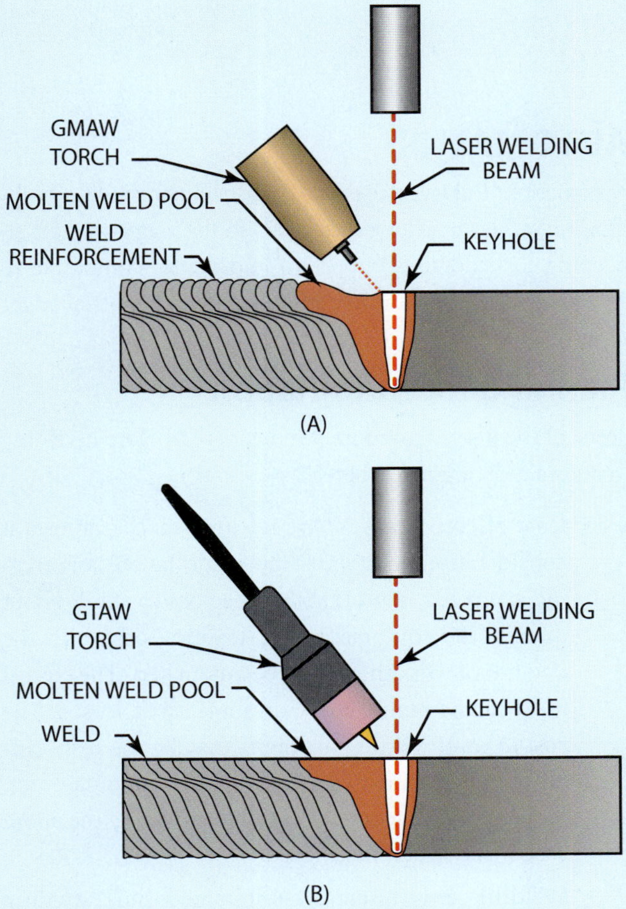

FIGURE 13-5 Hybrid laser welding with (A) GMAW and (B) GTAW.

(Continued)

- Modulated current GMAW—Uses a computer-controlled welding power supply to accurately control the size and shape of the molten weld pool.
- Metal cored FCAW—Uses a tubular filler wire much like the traditional FCA welding wire, but the filler is mostly metal powder that provides alloying elements and filler metal for an improved weld mechanical property.
- Submerged arc welding (SAW)—Has been used for years to weld longitudinal seams on pressure cylinders. Improvements in automation and in process controllers have resulted in SAW now being used to join pipe and fittings, **Figure 13-6**.

Larry Jeffus

FIGURE 13-6 Submerged arc beam welder.

The higher welding speeds that are capable with automated welding have some significant advantages other than just higher weld production. The higher speeds reduce the size of the heat-affected zone, which can be very important for some high-strength steels. Less heat input from the higher speeds also reduces the number of oxides that form on the hot pipe surface. Also, the reduction in heat input will result in a smaller amount of weld distortion.

AUTOMATION

Automation today is used to produce consistent, high-quality welds at a lower cost. In today's global market, keeping costs down is important in order for companies to stay competitive, and automation can give them that edge.

Planning for Automation

Some of the items that must be considered before investing in automation are listed here:

- **Cost-effectiveness**—One of the most important considerations that must be made before automating a process is profitability. Factors to think about include the equipment cost, training cost, and the amount of work that could be automated. The equipment can be purchased or, in some cases, leased. The cost to train welding operators may be the responsibility of the company, the equipment manufacturer, or both. The higher the volume of work, the more cost effective automating is.
- **Welding environment**—Automated pipe welding performed in a shop has significantly different requirements than automated pipe welding done in the field. The welding environment in a shop can be closely controlled, as opposed to field work, where wind, rain, dirt, and other contaminants are more difficult to contain. Welding tents are available to both provide shelter for the welder and control the welding environment.
- **Consistency**—The fit-up of the joints to be welded must be consistently accurate enough so that the automated process does not have to be tweaked between each weld. This is because having to make adjustments takes time and could void the welding procedure.

Property

The quality and type of welds that are needed for an application should be considered when deciding to automate welding. The important elements include:

- **Weld quality**—An automated welding system can repeatedly produce high-quality welds, thus reducing the rejection/repair rate typical of all manual welding. This can significantly reduce the total part cost while improving reliability.
- **High-strength and alloyed pipe**—Automated welding processes can produce the high-quality welds required for many of the new piping materials, which are more sensitive to welding.

SKILLED WELDERS

The demand for welders with the skills to perform pipe welds is very high, and the number of pipe welders available cannot meet the demand. Because of this, automation is often used in order to handle the workloads. Automating the pipe-welding process requires more welding skill and knowledge. Typically, companies that produce automated welding equipment recommend that experienced welders be trained to operate the automated equipment. This is because experienced welders can see (and sometimes hear) welding problems and correct them before they become too severe. This can help to produce more welds with fewer welders.

TYPES OF EQUIPMENT

The two major types of automated pipe-welding equipment are those where the welding torch is stationary and the pipe is rotated on a roller bed past the fixed welding torch, and also orbital welding, where the pipe is held stationary and the welding torch is mechanically moved all the way around the pipe. **Orbiting pipe welding** is an automated process where a track-mounted welding head moves around a fixed pipe or tubing joint to produce a weld that joins two sections.

Stationary Pipe-Welding Equipment

Advantages of rotating the pipe and having the welding head stationary are:

- The weld is being made in the flat position, and flat position welding is the most efficient technique and produces the highest-quality welds.
- Since the welding head does not have to move, it can be more robust, making it less likely to be damaged in the welding environment.
- The welding torch is always above the weld, so it is not subjected to as much spatter and thus will last longer.
- The coil of welding wire can be located well away from the welding spatter, and it can be larger since it does not have to be carried around the pipe on the welding head.
- Shielding gas lines are always above the welding spatter, so they are less likely to be damaged.
- A wide range of pipe sizes can easily be accommodated on the rollers by simply raising or lowering the weld torch, making it easier to change from one size of pipe to another.
- SAW can be used since the powdered flux will stay in place on the top of the pipe in the flat position.

Disadvantages of rotating the pipe and having the welding head stationary are:

- Not all pipes are round enough for the weld to be made without having to raise and lower the welding torch.
- **Lateral walking** (also called just "walking") describes what happens when the pipe moves horizontally as it is rolled because of unevenness in the pipe surface or misaligned rollers. Excessive walking can make it impossible to track the weld.
- Some slippage of the rollers can produce an uneven rate of weld travel.
- Pipe rollers, by their very nature, will have a maximum length of pipe that they can handle.
- Pipe rollers are not very portable, so they are typically only used in welding shops, not in the field.

Orbiting Pipe-Welding Equipment

Orbiting pipe-welding heads, sometimes called "bugs," can be closed or open. Closed orbital welding heads completely encircle the joint and enclose the GTAW torch. The welding tungsten is rotated around inside the closed head, where it and the joint are out of sight of the welder. They are often used to make small-diameter pipe or tubing welds ranging from a fraction of an inch up to around 8 in. (200 mm). Closed-head systems typically perform fusion welds without the addition of filler metal. Open-head orbital welding heads can be used with GTAW, GMAW, and FCAW processes and use a track, band, or chain attached to the pipe to both guide the head and provide traction for the drive wheels to pull the head around the pipe. The arc and joint are in full view of the welder. They are used on pipe of almost any size.

The welding heads also contain the necessary components to control the motion of the torch. Typical features of an orbital welding head include the rate of travel, arc length, torch oscillation, and dwell time.

- Travel speed can range from 4 to 60 in./min. (100 to 1500 mm/min.).
- The arc voltage is a function of the vertical height of the torch. Detection of any changes of the arc voltage is done with an automatic voltage control (AVC) sensor. The input from the sensor is used to make adjustments in the torch-to-work distance to maintain a preset voltage.
- The weave, oscillation, or both can be programed for the width, speed, and dwell time on the sides of the

weave. A typical oscillation rate ranges from 0 to 250 cycles per minute, **Figure 13-7**.

- The dwell time adds heat to the side of the weld to ensure groove face fusion.

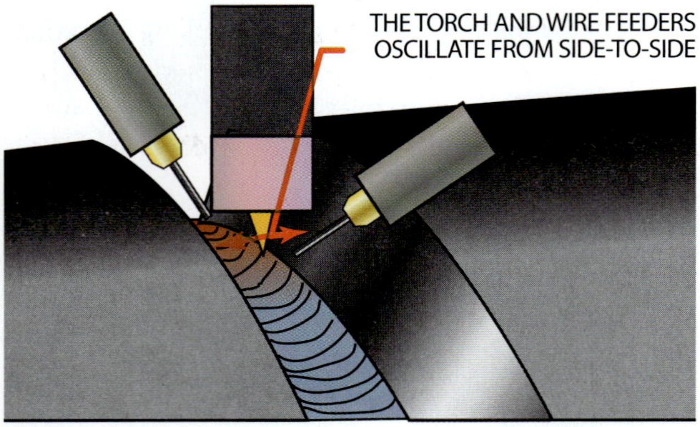

FIGURE 13-7 Orbiting GTAW twin wire feeder welder.

For GMAW, FCAW, and hot-wire GTAW, a wire feeder is also carried by the weld head. Most orbiting welding requires a shielding gas that is supplied through a gas hose. Some large-capacity orbital welding heads may also have cooling water circulating through hoses connected to the head. All orbiting heads have an electrode power cable attached.

Some orbital welding heads have feedback circuits that allow the program to make intelligent adjustments in the welding parameters to compensate for changes that might occur during the weld. These sensors can detect changes in the weld size, the pipe's temperature buildup, welding position, joint fit-up, joint tracking, arc force, and weld bead shape, and automatically make the necessary adjustments to produce an acceptable weld.

Advantages of orbital pipe welding heads where the pipe remains stationary are:

- There is no limit on how long a pipe section can be for an orbiting welding head to make pipe welds.
- The use of microprocessors have made orbiting welding heads relatively light, so they can easily be used both in the shop and in the field.
- Microprocessors have made it possible for preprogramed welding parameters to be downloaded to the orbital welding head, which makes it easier to repeatedly make high-quality welds.
- Low-profile orbital welding heads make it possible to make pipe welds on parallel, closely fitting pipes.
- An orbital welding head can be used on pipe in any fixed position, such as horizontal, vertical, or at any angle.
- One skilled welder can operate more than one orbital welding head at the same time.
- Multiple orbital weld heads can operate on the same pipe joint, leading to higher welding speeds and reduced distortion.

Disadvantages of orbital pipe welding heads where the pipe remains stationary are:

- The initial cost of the equipment can be prohibitive for small welding shops or when only limited work is being considered.
- Orbital welding requires accurate preparation and fit-up, so if that cannot be maintained, then orbital welding will not work.
- Even a slight misalignment of the orbital head's welding gun can result in lack of fusion.

Multiple Weld Heads

Two weld heads can be used at the same time, which allows the weld to be made faster and reduce weld distortion. The second head can be started after the first head has traveled a certain distance. It will travel on the opposite side of the pipe making a weld. Once the first weld head reaches a predetermined point, the welding with that weld head is stopped. The second head continues welding to tie into the first weld.

WELDING TORCH SETUP

There are four different types of welding torch setups that are used in automated welding. The first is the **single-wire torch**. This is the standard type of arrangement, where the welding torch has a single-wire electrode that is being fed. This would be the same as a standard GMAW machine.

The second type of arrangement is the **dual-torch** setup, which is when two single-wire torches are attached to a single weld head or bug. This setup requires two power sources, two feeders, and two welding torches.

The third type of welding torch arrangement is the **tandem-torch** setup, which requires two special power sources, two wire feeders, and a single torch that has two contact tips feeding wire out of each, but arranged in line out of a single welding torch. The tandem-torch setup has a lead arc or electrode and a trailing arc or electrode. Each welding electrode or arc is set independently from the other. The lead arc may employ a larger-diameter electrode and a different mode of metal transfer. The first electrode is responsible for establishing the depth of penetration and the weld pool. The second electrode is typically smaller and may use a pulse mode of metal transfer. Using the pulse mode will reduce the amount of heat input to the weld. The second wire is typically directed at the weld pool. The second electrode has several functions: to add filler metal to the weld pool and control the shape of the weld bead, sidewall penetration, and weld pool follow speed. On average, the leading arc contributes about 60% of the total weld metal that is deposited, and the trailing arc accounts for the remaining 40%.

The fourth type of welding torch configuration is the **dual-torch, tandem welding** configuration. It is a dual-torch setup but incorporates it into a tandem-torch configuration. A system of this nature would be used for larger-diameter pipe that would also have a thicker wall.

Wire Feeder

There are three different types of wire feeder systems that can be used. The GMAW and the FCAW systems use a filler wire, which is actually an electrode that carries welding current. The other system is referred to as a "cold wire," which is typically utilized with the GTAW process. A third type of feeder system is known as the "hot-wire feeder system." Hot-wire systems are typically used with either the GTAW or the PAW process, and they actually electrically preheat the filler wire prior to it entering the weld pool. By preheating the wire, it is possible to achieve deposition rates with the GTAW process that would be similar to that of GMAW.

ORBITAL WELDING HEAD SETUP

The weld head travels around the pipe using a band, track, or chain, each of which can serve to hold the head in place and provide it with a way to propel itself around the pipe. These devices are attached around the pipe and sit to one side of the weld joint. The offset distance is important since it determines where the welding torch will track. Check the head manufacturer's data sheet to determine the required offset for the orbital head that you are using.

The weld head attaches to the band and can engage the band with drive wheels to propel itself around the pipe. When necessary, the drive wheels can be disengaged so that it can freewheel around on the band for quick movement from one side of the pipe to the other. This feature is very beneficial when two weld heads are being used at the same time.

Many orbital weld heads are programmable. In some cases, the programing is done directly on the weld head. These systems often lack the ability to be programmed for multiple welds, so they may require additional setup time.

Fully programmable orbital welding heads can be set to multiple welding specifications. Depending on the equipment, the program or programs can be downloaded by using smart phones, tablets, laptop computers, or a network connection.

The following are the typical steps to set up an orbiting welding head:

1. Lock the head onto the track or band so it can circumnavigate the pipe.
2. Set the height and lateral alignment of the welding torch.
3. Confirm that the programming is correct by turning off the power and turning on the travel and oscillation, and watch the head as it travels around the pipe. Make sure that it is tracking correctly.
4. Check the travel speed by marking the position of the head and run the head for 6 seconds. Stop and measure the travel time between the original chalk mark and the current tip location. Multiply that by 10 to calculate inches (or millimeters) per minute.
5. Check the wire feed speed by running the wire feeder for 6 seconds, measuring the wire length and multiplying by 10 to get the inches (or millimeters) per minute.
6. If everything checks out, start the arc and allow the molten weld pool to be established before engaging in the travel.
7. Watch the progress and make necessary adjustments in travel speed, welding current, and horizontal tracking.

Summary

Over the years, there have been many changes in technology that have resulted in new types of welding power sources and orbital pipe welding systems. This has provided the industry with equipment that can produce consistent, high-quality welds that are repeatable and take less time to produce. This has helped to alleviate the shortage of skilled pipe welders.

Review Questions

1. Some form of pipe-welding automation has been around for approximately ____________ years.
2. How can automating pipe welding help with the shortage of highly skilled welders?
3. List three advantages of automation.
4. Name the two most common types of automated pipe welding equipment.
5. Explain what needs to be done so that consistently acceptable welds can be made with an automated process so that it is not necessary to tweak the settings for each weld.
6. On a tandem feeder system, the leading arc contributes an average of about _____% of the total weld metal that is deposited, and the trailing arc accounts for the remaining _____%.
7. What is the difference between a GMAW wire feeder system and a hot-wire feeder system?
8. In the pipe welding industry, the weld head is sometimes referred to as the _____.
9. Name the four different types of weld heads.
10. In orbital pipe welding, the welding head is propelled around the pipe via a(n) ____________________, ____________________, or ____________________.
11. How can the speed of the weld head be accurately checked?
12. How can the wire feed speed be accurately checked?
13. When setting up the welding head, why is important to check its offset?
14. The dual-torch, tandem-welding configuration is typically used for what diameter of pipe?

Chapter 14

Filler Metals

OBJECTIVES

After completing this chapter, the student should be able to:

- Describe how preheating and postheating pipe can affect a weld.
- Identify types of electrodes used in various weld passes.
- Describe how the type of shielding gas will affect a gas metal arc weld (GMAW).
- Explain how consumable inserts are used.
- Discuss the proper storage and handling of filler metals.
- Discuss how hydrogen embrittlement occurs and the problems that it causes.

KEY TERMS

Atomic hydrogen

Chromium-carbide precipitation

Consumable inserts

Hydrogen embrittlement

Interpass temperature

Molecular hydrogen

Preheating

INTRODUCTION

Manufacturers of filler metals may use any one of a variety of identification systems for their products. There is no mandatory identification system for filler metals. The most widely used numbering and lettering system is the one developed by the American Welding Society (AWS). Other numbering and lettering systems have been developed by the American Society for Testing and Materials (ASTM) and the American Iron and Steel Institute (AISI). A system of using colored dots has also been developed by the National Electrical Manufacturers Association (NEMA). Some manufacturers have produced systems that are similar to the AWS system, and most major manufacturers include both the AWS identification and their own identification on the box, on the package, or directly on the filler metal.

The AWS classification system gives the minimum requirements within a grouping. Filler metals manufactured within a grouping may vary but still be classified under that grouping's classification. However, a manufacturer may add elements to the metal or flux, such as more arc stabilizers. When one characteristic is improved, another characteristic may also change. The added arc stabilizer may make a smoother weld with less penetration. Other changes may affect the strength, ductility, or other welding characteristics.

Because of the variables within a classification, some manufacturers make more than one type of filler metal that is included in a single classification. This and other information may be included in the data supplied by manufacturers.

PREHEAT, INTERPASS, AND POSTHEAT TEMPERATURES

Preheating the pipe before starting the weld will reduce the stress caused by the weld and will help the filler metal flow. The most commonly used preheating temperature range is between 250°F and 400°F (120°C and 200°C) for most steel, **Table 14-1**. The preheating temperature can be as high as 1,200°F (650°C) when welding cast iron. Preheating is required anytime the metal to be welded is below 70°F (20°C) because the cold metal quenches the weld.

	Plate Thickness (in.)			
ASTM AISI	1/4 or Less	1/2	1	2 or More
A36	70°F	70°F	70°F	150°F to 225°F
A572	70°F	70°F	150°F	225 to 300°F
1330	95°F	200°F	400°F	450°F
1340	300°F	450°F	500°F	550°F
2315	Room Temp.	Room Temp.	250°F	400°F
3140	550°F	650°F	650°F	700°F
4068	750°F	800°F	850°F	900°F
5120	600°F	200°F	400°F	450°F

TABLE 14-1 Metal Types, Thicknesses, and Preheat Temperatures

The **interpass temperature** is the temperature of the pipe between weld passes. As welding progresses on a pipe joint, the heat of the arc can cause the pipe to heat up. Controlling the interpass temperature on smaller-diameter pipes is more problematic because they heat up faster than do larger-diameter pipes. The interpass temperature affects the mechanical and microstructural properties of the pipe and weld. A pipe's yield point (where the metal may first deform) and the pipe's tensile strength (the point where the metal fails) both can improve with the correct interpass temperature. Too high an interpass temperature can reverse these benefits. For most steel pipes, the interpass temperature should not exceed 500°F (260°C).

Postheating slows the cool-down rate following welding, which will prevent postweld cracking of some metals. Postheating also reduces weld stresses that can result in cracks forming some time after the part is repaired.

SMAW ELECTRODES

There is a variety of shielded metal arc welding (SMAW) electrode classifications and sizes available for the pipe welder. In addition, some manufacturers have more than one welding electrode in some AWS classifications. In these cases, the minimum specifications for the classification have been exceeded. These slight changes may alter the way the electrode is manipulated, provide faster or more efficient weld deposits, or allow the use of an additional welding current or polarity. The characteristics of each manufacturer's filler metals can be compared to one another by using data sheets supplied by the manufacturer. Most pipe welders are very selective and often are very loyal to a specific manufacturer's electrode.

E6010 AWS A5.1-04

Current DC DCEN These cellulose-based, flux-coated electrodes are used for vertical down-welding on large-diameter pipe. They can be used for root, hot, filler, and cover passes. Because this electrode's molten weld pool has excellent surface tension, it makes for a great open root pass and can be used to bridge gaps. **Figure 14-1** shows the minimum preheating and interpass temperatures for a range of pipe thicknesses.

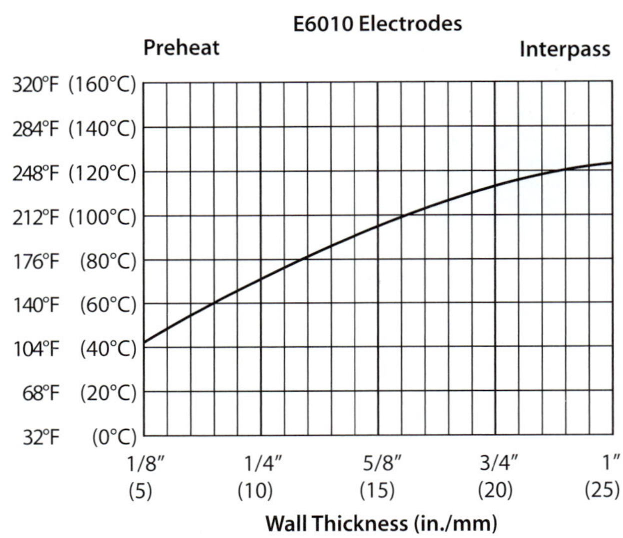

FIGURE 14-1 Minimum preheat and maximum interpass temperatures for E6010 electrodes.

E7010 AWS A5.5-96

Current DC These cellulose-based, flux-coated electrodes are used for vertical down-welding on large-diameter pipe. It can be used for hot, filler, and cover passes. Because of this electrode's penetrating arc and low slag, weld bead control is excellent in all positions for good x-ray quality welds. **Figure 14-2** shows the minimum preheating and interpass temperatures for a range of pipe thicknesses.

E8010 AWS A5.5-96

Current DC DCEN These cellulose-based, flux-coated electrodes are used for vertical down-welding on high-strength, large-diameter pipe. They can be used for hot,

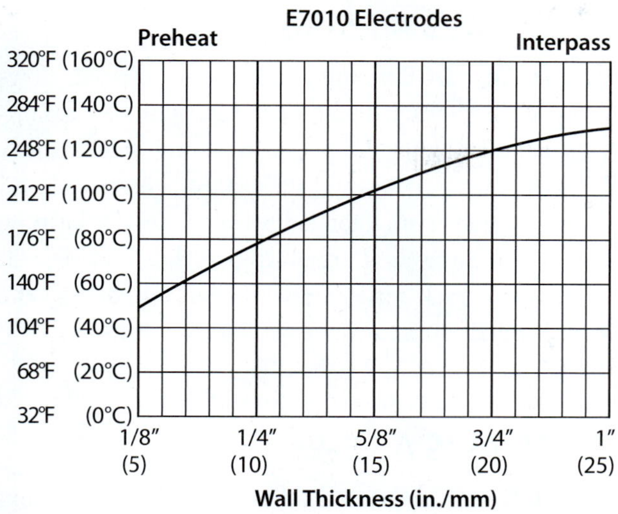

FIGURE 14-2 Minimum preheat and maximum interpass temperatures for E7010 electrodes.

filler, and cover passes and are excellent for cross-country pipelines. Because 0.6% nickel has been added to the electrodes, the weld deposit maintains the joint strength for temperatures as low as −40°F (−40°C). **Figure 14-3** shows the minimum preheating and interpass temperatures for a range of pipe thicknesses.

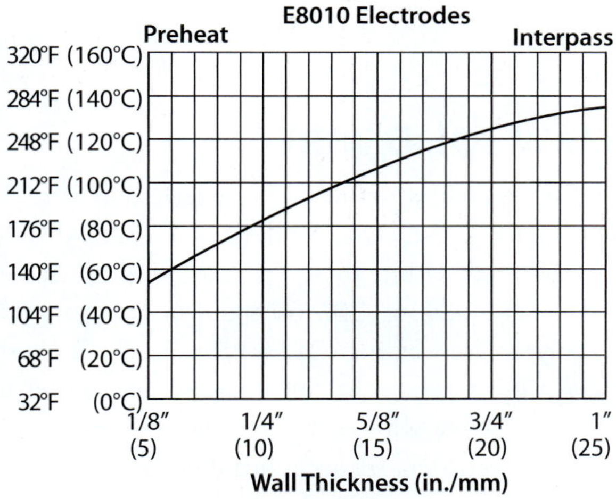

FIGURE 14-3 Minimum preheat and maximum interpass temperatures for E8010 electrodes.

E9010 AWS A5.5-96

Current DC DCEN These cellulose-based, flux-coated electrodes are used for vertical down-welding on high-strength, large-diameter pipe. They can be used for hot, filler, and cover passes. Because of the composition of the flux and core wire, there is excellent molten weld pool visibility, and the weld deposits have excellent metallurgical quality and soundness. They can be used in all positions and are resistant to porosity formation. **Figure 14-4** shows the minimum preheating and interpass temperatures for a range of pipe thicknesses.

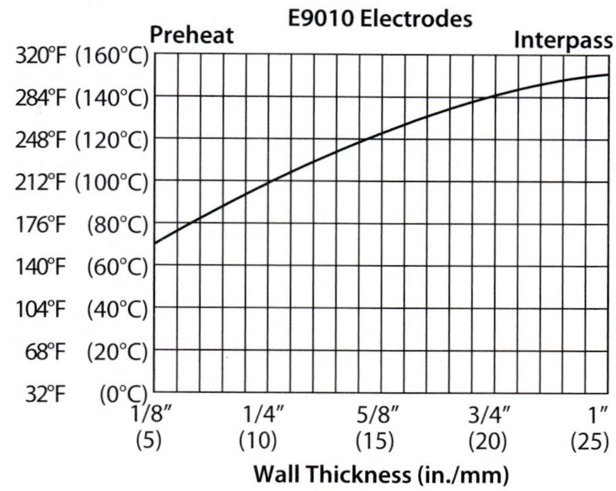

FIGURE 14-4 Minimum preheat and maximum interpass temperatures for E9010 electrodes.

NOTE

The weld joint design affects how well a cellulose-based electrode performs, **Figure 14-5**.

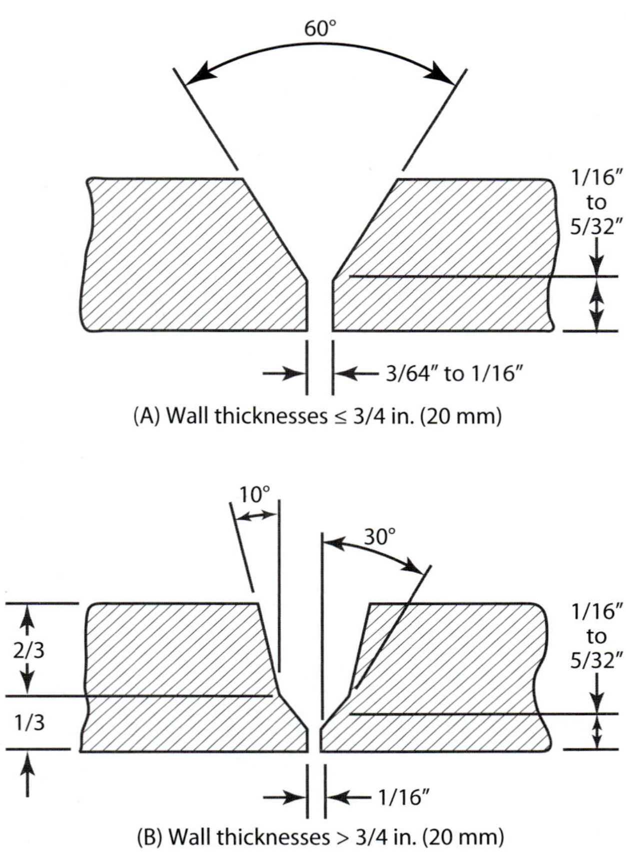

FIGURE 14-5 Recommended joint preparation for welding with cellulosic electrodes.

E7016 AWS A5.1-04

Current DC and AC This electrode is mineral-based, with the addition of rutile and silicates to the flux-coated electrode. DCEN works well on the root pass, and DCEP and AC can be used for filler and cover passes on pipe, tubing, and plate. The strong surface tension of the molten weld pool makes it easier to bridge gaps. The root gap should range from 3/32–1/8 in. (2–3 mm) with a root face ranging from 3/32–1/8 in. (2–3 mm). Postweld slag cleanup is easy.

E8016 AWS A5.5-96

Current DC and AC This is a mineral-based, low-hydrogen, flux-coated electrode. DCEN should be used for the root pass, and DCEP or AC can be used for filler and cover passes and for welding on plate. The optimum root gap should range from 3/32–1/8 in. (2–3 mm), with a root face ranging from 3/32–1/8 in. (2–3 mm). The weld bead has good impact properties for temperatures down to −40°F (−40°C).

E9016 AWS A5.5-96

Current DC and AC Mineral-based low-hydrogen flux-coated electrode. DCEN should be used for the root pass, and DCEP or AC can be used for filler and cover passes on pipe and tubing and for welding on plate. The optimum root gap ranges from 3/32–1/8 in. (2–3 mm), with a root face ranging from 3/32–1/8 in. (2–3 mm). The weld bead has good impact properties for temperatures down to −40°F (−40°C).

E8018 AWS A5.5-96

Current DC DCEN This is a mineral-based, low-hydrogen, flux-coated electrode that can be used for vertical up root filler and cover passes on large-diameter pipe. Its flux allows an easy arc strike and resists the formation of porosity at startup. It produces a weld that is extremely crack resistant and very tough. Because of the addition of 0.9% nickel to the electrode, the weld deposits maintain the joint strength and toughness for temperatures as low as −58°F (−50°C).

E9018 AWS A5.5-96

Current DC DCEN This is a mineral-based, low-hydrogen, flux-coated electrode for use with vertical up-filler and cover passes, large-diameter pipe, and structural steel. Its flux allows an easy arc strike and resists the formation of porosity on cover pass arc startups. The weld deposit is extremely crack resistant, with high toughness.

E10018 AWS A5.5-96

Current DC DCEN This is a mineral-based, low-hydrogen, flux-coated electrode for use with vertical up-welds on large-diameter pipe and structural steel. Its flux allows easy arc starting, and the electrode has very good welding characteristics for fieldwork. It resists forming porosity on cover pass arc startups. The weld deposit is extremely crack resistant, with high toughness. The recommended minimum interpass temperature is 212°F (100°C).

E11018 AWS A5.5-96

Current DC DCEN Mineral-based low-hydrogen flux-coated electrode for vertical up-welds on large-diameter pipe and structural steel. The weld deposit is extremely crack resistant, with high toughness. The recommended minimum interpass temperature is 230°F (110°C).

E12018 AWS A5.5-96

Current DC DCEN This is a mineral-based, low-hydrogen, flux-coated electrode used for vertical up-welds on large-diameter pipe and structural steel. The weld deposit is extremely resistant to cracking, with high toughness. The recommended minimum interpass temperature is 250°F (120°C).

GMAW SOLID WIRE

The AWS specification for carbon steel solid wire filler metals for gas shielded welding wire is A5.18. Filler wire classified within these specifications is used for gas metal arc welding (GMAW) and also can be used for gas tungsten arc welding (GTAW) and plasma arc welding (PAW) processes. The shielding gas selected for use with any GMAW wire will affect the welding process and the properties of the weld metal deposited. For that reason, the manufacturer's data sheet for a specific wire will contain the mechanical test results for weld metal deposited with each recommended shielding gas type or mixture.

ER70S AWS A5.28-96

This deoxidized mild steel filler wire is designed for high-quality pipeline welds. It can be used with 100% CO_2 or a mixture of argon and 15%–25% CO_2. The weld deposit is extremely crack resistant and maintains its toughness down to −58°F (−50°C). Welds can be made in all positions.

ER80S AWS A5.18-01

This deoxidized mild steel filler wire is designed for high-quality pipeline welds. It can be used with 100% CO_2 or a mixture of argon and 15%–25% CO_2. The addition of 0.9% nickel makes this an excellent choice for both on- and off-shore pipe welding. The weld deposit is extremely crack resistant and maintains its toughness down to −58°F (−50°C). Welds can be made in all positions.

ER90S AWS A5.18-01

This nickel molybdenum mild steel filler wire is designed for high strength and low creep at elevated temperatures. These properties make it a good choice for boilers, pressure vessels, and piping. It can be used with 100% CO_2 or a mixture of argon and 15%–25% CO_2. The weld deposit has excellent ductility and crack resistance. It maintains its toughness down to −75°F (−60°C).

GTAW FILLER METALS

Mild Steel Filler Metals

ER70S-6 AWS A5.18-01 This mild steel filler rod has high silicon content. It is used to join boiler pipes and tubing where high strength at elevated temperatures is required.

Stainless Steel Electrodes

The AWS specification for stainless steel bare filler rods is A5.9. Following the prefix ER, the AISI three-digit stainless steel number is used, which indicates the type of stainless steel in the filler metal. To the right of the AISI number, the letter L may be added to indicate a low-carbon stainless welding electrode. Keeping the carbon content low helps prevent chromium-carbide precipitation.

Chromium-carbide precipitation, often called "carbide precipitation," occurs in some stainless steels during welding when carbon combines with chromium. This reduces the chromium's ability to protect the pipe from corrosion. If carbide precipitation is going to form, it will occur at a temperature between 800°F and 1,500°F. Some stainless steels are less likely to form carbide precipitation during welding when the filler metal has less carbon.

ER308 and ER308L AWS A5.9 ER308 filler metals are used to weld 308 stainless steels that are used for chemical equipment piping, tanks, pumps, and stainless steel tubing.

ER309 and ER309L AWS A5.9 ER309 filler metals are used to weld 309 stainless steels that are used for high-temperature service, such as furnace parts and exhaust tubing.

ER316, ER316L, and ER316L-Si AWS A5.9 ER316 filler metals are used to weld on 316 stainless steels that are used for high-temperature service where high strength with low creep is desired. Molybdenum is added to improve these properties and to resist corrosive pitting. E316 filler metals are used for welding tubing, chemical pumps, filters, tanks, and furnace piping.

CONSUMABLE INSERTS

Consumable inserts are preplaced filler metals used for the root pass when consistent, high-quality welds are required. These inserts help to reduce the number of repairs or rejections when welding under less-than-ideal conditions, such as in a limited space. Although most inserts are used on pipe, they are available as strips for flat plate.

Inserts are classified by their cross-sectional shape (**Figure 14-6**) and listed as follows:

- Class 1, A-shape
- Class 2, J-shape
- Class 3, rectangular shape
- Class 4, Y-shape
- Class 5, rectangular shape (contoured edges)

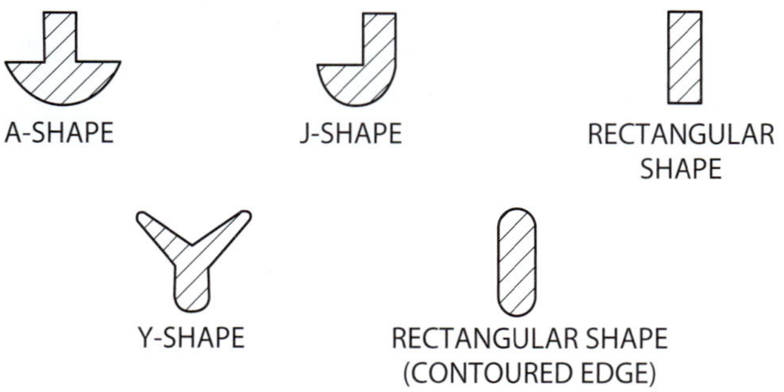

FIGURE 14-6 Common shapes for consumable inserts.

These are the most frequently used designs, **Figure 14-7**. Other shapes can be obtained from manufacturers upon request.

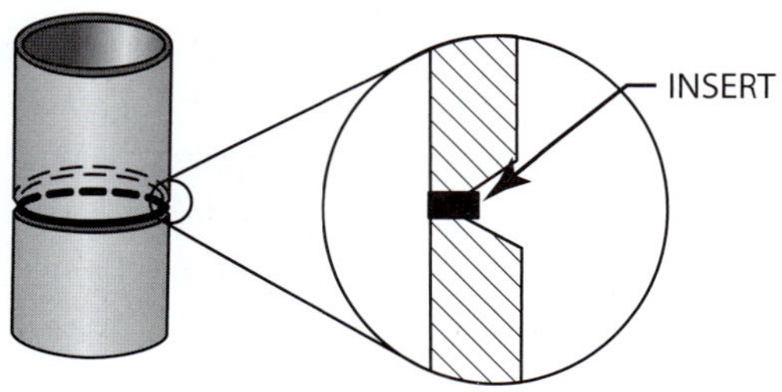

FIGURE 14-7 Placement of consumable inserts.

When ordering consumable inserts, the welder must specify the classification, size, style, and pipe schedule. Other information is available in the AWS Publication A5.30, "Specification for Consumable Inserts."

FILLER METAL STORAGE AND HANDLING

It is very important that filler metals are stored and handled properly. They must be stored in a clean and dry electrode hot locker to protect them from sources of hydrogen such as moisture, oil, grease, or other hydrocarbons, **Figure 14-8**. Electrodes can easily become contaminated with moisture, and this is one of the leading causes of hydrogen contamination of welds. Some electrode manufacturers package their electrodes in hermetically sealed metal cans to guard against moisture contamination during shipping and storage.

Larry Jeffus

FIGURE 14-8 An electrode hot locker is required to properly maintain electrodes.

Codes for some high-strength pipe welding require that the electrodes be kept hot until they are used and may limit the time that electrodes may be outside a hot electrode storage box before they are used.

Once some electrode types have been contaminated with moisture, they may not be used and must be discarded. These are primarily the cellulose-based, flux-coated electrodes. Some mineral-based, low-hydrogen, flux-coated electrodes can be used once they have been properly redried. The procedure for redrying varies with each type of electrode but generally requires them to spend 2 hours in an electrode hot locker at a temperature range from 570°F to 660°F (300°C to 350°C). See the electrode manufacturer's data sheet for specific requirements for redrying any damp electrodes.

When handling electrodes, it is important that they be protected from contamination. Some tips to achieve this are:

- Place electrodes to be used in a portable, thermal-type container when they are taken out of the heated electrode storage locker.
- Don't handle the electrodes with dirty, oily, or sweaty hands.
- Don't handle the electrodes with dirty, oily, or wet gloves.
- Don't open new electrode containers until you are ready to use them.
- Don't put electrodes in your pocket, **Figure 14-9**.

Larry Jeffus

FIGURE 14-9 The only electrodes you should ever put in your pocket are like the ones shown here—damaged or used and need to be disposed of properly.

HYDROGEN EMBRITTLEMENT

Hydrogen embrittlement is a potential problem for all steels, and it becomes a greater problem the higher the alloys in the steel are. A bonded pair of hydrogen atoms is **molecular hydrogen**, with the atomic symbol H_2, **Figure 14-10**. In nature, hydrogen atoms exist only if they are bonded together or to other types of atoms, such as oxygen, with which it forms water (H_2O). Hydrogen atoms are also present in many other substances, such as grease and oil. The heat of a welding arc can break the hydrogen bond, forming free, unbonded hydrogen atoms called **atomic hydrogen**, with the atomic symbol H. Atomic hydrogen is freely dissolved into the molten weld pool. As the weld metal cools, however, most of the dissolved hydrogen comes out of solution to form bubbles, much like CO_2 forms bubbles in a carbonated drink as it warms. Once the weld metal solidifies, these bubbles become porosity in the weld metal.

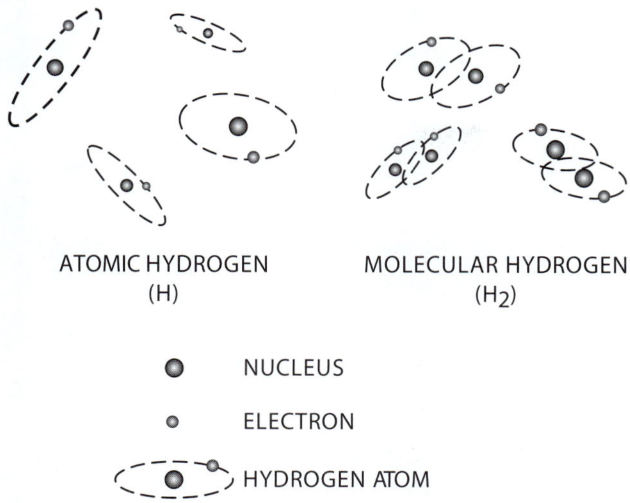

FIGURE 14-10 The two different forms of hydrogen.

Some hydrogen atoms stay dissolved in the weld metal as it solidifies and do not form porosity. Because hydrogen atoms are so small, they can move into the tiny spaces between the grains of solid metal. If enough of these free hydrogen atoms meet and combine to form molecular hydrogen, they can cause the tiny space to form an underbead crack, **Figure 14-11**. This hydrogen-caused cracking can appear hours, days, or weeks following welding. To control postweld hydrogen cracking-related problems, welding must be performed under the following conditions:

- All of the filler metals must be classified as low hydrogen, or a low-hydrogen process must be employed.
- Welding should not take place if the weld is wet or damp.

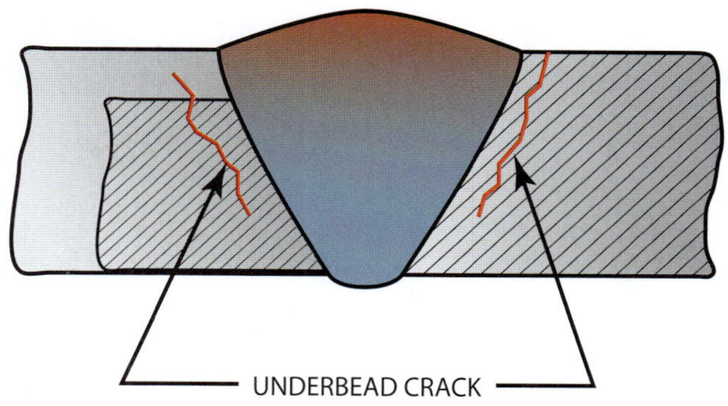

FIGURE 14-11 Underbead cracking caused by hydrogen embrittlement.

- All filler metals must be properly stored and handled.
- A wind block or shield should be used if there is a strong wind that may blow the shielding gas cloud away from the molten weld pool, resulting in atmospheric moisture contaminating the weld, **Figure 14-12**.

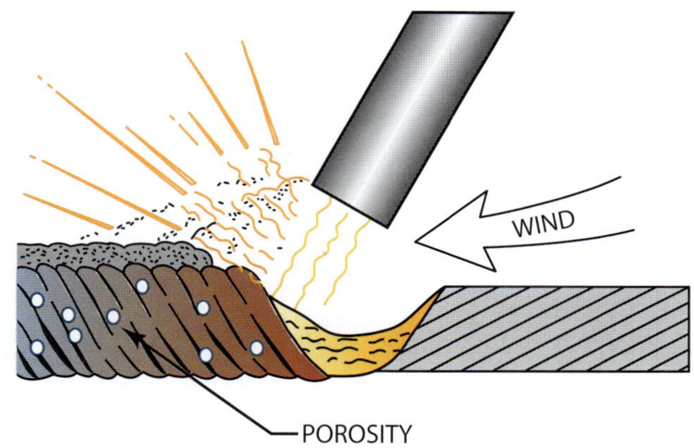

FIGURE 14-12 Shield the welding zone from wind.

Summary

At first glance, one might assume that selecting an electrode with a higher strength, such as E12018, would be a better choice than using an E7018. However, there are a number of reasons that would not be a good way to go. First, the higher-strength electrodes are usually a lot more expensive. Second, if the weld is too strong, it will be less flexible than the pipe, which could cause the pipe to crack next to the hard weld. Third, many of the higher-strength electrodes are harder to use, so it would take longer to make the weld. Fourth, and most important, in most cases, the type of electrode to be used on a pipe weld is mandated by code or the welding engineer.

Review Questions

1. What is the most widely used numbering and lettering system for filler metals?
2. What can be done before welding on pipe to reduce stress and help the filler metal flow?
3. What is the maximum temperature that the metal must be preheated to before starting to weld?
4. What is interpass temperature?
5. What can be affected by the interpass temperature?
6. What is the maximum interpass temperature to heat most steel pipes to?
7. What is the purpose of postheating following welding?
8. How can you compare one manufacturer's filler metal to that of another manufacturer?
9. Between what temperatures does carbide precipitation occur on stainless steels?
10. To reduce the possibility of carbide precipitation occurring on stainless steel, a welder can choose an electrode that has less of what element?
11. What are consumable inserts?
12. Under what welding conditions are consumable inserts used?
13. Why must filler metals be stored in a clean and dry electrode hot locker?
14. Why might an unused cellulose-based, flux-coated electrode need to be discarded?
15. Describe the general procedure for redrying electrodes.
16. List two things that can be done to protect electrodes from contamination.
17. Hydrogen embrittlement becomes a greater problem when the steel contains what?
18. What problems can hydrogen embrittlement cause?

Chapter 15

Welding Metallurgy

OBJECTIVES

After completing this chapter, the student should be able to:

- Explain why pipe standards are important.
- List the major mechanical properties of pipes.
- Explain the difference between tensile strength and yield strength.
- List and describe each of the 5 five most common types of pipe corrosion.
- Explain how to determine the corrosion allowance for pipe.
- List and explain the factors that affect the size of a weld's heat-affected zone.
- Describe how weld distortion occurs.

KEY TERMS

As-cast grain structure
Brittleness
Carbide precipitation
Compressive strength
Corrosion
Crevice corrosion
Ductility
Exfoliation corrosion
Galvanic corrosion
Hardness
Heat
Heat-affected zone (HAZ)
Localized or pitting corrosion
Pipe grades
Pipe standard
Shear strength
Temperature
Tensile strength
Thermal conductivity
Thermal expansion
Torsional strength
Toughness
Ultimate strength
Uniform or general corrosion
Yield point
Yield strength

INTRODUCTION

Metallurgy will help you to better understand the effect that heating, cooling, forming, alloys, and other elements have on a pipe's properties. Welding metallurgy will help you understand how welding may affect those properties because when the proper welding techniques are applied to a pipe, the weld will be as strong as or stronger than the original pipe.

Because carbon steel is the most commonly used piping material, and since some of the welding effects are similar to those of stainless, this chapter will focus primarily on carbon steel pipes.

ASTM, ASME, AND API

The American Society of Testing and Materials (ASTM) International, American Society of Mechanical Engineers (ASME), and American Petroleum Institute (API) all have developed a set of standards for pipe. Each **pipe standard** has a well-defined set of specifications covering the manufacturing process and chemical composition of pipes and fittings that fall within each standard. In addition, within some standards, such as A53, there are subgroups known as **pipe grades**, which further define the manufacturing process and chemical composition that give each grade its unique properties. Some pipe standards, such as A335, are specified to be manufactured as seamless pipe, **Figure 15-1**. A135 is available only as a welded pipe, **Figure 15-2**, while A53 pipe may be manufactured in seamless or welded form.

CARBON STEEL PIPE

Carbon steel pipe, which may be referred to as *steel pipe* or *black iron pipe*, is the most commonly used material for pipe because it is strong, ductile, machinable, reasonably corrosion resistant, weldable, and cheaper than other materials. Therefore, when carbon steel can meet the engineering requirement for any application, it should be the first choice. Carbon steel pipe is made from an alloy of iron (Fe) and carbon (C) and may contain small amounts of a variety of other alloys. Alloys are added to create a unique set of properties, such as added strength or corrosion resistance.

MECHANICAL PROPERTIES OF PIPES

The mechanical properties of a pipe can be described as those quantifiable properties that enable the metal to resist externally imposed forces without failing. The mechanical properties of all pipe standards and grades are known. As a result, a safe and sound system can be constructed.

The mechanical properties of the metal in pipes all interact with one another. Some properties are similar and complement each other, but others tend to be opposites. For example, a pipe cannot be both hard and ductile. Some pipes (such as A120) are hard and cannot be bent in close coils, while other pipes (such as A106) can be bent, flanged, or undergo another similar forming process. Probably the most outstanding property of metals is the ability to have their properties altered by some form of heat treatment. Pipes can be soft and then made hard, brittle, or strong by the correct heat treatment, yet other heat treatments can return the pipe to its original form.

The following sections describe some of the significant mechanical properties of pipes.

Hardness

Hardness may be defined as resistance to penetration. Most methods of testing hardness measure the depth that a small rounded ball penetrates the pipe's surface. The hardness in many pipes can be increased or decreased by heat-treating methods. Since hardness is proportional to strength, it is a quick way to determine strength. Harder

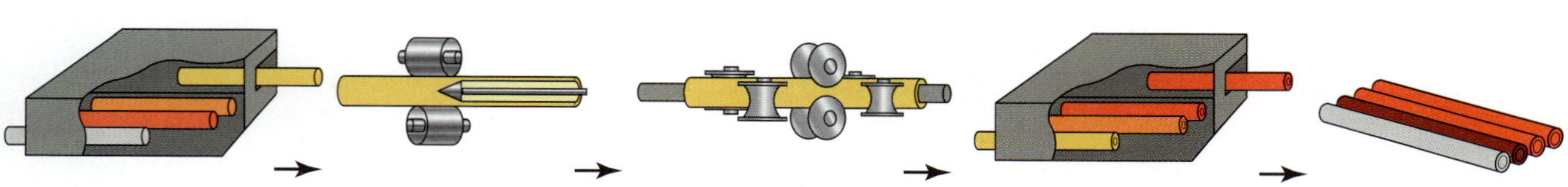

FIGURE 15-1 Manufacturing seamless pipe.

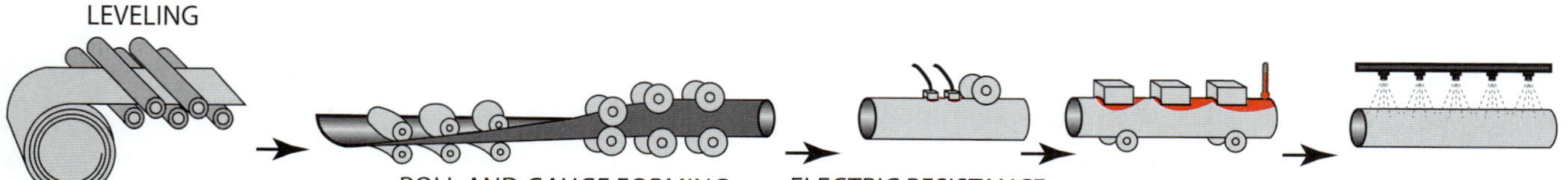

FIGURE 15-2 Manufacturing welded pipe.

metals tend to be stronger but more brittle, and softer metals tend to be more ductile. It is also useful to determine whether the metal received the proper heat treatment during manufacturing, during welding, or both. Hardness testing can be used to determine if a pipe was properly preheated and postheated during a weld. If the weld was overheated, it could be softer and more ductile but weaker; and if it was underheated, it may be harder and more brittle but stronger.

Ductility

Ductility is the ability of a pipe to be bent or formed without cracking or breaking. Ductility can be measured directly only by destructive tests such as tensile tests, where the percentage of elongation is measured, or bending testing. It can also be determined by a hardness test.

Brittleness

Brittleness is the ease with which a pipe will crack or break apart without any noticeable bending. Generally, as the hardness of a pipe is increased, the brittleness is increased as well. Brittleness is not measured by any testing method. It is the absence of ductility.

Toughness

Toughness is the property that allows a pipe to withstand forces, sudden shock, or bends without fracturing. Toughness may vary considerably with different methods of load application and is commonly recognized as resistance to shock or impact loading.

Toughness can be measured with the Charpy impact test. This test yields information about the resistance of a test specimen cut from a pipe to sudden loading in the presence of a severe notch. This test can be performed at room temperature, but because some metals become brittle and lose their toughness at low temperatures, some specimens are tested at a much lower temperature, such as −40° F (−40° C). A333 is a piping material that maintains its strength at low temperatures.

NOTE

A good example of a material that is very ductile when it is being pulled slowly under a light load but has little toughness under a sudden load and breaks with a brittle fracture is Silly Putty®.

Strength

Strength is the property of a metal to resist deforming. Common types of strength measurements are tensile, compressive, shear, and torsional, **Figure 15-3.**

- **Tensile strength** refers to the property of a pipe material that resists forces applied to pull metal apart. Tension testing has two stages: the yield strength and the ultimate strength. **Yield strength** is the amount of strain needed to deform a test specimen permanently. The **yield point** is the point during tensile loading when the pipe's metal stops stretching and begins to be made permanently longer by deforming. Like a rubber band that stretches and returns to its original size, metal that stretches before the yield point is reached will return to

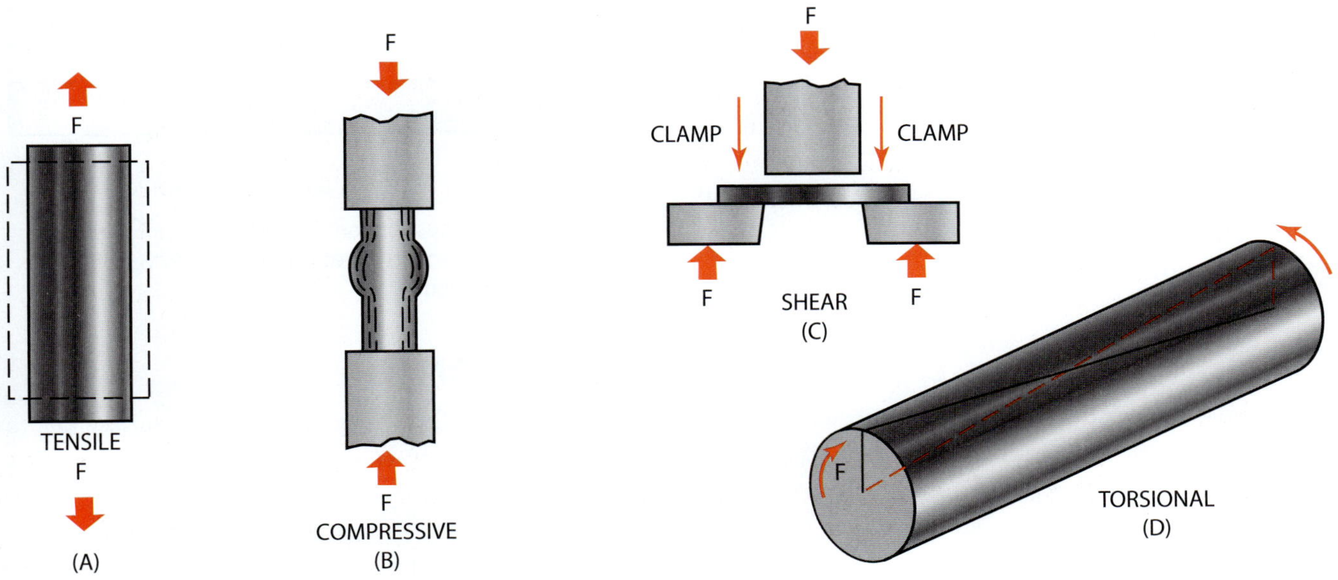

FIGURE 15-3 Stresses placed on metal.

its original shape. Pressure inside a pipe or pressure vessel causes it to expand, but so long as the pressure does not exceed the yield point, no permanent deformation will have occurred. After the yield point is reached, the metal is usually longer and thinner. Some metals stretch a great deal before they yield, and others stretch a great deal before and after the yield point. These metals are considered to have high ductility. **Ultimate strength** is a measure of the load that breaks a specimen. Some metals may become work-hardened as they are stretched during a tensile test. These metals will actually become stronger and harder as a result of being tested. Other metals lose strength once they pass the yield point and fail at a much lower force. Pipes that do not stretch much before they break are brittle. The tensile strength of a pipe's metal can be determined by a tensile testing machine.

- **Compressive strength** is the property of a material to resist being crushed. The compressive strength of cast iron, a rather brittle material, is three to four times its tensile strength.
- **Shear strength** of a material is a measure of how well a part can withstand forces acting to cut or slice it apart.
- **Torsional strength** is the property of a material to withstand a twisting force.

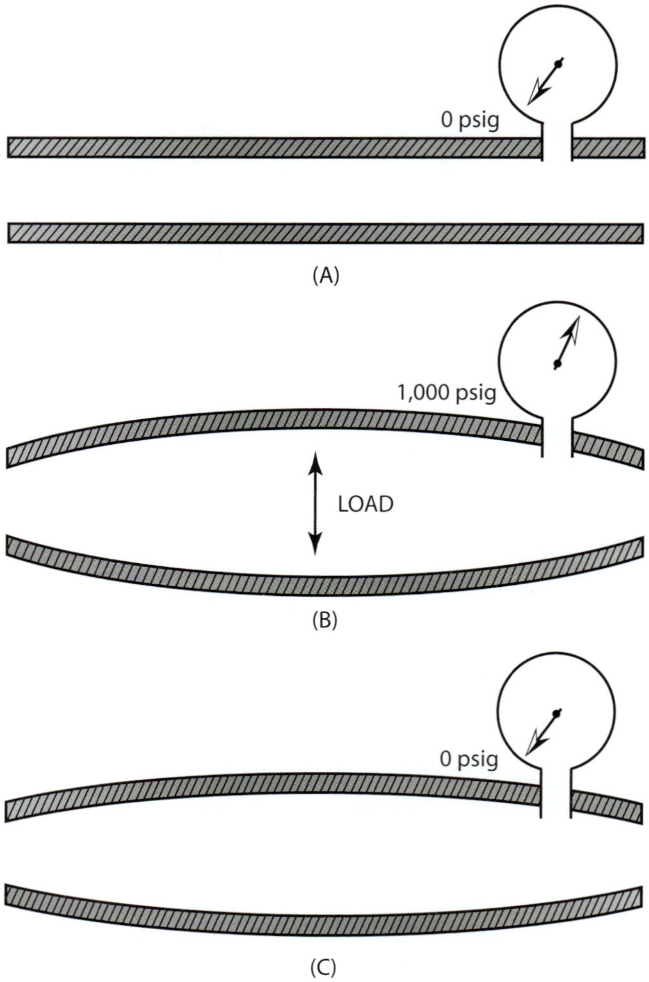

FIGURE 15-4 Stress versus strain.

Other Mechanical Concepts

Strain is deformation caused by stress. Some common stresses that might deform a pipe are overpressurizing and inadequate support for the pipe across a long span. The pipe shown in **Figure 15-4** is under stress and is strained (i.e., deformed) by internal or external loads. The deformation takes the form of a bend.

Elasticity is the ability of a material to return to its original form after a load is removed. The yield point of a material is the limit to which the material can be loaded and still return to its original form after the load has been removed, **Figure 15-5.**

Elastic limit is defined as the maximum load per unit of area to which a material will respond with a deformation directly proportional to the load. When the force on the material exceeds the elastic limit, the material will be deformed permanently. The amount of permanent deformation is proportional to the stress level above the elastic limit. When stressed below its elastic limit, the metal returns to its original shape.

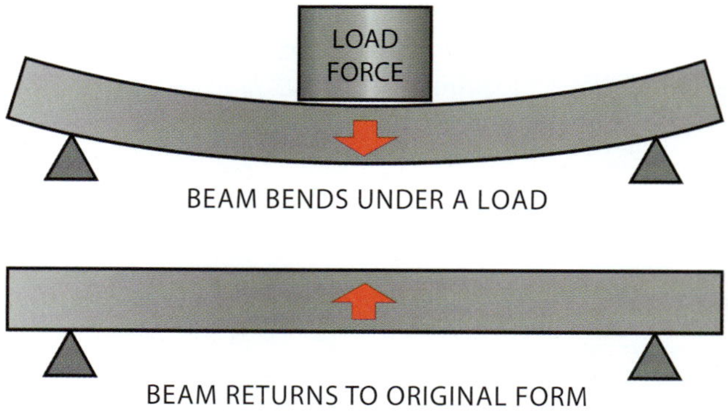

FIGURE 15-5 When a beam is loaded with a force less than its elastic limit, it will return to its original shape once the load is removed.

CORROSION

Corrosion is the slow removal of a pipe's surface due to oxidation. Oxidation is generally an electrolytic process where individual atoms of the metal are combined with oxygen. In the case of steel, the oxide that is formed this way is red and called *rust;* its molecular formula is Fe_2O_3.

Types of Corrosion

There are typically eight different classifications of corrosion. They are based on the type of corrosion and the environment the pipe is exposed to. The five most common types of corrosion that piping systems are affected by are uniform or general corrosion, galvanic corrosion, crevice corrosion, localized or pitting corrosion, and exfoliation corrosion, defined as follows:

- **Uniform or general corrosion** is the most common type of corrosion. It is what is seen covering the entire surface of an exposed plate, **Figure 15-6(A).**
- **Galvanic corrosion** occurs between two different types of metal that are placed together or within the same corrosive environment, **Figure 15-6(B).**
- **Crevice corrosion** occurs inside a crack or between two pieces of metal where moisture can be trapped, **Figure 15-6(C).**
- **Localized** or **pitting corrosion** is the type that does not occur evenly across the metal surface, leaving it covered with varying sizes of pits or dimples, **Figure 15-6(D).**
- **Exfoliation corrosion** occurs within the layers of metal causing pieces of the surface to become loose and possibly fall off, **Figure 15-6(E)**.

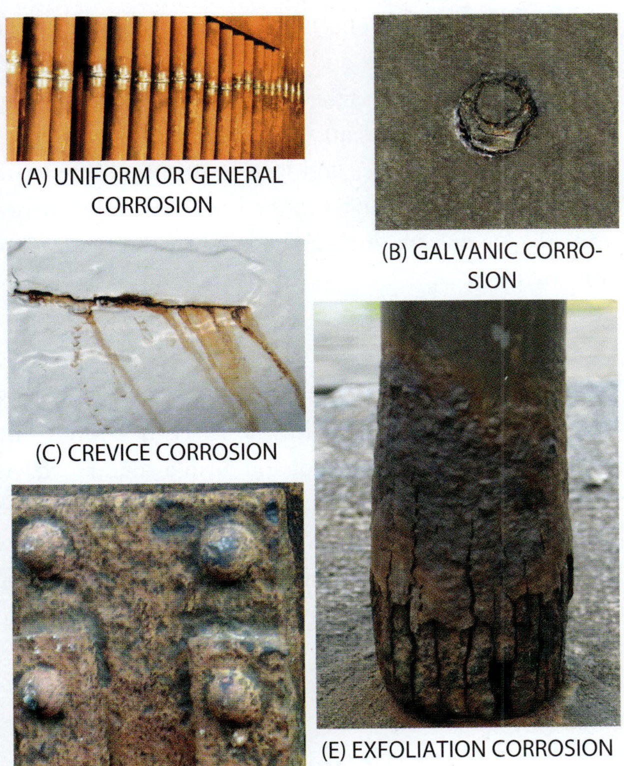

(A) UNIFORM OR GENERAL CORROSION

(B) GALVANIC CORRO-SION

(C) CREVICE CORROSION

(D) LOCALIZED OR PITTING CORROSION

(E) EXFOLIATION CORROSION

Larry Jeffus

FIGURE 15-6 Types of corrosion.

In many cases, a system can be attacked by more than one of these types of corrosion at the same time.

FACTORS THAT AFFECT CORROSION

A number of factors affect the rate that a pipe corrodes, including its composition, its environment, and any coating that may be applied. Stainless steel, aluminum, and copper are examples of piping metal whose composition resists corrosion.

The environment surrounding a pipe has the greatest potential effect on its corrosion. Since this type of corrosion is an electrolytic process, moisture and reactive chemicals in the moisture are a major factor. For example, salts, like those found in seawater or to some extent in soil, will accelerate corrosion, but extremely dry conditions will stop almost all corrosion.

Black iron pipe has a coating of black ferrous oxide (Fe_3O_4), which provides a limited amount of protection, while the bluish green epoxy coating on this natural gas line will give it years of protection even though it is being buried in damp soil, **Figure 15-7**.

Larry Jeffus

FIGURE 15-7 Corrosion protection on direct buried pipe.

NOTE

Black ferrous oxide is formed when steel is heated. In the case of pipe, it forms on the pipe as it is being manufactured. It provides some protection from rusting because an oxide cannot be oxidized again. However, if the black oxide is ground off, the pipe will quickly form red oxide if it is wet.

Allowing for Corrosion

All metals corrode, some faster than others. System designers know the required wall thickness of a pipe given the mechanical and/or pressure load. They also know the life expectancy of a piping system and the corrosion rate given its pressure, temperature, application, and environment. The standard safety allowance for corrosion is twice the expected depth of corrosion for the conditions. So if it is determined that the pipe wall thickness for the load is ¼ in. and that the depth of corrosion over the life expectancy is 1/16 in., the pipe selected should have a wall thickness of ¼ in. plus 2 × 1/16 in. (i.e., 3/8 in. thick).

HEAT, TEMPERATURE, AND ENERGY

Heat and **temperature** are both terms used to describe the quantity and level of thermal energy present. To better comprehend what takes place during a weld, you must understand the difference between heat and temperature. Heat is the quantity of thermal energy, and temperature is the level of thermal activity.

Although both heat and temperature are used to describe the thermal energy in a material, they are independent values. A material can have a large quantity of heat energy in it, but the material can be at a low temperature. Conversely, a material can be at a high temperature but have very little heat.

Example 1: Temperature versus Heat

A small spark from a grinder and a ball of slag flying off a weld are both at about the same temperature (around 2,700°F). But the small spark has so little heat that you might only feel a slight tingle if it hits your hand. However, the ball of slag has a lot more heat; if it lands on your hand, it will definitely burn you.

Example 2: Temperature versus Heat

If the point of a needle is held in a flame from a wooden match, the needle point will glow red hot. However, if you hold the flame under a coin, the coin will never get hot enough to glow. The difference is that the match flame does not have enough heat energy to raise the coin temperature high enough for it to glow.

Example 3: Heat versus Temperature

Suppose that you have two containers, and one is full of water at 32°F and the other is full of ice at 32°F, and you set them on a gas burner to put heat into each of them. Within a short time, the temperature of the container of water will begin to rise; but the container full of ice will stay around 32°F so long as there is ice remaining in it. The containers both started at the same temperature, but it took a lot more heat to raise the temperature of the container full of ice.

Example 4: Heat versus Temperature

If you have two steel blocks, one painted black and the other painted white, and place them in sunlight, the temperature of both will begin to rise, but the black block will get hotter faster. Both blocks are exposed to the same temperature, but the black block is picking up more heat. At some point, both blocks will reach the same temperature, and at that time, they will have the same level of heat.

Changes in the level of thermal activity can easily be read on a thermometer as a change in temperature. But the change in heat cannot be read directly with any test instrument.

The only way of determining a change in the heat of an object is by measuring its change in temperature and knowing its weight. In Examples 1 and 2, the temperature was the same, but the effect was different because of the size (mass) of the spark and needle point. In Examples 3 and 4, the heat input was the same, but its effect on changing the temperature was affected by something other than the size (mass) of the ice and black block. Both the spark and the needle had changes in the level of thermal energy, while the ice and black block had changes in the level of heat energy.

GRAIN STRUCTURES OF METAL

Molten metal from the smelting furnace is poured into molds, where it begins to cool and form a solid block. As the metal cools, it begins to solidify and form crystals, just like the crystals found in an ice cube. The crystals grow in random directions, so there is no real orientation. The term for this type of crystal is **as-cast grain structure**, **Figure 15-8**. Cast metal, like an ice cube, is hard and brittle.

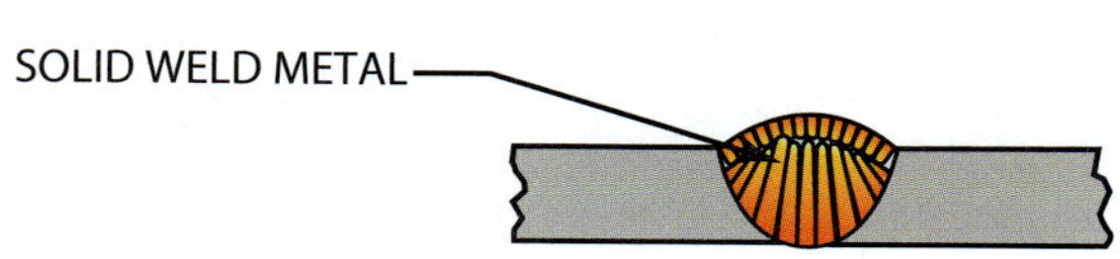

FIGURE 15-8 Grain growth in a weld begins along the unmelted edge of the base metal and grows inward toward the center of the weld.

To form pipe from a block of metal, it passes through many sets of rollers, where it is reshaped into a sheet to form welded pipe or into a blank to form seamless pipe. During the rolling process, the random crystals of the as-cast grain structure get broken down into smaller grain sizes and become oriented in the direction that they are being rolled, **Figure 15-9**. The breaking down of the grain structure is called *grain reduction,* and it makes metal stronger and less brittle.

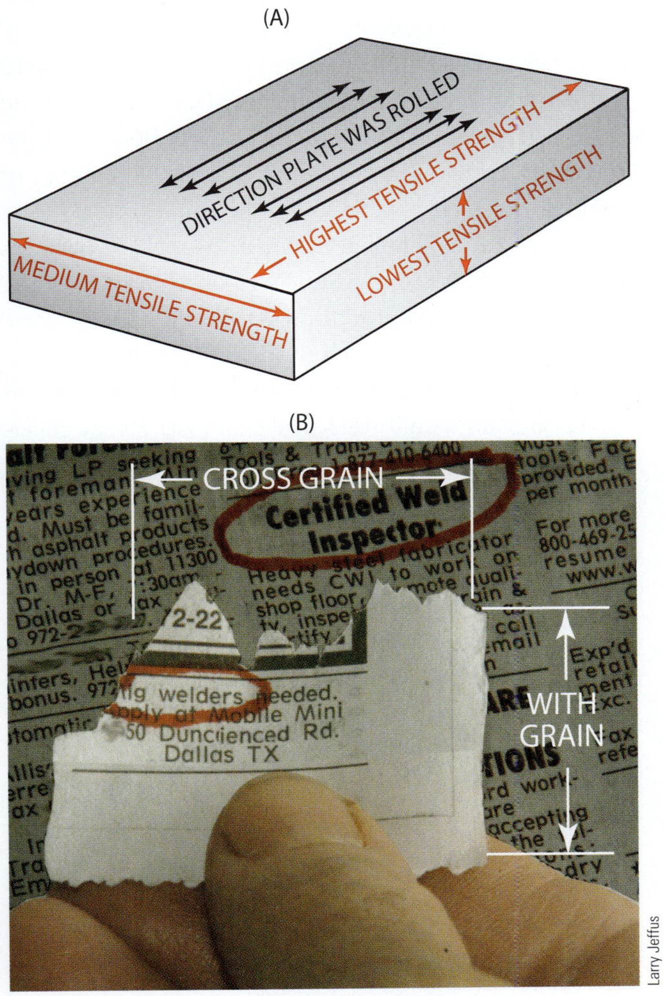

FIGURE 15-9 (A) Grain structure of a rolled plate. (B) Example of grain structure.

The tensile strength of metal differs depending on the direction from which the load is placed on it as compared to the rolling direction of the metal, **Figure 15-9(A)**. Wood is the most recognized material that has a visible grain structure; metal, much like wood, will break in one direction more easily than in another. In this aspect, steel is similar to a wooden board in that the direction of the rolled grain of the metal and the direction of the wood grain in a board both affect their strength. Many common materials have a grain; for example, when you tear a newspaper down the page, it tears fairly easily and straight. However, when you try to tear it across the page, the tear is much more jagged, **Figure 15-9(B)**.

Weld Metal Grain Structure

During arc welding, the grain structure of the pipe is melted along with the electrode forming the molten pool of metal. Conduction, convection, and radiation pull the heat out of the weld metal. This causes the temperature of the deposited weld metal to cool until it becomes solid. As the metal cools, it naturally forms an as-cast grain structure, which would result in a weak brittle weld. However, adding alloying elements to the welding rod causes the crystals to reduce in size and become more refined. This makes them much stronger; in fact, when done properly using the correct electrode and process, a weld will be stronger than the pipe itself.

Heat-Affected Zone

Within a few seconds, the weld and pipe go from being a solid at room temperature to a liquid and back to a solid near room temperature. This causes metallurgical changes in the heated region that is inevitable. The area where the grain structure was changed due to the weld is called the **heat-affected zone (HAZ)**, **Figure 15-10**. The lowest temperature at which any such changes occur in the pipe grain structure is defined as the edge of the HAZ. The exact size and shape of the HAZ are affected by a number of factors:

- *Type of metal or alloy:* Some metals are more easily affected by the welding process, while others are more resistant.
- *Method of applying the welding heat:* There is a large difference in the rate of heat input to the pipe by different welding and cutting processes. The faster the rate of heating, the less time that the heat can be conducted into the surrounding metal away from the weld or cut, **Figure 15-11**. Some heat sources, such

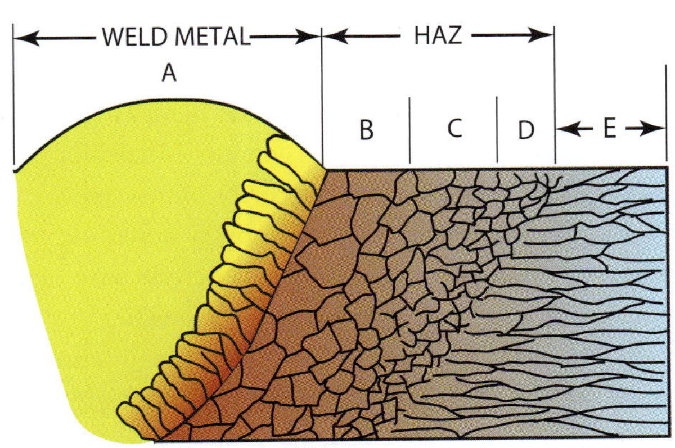

FIGURE 15-10 Heat-affected zone (HAZ).

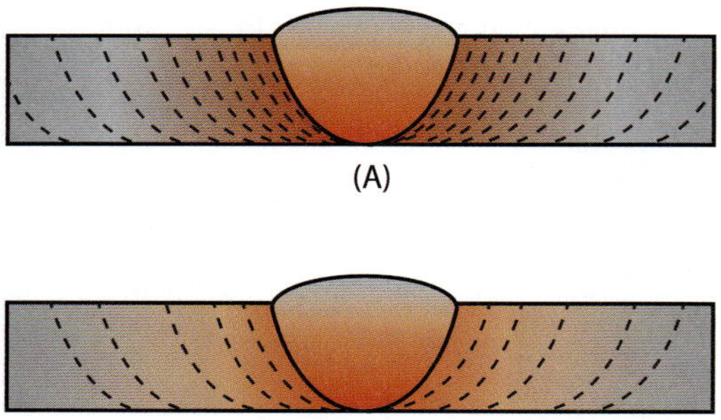

(A)

(B)

FIGURE 15-11 (A) Rapid heating concentrates the temperature close to the weld or cut. (B) Slower heating spreads the temperature farther across the base metal.

as plasma arc welding (PAW), are very concentrated. This high-intensity welding process can have a HAZ area that is only a few thousandths of an inch wide. Gas tungsten arc welding (GTAW) is a much slower process, and the resulting HAZ will be very large.

- *Weld bead size:* If a weld groove is filled with one large weld, it will have a wider HAZ than if the same groove is filled with a number of smaller welds.
- *Mass or size of the part:* The larger diameter and thicker the pipe being welded, the greater its ability to absorb heat without a significant change in temperature of the pipe, while small pipes may almost completely reach the melting temperature.
- *Preheating and postheating:* As the temperature of the base metal increases from preheating or postheating, the HAZ becomes larger. A cold pipe may have an extremely narrow HAZ.

The width of the HAZ is not a good or bad thing. Welds on some pipe alloys should have a narrower HAZ, while other pipe alloys should have a wider one. The controlling factor for the best width of the HAZ is the pipe alloy. The welding engineer for the pipe job will have established the welding procedure to ensure that the HAZ is the correct size.

STAINLESS STEELS

Stainless steels consist of four groups of alloys: austenitic, ferritic, martensitic, and precipitation hardening. The austenitic group is by far the most commonly used for piping. Its chromium content provides corrosion resistance, while its nickel content produces the tough austenitic microstructure. Stainless steel piping is relatively easy to weld, and a large variety of filler metals are available.

The most widely used stainless steels are the chromium-nickel austenitic types, which are usually referred to by their chromium-nickel content: 18/8, 25/12, 25/20, and so on. For example, 18/8 contains 18% chromium and 8% nickel, with 0.08% to 0.20% carbon. To improve weldability, the carbon content should be as low as possible. Carbon should not be more than 0.03%, with the maximum being less than 0.10%.

Keeping the carbon content low in stainless steel will also help reduce **carbide precipitation**, which occurs when alloys containing both chromium and carbon are heated. The chromium and carbon combine to form chromium carbide (Cr_3C_2).

The combining of chromium and carbon lowers the chromium that is available to provide corrosion resistance in the metal. This results in a metal surrounding the weld that will oxidize or rust. The amount of chromium carbide formed depends on the percentage of carbon, the time that the metal is in the critical range, and the presence of stabilizing elements.

Chromium carbides form when the weld is between 800°F and 1500°F (625°C and 815°C). The quicker the metal is heated and cooled through this range, the less time that chromium carbides can form. Since austenitic stainless steels are not hardenable by quenching, the weld can be cooled using a chill plate. The chill plate can be water-cooled for larger welds.

Some filler metals have stabilizing elements added to prevent carbide precipitation. Columbium and titanium are both commonly found as chromium stabilizers. Examples of the filler metals are E310Cb and ER309Cb.

In fusion welding, stainless austenitic steels may be welded by all of the methods used for plain carbon steels.

WELD DISTORTION

To make it easier to understand what is happening to the metal during welding, we must first define the terms used. Most dictionaries define the terms *distortion* and *warp* and the terms *distorted* and *warped* very similarly. However, in this discussion, the terms *distortion* and *warp* will be used as active and temporary events, such as "the part warps during welding" or "weld distortion affects the weldment." In these cases, it should be understood that once the welding is over, the metal will return to nearly its prewelded shape. The terms *distorted* and *warped,* as in "the weld distorted the plate" or "the weld warped the weldment," are past tense and refer to the fact that after it has cooled, the postwelded metal has been significantly misshapen as a result of the welding process.

All metals will distort by expansion when heated and will distort by contraction when cooled. Parts will return to their original shape when cooled if the heating is uniform and their shapes symmetrical. However, if the heating or the parts' shapes (or both) are not symmetrical, the parts will be distorted permanently to some degree as the result of the heating/cooling cycle. Almost every welding process involves some heat cycling. Most welding heat cycling is not symmetrical; as a result, the weldment will be distorted to some degree.

Thermal Expansion

The two factors that affect the degree to which a metal will distort and possibly remain distorted are its rate of thermal expansion and its rate of thermal conductivity, **Table 15-1**. **Thermal expansion** is the change in volume that metal has as the result of being heated. Basically, the higher the coefficiency of thermal expansion, the more the metal distorts. From Table 15-1, we see that tungsten has the smallest coefficiency of expansion and zinc has the largest. What this means is that if the same size pieces of tungsten and zinc are heated to the same temperature, the zinc will expand a lot more. Use the following linear expansion formula:

$$\Delta L = \alpha L_O \Delta T$$

where ΔL is the change in length, α is the coefficiency of linear expansion, L_O is the original length, and ΔT is the change in temperature, which is the difference between the beginning temperature and the ending temperature ($T_1 - T_2$). (A change in temperature is always expressed as a positive number whether the temperature was increased or decreased.)

Using this formula, calculate the change in length of a 100-in.-long tungsten bar and the change in length of a 100-in.-long zinc bar. The beginning temperature for both bars is 70°F, and the ending temperature is 570°F. Use Table 15-1 to determine the coefficiency of thermal expansion for each metal.

From Table 15-1: α (alpha) for tungsten is 2.4×10^{-6} or 0.0000024, and for zinc, it is 22.1×10^{-6} or 0.0000221.

Tungsten

$$\Delta L = \alpha L_O \Delta T$$
$$\Delta L = 0.0000024 \times 100 \times 500°F$$
$$\Delta L = 0.12 \text{ in.}$$

Zinc

$$\Delta L = \alpha L_O \Delta T$$
$$\Delta L = 0.0000221 \times 100 \times 500°F$$
$$\Delta L = 1.105 \text{ in.}$$

As you can see, the α for zinc at 22.1 is about 10 times greater than the α for tungsten at 2.4; and the expansion in length for the zinc bar was about the same—10 times

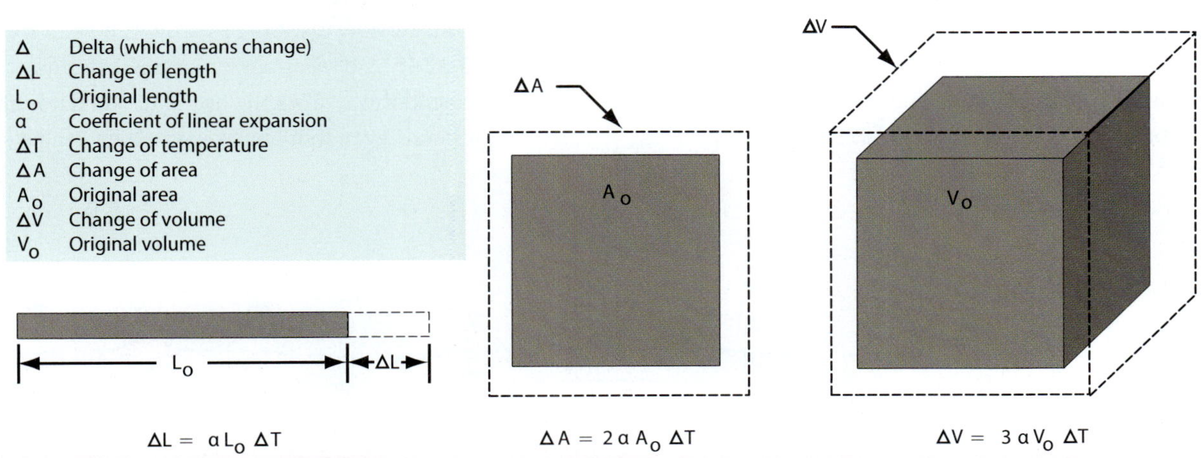

Properties of Metals

Type of Metal	Coefficient of Linear Expansion X 10^6 Per degree in./°F	Coefficient of Linear Expansion X 10^6 Per degree mm/°K	Relative Thermal Conductivity Copper = 1	Type of Metal	Coefficient of Linear Expansion X 10^6 Per degree in./°F	Coefficient of Linear Expansion X 10^6 Per degree mm/°K	Relative Thermal Conductivity Copper = 1
Aluminum	13.0	24.0	0.52	Steel, low carbon	6.7	12.1	0.17
Brass	11.0	19.0	0.28	Steel, medium carbon	6.7	12.1	0.17
Bronze	16.6	29.9	0.15	Steel, high carbon	7.2	13.0	0.17
Copper	9.4	17.0	1.00	Steel, stainless			
Gold	8.2	14.2	0.76	Austentic	9.6	17.3	0.12
Gray cast iron	6.0	10.8	0.12	Matensitic	9.5	17.1	0.17
Inconel	7.0	12.6	0.04	Ferritic	5.5	9.9	0.17
Lead	15.1	28.0	0.08	Tantalum	3.6	6.5	0.13
Magnesium	14.3	25.7	0.40	Tin	13.0	23.4	0.15
Monel	7.5	13.5	0.07	Titanium	4.0	7.2	0.04
Nickel	7.4	13.3	0.16	Tungsten	2.4	4.3	0.42
Silver	10.0	18.0	1.07	Zinc	22.1	39.8	0.27

TABLE 15-1 Thermal Expansion

greater than the tungsten bar. The length of the heated tungsten bar would be 100.12 in., and the length of the heated zinc bar would be 101.105 in.

Some metals have high values of coefficients of expansion, and others have much lower values. The metals with the higher values tend to distort more during welding, and those with lower values distort less.

Thermal Conductivity

Thermal conductivity is another factor that affects the degree of distortion experienced during welding. The more uniform a part is heated, the more uniform the expansion and contraction, which can result in the metal being less distorted following the heating/cooling cycle. The conduction and thermal expansion of copper is used as the standard, and it is given the value of 1.0. Other metals are compared to copper and given a value based on 1. Most metals conduct heat a little less quickly than copper. Silver at 1.07 conducts heat a little faster than copper. So when looking at Table 15-1, you see that all the metals listed have values less than 1.0 and conduct heat slower than copper. Metals with low thermal conductivity, like austenitic stainless steel that has a conductivity value of 0.04 and a higher coefficiency of expansion value at 6.7, and other metals with low conductivity numbers and high expansion numbers tend to be distorted more during welding.

You usually cannot change from one metal that is known for its problem with distortion to one that is less likely to be distorted. If the specifications call for the part to be made of stainless steel or any other metal that tends to be distorted from welding, that is the metal you have to use. But there are ways of controlling distortion to keep the distorted part within tolerance. Weld shrinkage is the primary reason that welds cause metal to be distorted.

If you take a flat plate and begin to heat it in the center to make a weld, it begins to expand away from the heat, **Figure 15-12(A)**. Initially, the expansion is uniform in all directions and the plate is bent away from the heat, **Figure 15-12(B)**. But as the metal gets even hotter, it becomes softer, and it reaches the molten state where it becomes fluid. It no longer is strong enough to bend the plate, so it begins to expand outward from the plate, **Figure 15-12(C)**. The result is that the hot metal has actually expanded outward more than it has expanded inward toward the cooler surrounding metal. At this time, the bending of the plate stops. Once the heat is removed, the metal begins to cool and shrink uniformly, **Figure 15-12(D)**. This uniform shrinkage causes the plate to bend toward the weld, **Figure 15-12(E)**. Because the heat caused the metal to become a little thicker where it was heated, there is more metal there to shrink. As the plate cools, it bends toward the heated side, resulting in a slight bow in the plate, **Figure 15-12 (F)**. To recap the process: As metal is heated, it initially bends away from the heat, but when it is allowed to cool, it bends even more back toward the heated spot.

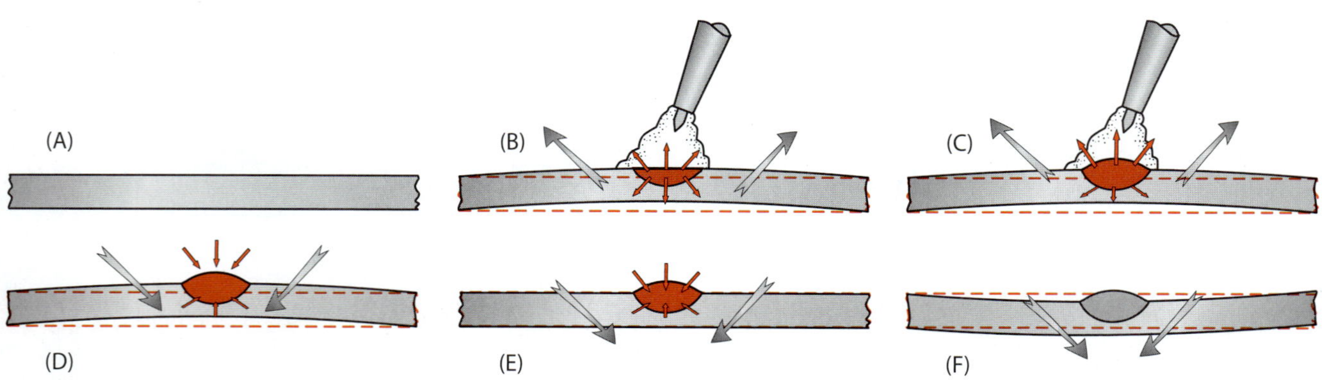

FIGURE 15-12 Thermal cycling causes metal distortion.

Summary

Thermal cycling occurs every time a pipe is welded, and this causes stresses to be applied to pipes and changes to their physical and mechanical properties. Often, as part of a postweld procedure, pipe welds will have the joint or part heat-treated. Computer programs are available to engineers to determine the optimal thermal cycling that will allow for the greatest strength in the piping system.

You do not have to have as detailed an understanding of the thermal effects on metals as a welding engineer does, but you must know the importance of controlling temperature cycles during welding. Pipe failures may be a result of welder-created problems if the welder fails to follow the required procedure of preheating and postheating.

Review Questions

1. Which organizations have developed standards for pipes?
2. What are some other names for carbon steel pipe?
3. What are some of the reasons that carbon steel pipe is the most commonly used piping?
4. Define the following terms as they relate to pipe:
   (a) hardness
   (b) ductility
   (c) brittleness
   (d) toughness
5. Define the following terms as they relate to pipe:
   (a) tensile strength
   (b) yield strength
   (c) yield point
   (d) ultimate strength
6. Define the following terms as they relate to pipe:
   (a) compressive strength
   (b) shear strength
   (c) torsional strength
7. What process can slowly remove a pipe's surface by oxidation?
8. List the five most common types of piping system corrosion.
9. Where does the term *black iron pipe* come from?
10. How thick would a pipe wall have to be to allow for corrosion if it has to be 0.5 in. mechanically and it is expected to lose 0.025 to corrosion over its lifetime?
11. What term refers to the quantity of thermal energy?
12. What term refers to the level of thermal activity?
13. What would the mechanical properties of an as-cast pipe have?
14. How does a block of steel become a pipe?
15. What keeps the weld metal from forming as-cast grain structure when it cools?
16. What defines the edge of a HAZ?
17. What effect on the width of the HAZ would a large weld have?
18. What effect does chromium and nickel have on austenitic stainless steel?
19. What temperature range might chromium carbides form in stainless steel?
20. What is meant by the term *thermal expansion?*
21. If metal is heated, does it bend toward or away from the heat?

Chapter 16

Weld Discontinuities and Defects

OBJECTIVES

After completing this chapter, the student should be able to:

- Explain the difference between discontinuities and defects.
- Describe discontinuities common to pipe welding.
- Discuss the causes of pipe-welding discontinuities.
- Explain how to avoid causing pipe-welding discontinuities.
- Describe the difference between lamination and delamination.
- Describe weld-related cracks.

KEY TERMS

Clustered porosity
Cold cracks
Cold lap
Cracks
Crater cracks
Cylindrical porosity
Defect
Delaminations
Discontinuity
Fit for service
Flaw
Hot cracks
Inclusions
Incomplete fusion
Interpass cold lap
Lack of sidewall fusion
Lamellar tears
Laminations
Linear porosity
Linear slag inclusions
Nonmetallic inclusions
Overlap
Piping porosity
Porosity
Root cracks
Scattered inclusions
Spherical porosity
Throat cracks
Toe cracks
Underbead cracks
Underfill
Uniformly scattered porosity
Weld tolerance
Wormhole

INTRODUCTION

The term **discontinuity** refers to anything that is not exactly like its surroundings—in short, a lack of uniformity. A knot in a wooden board and a cloud in a blue sky are both examples of discontinuities, but neither would be considered a defect in every case. However, a knot in a board would be a defect if it were so large that it caused the board to break. The term "defect" refers to any discontinuity that is so large or in such a location that it would cause the part not to meet the applicable standard; i.e., it might cause the pipe to break. **Flaw** is another word used to describe a discontinuity in welds. All welds have discontinuities and flaws, but they are not necessarily defects.

The American Welding Society (AWS) defines **defect** as "a discontinuity or discontinuities that by nature or accumulated effect render a part or product unable to meet minimum applicable acceptance standards or specifications. This term designates rejectability." Examples of defects are total porosity or length of slag inclusions.

In this chapter, the various discontinuities that may be found in pipe welds will be discussed. However, whether a discontinuity is a defect or not will be discussed in Chapter 18, "Testing and Inspecting Welds."

DISCONTINUITIES

Ideally, a weld should not have any discontinuities, but that is practically impossible. So most piping systems have welds that contain some discontinuities, and that is fine so long as these discontinuities fall within acceptable limits. The acceptable limits for discontinuities are determined by the code or standard that is being used. The code or standard is based on what type of weld would be fit for service. The term **fit for service** means that there is a very high likelihood that the weld will never fail during the expected life of the piping system. The term **weld tolerance** means that taken in its entirety, all the discontinuities fall within the limits set by the code or standard.

The reason that codes and standards based on a weld's fitness for service and tolerances have been established for welding is that the more perfect you try to make each weld, the more it will cost. That is, if every porosity bubble and spot of undercut were removed and rewelded on every pipe joint, untold hours would be needed even though the original weld would have been fit for service. It is important to note that welders try to make every weld flawless. Some spots of discontinuities would be acceptable individually, but when the total number of discontinuities in a weld passes the acceptable limit, they can become a defect.

So the only difference between a discontinuity and a defect is when the discontinuity becomes so large, or when there are so many small discontinuities, that the weld is not acceptable under the standards for the code for that product. Some codes are more strict than others, so that the same weld might be acceptable under one code but not another. But piping systems cannot have welds that contain defects because the system is no stronger than its weakest weld.

TYPES OF DISCONTINUITIES

The nine common types of weld discontinuities that may be found in pipe welds are listed in **Table 16-1**.

Porosity

Porosity refers to the small bubbles of gas that appear in solid weld metal. It results from gas that was dissolved in the molten weld metal, forming bubbles that are then trapped as the metal cools and becomes solid. The molten weld pool does not cool evenly. Much of the weld heat is conducted in the surrounding pipe and the remainder radiates off the top surface. This results in most of the weld grains growing from the sides and bottom of the weld, with smaller grains growing from the surface, **Figure 16-1**.

The bubbles that make up porosity cannot be seen because they most often form below the weld metal as it cools. Examples of gases forming bubbles as the temperature changes are the CO_2 bubbles that form in a carbonated drink as it warms and the air bubbles that collect to form a white mass in the center of an ice cube as the water freezes.

When porosity forms, it can be either **spherical porosity**, which is ball-shaped, or **cylindrical porosity**, which is tube-shaped. The cylindrical porosity is often called a **wormhole** because it can travel inside the weld as the weld moves along the joint. It is the most likely type of porosity to reach the weld surface and be seen. The rounded edges tend to reduce the stresses around them. Therefore, unless porosity is extensive, there is little or no loss in strength.

Discontinuity	Shielded metal arc welding (SMAW)	Gas metal arc welding (GMAW)	Flux cored arc welding (FCAW)	Gas tungsten arc welding (GTAW)	Oxyacetylene welding (OAW)	Oxyhydrogen welding (OHW)	Submerged arc welding (SAW)	Laser beam welding (LBW)	Plasma arc welding (PAW)	Electron beam welding (EBW)	Carbon arc welding (CAW)	Pressure gas welding (PGW)	Electroslag welding (ESW)	Thermite welding (TW)
Porosity	X	X	X	X	X	X	X	X	X	X	X	X	X	X
Inclusions	X	X	X				X				X		X	X
Inadequate joint penetration	X	X	X		X		X		X	X	X		X	
Incomplete fusion	X	X	X	X	X	X	X	X	X	X	X	X	X	X
Arc strikes	X	X	X	X										
Overlap (cold lap)	X	X	X	X	X	X	X	X	X	X	X	X	X	X
Undercut	X	X	X	X	X	X	X		X		X			
Crater cracks	X	X	X	X	X	X	X		X		X			
Underfill	X	X	X	X	X	X	X		X		X			

TABLE 16-1 Common Discontinuities and the Joint Types Where They Might Be Found

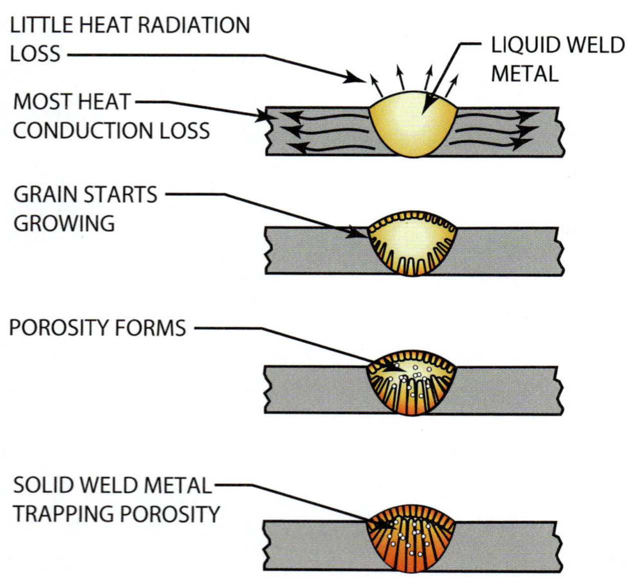

FIGURE 16-1 The weld begins cooling from the sides and top, so the center is the last to cool. This results in any impurities being forced to the center, where they remain in the solidified weld bead.

Causes of Porosity The two most common causes of porosity include improper welding techniques and contamination:

- **Improper welding techniques**—Some filler metals and some welding processes' shielding gas coverage of the molten weld pool may be lost if the proper techniques are not used. Examples of filler metals that can be affected by improper welding techniques are the many mineral-based, low-hydrogen electrodes such as E8016, E9016, E8018, etc. They should not be weaved wider than two-and-one-half times the electrode diameter because they produce very little shielding gas; and if the weld is not adequately covered, nitrogen from the air dissolves in the weld pool. As the weld cools, the nitrogen will escape the solidifying weld metal to form porosity.

NOTE

Even though the electrode is only weaved two-and-one-half times its diameter, the weld pool can be much wider than the weave.

An example of a welding process that can produce porosity if improper welding techniques are used is gas metal arc welding (GMAW). If the shielding gas coverage is lost due to high winds, improper welding torch manipulation, improper shielding gas flow rates, or a combination, porosity can be formed.

- **Contamination**—A weld can be contaminated by impurities left on the pipe surface or by improper storage and handling of filler metals. Some of the surface impurities that can contaminate a weld and cause porosity are paint, oil, and moisture. The intense heat of the weld arc causes these materials to form a vapor or burn, emitting smoke that can be dissolved into the molten weld metal. To prevent surface impurities from contaminating a weld, the surrounding pipe surface should be ground or wire-brushed clean within 1 in. (25 mm) of the weld groove face, **Figure 16-2**.

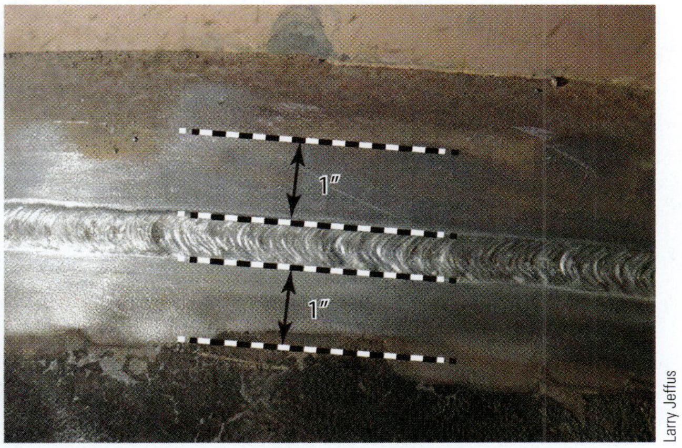

FIGURE 16-2 Cleaning the pipe surface at least 1 in. from the weld will prevent any impurities on the pipe from entering the weld.

One of the most common contaminants caused by improper storage and handling of filler metals is hydrogen. If filler metals are allowed to get wet (or even damp from atmospheric humidity), the weld arc will break water into free hydrogen and oxygen. The electrode's gaseous cloud will eliminate the oxygen, but the hydrogen will become dissolved into the molten weld metal where it will form porosity later, **Figure 16-3**. Electrodes can also be contaminated if they are handled with oily gloves.

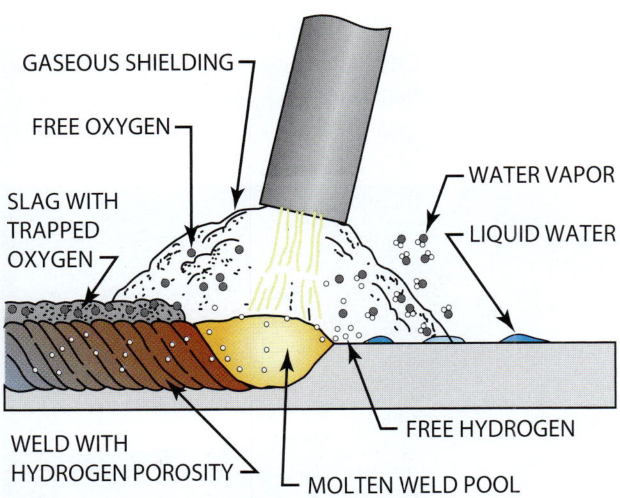

FIGURE 16-3 Water decomposes in the presence of the welding arc. The oxygen combines with the gaseous cloud, but some of the hydrogen is dissolved into the weld pool.

Types of Porosity Porosity can be grouped into four major types according to what causes it, as follows:

- **Uniformly scattered porosity** is most frequently caused by poor welding techniques or faulty materials, **Figure 16-4.**

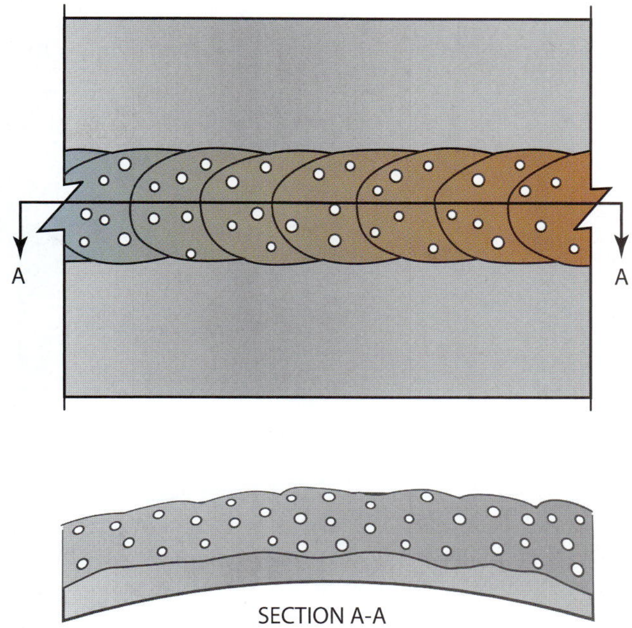

FIGURE 16-4 Uniformly scattered porosity.

- **Clustered porosity** is most often caused by improper starting and stopping techniques, **Figure 16-5.**

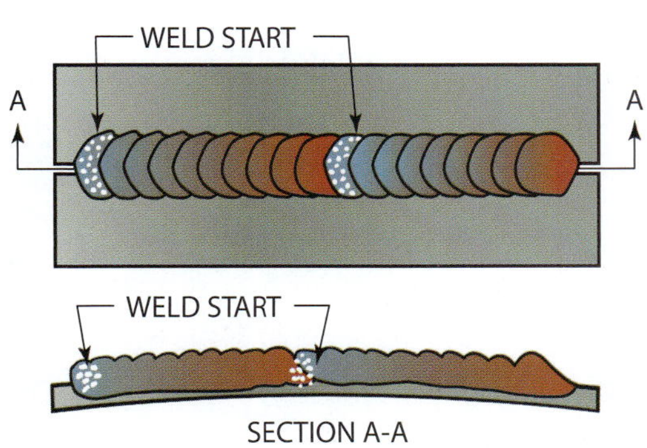

FIGURE 16-5 Clustered porosity.

- **Linear porosity** is most frequently caused by contamination within the joint, root, or interbead boundaries, **Figure 16-6.**

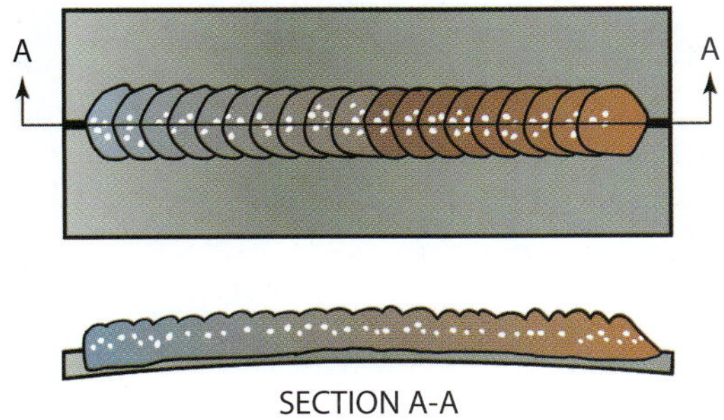

FIGURE 16-6 Linear porosity.

- **Piping porosity**, also known as "wormhole," is most often caused by contamination at the root, **Figure 16-7.** This porosity is unique because its formation depends on gas escaping from the weld pool at the same rate as the pool is solidifying.

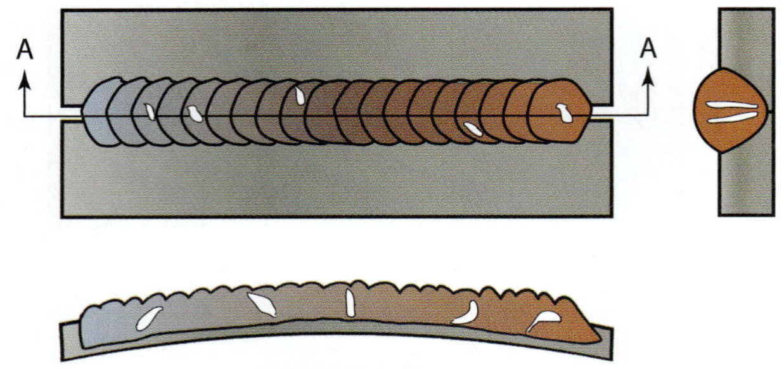

FIGURE 16-7 Piping (wormhole) porosity.

Inclusions

Inclusions are foreign objects not melted by the weld arc that are left in the weld groove as weld metal is deposited over them, **Figure 16-8**. There are two types of foreign objects that can be inclusions in pipe welds–nonmetallic and metallic materials. Examples of **nonmetallic inclusions** are slag and oxides. The only metallic (foreign metal) inclusion found in pipe welding is tungsten from the gas tungsten arc welding (GTAW) process.

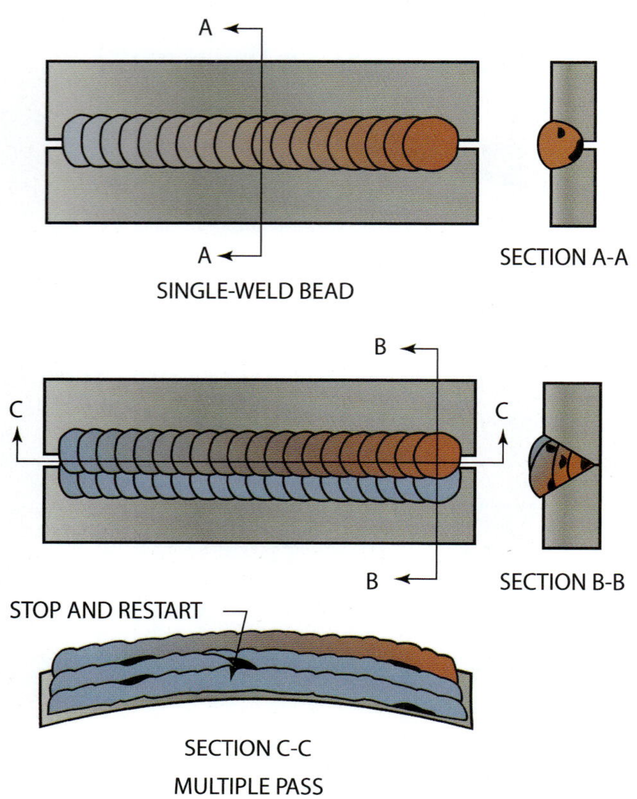

FIGURE 16-8 Nonmetallic inclusions.

Causes of Inclusions There are four common causes of weld inclusions–improper joint cleaning between weld passes, improper groove preparation, joint design, and improper welding technique:

- **Improper joint cleaning between weld passes**—Nonmetallic inclusions get trapped between the layers of welds when multiple weld passes are required to fill a joint and the previous weld pass was not properly cleaned.

Slag can be trapped alongside a weld if it was not cleaned properly before the next weld pass was made. Some welders don't completely chip, wire-brush, or grind the slag off a weld before making the next weld pass. They believe that the slag will be melted away; however, the melting temperature of slag is much higher than the temperature of the molten weld pool. Depending on the slag and base metal, there may be more than a 1,000°F (540°C) temperature difference between the melting temperature of the slag and the base metal. It would be much like trying to melt sand by pouring hot water on it. However, in some cases, a hot pass can "burn out" small amounts of slag if it is hot enough to melt all the base metal around the slag. The problem is that if it was not all burned out, the weld may not pass inspection and have to be removed and the joint rewelded.

Oxides such as rust or corrosion can become an inclusion if they are not removed before welding begins. Like slags, mill scale and rust have much higher melting temperatures than does the base metal. Some of the sources of oxides are the ones left on a thermal cut or rust that might form overnight, **Figure 16-9**.

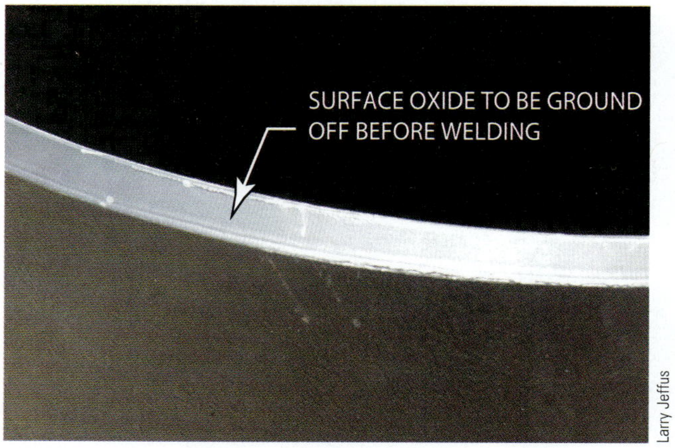

FIGURE 16-9 Grind down the surface oxide left from a thermal cut.

- **Improper groove preparation**—Unevenness of the weld groove face or of the previous weld surface can result in defects and discontinuities, where weld metal can get deposited over foreign material such as slag or rust, trapping it under the weld.

- **Joint design**—If there is not sufficient room for the correct manipulation of the electrode and when slag is allowed to flow ahead of the arc due to limited electrode movement, then weld slag entrapment can occur.
- **Improper welding technique**—Metallic inclusions can occur when the tip of the tungsten electrode is accidently touched to the molten weld pool and gets broken off due to improper welding technique.

Types of Inclusions Inclusions can be grouped into two major types:

- **Linear slag inclusions** are long strings of slag that lie along one side or both sides of a previous weld pass. Often, welds that have a very high crown trap slag along the sides, which requires extensive grinding to remove. On an radiograph (x-ray,) linear slag inclusions may or may not show as continuous lines alongside the weld and are referred to as "wagon tracks" because of their appearance.
- **Scattered inclusions** can resemble porosity, but unlike porosity, they are generally not spherical. These inclusions can also result from inadequate removal of earlier slag deposits and poor manipulation of the arc. Additionally, heavy mill scale or rust serves as their source, or they can result from unfused pieces of damaged electrode coatings falling into the weld. In radiographs, some detail will appear.

Inadequate Joint Penetration

Inadequate joint penetration is when the weld metal does not get fused all the way to the root of the weld joint, **Figure 16-10.** On some larger-diameter pipes, the welder may use a light and a mirror to look at the root of a weld pass to see that complete joint penetration was achieved, **Figure 16-11**.

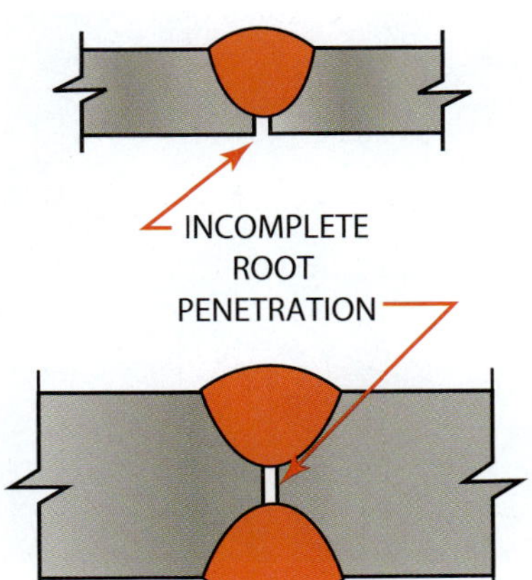

FIGURE 16-10 Incomplete root penetration.

Larry Jeffus

FIGURE 16-11 Pipe welders often visually inspect the root surface for proper reinforcement.

Causes of Inadequate Joint Penetration The major causes of inadequate joint penetration are:

- **Improper welding technique**—The most common cause is a misdirected arc, or where the molten weld pool is well established before the weld is started along the joint.
- **Not enough welding current**—Pipes that are thick or have higher thermal conductivity may require higher-current settings to produce a weld pool that is large enough to fully fuse the entire depth of the root of the joint. Sometimes preheating can be used where the weld heat would be drawn away so quickly by the surrounding metal that the weld cannot penetrate the joint.
- **Improper joint fit-up**—This problem results when the pipe joints are not prepared or fitted accurately. Too small a root gap or too large a root face will keep the weld from penetrating adequately.

Types of Inadequate Joint Penetration Most inadequate joint penetration in pipe welding occurs along the root of the weld. In some pressure vessel welding where both sides of the joint can be welded, inadequate joint penetration may occur between the inside and outside welds.

Incomplete Fusion

Incomplete fusion is when one weld pass does not melt the surface of a previous weld pass, or when it does not melt the side of the weld groove so that they can flow together, **Figure 16-12**.

Cause of Incomplete Fusion There are five major causes of incomplete fusion in pipe welding:

- **Inadequate agitation**—Lack of weld agitation to break up oxide layers. The base metal or weld filler

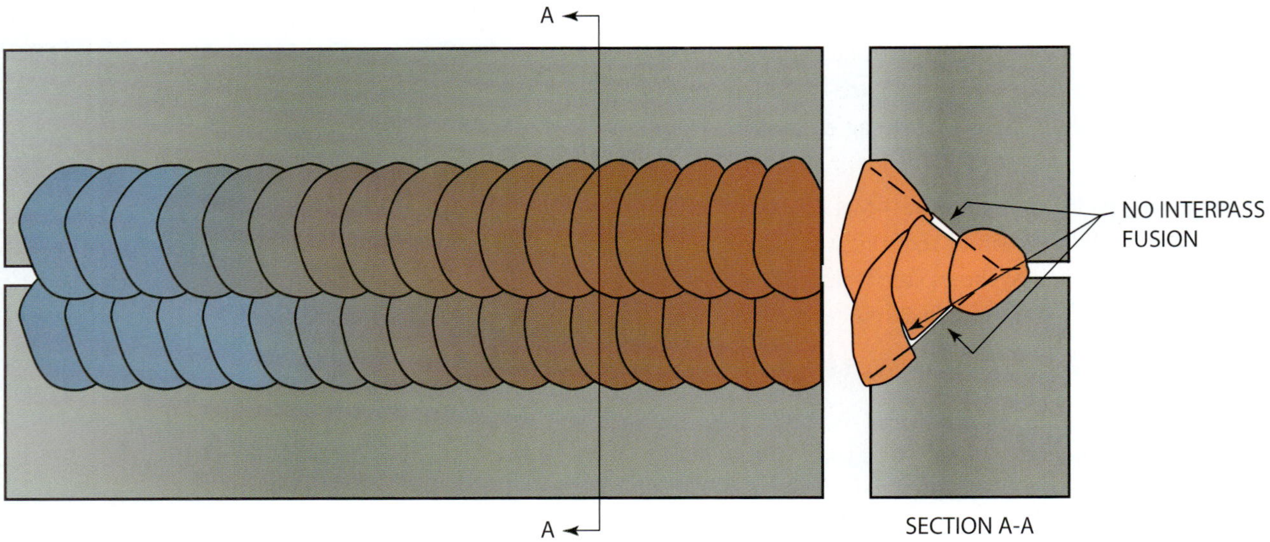

FIGURE 16-12 Incomplete fusion.

metal may melt, but a thin layer of oxide may prevent coalescence from occurring.

- **Improper welding techniques**—Examples include poor manipulation, such as moving too fast or using an improper electrode angle.
- **Improper edge preparation**—Any notches or gouges in the edge of the weld joint must be removed. For example, if a flame cut plate has notches along the cut, they could result in a lack of fusion in each notch, **Figure 16-13.**

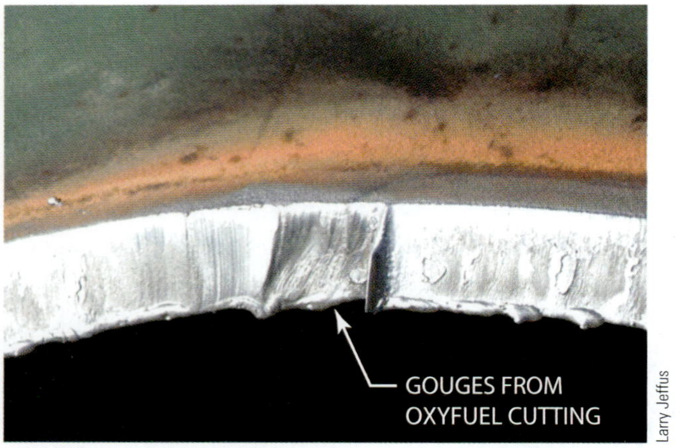

FIGURE 16-13 Remove gouges along the joint surface before welding.

- **Improper joint design**—Incomplete fusion may also result if there is not enough heat to melt the base metal, or too little space was allowed by the joint designer for correct molten weld pool manipulation.
- **Improper joint cleaning**—Failure to clean surface oxides left behind from oxy-fuel cutting of the pipe, or failure to remove slag from a previous weld, can result in defects and discontinuities.

Types of Incomplete Fusion There are two major types of incomplete fusion. The lack of fusion between the filler metal and previously deposited weld metal is called **interpass cold lap.** The lack of fusion between the weld metal and the joint face is called **lack of sidewall fusion.** Both of these problems usually occur along all or most of the weld's length.

Arc Strikes

Arc strikes become defects when they accidentally occur outside the weld groove, where they can cause localized hardening, which can cause a future pipe-cracking problem, **Figure 16-14**. The localized hardening of an arc strike is the result of the quenching effect of it by the surrounding metal. On some higher-alloyed pipes, the hard brittle spot can form a crack almost instantly.

FIGURE 16-14 Arc strikes outside the weld groove create hardness zones.

Cause of Arc Strikes Trying to strike an arc in the weld groove can be difficult at times because of limited access or limited visibility, or if it was happening as you were trying to get in position to start welding. Trying to get in a position where you can see the bottom of a pipe, and not being in a spot where all the sparks are going to land on you, can be difficult and could lead to an accidental arc strike, **Figure 16-15**. Also, a faulty ground connection can cause an arc strike if it arcs on the pipe.

FIGURE 16-15 Welders often brace themselves against the pipe and use both hands to guide the electrode in the groove to prevent accidental arc strikes.

Even if an arc strike is ground smooth, the underlying hardness zone will remain. Often, the only way to completely remove the damage caused by an arc strike is to grind out the hardened zone and reweld it using the same welding procedure that is being used to weld the pipe.

Overlap

Overlap, also called **cold lap,** is the area along the side of a weld where the weld metal flowed out on top of the pipe without fusing to it, **Figure 16-16**.

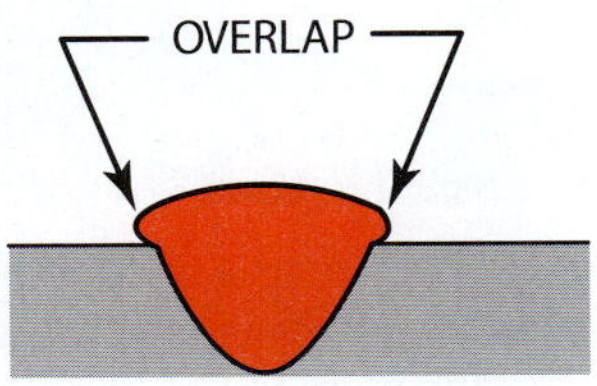

FIGURE 16-16 Rollover or overlap.

Causes of Overlap The two most common causes of overlap are oversizing of a weld pass and misdirecting of the arc. Making a number of stringer beads as opposed to a wide weave bead will help eliminate overlap caused by too large a weld bead. Moving the arc closer to the sides of a GMAW can put enough heat into the toe of the weld to ensure complete fusion.

Overlap generally occurs on the horizontal leg of a horizontal fillet weld around a flange, or on the flat portion of a saddle of a tee joint. It can also occur on both sides of the vertical portion of the capping passes. Overlap is more of a problem with GMAW than with the other processes used for pipe welding.

Undercut

Undercut is the area alongside a weld that was melted away by the arc but was not filled by weld metal, **Figure 16-17**.

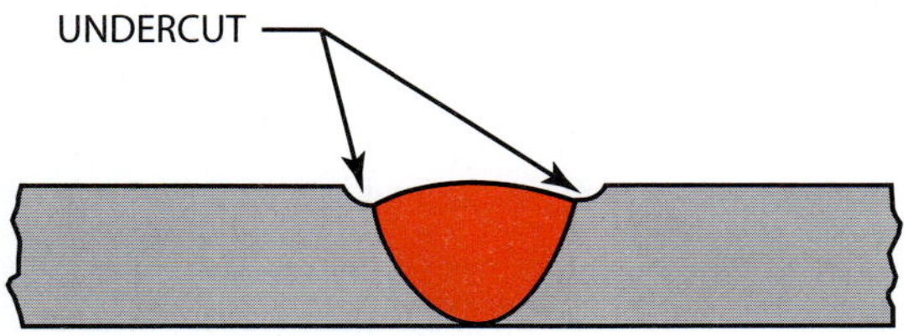

FIGURE 16-17 Undercut.

Causes of Undercut The major causes of undercutting are too high a welding current, improper welding technique, and with GMAW, the wrong shielding gas. In vertical, overhead, and horizontal positions, the larger weld beads caused by high welding currents allow the filler metal to pull away from the edge of the weld as it flows toward the center of the weld.

On a vertically up weld, a small spot of undercut will occur as the base metal is melted away along the leading edge of the side of the weld. So the proper welding technique is to hold the electrode slightly above the undercut for a sufficient period for it to be filled.

GMAW pools are not as fluid as those created by shielded metal arc welding (SMAW) unless the proper shielding gas mixture is used. The oxygen percentage in the shielding gas is the major controlling factor in the fluidity of the GMAW pool. The higher the oxygen percentage, the more fluid the pool will be. A more fluid weld pool will help to eliminate undercutting alongside the GMAW.

Underfill

Underfill is an area of depression of the face or root of a weld where the filler metal did not completely fill the weld groove, so it is level with the surface of the pipe, **Figure 16-18**.

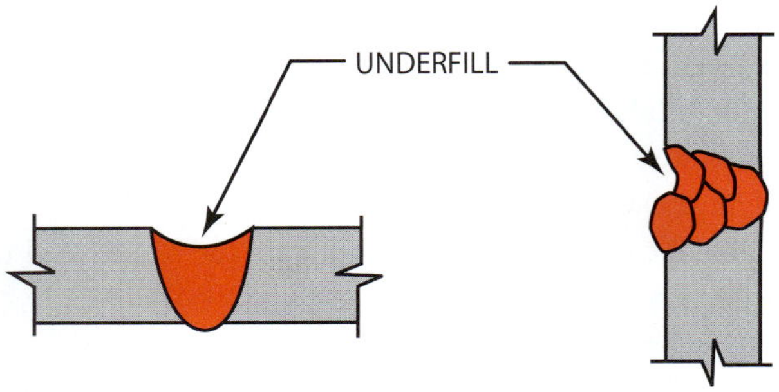

FIGURE 16-18 Underfill.

Causes of Underfill The face of a weld will be underfilled if the weld is being made so fast so that not enough filler metal is being deposited to fill the weld groove completely, **Figure 16-19**. This is a more serious problem when making a vertically down cover pass. Notice the slight underfill in this portion of the downhill cover pass, **Figure 16-20**, which could have been prevented if multiple stringer beads were used on the cover pass, **Figure 16-21**. Underfill can also occur on the top edge of a horizontal weld even if stringer beads are used.

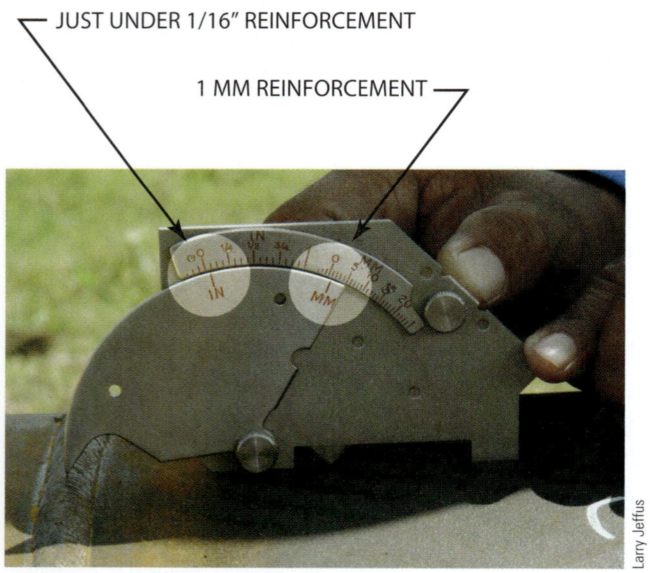

FIGURE 16-19 Undercut can be measured accurately with a bridge cam or other similar inspection tool.

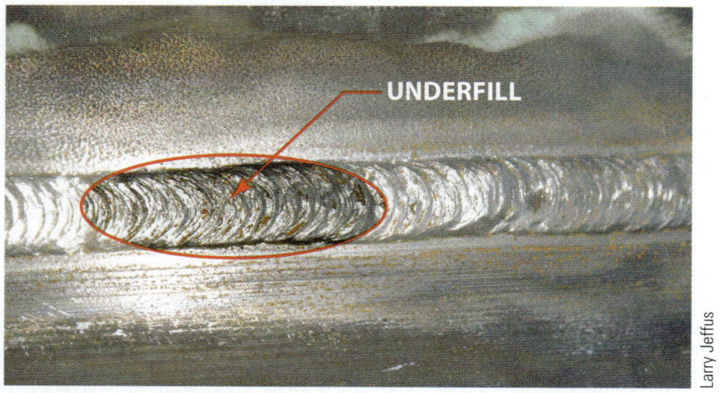

FIGURE 16-20 Underfill alongside a single cover bead.

FIGURE 16-21 Multiple weld passes can help prevent undercut on cover beads.

On a fillet weld, underfill occurs when the face of the weld is more concave than the code allows. The concavity of a fillet weld is measured from a straight line drawn between the toes of the weld and measured in the center of the face of the weld, **Figure 16-22**.

(A)

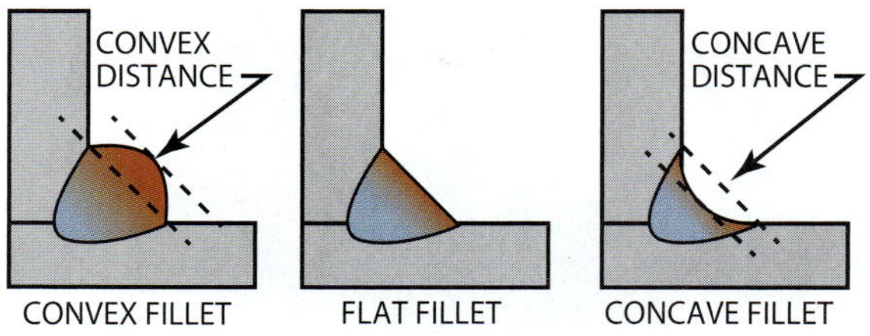

(B)

FIGURE 16-22 (A) Do not overfill or underfill fillet welds. (B) This roller coaster's base plate is subjected to a great deal of stress, so having a properly sized fillet weld is critical to its safe operation.

Underfill of the root face of a weld is also called "suck back" because it is the surface tension of the molten weld metal that draws the root face back into the weld, **Figure 16-23**. Increasing the size of the root face can reduce or eliminate suck back; however, too large a root face may cause a lack of root fusion, **Figure 16-24**.

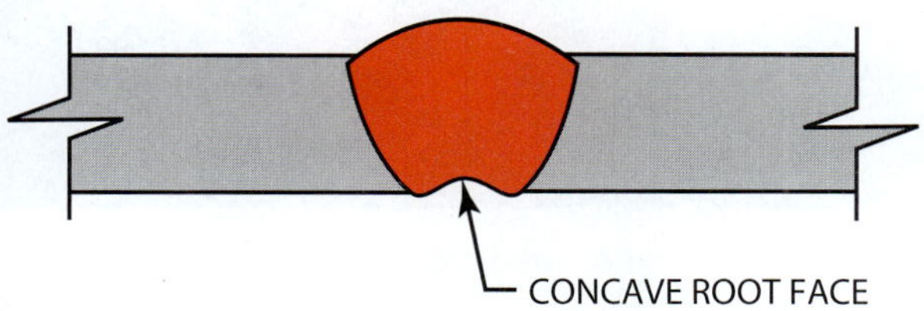

FIGURE 16-23 Root face concavity caused by weld shrinkage or suck back.

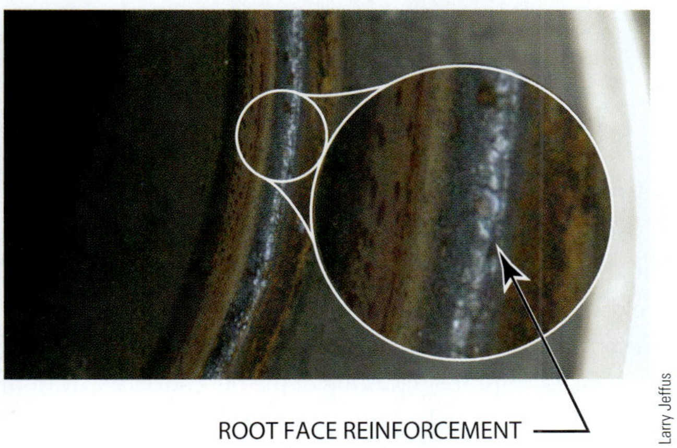

FIGURE 16-24 Proper weld bead root face reinforcement.

WELD PROBLEMS CAUSED BY INHERENT PIPE DISCONTINUITIES

Not all welding problems are caused by weld metal, the process, or your lack of skill in depositing that metal. The pipe being fabricated can be at fault, too. Some problems result from internal defects that you cannot control. Steel producers try to keep their steels as sound as possible, but internal discontinuities can often occur, even with the best steelmaking practices. These types of discontinuity are described in the next sections.

Laminations

Laminations are layers of nonmetallic contamination, such as slag and oxides, that exist before steel is rolled out to form a pipe that remains tightly pressed together. During the rolling process, the slag and oxides are pressed into thin layers. Laminations are usually found within the walls of the pipe. They are not always a basis for rejecting the pipe, so long as they can be sealed during the welding, **Figure 16-25**.

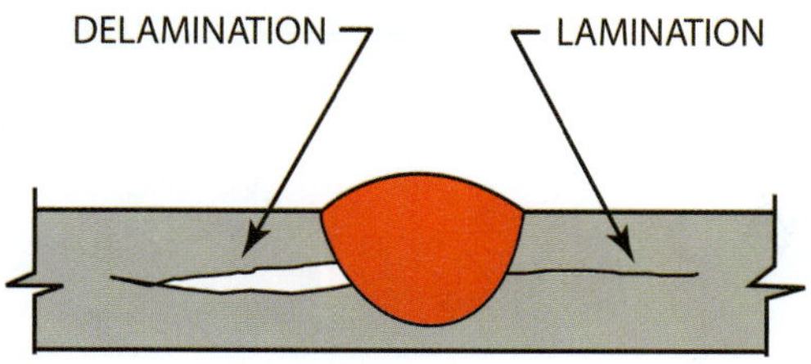

FIGURE 16-25 Lamination and delamination caused by base metal imperfections.

Delaminations

Delaminations are laminates that have separated and now have formed a gap between the layers of pipe, Figure 16-25. When laminations intersect a joint being welded, the heat and stresses of the weld may cause some of them to become delaminated. Contamination of the weld metal may occur if the lamination contains large amounts of slag, mill scale, dirt, or other undesirable materials. Such contamination can cause wormhole porosity or lack-of-fusion defects.

CRACKS

Cracks are the result of internal stresses within the weld or surrounding metal that exceed the ultimate strength of the metal. Not all cracks are weld discontinuities, but they may occur as a result of welding. Cracks can be identified as being hot cracks or cold cracks. **Hot cracks** develop following the solidification of the weld and subsequent cooling of the weld and surrounding metal. **Cold cracks** develop sometime after the weld has reached room temperature. Some cold cracks may develop even days after the weld; that is why weld testing may be delayed for a week or so.

Types of Cracks

Crater cracks are tiny cracks that develop in a weld crater because it is shrinking as it cools. All metals expand when heated and shrink as they cool. During the weld pool's cooldown cycle, it is much weaker than the surrounding weld and base metal, which is also cooling and shrinking at the same time. Since the weld crater is the weakest point, it can get pulled apart, forming a star-shaped crack in its center, **Figure 16-26**.

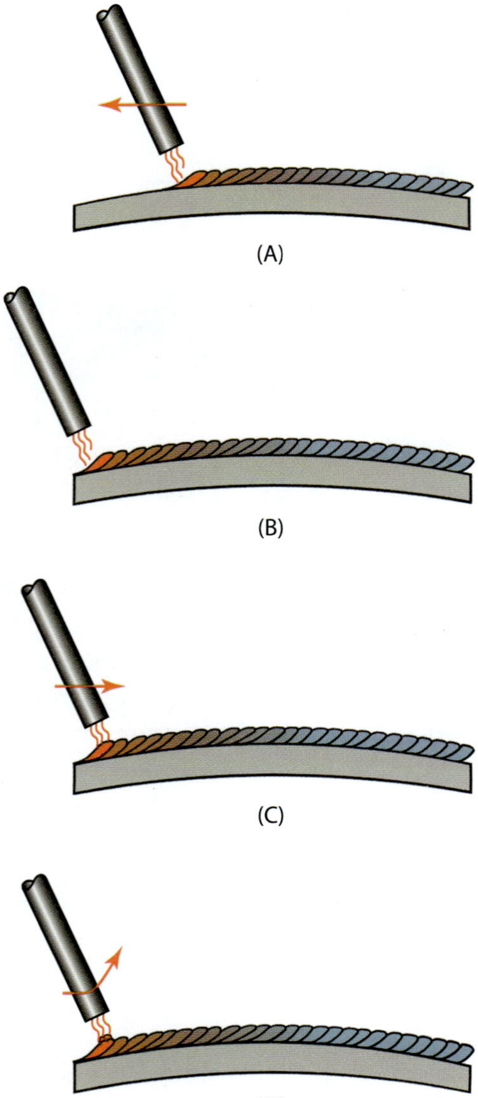

FIGURE 16-26 When ending a weld (A), pause for a moment to fill the weld crater (B) before gradually moving the arc back over the top of the weld bead (C) and then quickly breaking the arc (D).

Crater cracks can be minimized, if not prevented, by not stopping the arc quickly at the end of a weld. This allows the arc to lengthen, the current to drop gradually, and the crater to fill and cool more slowly. Some GMAW equipment has a crater-filling control that automatically and gradually reduces the wire feed speed at the end of a weld. For most welding processes, an effective way of preventing crater cracking is to pull the weld slightly back over the top of the end of the weld. This allows the pool to end up on top of the weld, where cracking can be minimized, **Figure 16-27**.

Toe cracks are longitudinal cracks that run along beside the toe (edge) of the weld in the base metal, **Figure 16-28**. They are open to the surface and are easily detected. If they are hot cracks, they may be eliminated

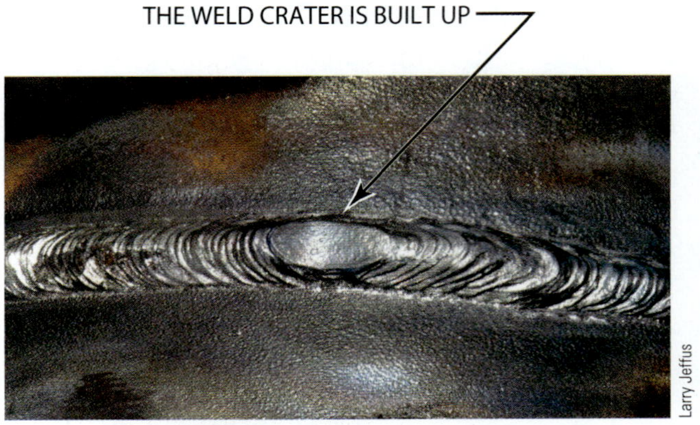

FIGURE 16-27 Proper weld crater filling.

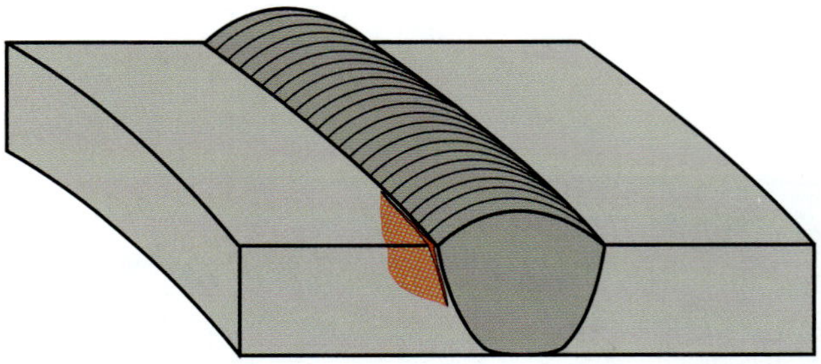

FIGURE 16-28 Toe crack.

in the future by preheating and postheating. If they are cold cracks, they are usually caused by hydrogen entrapment, which can occur because of improper handling of electrodes, improper cleaning of the pipe surfaces, or from other sources of hydrogen.

Throat cracks are longitudinal cracks that appear in the face of the weld, **Figure 16-29**. They are open to the surface and are easily detected. They are hot cracks, and they may be eliminated in the future by preheating and postheating.

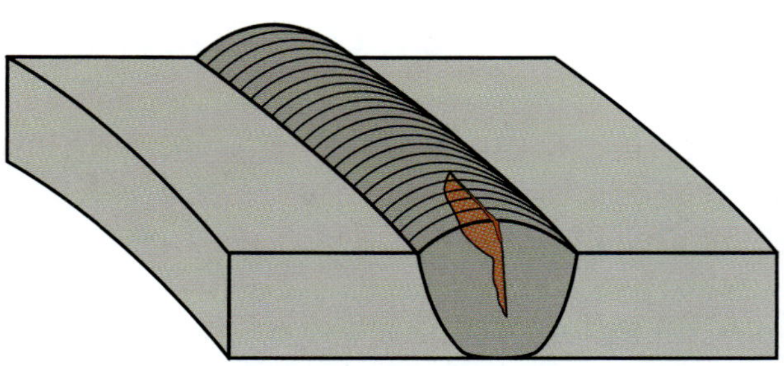

FIGURE 16-29 Centerline crack.

Root cracks are longitudinal cracks that appear in the root face of the weld, **Figure 16-30**. They are open to the surface and are easily detected. They can be either hot cracks or cold cracks. Hot cracks may be eliminated by making a larger root pass or by preheating and postheating. Cold cracks also can be controlled by making a larger root pass and by controlling hydrogen embrittlement.

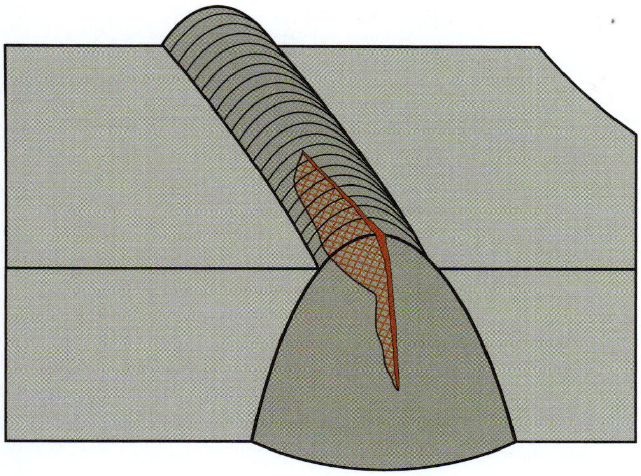

FIGURE 16-30 Root crack.

Underbead cracks are longitudinal cold cracks that appear in unmelted base metal, **Figure 16-31**. They are caused by hydrogen embrittlement. They can be prevented by using low-hydrogen electrodes.

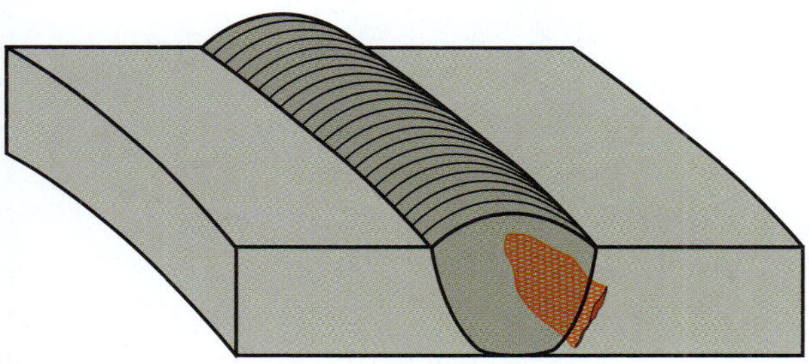

FIGURE 16-31 Underbead crack.

Lamellar tears are cracks that tear open parallel to and under the steel surface. In general, they are not in the heat-affected zone, and they have a steplike configuration. They result from thin layers of nonmetallic inclusions that lie beneath the plate surface and have very poor ductility. So the layers do not bend; rather, they pull apart under the welding stresses. Although barely noticeable, these inclusions separate when overly stressed, producing laminated cracks. These cracks are evident if the plate edges are exposed. Changing the way that the pipe joint intersects so the internal forces are parallel to the nonmetallic layers will prevent these cracks from appearing, **Figure 16-32.**

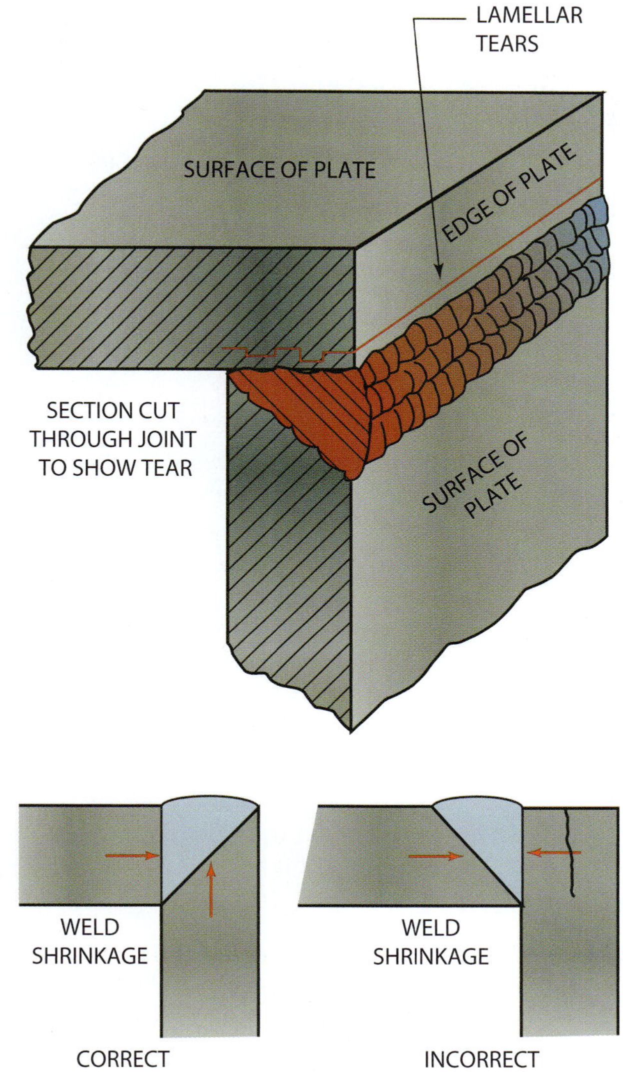

FIGURE 16-32 Preventing lamellear tears.

Summary

It is important to try and make every weld as perfect as possible. Watching the molten weld pool carefully, understanding the stresses a weld has on the joint, and skillfully setting up and performing the weld will help you make the best weld possible. Welding is often referred to as being more of an art than a science, but in reality, it is both. Laying a weld bead on a pipe joint takes as much skill as when an artist makes a brushstroke on a canvas, but you must also understand the underlying physics and chemistry of a weld, too.

Becoming a highly skilled welder is a lifelong learning exercise; you should learn something from each weld and welding job. After each weld, take note of the results and what factors contributed to them. For example, "How did that undercut happen?" or "Why did that crack form?" But most important, watch other welders and learn from them as well, so someday every weld you make will be discontinuity free . . . perfect.

Review Questions

1. What is the difference between a discontinuity and a defect?
2. How is the acceptable limit for discontinuities in a weld determined?
3. What does the term "fit for service" mean?
4. What term means that all of the discontinuities in a weld fall within the limits set by the code or standard?
5. Describe how porosity forms.
6. Name two common causes of porosity.
7. How can a weld become contaminated?
8. How can surface impurities be prevented from contaminating a weld?
9. How can filler metals be contaminated?
10. Name the four major types of porosity and describe an example of each.
11. What are weld inclusions?
12. What are four common causes of weld inclusions?
13. What causes inadequate joint penetration?
14. Describe a weld with incomplete fusion.
15. Which is more likely to be a result of incomplete fusion after a root pass–interpass cold lap or lack of sidewall fusion?
16. What is the difference between an interpass cold lap and a lack of sidewall fusion?
17. What problem can happen when an arc is struck outside the weld groove?
18. What is a synonym for overlap?
19. Compare undercut to underfill.
20. What is the difference between laminations and delaminations?
21. Why are some cracks called "cold cracks"?
22. List the types of cracks that might form in or under a pipe weld.

Chapter 17

Pipe Weld Repairs

OBJECTIVES

After completing this chapter, the student should be able to:

- List common types of defects that have to be repaired.
- Describe some of the causes of weld cracking.
- Identify commonly used tools in the repair of pipe welds.
- Explain the steps used in the repair of pipe welds.

KEY TERMS

Complete joint penetration

Incomplete joint penetration

Liquid penetrant

Magnetic particle

Nonmetallic inclusions

Wagon tracks

Weld toe

INTRODUCTION

Weld repairs on pipe can be extremely expensive to perform and result in the job completion being delayed, which can incur even more costs.

Regardless of the cause, it is sometimes necessary to repair some piping, especially with aging systems. Some welds have to be repaired due to unavoidable events such as welding power fluctuation, loss of shielding gas due to a sudden strong wind, or an unexpected event. For instance, an unexpected event occurs on the north slope of Alaska when a bear enters the work camp and an alarm sounds, so welders must stop immediately. Occasionally, an experienced welder may be asked to repair a weld because of poor work by another less-skilled welder.

In many cases, these repairs are a result of not following proper welding procedures and practices. Improper preparation often results in discontinuities. One example is not properly cleaning the material prior to welding, which can result in defects that must be repaired. Failure to clean slag and silicon in between weld passes can cause inclusions. Porosity can result from poor technique that results in improper shielding of the weld pool, **Figure 17-1**. In some cases, fixes may be needed as a result of improper preheating or postweld heat treatment

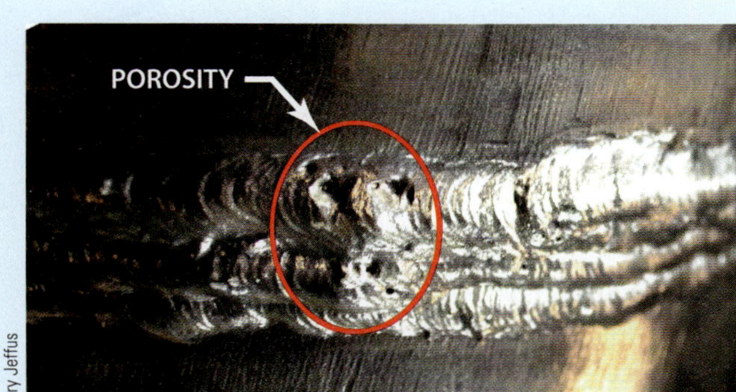

Larry Jeffus

FIGURE 17-1 Weld porosity.

(Continued)

that can result in hydrogen cracking. The other types of repairs that must be done are as a result of wear related to service, such as erosion, corrosion, cyclical loading in service, or accidental damage.

For the purposes of this chapter, we will discuss repairs based on workmanship flaws. This includes the following defects: nonmetallic inclusions, lack of fusion, lack of penetration, and porosity.

REPAIR CONSIDERATIONS

Things to take into account when making repairs are the location, the length of the defect, and how deep it is within the weld. This can often be determined by a nondestructive testing method. Inspection processes that can be used include x-rays, ultrasound, liquid penetrants, and sometimes visual inspection. The tools that are commonly used in the repair of welds include angle grinders, **Figure 17-2**, die grinders, **Figure 17-3**, oxy-fuel cutting/gouging torches, plasma arc gouging torches, and air carbon arc gouging torches.

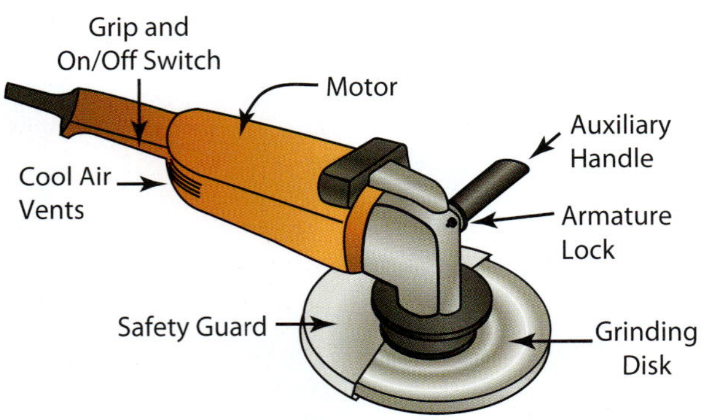

FIGURE 17-2 Angle grinder.

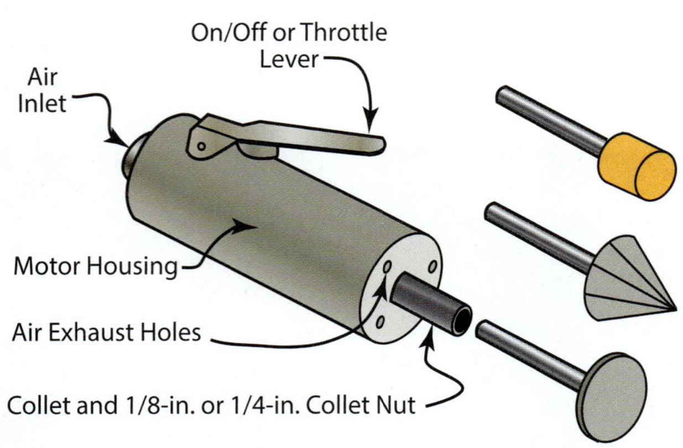

FIGURE 17-3 Die grinder.

The tool that you use is largely determined by the type of material that the pipe is made of, location of the defect, and the length of the defect to be repaired. In cases where the defects are short or are located near the surface, they can be removed easily.

Angle grinders range in size from 4½ in. (115 mm) to as large as 9 in. (230 mm). The smaller grinders are easier to use and better for controlling how much metal to remove at a time. The larger grinders are good when you need to remove a lot of metal quickly. Fatigue can be an issue when using a larger grinder. A larger grinder is also harder to handle, making control of the tool and precise removal of the defect more difficult. It is also easier to be injured trying to use them.

Care should always be taken when using grinders. It is always important to make sure that you read and follow all manufacturer guidelines and instructions when operating them. First and foremost, make sure that you have selected the proper size and type of grinding disk for the job. Refer to the manufacturer's suggestions for the correct disk to use. Some disks are designed for grinding at a 30-degree angle, while others can be used at a 90-degree angle. For grinding and repairing pipe, it is best to select a disk that can be used at a 90-degree angle. This allows the grinder to be tipped up on the edge and grind down in the weld groove. This is important in order to grind the root and do repair work on welds that are in the weld groove.

NONMETALLIC INCLUSIONS

Nonmetallic inclusions are typically encountered with processes that use a flux or leave a slag or residual silicon after welding. Nonmetallic inclusions are pockets of material in a weld that are not metal. The inclusion will typically be located between weld passes. They are also located along the edge of the weld, such as an element called **weld toe**. The weld toe is the junction between the edge of the weld and the base metal. When they are located along this region, they are usually referred to as **wagon tracks**. The reason for this term is that they look similar to wagon tracks on an x-ray. Typically, wagon tracks would be seen on the root weld pass. This is common because when running a cellulosic rod downhill on the root pass, the result is a convex weld. This makes it difficult to remove the slag along the toe. This is why it is important to remove this slag by chipping, wire-brushing, or by using a needle scaler, **Figure 17-4**, die grinder, or angle grinder to clean the root pass prior to making the hot pass. The hot pass can help to eliminate the slag by melting the surrounding base metal and floating the slag to the top of the weld pool, but this does not always happen.

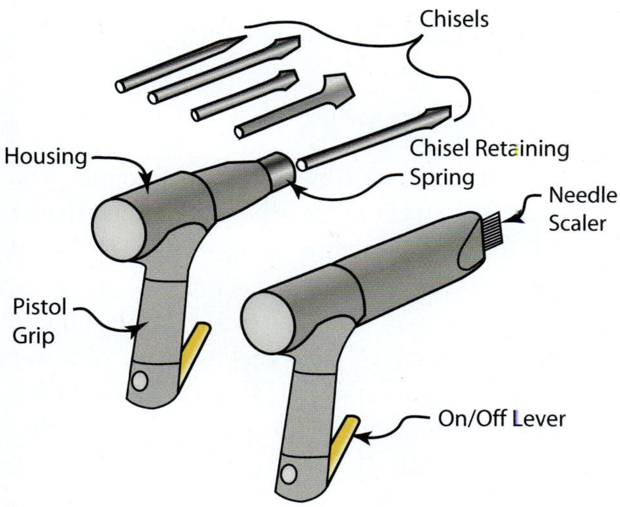

FIGURE 17-4 Pneumatic chisel and needle scaler.

PRACTICE 17-1

Slag Removal

1. To remove the slag, start by using a chipping hammer or file modified for removing slag. Eliminate the large pieces first, **Figure 17-5**.

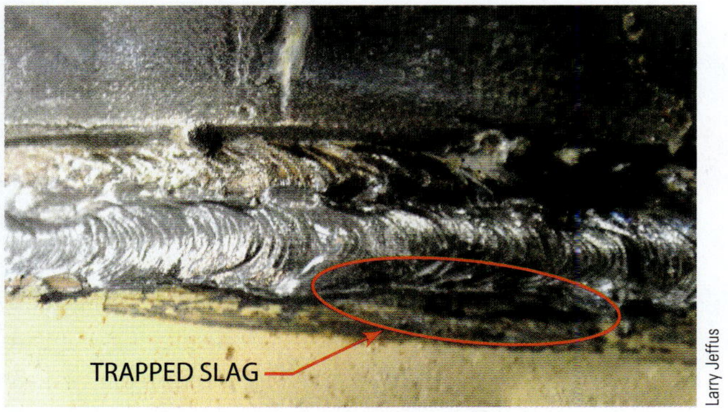

FIGURE 17-5 Trapped slag.

2. Next, you can use a grinder with a twisted wire wheel. Depending on the weld joint geometry, the wire wheel should remove a large amount of the trapped slag. If this does not eliminate all the slag, remove the wire wheel from the grinder and install a grinding wheel designed to be used at a 90-degree angle to the work surface.

> **NOTE**
>
> The pipe size, schedule, and geometry will determine the size of grinding disk you should use. In most cases, using a 4½–5-in. (115–130-mm) grinder and matching wheel is a good choice.
>
> While grinding, it is often necessary to get in an awkward position. A smaller grinder is easier to control and reduces fatigue, making grinding safer.
>
> A 1/8-in. (3-mm) wheel thickness for grinding the root is probably the best choice. This will help you control how much metal is removed at the root of the weld.

3. You should grind the entire length of the inclusion. Gradually increase the depth as you work your way around the length of the inclusion.
4. The depth of the groove should be relatively uniform. Never force the grinding disk into the work. Doing so may cause the wheel to bind, which can cause it to kick back, risking injury.
5. When grinding out the weld area, take care not to cut or gouge into the base metal.
6. You should maintain the integrity of the original joint design as much as possible.
7. When grinding a weld in this manner, always make sure that the stone is turning into the direction that the grinder is traveling. This will help prevent the grinding disk from kicking back.
8. You should grind down to bright, shiny metal to make sure that all the slag is removed from the weld.
9. Once you have removed all the slag, have your instructor inspect the groove before welding.

> **NOTE**
>
> The instructor will be looking to see that all the slag has been removed, as well as making sure that the weld groove face has maintained the original angles. It is important that the area of the groove near the root not be back-beveled. Back-beveling results from grinding at an angle that is opposite the bevel angle. Because of a combination of the back-beveling angle, the welding electrode will not be able to be directed at the opening. A void will likely be left, and it may become filled with slag.

10. Turn off all equipment, return all tools and supplies, discard all scrap, and clean up the welding booth. ◆

PRACTICE 17-2

Welding Repair

Once the instructor has checked your preparation, you can start welding.

1. Use an E7018 or similar low-hydrogen electrode; the type of current should be direct current electrode positive (DCEP).
2. Be sure to check your current settings to make sure that they are within the range specified by the welding procedure or recommended by the electrode manufacturer.
3. It is also important to make sure that you have a good secure ground.
4. Preheat if required; it is important to check and make sure that the material is within acceptable limits.
5. When starting to weld, make sure that the arc is started within the weld zone and slightly ahead of the direction that the weld will progress in.
6. As you are striking the arc, move the rod back away from the direction of travel just to the edge of where the grinding feathers and ends. The distance should be 1–1.5 in. (25–38 mm) from where the end of the feather is.
7. When you reach this point, the arc should be stable; you should now tip the rod into the direction of travel.
8. Weld metal will be deposited once you reach this point and the direction of travel changes.
9. As the weld progresses, the area where the arc was started will be welded over, remelted, and fused.
10. This motion should be rather quick and will prevent slag from being deposited. This will result in a nice, smooth tie-in and transition between the original weld and the weld that was reworked. It takes time and practice to make smooth restarts as well as stops.
11. Have your instructor inspect your work for uniformity.
12. Turn off all equipment, return all tools and supplies, discard all scrap, and clean up the welding booth. ◆

INCOMPLETE JOINT PENETRATION

Incomplete joint penetration is a discontinuity associated only with groove welds. It occurs when weld metal does not extend completely through the thickness of the joint when specifications require **complete joint penetration**, which is when the weld extends completely through the thickness of the joint. In some instances, incomplete joint penetration is allowable when the specifications say so. In this case, it would be referred to as a partial joint penetration weld.

Welds performed on pipe that are butt-type joints that form groove joints often will require complete joint penetration. Socket-type welds form fillet joints that typically do not penetrate the joint completely.

REPAIR OF CRACKS

Cracks are often caused by stresses within the weld. Some of these are the result of making tack welds that are too small for the joint that is being welded. The wrong type of filler metal selected for making tack welds can also result in cracks in welds and tack welds. Insufficient preheating can result in rapid cooling of the tack weld or weld, resulting in cracking. In some cases, higher-strength steels require proper preweld and postweld heat treatment to prevent hydrogen cracking that can take place hours and in some cases days after welding. The important thing to remember is to take steps prior to welding that will help ensure that these types of defects are prevented.

In order to repair cracks, it is important to properly locate where the crack is. The length of the crack and the depth of the crack must be determined so that it can be removed completely prior to rewelding. Surface cracks can be located by using either a **liquid penetrant** or **magnetic particle** inspection process. Magnetic particle testing requires that the metal being tested be magnetic in nature. Both of these inspection methods are nondestructive tests and are good for locating surface defects. Magnetic particle can also locate defects that are near the surface.

Liquid penetrant uses a solution that is applied to the area that is to be inspected. After a period of time, it is wiped off the surface, and a developer is applied that will show any surface cracks. The magnetic particle process uses a magnetic field that is set up around the area to be inspected. A small amount of iron particles are sprinkled on the surface. If there are imperfections at or near the surface, it will disrupt the field and cause the magnetic particles to gather in the area of the discontinuity revealing the flaw. Once the discontinuities are located, the same technique used in repair of nonmetallic inclusions should be used when repairing cracks. So the first step is to determine the length of the crack. Once the length of the crack is determined, it is important to drill the ends of the crack to make sure that the crack does not continue when welding is resumed. If it is not possible to drill the crack, it is necessary to remove the crack completely by grinding. Next, the crack should be ground along its entire length and depth to make sure that it is completely removed. It is important to remember to grind the ends of the crack to a featheredge, which will help to make a smooth transition from the previous weld pass to the next.

Summary

No mater how experienced a welder is or how well a job is,laid out, there will be occasions where welds must be repaired. As you are learning, it is a good practice to repair some of your welds that might not meet the acceptance criteria. This will give you invaluable experience that will have an immediate impact on your welding future. Welders who can fix their or other welders' defects are regarded as valuable assets on a pipeline team.

Review Questions

1. When making repairs on welds, what are some of the things that should be considered?
2. What are some of the nondestructive testing methods that could be used to identify defects in pipe welds?
3. List some of the disadvantages of the liquid penetrant and magnetic particle tests.
4. *True or False:* When grinding out a weld defect, it does not matter if you gouge into the base metal, so long as you remove the defect.
5. What tools would commonly be used in order to repair weld defects in pipe?
6. Give an example of a nonmetallic inclusion.
7. Name one cause of porosity.

Chapter 18

Testing and Inspecting Welds

OBJECTIVES

After completing this chapter, the student should be able to:

- State the purpose of quality assurance (QA) and quality control (QC).
- Name the three main organizations that have pipe welding codes and standards.
- Discuss the purpose of a Welding Procedure Specification (WPS) and what it contains.
- Compare destructive testing (DT) to nondestructive testing (NDT).
- Explain what results can be obtained from DT and NDT.
- Give methods of protecting buried carbon steel and galvanized pipe from corrosion.

KEY TERMS

Codes

Destructive testing (DT)

Nondestructive testing (NDT)

Standards

Welding Procedure Specification (WPS)

INTRODUCTION

Welding codes and standards have been developed over many years to aid in the production of quality welded piping systems. The committees that have developed these documents have had input from every aspect of the welding industry including welding engineers, industrial fabricators, and end users. Local, state, and federal governments have also influenced them as they relate to public safety and reliability.

Welding engineers rely on codes and standards when they are designing a system; fabricating companies consult and follow codes and standards during fabrication; and inspectors refer to codes and standards as they check the work in every phase of the fabrication process. As a pipe welder, you should be familiar with all the codes and standards that might apply to any project you are welding on to ensure that all of your work is acceptable.

QUALITY ASSURANCE (QA) AND QUALITY CONTROL (QC)

Quality assurance (QA) focuses on preventing defects through strict adherence to the process and **quality control (QC)** focuses on identifying defects in the finished welds and piping systems. The goal of a QA program is to constantly improve the quality of welds by changing the procedures when needed to reduce the number of defects and to reduce construction costs. The goal of a QC program is to identify all the defects before the piping system is released to the owner or operator.

CODES AND STANDARDS

Codes and standards may be called by a number of different names such as "guides," "recommended practices," "rules," "regulations," or "specifications." **Codes** are written documents adopted by jurisdictions that become part of their statutory law and are enforceable under their legal authority. Therefore, they must be consulted and followed for any jurisdiction where a piping system or pipeline is being constructed. **Standards** are written details that contain specifications, procedures, and tests to be used by welders, QC, and manufacturers to provide a well-defined basis to ensure the overall quality for the construction of the pipeline or piping system. Although codes and standards are different, you cannot have one without the other. That is why you almost always see the term "codes and standards."

Codes and Standards Organizations

The three main organizations that have codes that may be used in pipe welding are the American Welding Society (AWS), the American Society of Mechanical Engineers (ASME), and the American Petroleum Institute (API), **Table 18-1**. The AWS is the largest publisher of welding codes and standards covering all aspects of welding. ASME publishes many codes and standards that cover all the aspects of boiler and pressure vessel design, construction, and testing. The API focuses primarily on the petrochemical industry's piping requirements

Welding engineers select codes based on the requirements of the controlling governmental jurisdiction (i.e., city, county, state, or federal government, the end user, or the insurance company covering the piping system). The standards can be written by the designing engineers, or they can be selected from the many prequalified industrial standards available from the code organizations or other sources.

Welding Procedure Specification (WPS)

Codes and standards have specific requirements that each piping system's details are based on. Often, a **Welding Procedure Specification (WPS)** will be prepared, which is a set of written instructions that must be followed to ensure that the process and end product are in compliance with the applicable code. Following these written instructions will help to ensure that a sound weld is made. They should list parameters for welding process, technique, electrode or filler metal, current, amperage, voltage, preheating, and postheating. Although the format of a WPS document is not dictated by codes or standards, they generally are very similar.

Code	Application Area
	API
Standard 1104	Welding of Pipelines and Related Facilities
Standard 1160	Managing System Integrity for Hazardous Liquid Pipelines
Specification 5L	Specification for Line Pipe
Recommended Practice 1161	Guidance Document for the Qualification of Liquid Pipeline Personnel
	AWS
D1.1	Structural Welding Code for Steel
D10.1	Guide For Welding Mild Steel Pipe
D10.10	Recommended Practices For Local Heating Of Welds In Piping And Tubing (Historical).
D10.11	Recommended Practices For Root Pass Welding Of Pipe Without Backing (Historical)
D10.12	Guide For Welding Mild Steel Pipe
D10.18	Guide For Welding Ferritic/Austenitic Duplex Stainless Steel Piping And Tubing
D10.4	Recommended Practices for Welding Austenitic Chromium-Nickel Stainless Steel Piping and Tubing
D10.6	Recommended Practices For Gas Tungsten Arc Welding Of Titanium Piping And Tubing
D10.7	(Historical) Guide For Gas Shielded Arc Welding Of Aluminum And Aluminum Alloy Pipe
D10.8	Welding Of Chromium-Molybdenum Steel Piping And Tubing
D18.1	Specification For Welding Of Austenitic Stainless Steel Tube And Pipe Systems In Sanitary (Hygienic) Applications
	ASME
B31.3	Contains requirements for piping typically found in petroleum refineries; chemical, pharmaceutical, textile, paper, semiconductor, and cryogenic plants; and related processing plants and terminals.
B31.4	This Code prescribes comprehensive solutions for materials, design, fabrication, assembly, erection, testing and inspection. It also serves as a companion to ASME's other B31 codes on piping systems. Together, they remain essential references for anyone engaged with piping.
BPE MP	Discoloration Acceptance Criteria for Weld Heat-Affected Zones on Mechanically Polished 316L Tubing
B31G	Manual for Determining the Remaining Strength of Corroded Pipelines

TABLE 18-1 Testing Agencies and Codes

WPS Form

A WPS form has the following elements:

- **Title**—Lists the welding process and material type, size, and thickness so that the welder can quickly see if this is a WPS that might apply to a specific job
- **Scope**—Gives more details regarding the process than are in the title, such as joint type, welding position, and any limitations that may be applicable
- **Base metal**—Gives the type, grade, and other specifications of the pipe to be welded
- **Filler metal**—Lists the classification and diameter of the filler metal
- **Joint design and tolerance**—Could be a drawing of the joint geometry, or a reference to a location where there is such a drawing
- **Qualifications**—Gives the test requirements of the welder, which are typically the same standards that were used to qualify the WPS; may include the requirements to maintain the welding qualification or the requalifying test procedure
- **Fabrication**—May include welding technique, shielding gas type and flow, backing gas requirements, preheating and postheating, and other details that would be required to complete the weld
- **Inspection**—All information required to test the weld to ensure that it meets the specifications; included are details as to what is acceptable for the tolerances of the finished weld, such as reinforcement, and for discontinuities, such as undercut, underfill, overlap, cracks, porosity, burnthrough, etc.

WELD TESTING

The two methods used to test the quality of a weld are destructive testing (DT; also known as "mechanical testing") and nondestructive testing (NDT). These methods can be used individually or in combination. DT methods, except for hydrostatic testing, result in the product being destroyed. NDT methods do not destroy the part being tested.

"Mechanical testing" and **destructive testing (DT)** both refer to the commonly used methods to qualify you or the welding procedure you are using. It can be used in a random sample testing procedure in mass production. In many cases, a large number of identical parts are made, and a chosen number are destroyed by DT. The results of such tests are valid only for welds made under the same conditions because the only known weld strengths are the ones resulting from the tested pieces. It is then assumed that the strengths of the nontested pieces are the same.

Nondestructive testing (NDT) is used for weld qualification, welding procedure qualification, and product QC. Since the weldment is not damaged, all the welds can be tested, and the part can actually be used for its intended purpose. Because the parts are not destroyed, more than one testing method can be used on the same part.

DESTRUCTIVE TESTING (DT)

Tensile Testing

Tensile tests are performed with specimens prepared as round bars or flat strips. The simple round bars are often used for testing only the weld metal, sometimes called "all weld metal testing." This testing can be used on thick sections where base metal dilution into all the weld metal is not possible. Round specimens are cut from the center of the weld metal. The flat bars are often used to test both the weld and the surrounding metal. Flat bars are usually cut at a 90-degree angle to the weld, **Figure 18-1**. **Table 18-2** shows how a number of standard smaller bars can be used, depending on the thickness of the metal to be tested. Bar size also depends on the size of the tensile testing equipment available.

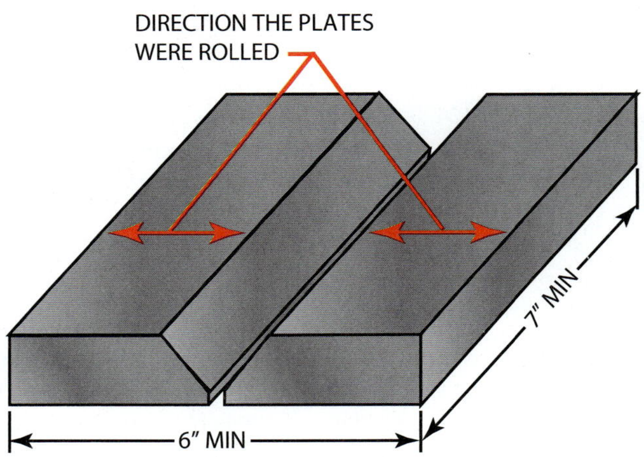

FIGURE 18-1 The direction that plate was rolled during its manufacture affects the test plate's ability to withstand a bend test.

Two flat specimens are commonly used for testing thinner pipe wall sections. When testing welds, the specimen should include the heat-affected zone and the base metal. If the weld metal is stronger than the pipe metal, failure occurs in the pipe; if the weld is weaker, failure occurs in the weld.

After the weld section is machined to the specified dimensions, **Figure 18-2**, it is placed in the tensile testing machine and pulled apart.

Specimen Width C	Dimensions of Specimen						
	in./mm A	in./mm B	in./mm C	in./mm D	in./mm E	in./mm F	in./mm G
C-1	0.500/12.7	2.0/50.8	2.25/57.1	0.750/19.05	4.25/107.9	0.750/19.05	0.375/9.52
C-2	0.437/11.09	1.750/44.4	2.0/50.8	0.625/15.8	4.0/101.6	0.750/19.05	0.375/9.52
C-3	0.357/9.06	1.4/35.5	1.750/44.4	0.500/12.7	3.500/88.9	0.625/15.8	0.375/9.52
C-4	0.252/6.40	1.0/25.4	1.250/31.7	0.375/9.52	2.50/63.5	0.500/12.7	0.125/3.17
C-5	0.126/3.2	0.500/12.7	0.750/19.05	0.250/6.35	1.750/44.4	0.375/9.52	0.125/3.17

TABLE 18-2 Tensile Testing Specifications

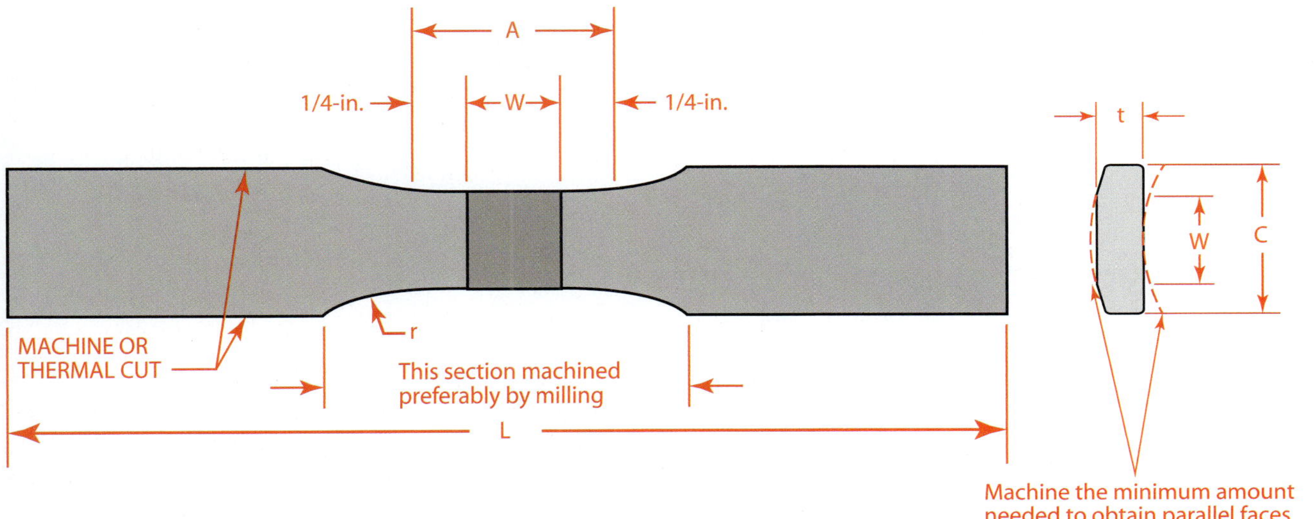

(A)

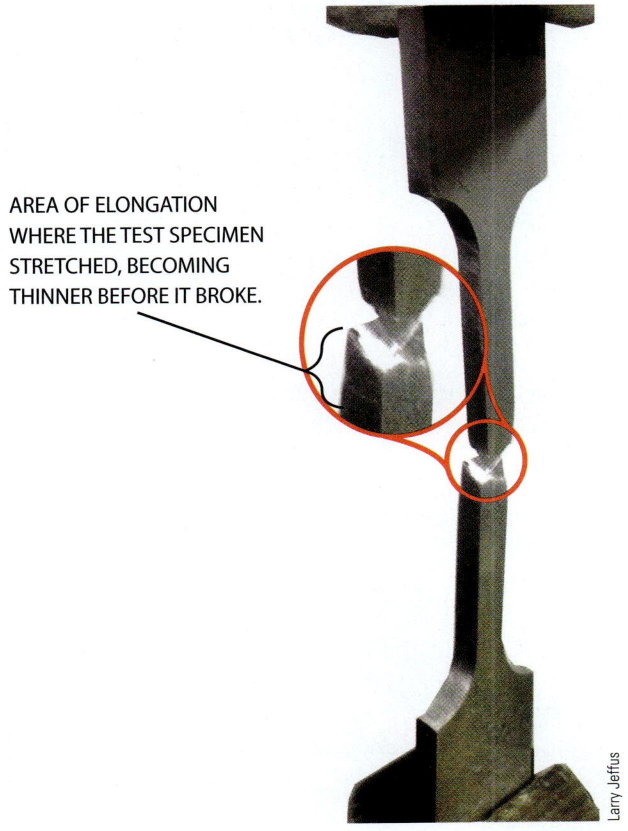

(B)

FIGURE 18-2 (A) Tensile test specimen. (B) Specimen being tensile tested.

Nick-Break Test

A specimen for this test is prepared as shown in **Figure 18-3A**. The specimen is supported as shown in **Figure 18-3B**. A force is then applied, and then the specimen is ruptured by one or more blows by a hammer. Theoretically, the rate of application could affect how the specimen breaks, especially at a critical temperature. Generally, however, there is no difference in the appearance of the fractured surface due to the method of applying the force. So for all practical purposes, striking the specimen with a small hammer, swung rapidly, would not affect the results any more or less than a strike with a much heavier hammer, swung more slowly. The force may be applied slowly or suddenly. The surfaces of the fracture should be checked for soundness of the weld.

Guided-Bend Test

To test your ability to make grooved butt joints on pipe that have a wall thickness of 3/8 in. (10 mm) or less, two specimens are prepared and tested—one face bend and one root bend. If the welds pass this test, you are qualified to make groove welds on pipe having a wall thickness range of from 3/8 in. to 3/4 in. (10 mm to 19 mm). These welds need to be machined and tested as shown in **Figure 18-4**. The test specimens will be bent in a U-shape in a testing machine

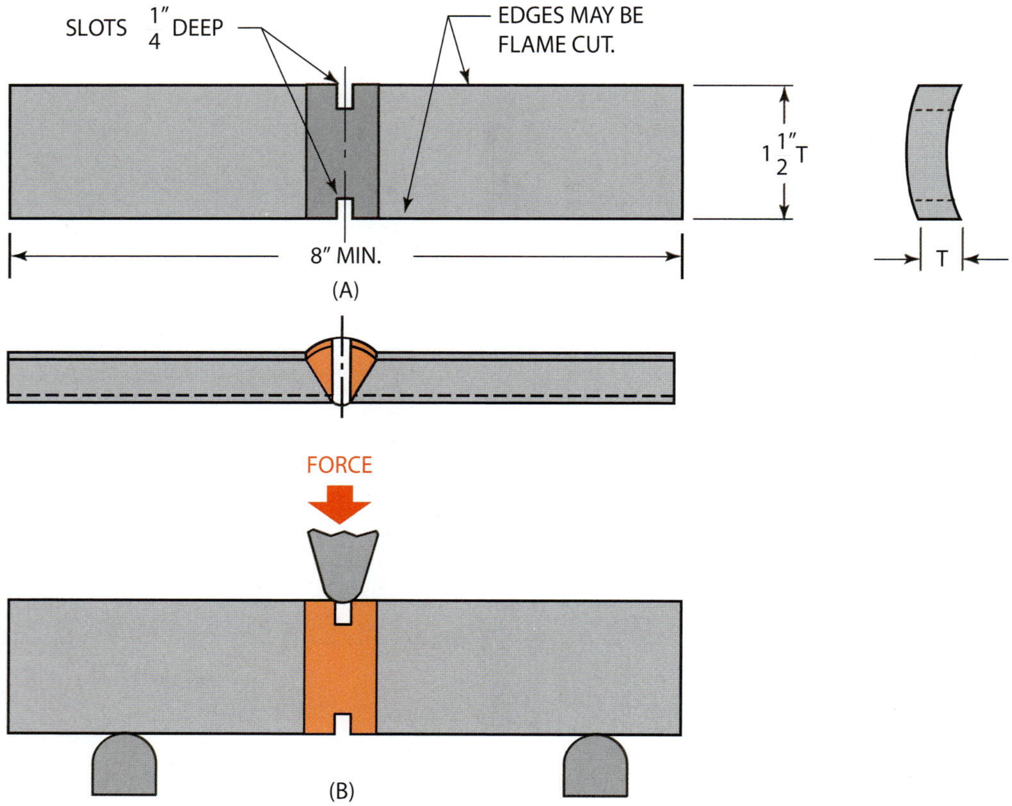

FIGURE 18-3 Nick-break specimen.

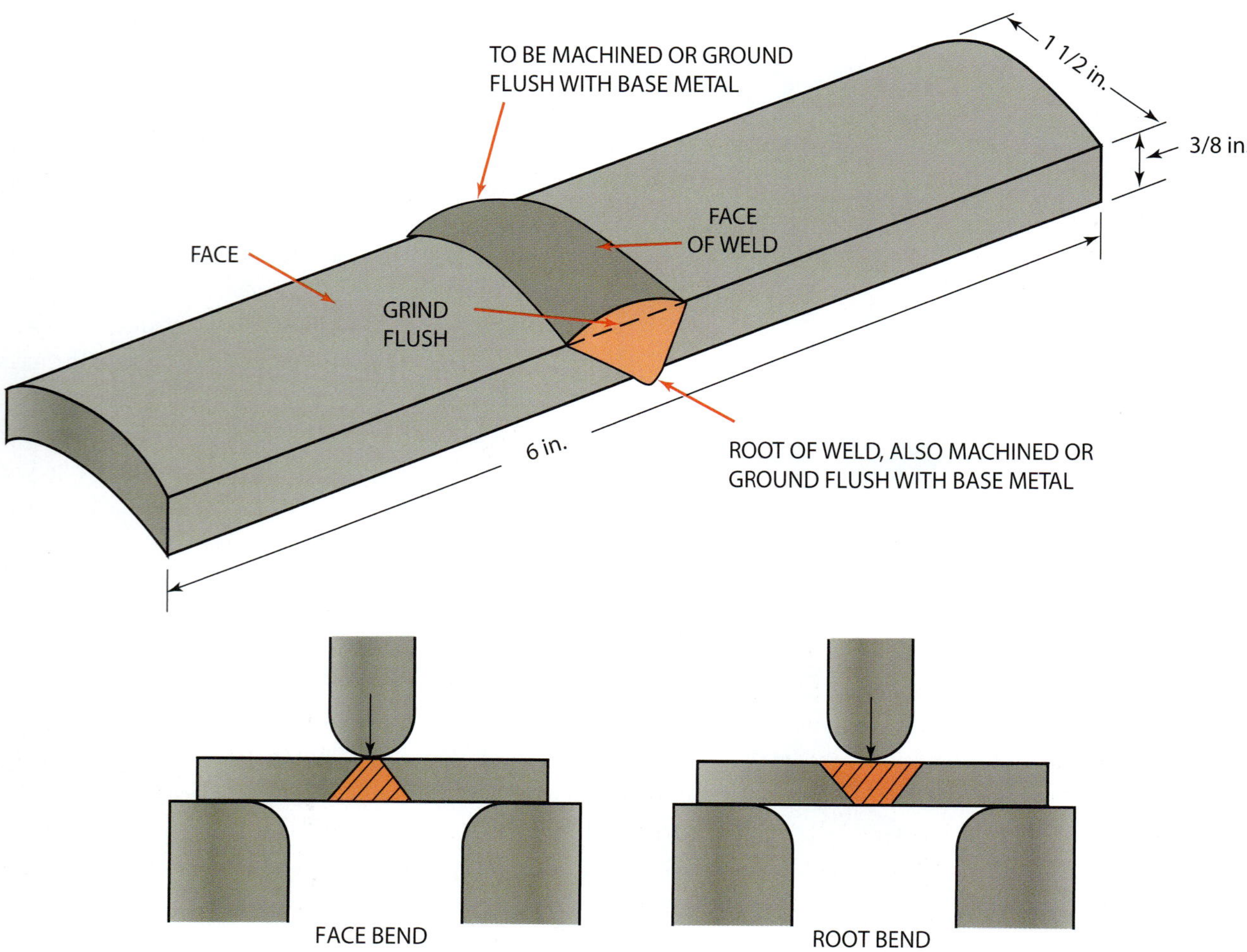

FIGURE 18-4 Root and face bend and side bend specimen.

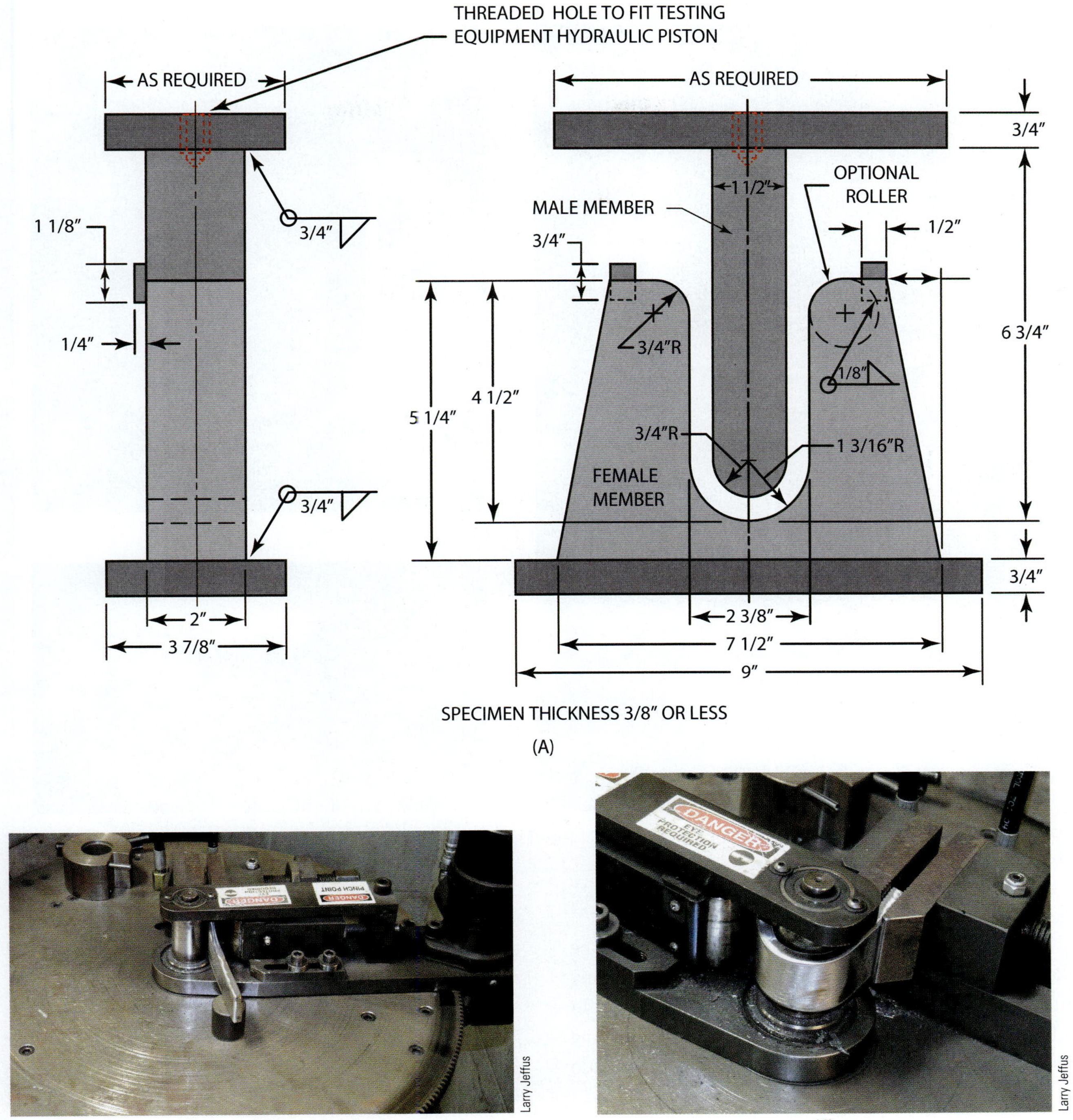

FIGURE 18-5 (A) Plunger type guided bend tester. (B) Bending guides are available for bending machines that can be used for both testing welds and for forming steel shapes for welding projects. (C) Successful bend test.

specifically designed for guided bend testing, **Figure 18-5A**, or in a bending machine with a die designed for guided bend testing, **Figure 18-5B–C**. If these specimens pass, you will also be qualified to make fillet welds on materials of any (unlimited) thickness. For grooved butt joints on thicker-walled pipe greater than 1/2 in. (13 mm) thick, two side bend specimens are prepared as shown in **Figure 13-6**, and tested as shown in **Figure 18-7**. If the welds pass this test, you are qualified to weld on metals of unlimited thickness.

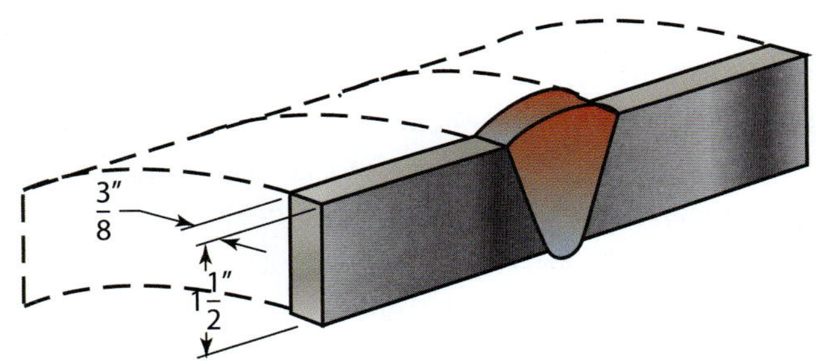

FIGURE 18-6 Side bend specifications.

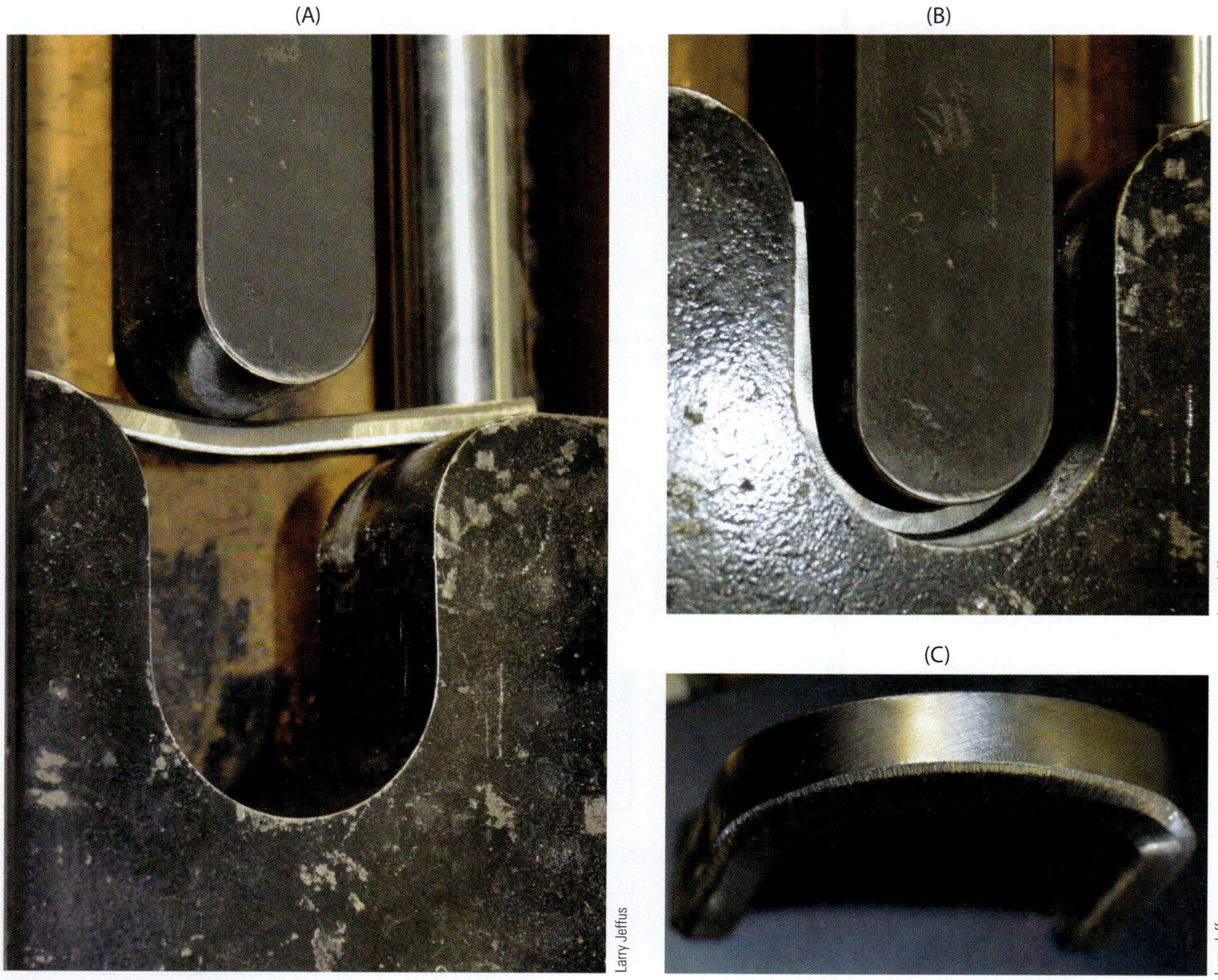

FIGURE 18-7 (A) Side bend specimen set for testing. (B) Bend specimen pushed all the way to the bottom of the test jig. (C) Side bend test specimen.

When the specimens are prepared, caution must be taken to ensure that all grinding marks run longitudinally to the specimen so they do not cause stress cracking. In addition, the edges must be rounded to reduce cracking that tends to radiate from sharp edges. The maximum ratio of this rounded edge is 1/8 in. (3 mm).

Guided-Bend Test Procedure

The jig shown in Figure 18-5A is commonly used to bend most specimens. Not all guided-bend testers have the same bending radius. Codes specify different bending radii depending on material type and thickness. Place the specimens in the jig with the weld in the middle. Face bend specimens should be placed with the face of the weld toward the gap. Root bend specimens should be positioned so that the root of the weld is directed toward the gap. Side bend specimens are placed with either side facing up. The guided-bend specimen must be pushed all the way through open (roller-type) bend testers and within 1/8 in. (3 mm) of the bottom on fixture-type bend testers.

Once the test is completed, the specimen is removed. The convex surface is then examined for cracks or other discontinuities and judged acceptable or unacceptable according to specified criteria. Some surface cracks and openings are allowable under codes.

Free-Bend Test

The free-bend test is used to test welded joints in pipe. A specimen is prepared as shown in **Figure 18-8**. Note that the width of the specimen is one-and-a-half times the thickness of the specimen. Each corner lengthwise should be rounded in a radius that does not exceed 1/10 the thickness of the specimen. If the surfaces are ground, hold the grinder so the grinder marks on the pipe run lengthwise of the specimen. Grinder marks that run across the specimen can start cracks that might not have occurred with lengthwise grinder marks.

Conversion Table							
in.	mm	in.	mm	in.	mm	in.	mm
1/4	6.35	9/16	14.2	1 1/8	28.5	2	50.8
3/8	9.52	3/4	19.05	1 1/4	31.7	6	152.4
1/2	12.7	15/16	23.8	1 1/2	38.1	8	203.2
5/8	15.8	1	25.4	1 7/8	47.6	9	228.6

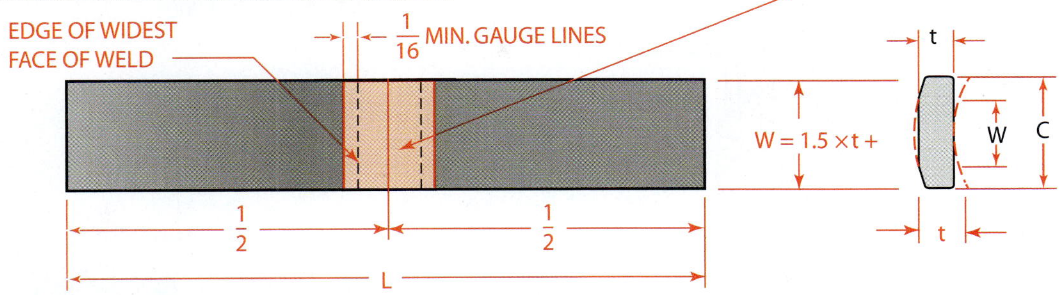

Dimensions							
T, inches –	1/4	3/8	1/2	5/8	3/4	1	1 1/4
W, inches –	3/8	9/16	3/4	15/16	1 1/8	1 1/2	1 7/8
L, min, inches –	6	8	9	10	12	13 1/2	15
B, min. inches –	1 1/4	1 1/4	1 1/4	2	2	2	2

FIGURE 18-8 Free bend test specimen specifications.

Gauge lines are drawn on the face of the weld, **Figure 18-9**. The distance between the gauge lines is 1/8 in. (3.17 mm), and the lines are 1/8 in. (3.17 mm) shorter on both ends than on the face of the weld. The initial bend of the specimen is completed in the device illustrated in **Figure 18-10**. The gauge line surface should be directed toward the supports. The weld is located in the center of the supports and loading block.

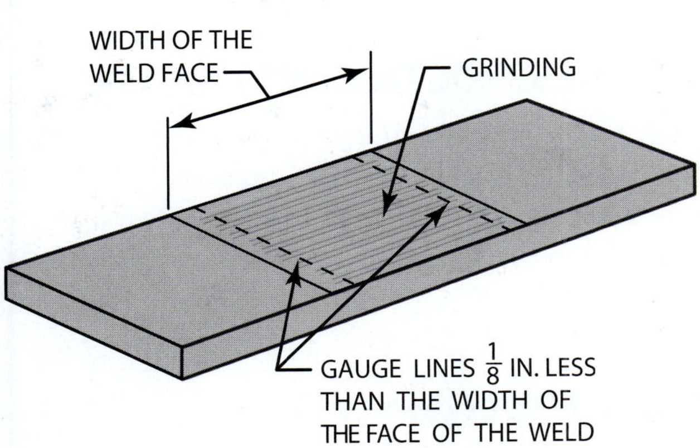

FIGURE 18-9 Layout of guidelines on a free bend specimen.

Alternate Bend

The initial bend may be made by placing the specimen in the jaws of a vise with one-third of its length protruding from the jaws. The specimen is then bent away from the gauge lines through an angle of 30 degrees to 45 degrees by blows of a hammer. The specimen is then inserted into the jaws of a vise, and pressure is applied by tightening the vise. The pressure is continued until a crack or depression appears on the convex face of the specimen. The load is then removed.

The elongation is determined by measuring the minimum distance between the gauge lines along the convex surface of the weld to the nearest 0.01 in. (0.254 mm) and subtracting the initial gauge length. The percent of elongation is obtained by dividing the elongation by the initial gauge length and multiplying by 100.

Impact Testing

A number of tests can be used to determine the impact capability of a weld. The Izod impact test determines the toughness of a specimen by measuring the foot-pounds it takes to fracture it, **Figure 18-11**. The specimen is accurately machined to size according to the ASTM specifications, and a sharp notch is cut into one side. One end of the specimen is placed in the specimen vise (anvil), and the swinging pendulum (hammer) is raised to a specific height that creates a known force measured in foot-pounds. When the pendulum is released, it swings downward striking the specimen. The tougher the specimen the more force (foot-pounds of energy) it takes to break it. Tougher specimens require more energy to break, and the pendulum would continue swinging only

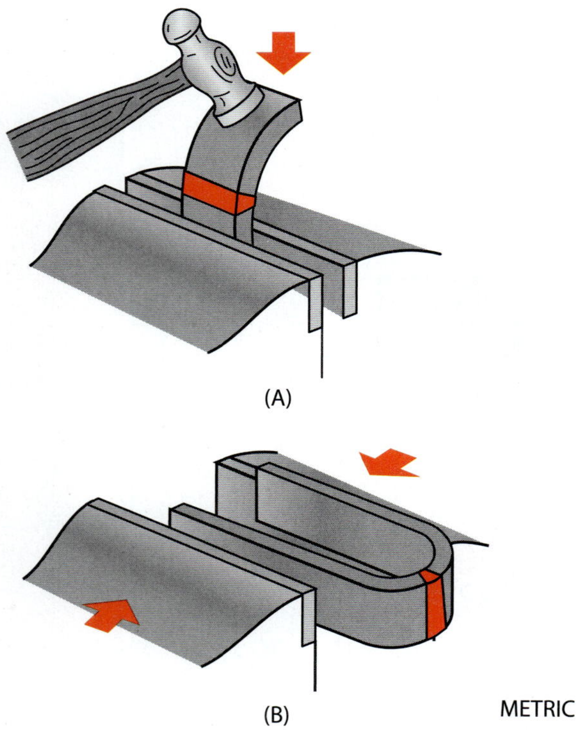

Conversion Table	
mm	in.
12.7	0.500
20.0	0.787
32.0	1.25
76.0	3.000

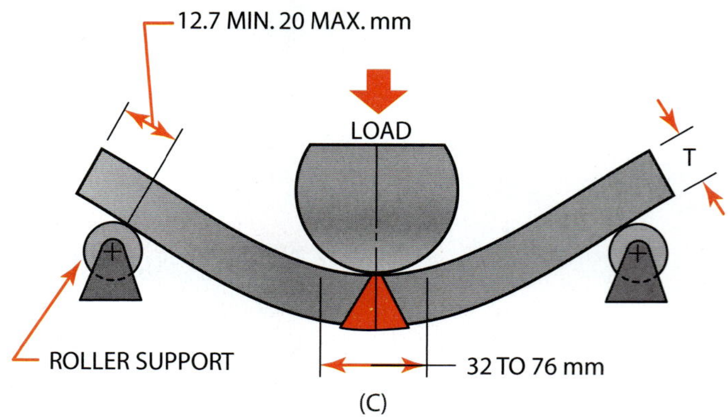

FIGURE 18-10 Free bend test.

a short distance upward after striking the specimen before stopping its upward motion (**Figure 18-11A**). A brittle specimen would absorb less energy from the swinging pendulum so that it would continue its upward swing to a much higher position (**Figure 18-11B**).

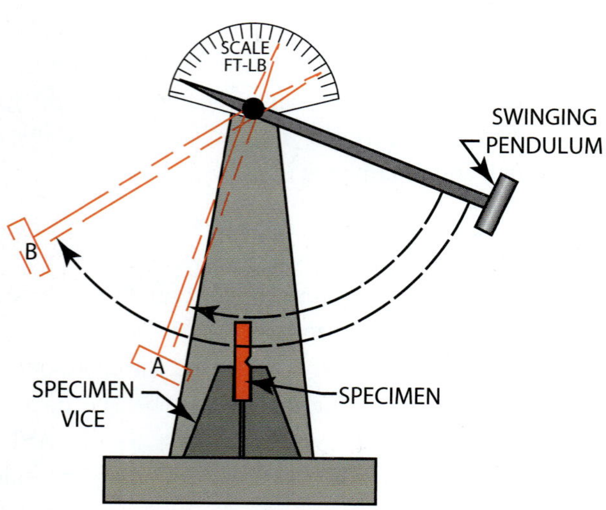

FIGURE 18-11 Impact test.

Another type of impact test is the Charpy test. The major differences between this and the Izod test are that the Izod test specimen is gripped on one end and is held vertically and usually tested at room temperature, and the Charpy test specimen is held horizontally, supported on both ends, and is usually tested at a very specific temperature. The Charpy test specimen temperature is important because some base metals and filler metals become brittle at low temperatures. Normal specimen temperatures can range from 20°C to −20°C, but the Charpy test can be performed on specimens at temperatures ranging from 280°C to −250°C.

NONDESTRUCTIVE TESTING (NDT)

NDT is a method used to test weld materials for surface defects such as cracks, arc strikes, undercuts, and lack of penetration. Internal or subsurface defects can include slag inclusions, porosity, and unfused metal in the interior of the weld.

Visual Inspection (VT)

Visual inspection (VT) is the most frequently used NDT method and is the first step in almost every other inspection process, **Figure 18-12**. The majority of welds receive only visual inspection. In this method, if the weld looks good, it passes; if it looks bad, it is rejected. This procedure is often overlooked when more sophisticated NDT methods are used. However, it should not be overlooked.

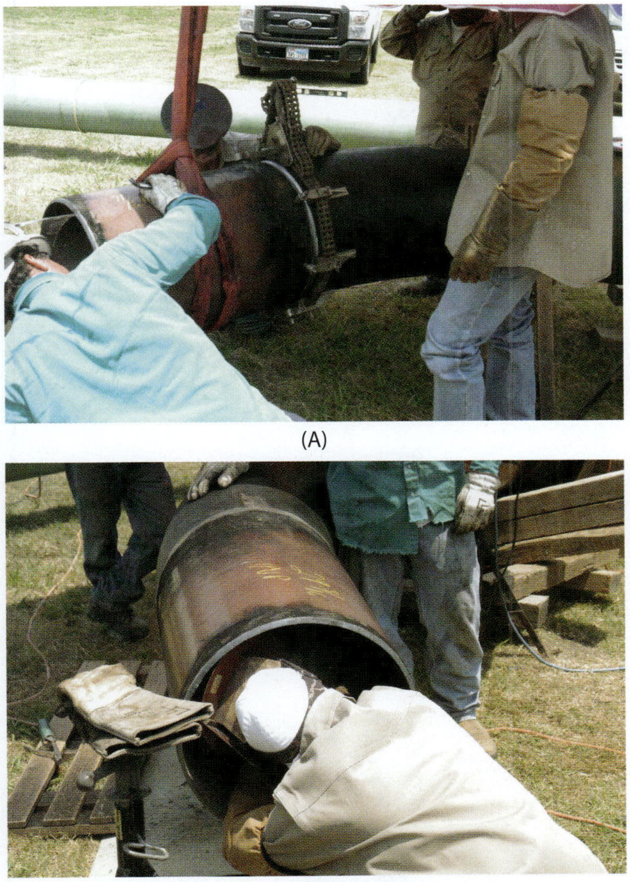

(A)

(B)

Larry Jeffus

FIGURE 18-12 Visually inspecting the root of pipe welds.

An active VT schedule can reduce the finished weld rejection rate by more than 75% because defects are identified and corrections can be made. Visual inspection can easily be used to check for fit-up, interpass acceptance, technique, and other variables that will affect the weld quality. Minor problems can be identified and corrected before a weld is completed. This eliminates costly repairs or rejection.

Visual inspection should be used before any other NDT or mechanical testing are used to eliminate (i.e., reject) the obvious problem welds. Eliminating welds that have so many surface discontinuities that they will not pass the relevant code or standards saves preparation time.

Penetrant Inspection (PT)

Penetrant inspection (PT) is used to locate discontinuities that are open to the surface, such as cracks and porosity. Two types of penetrants are now in use—color contrast and fluorescent. Color-contrast penetrants contain a colored dye (often red) that shows under ordinary light, **Figure 18-13**. Fluorescent penetrants contain a more effective fluorescent dye that shows under black light.

Larry Jeffus

FIGURE 18-13 Welds that have been penetrant inspected.

PRACTICE 18-1

PT Procedure

The following steps outline the procedure to be followed when using a penetrant, **Figure 18-14**:

1. Make sure that the test surface is clean and dry. Suspected flaws must be clean and dry so they are free of oil, water, or other contaminants.
2. Cover the test surface with a film of penetrant by dipping, immersing, spraying, or brushing.
3. Wipe, wash, or rinse the test surface gently with a cleaner of excess penetrant, and then dry it with cloth or hot air. Die penetrant will remain in any discontinuities that are open to the surface.
4. Apply a developing powder to the test surface as a blotter to pull the penetrant out of any flaws open to the test surface.
5. Depending upon the type of penetrant applied, make a visual inspection under ordinary light or ultraviolet (black) light. The penetrant fluoresces to a yellow-green color, which clearly defines the defect. ◆

PRACTICE 18-2

Magnetic Particle Inspection (MT)

Magnetic particle inspection (MT) is a low-cost testing method that can be used to locate surface and near-surface discontinuities. It uses very fine iron (ferromagnetic)

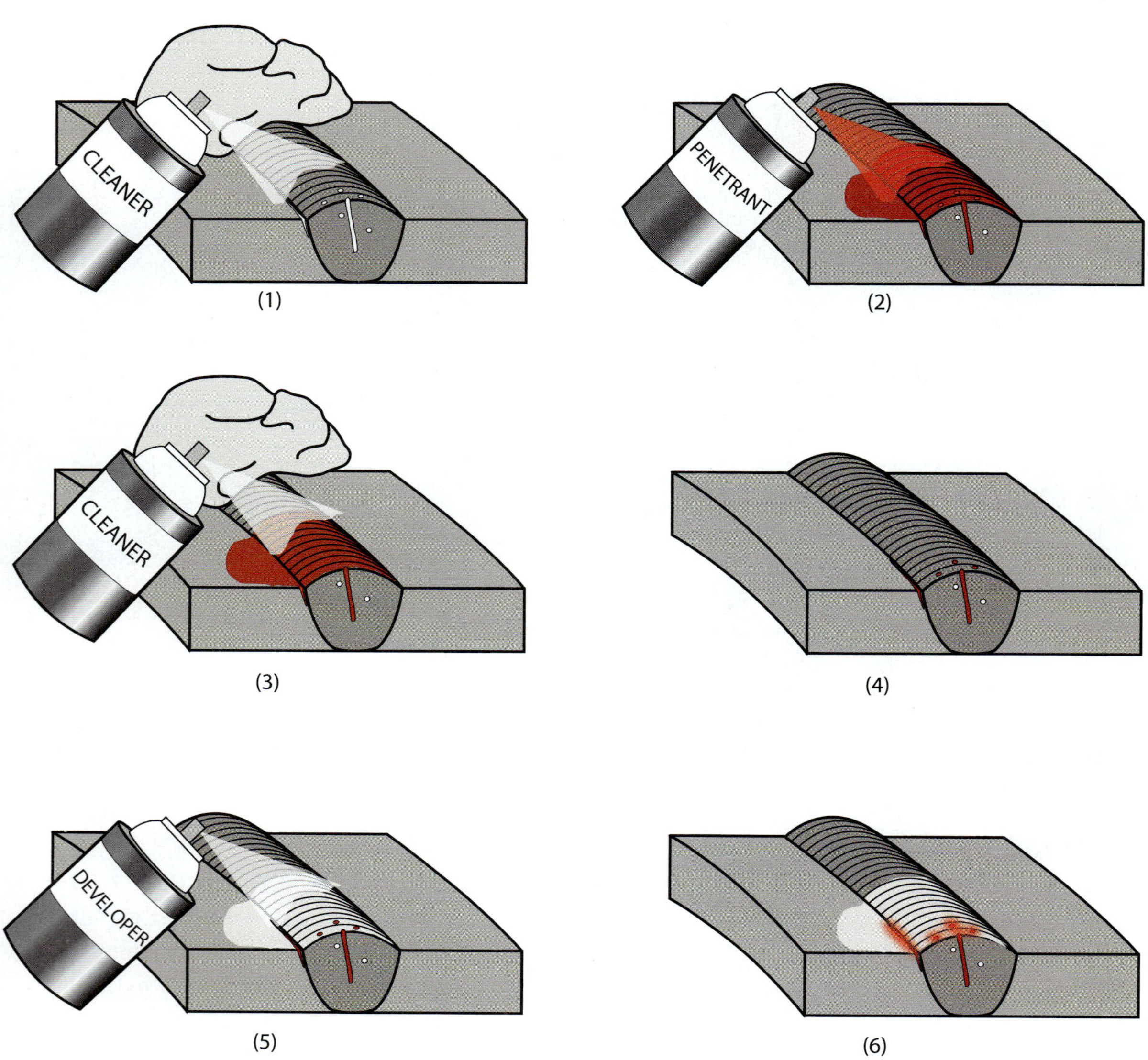

FIGURE 18-14 Penetrant inspection.

powder to indicate discontinuities open to the surface or just below the surface on magnetic materials. Powders are available in colors to have a stronger contrast with the surface being tested, and they are also available as fluorescent powders that are more visible under ultraviolet light.

When a magnetic field is induced in a part by passing an electric current through it, or a magnet is used to induce a field and a fine iron powder is applied, any discontinuities on or near the surface that are at a right angle to the magnetic field will attract the powder because the discontinuities cause a disruption in the magnetic field, **Figure 18-15A**. However, if the discontinuities are too far from the surface, the magnetic field will not show a deflection on the surface, **Figure 18-15B**.

The magnetic field can be induced in the pipe by passing a high-amperage, low-voltage current through it using a set of probes, **Figure 18-16**. Switches on the probes allow the current to be switched on and off so

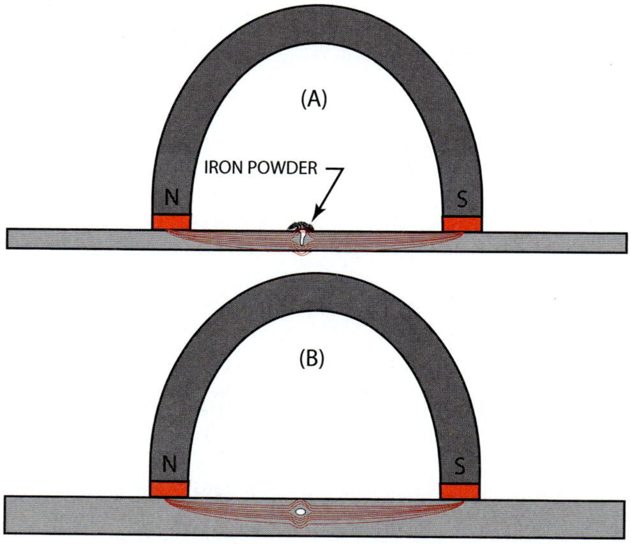

FIGURE 18-15 Flaws close to the surface cause the magnetic lines of flux to be deflected (A), while flaws farther below the surface do not deflect the lines of flux (B).

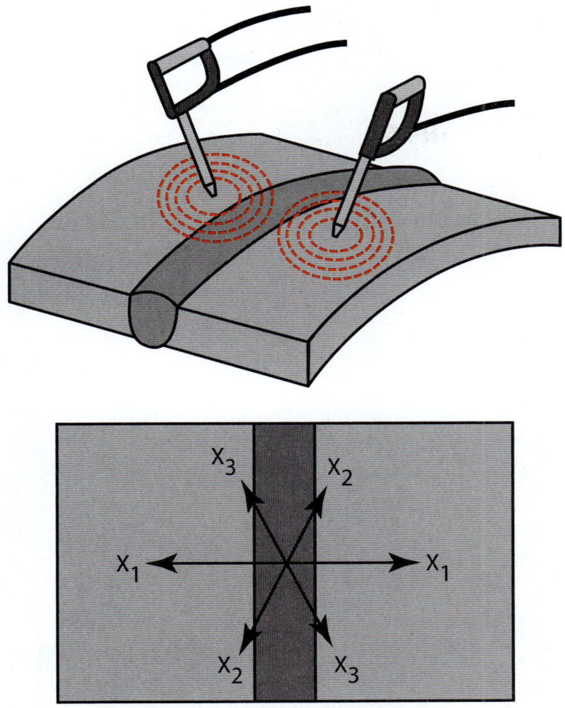

FIGURE 18-16 The probes are placed in a pattern all along the joint being tested.

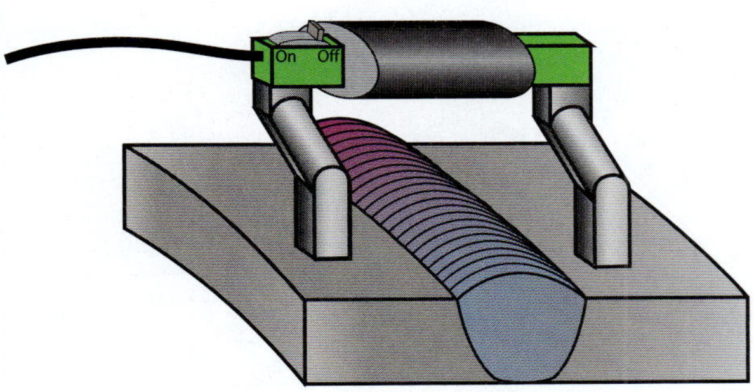

FIGURE 18-17 Yoke magnetic particle tester.

that no arc strikes occur between the probe and pipe. A yoke-type magnetic tester can also be used to test welds, **Figure 18-17**. ◆

MT Procedure

The following steps outline the procedure to be followed when using a yoke-type MT tool, **Figure 18-18**:

1. Clean the surface of any dirt, slag, oil, grease, or other material that might catch attract and collect the testing powder.
2. Place the probe at a right angle to the weld outside the heat-affected zone.
3. Turn on the power.
4. Using a spray can or a squeeze bulb, lightly dust the surface with the testing powder.
5. Inspect the powder visually to see if it has been attracted to any discontinuities.
6. Rotate the test probes as close to the weld as possible and repeat the test procedure to locate any discontinuities that may have been transverse to the weld.
7. Rotate the test probes in the opposite direction, and repeat the test procedure to ensure that all the discontinuities have been located.

Radiographic Inspection (RT)

Radiographic inspection (RT) uses the very short wavelength rays developed by x-ray machines or radioactive isotopes (gamma rays). Both of these types of rays can be used to see through metals to detect discontinuities anywhere inside weldments that are generally pointed toward the source or that are very localized, **Figure 18-19**. Unlike light waves, which are reflected from a material's surface, x-rays and gamma rays are so short that they can weave their way between the atoms that make up most materials. RT can be best used to detect discontinuities that are vertical or nearly vertical, but those that are uniform over a large area may not be seen on the x-ray film, **Figure 18-20**.

NOTE

Finding discontinuities with RT is much like trying to see a very clean, clear plate of glass. Unless there is a reflection from the surface of the glass, it may be invisible when viewed straight on; but it becomes more visible the more that it is rotated. When the edge is directly pointed in your direction, it becomes the most visible.

CAUTION

Follow all posted warning signs any time RT inspection is being performed because overexposure to x-rays is harmful to people and animals.

The power produced by most x-ray machines can penetrate only thin sections of metal, so the more powerful gamma rays emitted from radioactive isotopes are the most common source for RT inspection. Lead containment boxes are used to hold these isotopes safely because lead is so dense that radiation waves cannot pass through it.

For years, film was the only way of capturing the RT inspection images of welds. Just like most photography today, the RT process has been moving into

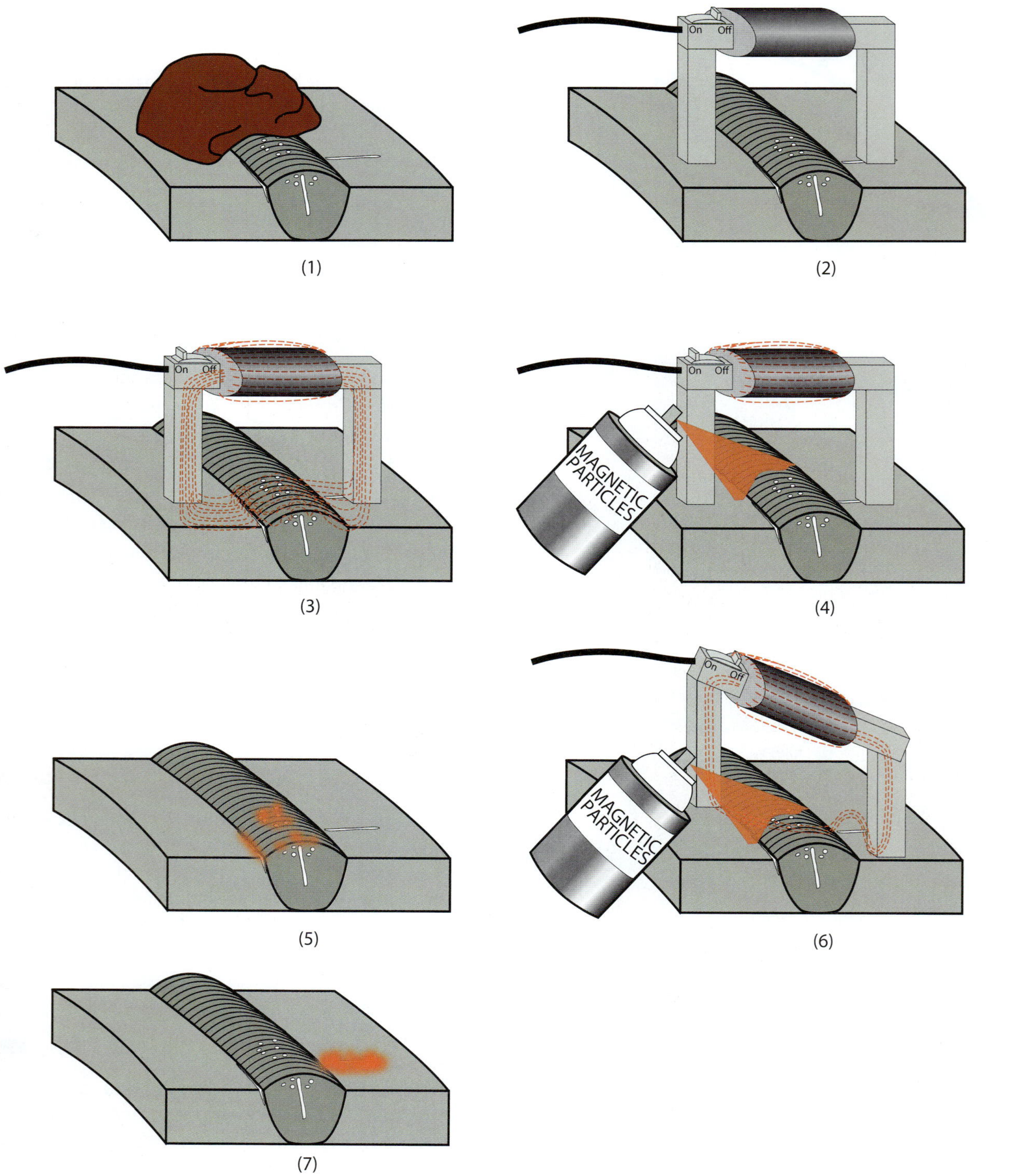

FIGURE 18-18 Magnetic particle testing procedure.

the digital age. In digital imaging, a detector plate is used instead of film to record any discontinuities in the weld. Digital imaging can provide an immediate look inside the weld.

Figure 18-21 shows film wrapped around one side of the pipe weld, and the gamma source is placed on the opposite side of the pipe. This places the film closest to the weld to minimize the distortion of the image of any flaws in the weld, **Figure 18-22**. However, there is still some distortion of the size of the flaw image on the x-ray film due to the flaw's distance from the center line of the film just as if a photo were out of focus. The farther the flaw is from the center line and the film, the larger and fuzzier the image appears.

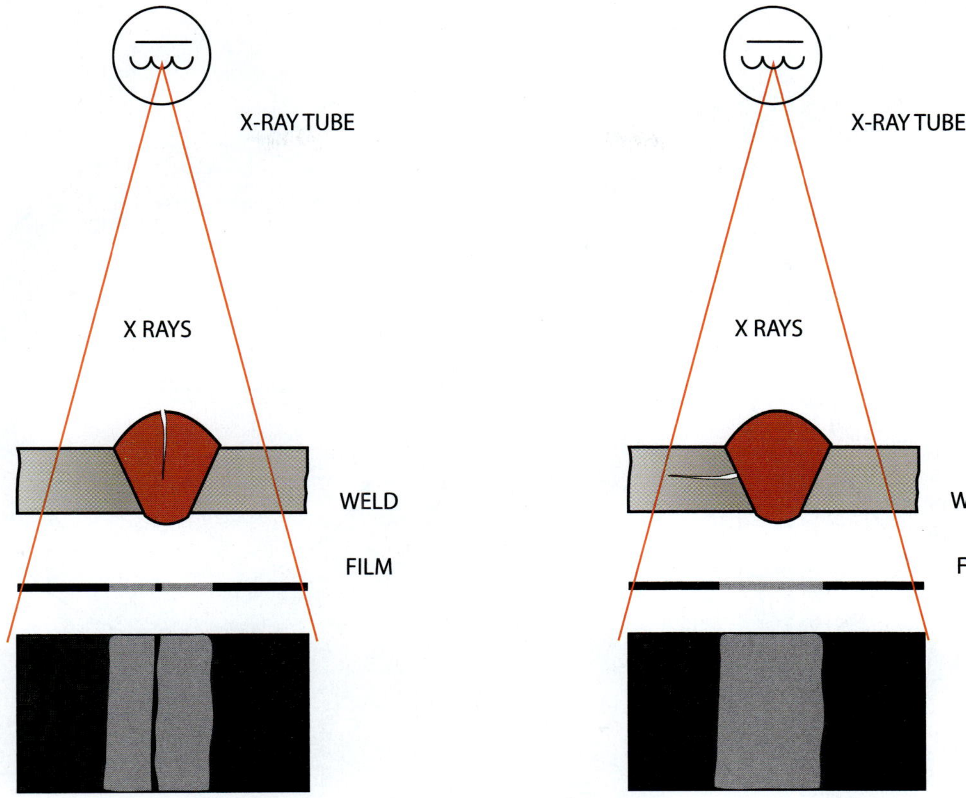

FIGURE 18-19 Not all flaws show up as easily on radiographic inspections.

Larry Jeffus

FIGURE 18-20 Radioactive source.

Larry Jeffus

FIGURE 18-21 X-ray film being exposed.

Hydrostatic Testing

Hydrostatic testing is a way of determining the soundness and strength of a pipeline, piping system, or pressure vessel by filling it with water and then applying a pressure higher than the system's design pressure to locate leaks and to ensure that it exceeds the strength requirement. The system is sealed, filled with water or another noncompressive liquid, a gauge is attached, and a specific pressure is applied, **Figure 18-23**. Because a noncompressive liquid is used, very little liquid needs to be added to achieve the desired pressure. Also, if the system fails during the test, only the small quantity of additional liquid would escape; however, if a compressive gas such as air were used, the pipe would fail, and there would be an explosion—much as a balloon filled with air pops while

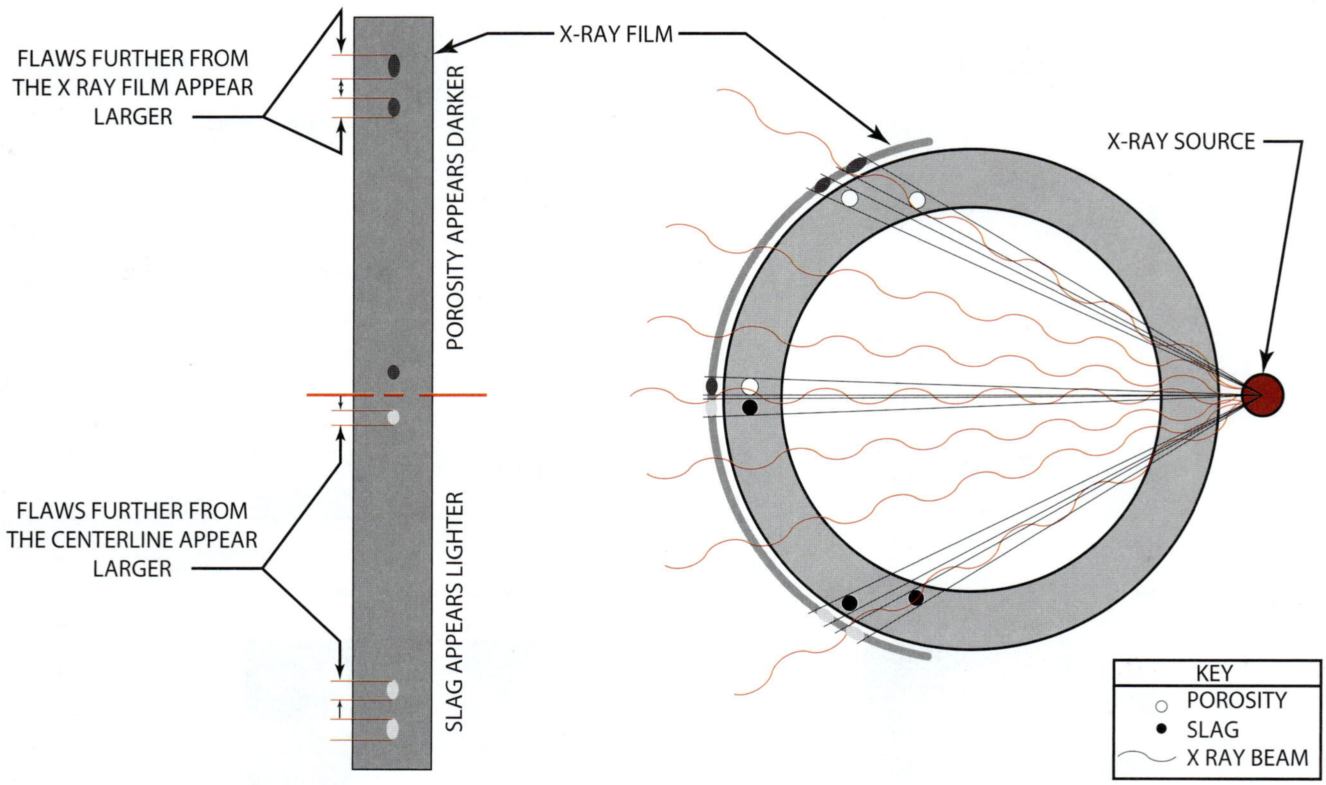

FIGURE 18-22 How flaws may appear on an x-ray film.

FIGURE 18-23 Hydrostatic testing.

one filled with water does not. Water is the liquid used most often because it is noncompressible, readily available, and inexpensive.

Depending on the code requirements, the maximum test pressure ranges from 125% to 166.66% of the designed maximum operating pressure. If the system is sound when it is isolated from the pressure pump, the test pressure will hold. However, if it has any leaks, the pressure will drop. If the system exceeds the strength requirement, it will not rupture. However, if more liquid, based on the known volume of the system, is required to obtain the desired test pressure, that could indicate a weak spot that is bulging out.

Ultrasonic Inspection (UT)

Ultrasonic inspection (UT) is fast and uses few consumable supplies, which makes it inexpensive to use, **Figure 18-24**. This inspection method employs electronically produced sound waves. These sound waves can range from a low frequency of roughly 1/4 cycles per second to a high frequency of around 25 million cycles per second, which penetrate metals and many other materials at speeds of several thousand feet or meters per second. A portable UT unit is shown in **Figure 18-25**.

Pulse-echo ultrasonic detection systems use sound generated in short bursts or pulses. Since the high-frequency sound is at a relatively low power, it has little ability to

FIGURE 18-24 Ultrasonic testing of welds on a pipeline.

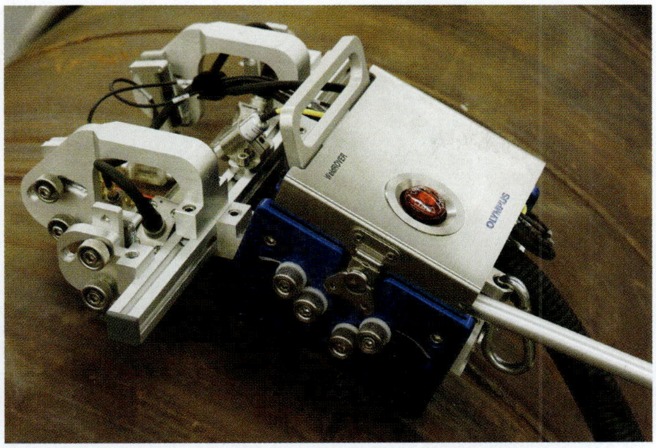

Courtesy of Olympus NDT

FIGURE 18-25 Portable ultrasonic tester.

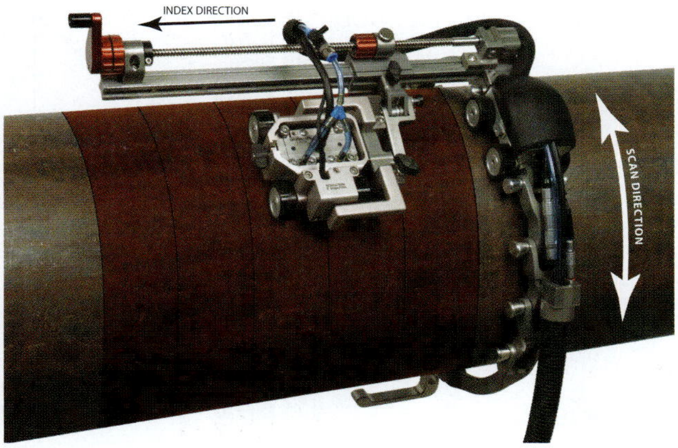

Courtesy of Olympus NDT

FIGURE 18-27 Automatic ultrasonic testing machine.

travel through air, so it must be conducted from the probe into the part through an acoustic medium.

The device that generates and receives the ultrasonic signal is called a "transducer." They can be either straight or angled. Because of the irregular surface of the weld, an angled transducer is used for most weld inspections. The speed that the sound waves travel through pipe is a known quantity, so the internal computer program converts the time into a precise measurement that is displayed on the detector screen. When the sound waves strike a flaw, they are reflected back to the transducer. The technician moves the transducer in a zigzag pattern to accurately measure the size of any flaws, **Figure 18-26**.

Courtesy of Olympus NDT

FIGURE 18-26 Manual ultrasonic testing of a pipe joint.

Some self-propelled testers can travel around a pipe weld, which eliminates many operator errors that can occur when a zigzag motion by hand is being used, **Figure 18-27**. In addition, phase-arrayed transducers contain multiple sending and receiving acoustic transmitters so that the entire depth of the weld can be viewed on multiple tracks displayed on the detector.

Eddy Current Inspection (ET)

Eddy current inspection (ET) uses magnetic induction to induce a current called an "eddy current" within the material being tested. An eddy current is an induced electric current circulating wholly within a mass of metal. This method is effective in testing nonferrous and ferrous materials for internal and external cracks, slag inclusions, porosity, and lack of fusion that are on or very near the surface. Eddy currents cannot locate flaws that are not near the surface.

CORROSION PROTECTION

All buried carbon steel and galvanized pipe, valves, fittings, and all other piping system components must be protected from corrosion. They can be double wrapped or epoxy coated and may have a sacrificial cathode buried alongside the pipe for additional protection, **Figure 18-28**. Some pipe comes precoated with a corrosion-protection material, **Figure 18-29**. Care must be taken when handling these precoated pipes so as not to damage the corrosion-protective surface.

Pipe Preparation

The surface of the pipe and other devices must be free from rust, mill scale, weld spatter, slag, dirt, oil and grease, or any other foreign materials. Shot blasting (sand blasting) of all the surfaces to be coated must be done immediately before the protective coating is applied to reduce the possibility of recontaminating the pipe surface, **Figure 18-30**.

The protective coating must be applied in accordance with the manufacturer's recommended application procedures and in accordance with any applicable codes or standards. Read and follow all the product's safety instructions

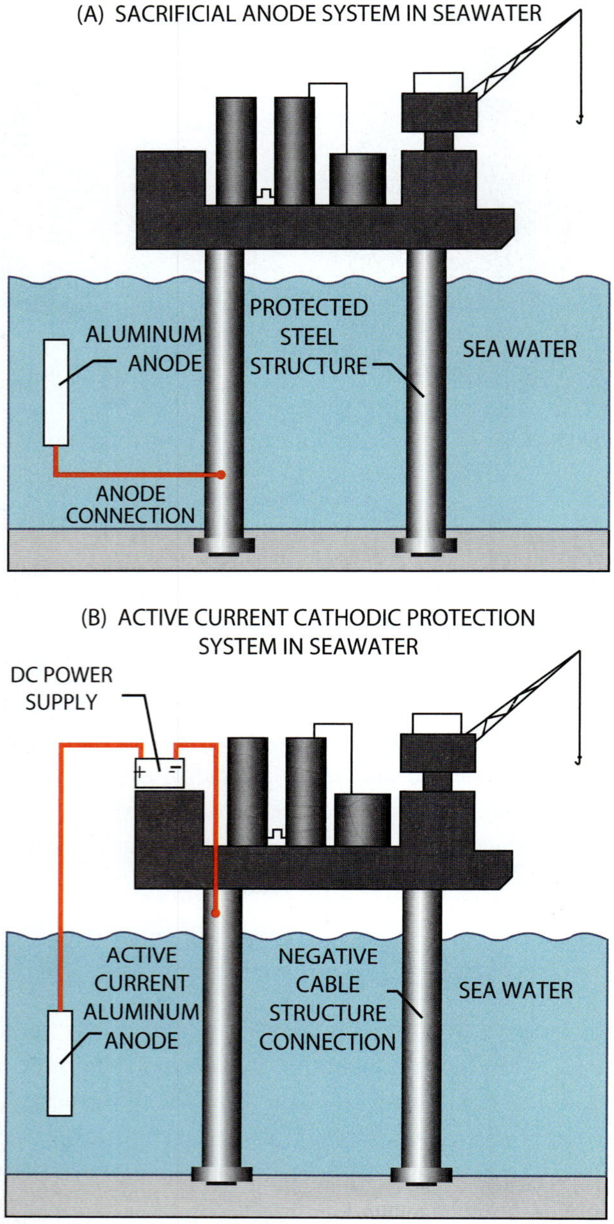

FIGURE 18-28 Two methods of anode protection for piping.

FIGURE 18-29 Precoated pipe sections.

FIGURE 18-30 Shot-blasting pipe before a protective coating is applied.

regarding the application, handling, storage, and disposal of materials.

Corrosion Protection Inspection

In a corrosion protection inspection, 100% of the pipe shall be visually inspected and 100% electronically tested for complete coverage with a Holiday Pipe Tester, **Figure 18-31**. Any area that does not pass both tests shall be marked and repaired according to the coating manufacturer's repair instructions.

FIGURE 18-31 Testing the thickness of the protective coating.

Larry Jeffus

FIGURE 18-32 Open storage area for pipe fittings on a job site.

Summary

Quality cannot be tested just in a weldment; quality is everyone's responsibility and must be part of the entire fabrication process. During the planning stage, the design must take into consideration the fabrication environment, weld joint accessibility, and qualifications of the welders. When the pipe, fittings, and valves arrive on the job site, care must be taken to prevent damage to the corrosion-protective surface material, **Figure 18-32**. As the joints are aligned, care must be taken to ensure that the fit-up is within the acceptable tolerance because errors in the fit-up will never get better and cannot be fixed by even the best welder.

When a job has to be redone to fix a problem, it costs you and your company five times: once for the time it took to do it wrong, once for the time it took to take it out, once for the time it took to redo it, and two times for the other work you could have done while redoing the problem. Take the time to check everything before starting to weld so you can minimize the number of problems that need to be fixed.

Review Questions

1. What is the difference between quality assurance (QA) and quality control (QC)?
2. What are some of the other terms that codes and standards are called?
3. What are the three main organizations that have codes used in pipe welding?
4. What is a welding procedure specification (WPS)?
5. What items are included on a WPS form?
6. What are the two methods used to test the quality of a weld?
7. What is the major assumption made when mechanical testing is used?
8. When tensile testing, why must the specimen include both the heat-affected zone and the base metal?
9. What should you check for on the fractured surface of a nick-break test specimen?
10. What determines the required bending radius of a guided bend testing jig?
11. What results are obtained by an alternate bend test?
12. Describe the two impact tests.
13. What are some surface defects that can be detected by NDT?
14. What type of inspection should be the first step in almost every inspection process?
15. What types of defects can penetrant inspection reveal?
16. What test uses short-wavelength rays to detect discontinuities anywhere inside weldments?
17. Hydrostatic testing can be used to determine what two things about a pipeline, piping system, or pressure vessel?
18. What is the name of the device that transmits and receives an ultrasonic signal for ultrasonic inspection?
19. How can buried carbon steel and galvanized pipe be protected from corrosion?

Chapter 19

Pipe Welding Certification—Welding Procedures

OBJECTIVES

After completing this chapter, the student should be able to:

- Identify the publishers of the most commonly used pipe welding codes.
- Explain what it means to be a certified welder.
- Discuss the purpose of the American Welding Society (AWS) SENSE program.
- Describe what can be measured by visual weld inspection tools.
- Explain what can cause a weld to be unacceptable.
- Compare discontinuities and defects as they relate to acceptability.

KEY TERMS

Defect

Discontinuity

Notch-sensitive

SENSE

Stress riser

INTRODUCTION

Welding codes related to piping and pressure vessels were first published around 1925 by the American Society of Mechanical Engineers (ASME) and American Welding Society (AWS). After the codes were established, welders were asked to demonstrate that they could produce welds that met these standards. This evolved into today's practice of requiring that welders be certified. The most common welding codes used in the piping industry today are published by the American Petroleum Institute (API 1104), the AWS (AWS D1.1, AWS D10.12, AWS D10.18, and AWS D18.1), and the ASME (ASME B31.3).

Welders in the industry are often required to demonstrate their welding skills by passing a weld test based on these codes. Once the test is passed and documented in writing, the welder is said to be "certified." The AWS SENSE program is the most commonly used certification test used by welding schools to certify students. The major difference between the certification tests taken by production welders and students is the cost, not the acceptance criteria. All welder certifications have time limits and can expire within a few weeks or months, depending on the test and industry standards. The time period that a certification is good for is often based on whether the welder is using the certification welding skills on a regular basis. So the expiration date can be based on the last time you made a particular pipe weld covered by the certification you hold.

So what does being a certified welder actually mean? It means that at the time you were tested, you had the skills to pass the certification weld specimen required by a specific code and that fact was documented. Unlike certifications in almost every other industry, when applying for a new job, you will most likely be asked to take another certification test, no matter how much

(*Continued*)

(or how little) time passed since the last test you took to demonstrate your current skill level. In fact, not even surgeons are asked to demonstrate their skills the way that welders must.

Pipe-welded workmanship specimens produced according to the practice welds in Chapters 9, 10, 11, and 12 of this book can be used for this chapter's certification test. Alternatively, workmanship specimens can be made in accordance with the welding instructions in the appropriate chapter that covers the process on which you wish to be tested.

QUALIFIED VERSUS CERTIFIED WELDERS

A qualified welder usually only has to pass a company's welding proficiency test. The company maintains the records of a welder's qualification. Also, there may be a requirement for the welder to complete a company-sponsored training program.

To become an AWS certified welder, you must pass a welding proficiency test at an AWS Accredited Test Facility (ATF). There are very strict guidelines for ATFs to follow in order to certify welders. Once the test is passed, the results are forwarded to the AWS, which then issues the welder a wallet card. AWS welder certifications are good for six months and can be renewed by proving current and ongoing welding activities in accordance with the certification.

AWS SENSE

The acronym **SENSE** stands for "Schools Excelling (through) National Skills (standards) Education." It was first published in the mid-1990s and was recently updated. This update combined some of the elements of the Level III certification with Level II and eliminated the Level III certification altogether. The current SENSE publications available from the AWS are EG2.0 for Level I (Entry Welder) and EG3.0 for Level II (Advanced Welder). This chapter concentrates on the skills in Level II.

There is a one-time fee for a school to join the SENSE program. Schools that are SENSE members can submit their students' names who have passed a certification test to be listed in the AWS national database, and companies looking for welders can access this database to recruit skilled welding students for jobs. Because of the nearly 20 years of providing excellence in certifying students through the SENSE program, many companies in the welding industry recognize a student's SENSE welding certification as a stamp of approval. Like any other skilled welder applying for a job, you will still have to be tested; but since you may have already passed the same test in school or on the job, your welding skills should be up to the challenge.

SENSE Certification

There are a number of individual SENSE certifications that can be earned by welding students. The fabrication workmanship standards require only a visual inspection, while the welding workmanship standard requires both an initial visual inspection, followed by a guided bend test. The AWS publications AWS C4.1, "Criteria for Describing Oxygen Cut Surfaces"; AWS B4.0, "Standard Methods of Mechanical Testing of Welds"; and AWS B2.1, "Standards for Welding Procedure and Performance Qualification" can provide more information about tests not included in this chapter.

SENSE Visual Inspection Tools

The basic visual inspection process requires tools to measure the joint geometry before the weld is produced and to measure the various aspects of the weld when it is completed. There are a number of specially designed weld gauges that can be used to measure pipe wall thickness and diameter, joint groove angles, root face, and root opening before welding; and most perform the postweld measurements as well. Items that must be measured after the weld is produced include length and depth of undercut; size and frequency of porosity; length of cracks; and weld width, length, and reinforcement. Some of these weld gauges are shown in **Figure 19-1**.

Inspection kits are available that contain a caliper, micrometer, magnifying glass, steel rule, and weld gauges that can be used to measure the wall or plate thickness, **Figure 19-2**. A steel rule or tape measure can be used to measure the length and diameter of the pipe workmanship specimen.

Sometimes a magnifying glass and a flashlight can be helpful to see the reading on a weld gauge.

SENSE Visual Inspection Testing

A common misconception is that only bend specimens that are free of any discontinuities can pass. *This is not true.* No weld is perfect. It is important to understand the difference between a discontinuity and a defect. A **discontinuity**

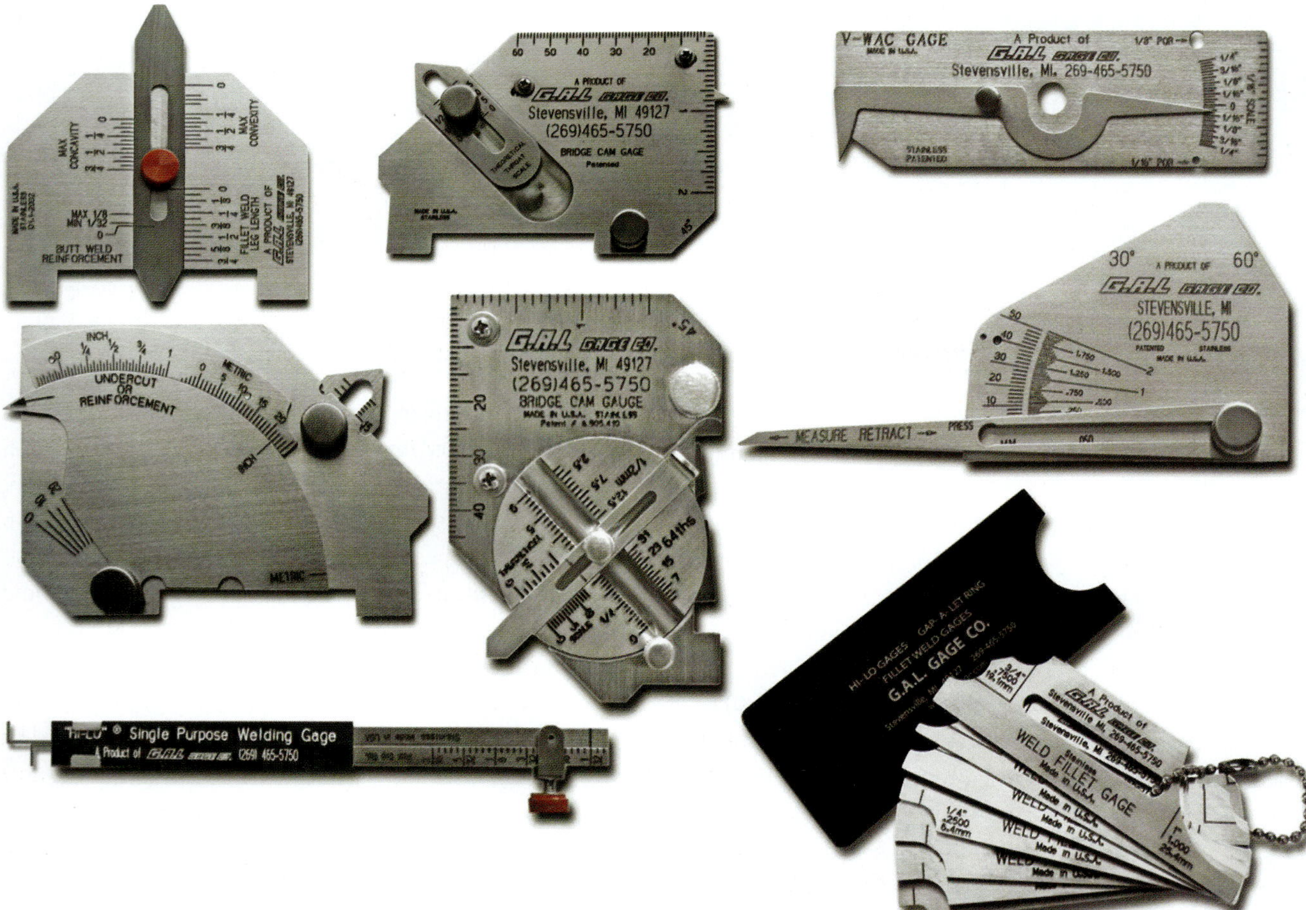

G.A.L. Gage Company

FIGURE 19-1 There are a number of different specialty gauges for measuring weld joints and welds.

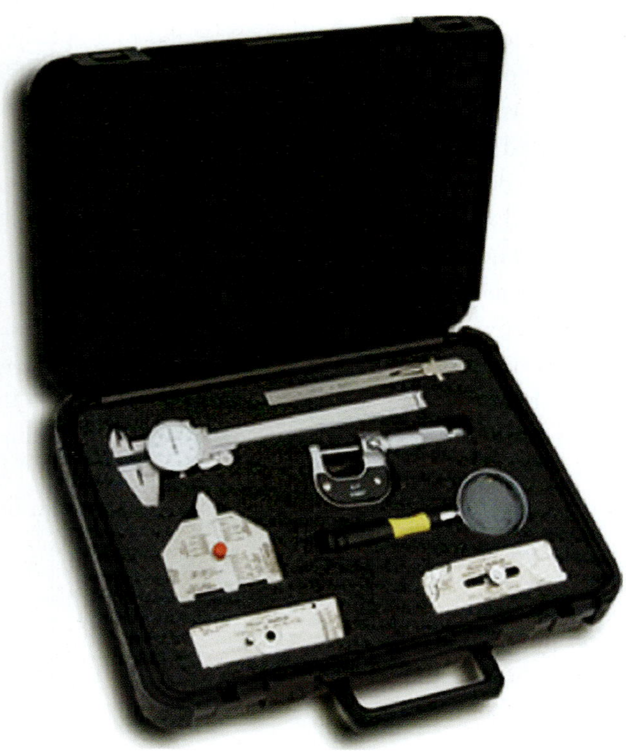

G.A.L. Gage Company

FIGURE 19-2 Weld visual inspection kit.

is an interruption of the typical structure of a material, while a **defect** is a discontinuity or discontinuities that by nature or accumulated effect render a part or product unable to meet the minimum acceptable standards or specifications.

The major difference between discontinuities and defects is their size and location and the applicable code or standard. For example, a single porosity with a 1/16-in. (1.5 mm) diameter that is open to the surface on the outside of a pipe weld would not be a defect in most codes and standards. But that same small porosity on the root face of a weld on piping used to manufacture cosmetics would be a defect because bacteria could collect in it and contaminate the cosmetics.

A **stress riser** is a point of stress concentration usually caused by an abrupt change in the thickness or shape of a material or the presence of a sharp groove or crack on the material's surface. An example of a stress riser that can cause a failure is the point on a bolt where the threads end and the shank begins. If a bolt is clamped in a vise and bent back and forth, it will most often break either at that point or at the top edge of the vise jaws.

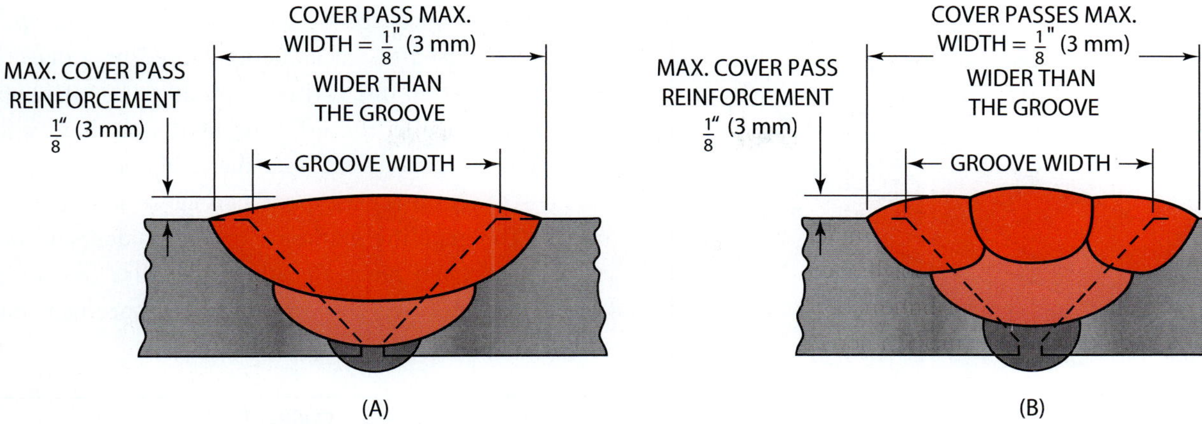

FIGURE 19-3 The weld contour must be the same whether the cover pass is made with one weld pass or several weld passes.

Both places cause stress risers. Stress risers are most detrimental when they are located on the surface. For example, the small groove around the pop top on a canned drink forms a stress riser that allows the can to be easily opened. It is important to note that by having the groove on the outside of the can, the internal pressure does not cause it to pop open on its own. Because stress risers are more detrimental on the surface, a flaw on the surface of a weld is more critical than the same flaw on the inside of a weld.

All metals have some level of notch sensitivity, but some metals are more susceptible to failure than others due to stress risers. **Notch-sensitive** metals are the ones that are more susceptible to stress riser failure.

SENSE Visual Inspection Test Results

The primary purpose of a visual inspection is to identify flaws (stress risers) that might pose a threat to the pipe's integrity. To prevent stress risers at the edge of the weld, the face of the weld must have a smooth transition between the weld and the pipe surface, **Figure 19-3**. The reinforcement of the weld is restricted to a maximum height of 1/8 in. (3 mm) to prevent the weld from excessively restricting the pipe's expansion and contraction due to temperature changes or pressure changes and to allow a more uniform flexing of the pipe during construction and shifting soil.

The following are the AWS SENSE requirements for welds to pass a visual examination:

- The weld must be uniform in appearance, with no roughness or abrupt changes in shape or size.
- The weld must be free of cracks and incomplete fusion, and the root fusion must be 100% around the entire pipe.
- Undercut cannot exceed 10% of the wall thickness or 1/32 in. (0.793 mm), whichever is less. For example, the allowable undercut on an 8-in. (200-mm) diameter Schedule 40 pipe with a wall thickness of 21/64 in. (8.3344 mm) would be 1/32 in. (0.7938 mm). However, the allowable undercut on an 8-in. (200-mm) diameter tubing with a wall thickness of 1/4 in. (6.350 mm) would be 3/128 in. (0.5953 mm).

SENSE Guided Bend Test Specimen Selection

There are very exacting specifications for selecting the pipe weld specimens for a guided bend test, **Figure 19-4**. In addition, there are very exacting specifications concerning the size of the specimens, **Figure 19-5**. It is important to identify the top of the pipe specimen for 5G, 6G, and 6GR workmanship samples since the location of the test specimens is determined from the top of the pipe as it was welded.

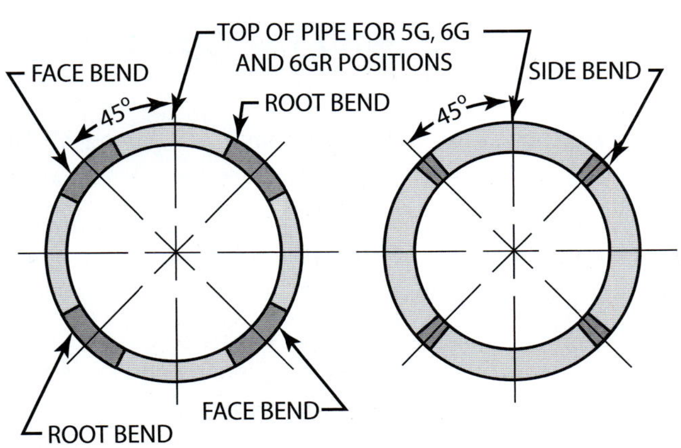

FIGURE 19-4 Weld specimen location.

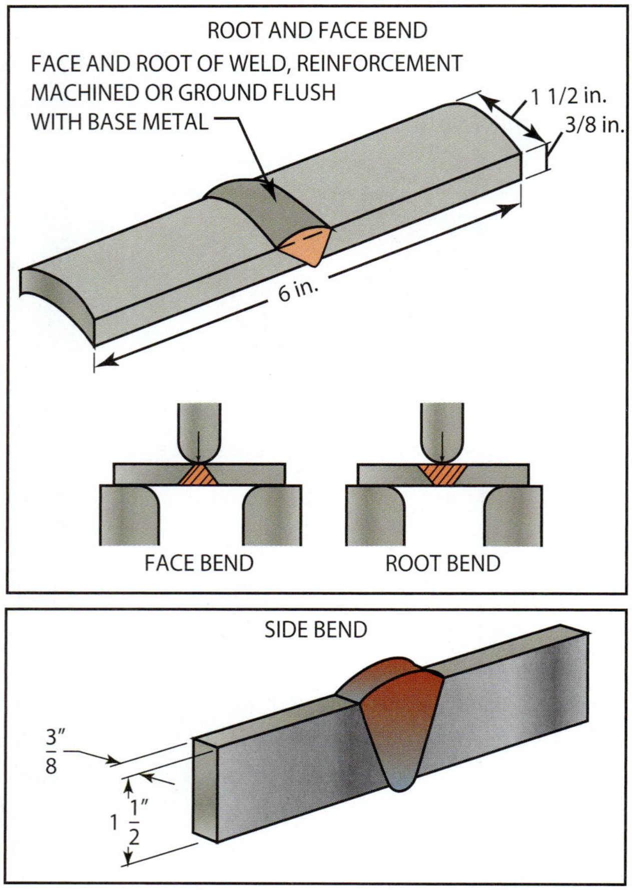

FIGURE 19-5 Weld specimen preparation.

The specimens can be thermally or mechanically cut from the finished workmanship sample. If they are going to be thermally cut out, they should be laid out wider than the finished size so that the effects of the thermal cutting can be removed and smoothed with a grinder before testing. Specimens that are mechanically sawed or cut with an abrasive wheel can be cut to the finished width since little or no thermal shock should occur during these processes.

SENSE Guided Bend Test Specimen Preparation

Only workmanship specimens that have passed all of the visual inspections can be bend-tested. Proper weld specimen proportion is essential to making an acceptable bend test. The specimen length, thickness, and width must all be within the tolerances as specified by the testing procedure.

Cracks are the major cause of a guided bend test specimen failing because they form stress risers. All of the following items are designed to reduce the possibility of cracks being caused by the preparation of the specimen, as opposed to cracks being caused by a defective weld. The bottom line is that if you do not prepare your specimen properly, the grinding marks or sharp edges can cause a good weld specimen to crack just as a small scratch on a glass plate can cause it to crack or break. The grinding marks on the top, sides, and back of the specimen must all run in the direction of the bend, **Figure 19-6**. The corners of the specimen must

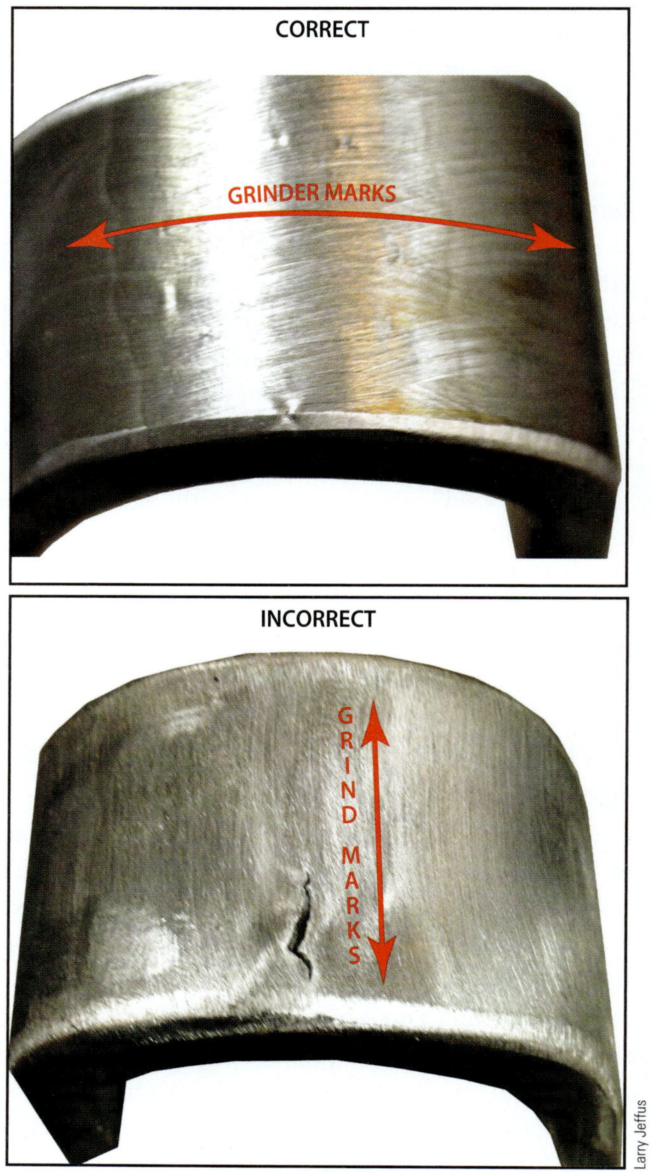

FIGURE 19-6 The grind marks must run parallel to the bend test specimen. The two small dimples on the edge and face of the top bend specimen would have formed a crack if the grind marks were incorrect or if the edge of the specimen had not been rounded. This specimen passed because it was prepared correctly. The bottom bend specimen may have passed as well, if it was ground correctly.

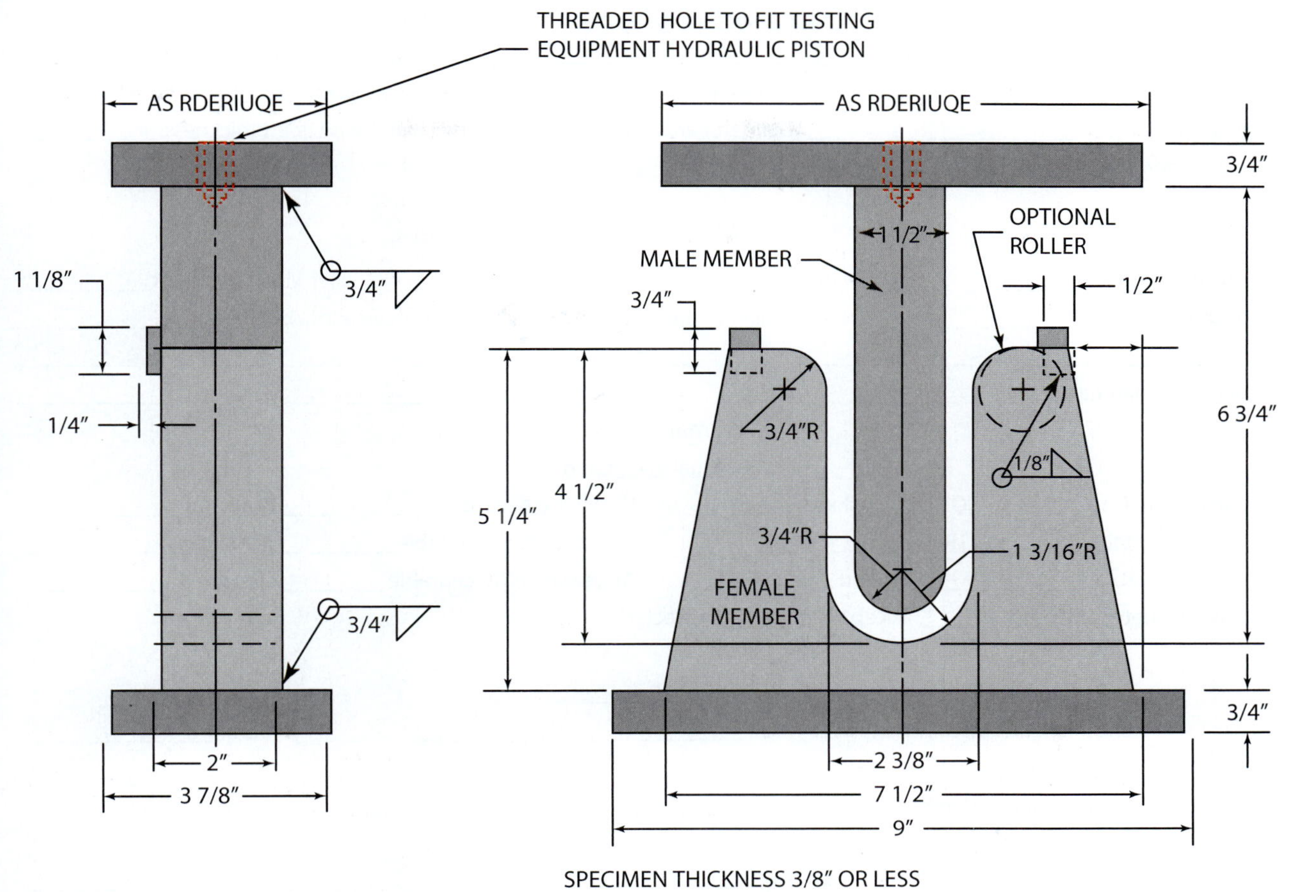

FIGURE 19-7 Typical guided bend test machine.

be slightly rounded. The best way to round the edges is with a hand file.

SENSE Guided Bend Test Jig

Guided bend testers may have rollers on the top edge of the female die or rounded shoulders, **Figure 19-7**. The guided bend test is designed to place a bending force on the face, root, or side of a weld. However, if the rollers do not turn freely or the rounded shoulder is not smooth and lubricated, the weld can be subjected to a strong pulling force that can rip apart even the best welds.

SENSE Guided Bend Test Results

The center of the weld should be near the center of the convex surface once the specimen has been bent. No opening shall be larger than 1/8 in. (3 mm), as measured in any direction. The only exception is small cracks that may start at the edge of the specimen, but only if they were not caused by an underlying discontinuity such as slag, lack of fusion, porosity, etc.

Occasionally, the surface of a bend will have an area that stretches differently from the surrounding surface as the result of a discontinuity just below the surface. Unless the discontinuity opens up to the surface, these areas are not to be considered in the evaluation of the bend.

Certification Records

Before the weld workmanship sample is visually inspected, the top part of the form Welder and Welding Operator Qualification Test Record (WQR), **Figure 19-8**, must be filled out. The form contains information regarding the weld. If it passes visual inspection, then the weld bend test coupons can be prepared for testing.

The bottom part, containing the record of the test results, will be filled out by the designated individual (usually the welding instructor) once all of the tests are completed. To pass the certification test, your weld must pass both the visual inspection and the entire bend test.

Welder and Welding Operator Qualification Test Record (WQR)

Welder or welding oporator's name:____________________Identification no.__________

Welding process: _________, Manual_____, Semiautomatic_____, Machine_____, Position: Groove________, Fillet________

Material specification:_________________Diameter and wall thickness (if pipe) - otherwise, joint thickness:__________

Thickness range this qualifies:____________to____________

Filler Metal

Specification no.:____________, Classification:____________, F-number:__________, Filler Metal Diameter__________

Trade name:__________________Describe filler metal (if not covered by AWS specifications):________________

____

Backing strip material (if any):____________________, Shilding gas type:__________,CFH:____________

Workmanship Sample

Visual Inspection

Reinforcement: Under_______,OK_______,Excessive_______ Undercut: Acceptable__________,Rejected__________

Overlap: Acceptable________,Rejected________ Penetration: Acceptable_________, Rejected__________

Cracks: Acceptable __________, Rejected _________ Appearance: Acceptable ________, Rejected __________

Porosity: Diameter of largest_______, Acceptable_______, Rejected________

Face and Root Bend Test Results

(1) Face Bend: Length of each discontinuity over 1/32" ________ ________ ________ ________ Sum: ____________

Accept ________, Reject________

(2) Face Bend: Length of each discontinuity over 1/32" ________ ________ ________ ________ Sum: ____________

Accept ________, Reject________

(3) Root Bend: Length of each discontinuity over 1/32" ________ ________ ________ ________ Sum: ____________

Accept ________, Reject________

(4) Root Bend: Length of each discontinuity over 1/32" ________ ________ ________ ________ Sum: ____________

Accept ________, Reject________

Pass:____________, Fail ________

Inspector's name:________________________ Date:________________

FIGURE 19-8 Sample form for recording a welder's test.

Summary

Welders become professional test takers, so it is a good idea to make every weld as if it was being made to pass a certification test. As you develop your skills making quality welds on a regular basis, taking and passing certification tests will become much easier.

Specific information regarding SENSE certification and registration is available through the AWS website (www.AWS.org) or by calling 800-443-9353. A sampling of the AWS SENSE pipe workmanship samples are provided in the appendix.

Review Questions

1. What are the names of three organizations that publish pipe welding codes?
2. What AWS program can be used to certify student welders?
3. What does being a certified welder mean?
4. What measurements can be made during a visual inspection of a weld?
5. What is the name of a weld flaw that makes it unable to meet minimum acceptable standards or specifications?

(Continued)

6. What is the major difference between discontinuities and defects?
7. What can cause a stress riser?
8. What is the purpose of a visual inspection of a weld?
9. Why is the reinforcement of a weld restricted to a maximum height of 1/8 in. (3 mm)?
10. What is the maximum allowed undercut on a weld?
11. Why is it important to lay out a thermally cut weld specimen wider than the finished size needed for the test?
12. What must be done before a workmanship sample can be bend-tested?
13. When preparing a specimen for a bend test, what should be done to ensure that the specimen does not crack as a result of inadequate preparation, rather than because of a weld defect?
14. The top part of what form must be filled out prior to visual inspection of a weld?
15. What must pass on the weld in order to pass the certification test?

Chapter 20

Pipe Threads

OBJECTIVES

After completing this chapter, the student should be able to:

- List the names of screw thread standards.
- Explain the primary purpose of bolts and nuts, power threads, and pipe threads.
- Give examples of standard bolt and nut threads.
- Describe the three parts of a thread groove.
- Explain the difference between bolts, machine screws, and studs.
- Discuss the purpose of using cutting oils for threading operations.
- Explain how pipes are threaded.
- Demonstrate how to thread a pipe.
- Demonstrate how to thread a drilled hole.
- Discuss problems that can occur if threads are overtightened.
- Explain the purpose of using pipe dope or Teflon® tape on pipe threads.

KEY TERMS

Acme
Bolt
Buttress
Crest
Cutting oils
Machine screw
Major diameter
Minor diameter
Pitch
Pitch diameter
Profile angle
Root
Square threads
Tap drill
Thread face
Thread flank
Unified National Coarse (UNC)
Unified National Fine (UNF)

INTRODUCTION

Archeologists have found screw threads on implements that were made as early as 400 BC. These screws were made of wood and were used to squeeze grapes and pump water. Before thread spacing, depth, and shapes were standardized, blacksmiths all made their own threads. As a result, it was impossible for parts to be interchanged or to replace a broken bolt unless you went back to the artisan who made the tool, plow, gun, or other implement.

In the 1770s, the first screw-cutting lathes were manufactured, which allowed screw threads to be made that were consistent in shape and threads per inch. However, there still were no standard screw threads until the 1860s. English engineer Joseph Whitworth collected a large number of thread samples from workshops all across Britain. After a number of years, he created a standard for threads which bear his name today—British Standard Whitworth (BSW) threads. An American screw thread standard was developed in 1865. The American Standard had two different standard thread types, National Coarse (NC) and National Fine (NF). Another standard was developed in Europe as that continent adopted the SI (Système International, better known as "metric") system of measurement.

Following the end of World War II, the United States, Canada, and Great Britain got together to create a standard thread form that would eliminate the problems caused during the war with bolts and nuts not fitting. The result was the creation of the Unified National Screw Threads **Unified National Coarse (UNC)** and **Unified National Fine (UNF)**. The M-series screw standard threads were created by a multinational group of nations that all use the metric system.

It can be argued that the Industrial Revolution and the Age of Steam were spurred on by the standardization of threads, which made it possible to begin the mass production of steam engines to power factories and transportation.

Back in the late 1700s, pipes were joined mechanically with a bell and spigot or they were riveted, **Figure 20-1**. By the early 1800s, pipes were being joined with threads; however, just like the early threads on bolts, there was not a standard shape or spacing. In 1820, Robert Briggs began working on standardizing piping threads. By the late 1880s, most major manufacturers had adopted Briggs's pipe threads as their standard.

In 1918, the American Standards Association (ASA) was formed with the help of engineering societies and government agencies for the purpose of developing a standardized national piping thread specification. This group created a pipe thread standard based on Briggs Standard Pipe Thread and named theirs National Pipe Taper (NPT). Later, to meet industrial needs, they created the National Pipe Taper Fuel (NPTF) pipe thread standard.

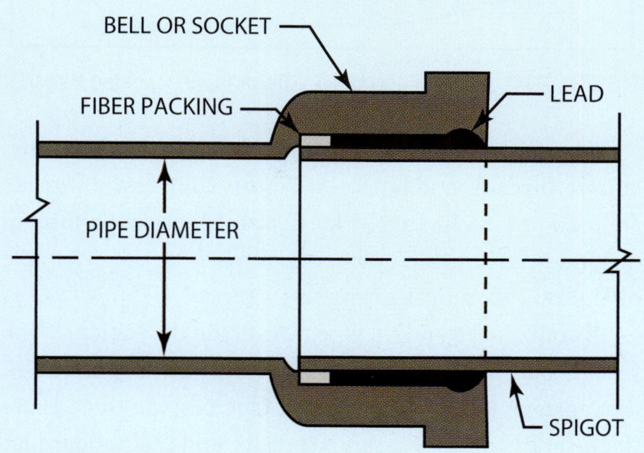

FIGURE 20-1 Bell and spigot pipe joint.

TYPES OF THREADS

Threads can be grouped into three broad categories based on their primary function. Bolts and nuts are primarily used to hold things together; power threads are primarily used to transfer power, apply force, or provide movement of parts; and pipe threads are used primarily to connect pipes to pipes or to connect pipes to machinery or valves.

V-threads are primarily used to hold things together and are available as UNC and UNF. UNC threads are less likely to become nicked or damaged and are less likely to be cross-threaded than fine threads. UNF threads are less likely to loosen as the result of vibration and have a slightly higher tensile and shear strength than the same size and type of coarse thread bolts or screws, **Figure 20-2A**.

Round threads, sometimes called "knuckle threads," are used on ceramic parts, lightbulb bases, and on materials where sharp threads might chip or break easily, **Figure 20-2B**. They also can be rolled into sheet metal. An example of SI bolt and nut thread is the M-series.

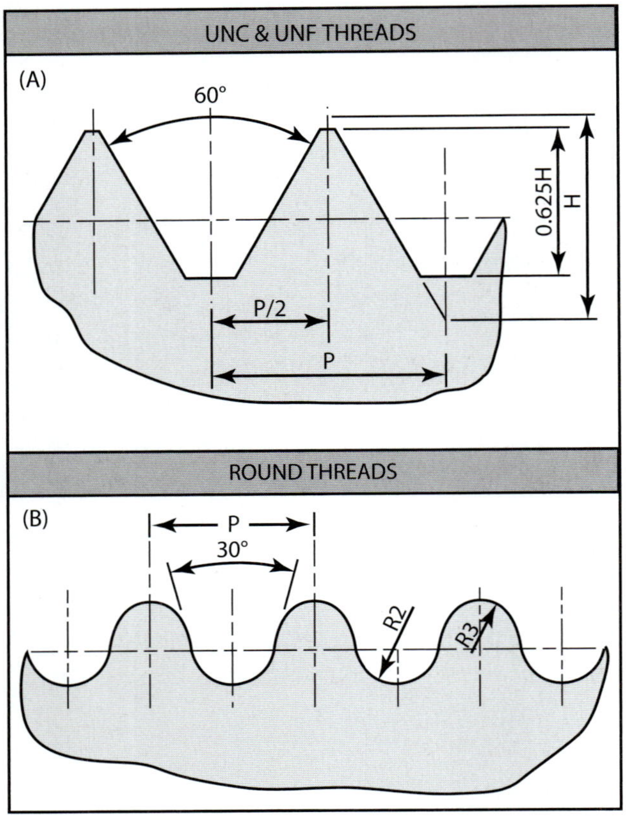

FIGURE 20-2 (A) UNF and UNC V-threads, (B) rounded threads.

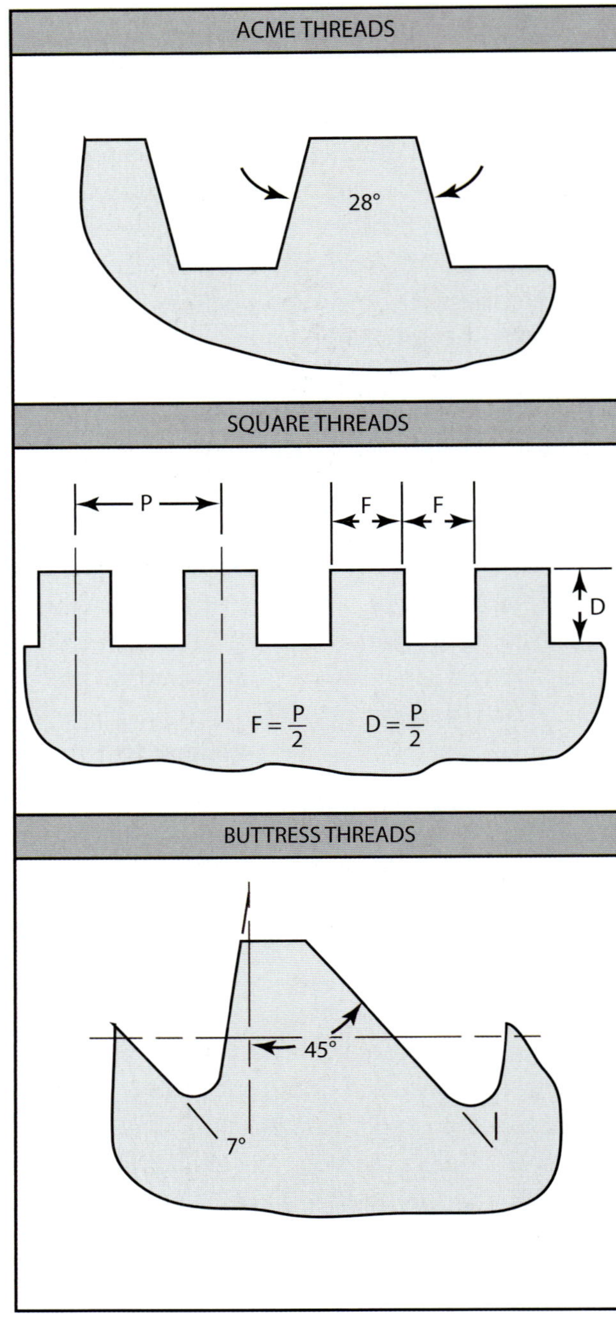

FIGURE 20-3 Examples of power threads.

Examples of standard power threads are the **acme**, **square threads**, and **buttress**, **Figure 20-3**. Power threads are used on bench vises, jacks, C-clamps, drives on milling machines, lathes, and other similar machines. An example of SI power threads is the Tr-series thread.

Examples of standard pipe threads are the National Pipe Thread and the National Pipe Thread Fuel, **Figure 20-4**. Examples of the British standard tapered pipe threads are the R-series and the Rc-series threads, and the standard for the parallel threads is the G-series threads. Most European and Asian countries on the metric system use the British inch measurement system for pipe and tubing.

THREAD SPECIFICATIONS

The various parts of all threads use the same nomenclature for identification. **Figure 20-5** shows the three parts of a thread groove. The top of the thread is called the **crest** (1), the bottom of the thread is called the **root** (2) and the side of the thread is called the **thread face** or **thread flank** (3). The distance from any point on one thread to the same point on the next thread is called the **pitch** (4). The pitch measurement is given as threads per inch for standard measurements and as millimeters in the SI system. The angle formed by the flank or sides of the thread is called the **profile angle** (5).

Although the terminology for the threads on bolts and machine screws is the same for pipe threads, the way the diameter is expressed is different. The diameters of bolts and machine screws are expressed as the outside diameter (OD) of the threads. A pipe's diameter is expressed as the inside diameter (ID) of the pipe.

Figure 20-6 identifies the terms for dimensioning a common bolt. Three terms are used to describe it—**major diameter**, **minor diameter**, and **pitch diameter**. The major diameter is the dimension that is used to identify the size of a bolt and nut. The minor diameter is the distance between the thread roots. The pitch diameter is measured at the center line between the major and minor diameters.

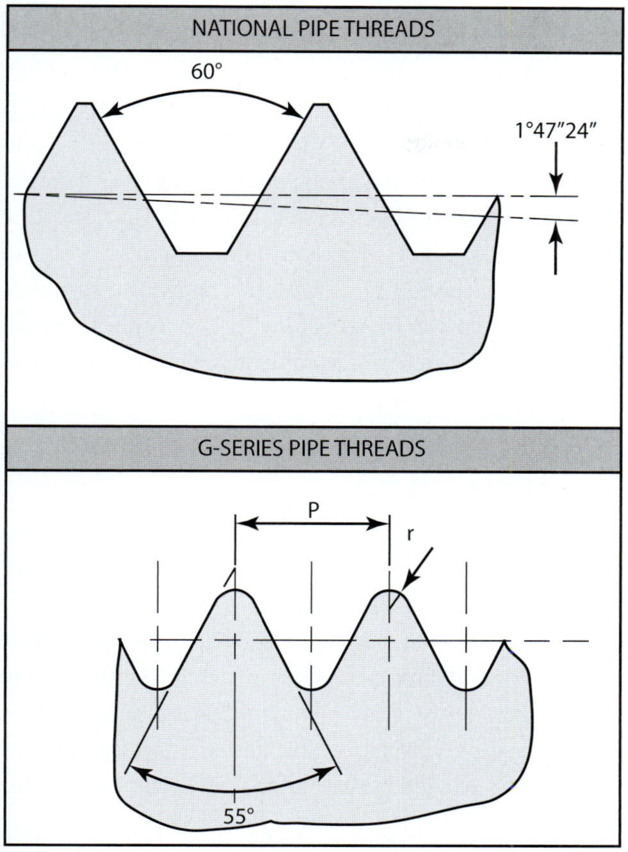

FIGURE 20-4 Examples of pipe threads.

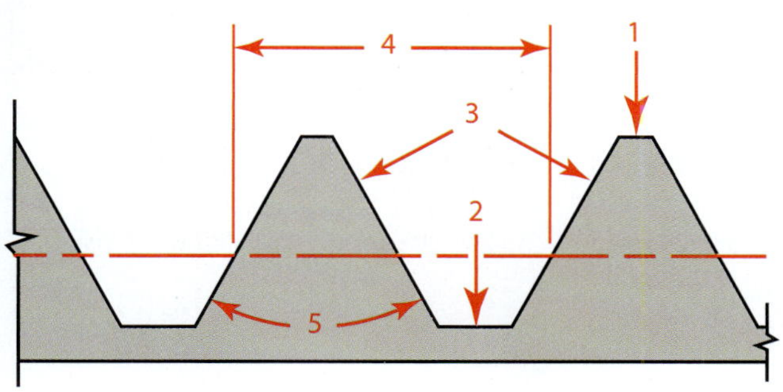

FIGURE 20-5 Thread nomenclature.

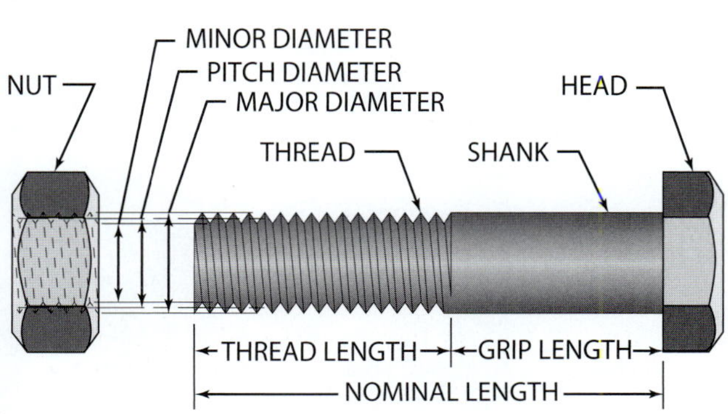

FIGURE 20-6 Bolt and nut nomenclature.

NOTE

Generally, the size of the drill bit that is used to drill the hole is the same as the minor diameter of the threads.

Thread Gauges

Thread gauges are available for all types and sizes of threads and for both internal and external threads. Some thread gauges are made of plastic, with a wide range of thread sizes and pitches. Metal pitch gauges have multiple blades that have teeth cut at varying spacing called "pitch," as noted previously. These gauges will provide the number of threads per inch, but not the diameter. Thread plugs and thread rings are precision-machined test instruments that can provide very accurate information about threads and whether they would meet the specifications, **Figure 20-7**.

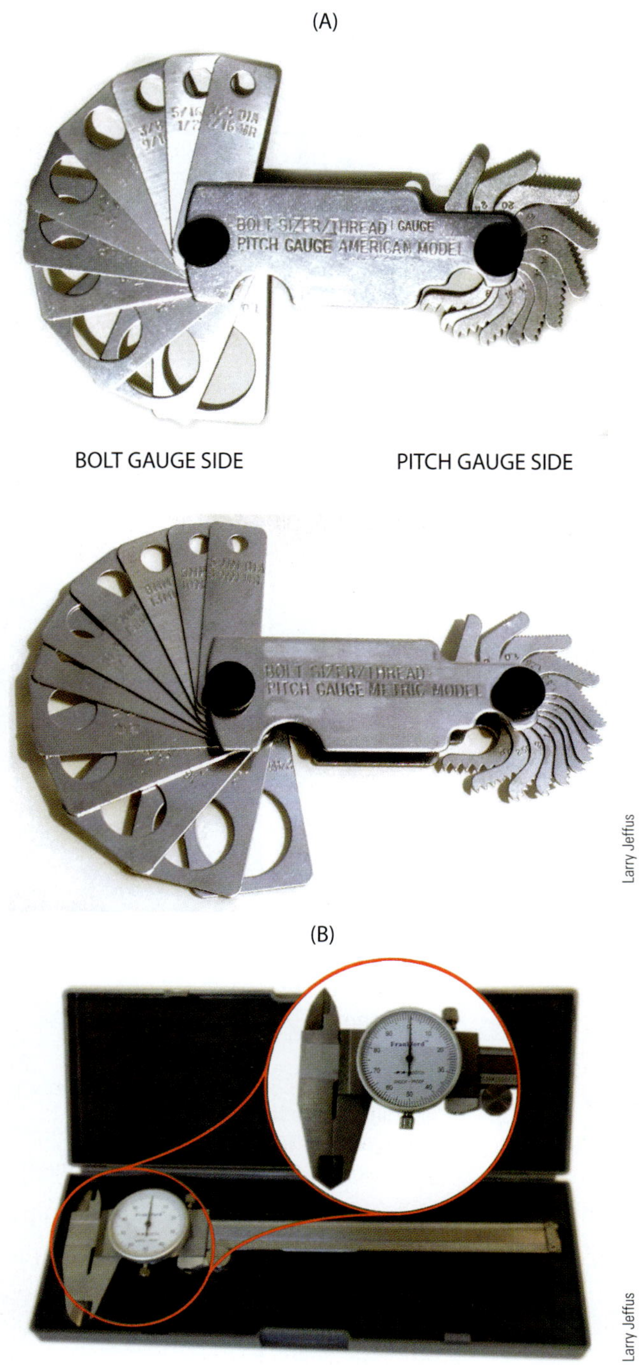

FIGURE 20-7 (A) Thread gauges, (B) Vernier caliper.

Bolts, Machine Screws, and Studs

The terms **bolt** and **machine screw** are often used interchangeably, but the primary difference is that bolts are tightened by turning a nut and machine screws are tightened by turning the heads of the screws. Bolts are used with nuts and may be fully threaded or partially threaded with an unthreaded shank. Machine screws are usually threaded all the way up to the head and are ¼-in. (6 mm) in diameter or smaller. Studs do not have a head, are threaded the full length, and may have different thread types on opposite ends, **Figure 20-8**.

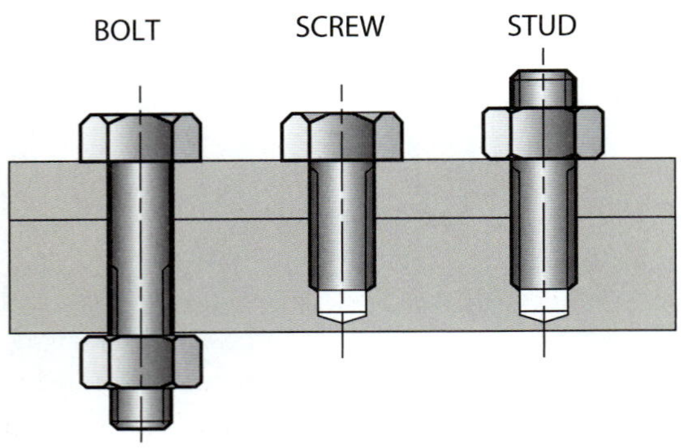

FIGURE 20-8 Examples of bolts, screws, and studs.

Bolt Grades

Of the 50 or more SAE International (SAE) and American Society of Testing and Materials (ASTM) fastener grades, there are three standard bolt grades and two SI grades commonly used in industrial applications, **Table 20-1**. Grade 2 has no markings on its head and sometimes is referred to as an "ungraded bolt." There are more than 30 SAE and ASTM grades that do not have head markings, so unless you know from the manufacturer which of these grades you have, there can be some confusion. Grades 5 and 8, as well as the SI bolt heads, are marked.

CUTTING OILS

Cutting oils that are used for hand-threading operations are made from petroleum distillates or synthetic or semisynthetic compounds. Petroleum distillate cutting oils are referred to as "mineral cutting oils." They can range from heavy, dark, and strong-smelling, which are sulfur-rich, to lightweight, almost clear oil. Synthetic or semisynthetic oils are most often water soluble and are used in machine threading to provide both cooling and lubrication for the cutting tool.

Head Marking	Grade and Material	Nominal Size Range	Mechanical Properties		
			Proof Load	Min. Yield Strength	Min. Tensile Strength
No Markings	**Grade 2** Low or medium carbon steel	1/4″ through 3/4″	55,000 psi	57,000 psi	74,000 psi
		Over 3/4″ through 1-1/2″	33,000 psi	36,000 psi	60,000 psi
3 Radial Lines	**Grade 5** Medium Carbon Steel, Quenched and Tempered	1/4″ through 1″	85,000 psi	92,000 psi	120,000 psi
		Over 1″ through 1-1/2″	74,000 psi	81,000 psi	105,000 psi
6 Radial Lines	**Grade 8** Medium Carbon Alloy Steel, Quenched and Tempered	1/4″ through 1-1/2″	120,000 psi	130,000 psi	150,000 psi
8.8	**Class 8.8** Medium Carbon Steel, Quenched and Tempered	All Sizes below 16 mm	580 MPa	640 MPa	800 MPa
		16 mm through 72 mm	600 MPa	660 MPa	830 MPa
10.9	**Class 10.9** Alloy Steel, Quenched and Tempered	5 mm through 100 mm	830 MPa	940 MPa	1040 MPa

TABLE 20-1 Classification of Bolt Strengths

All cutting oils or fluids are formulated to reduce the wear and heat buildup on the tap or die and improve the accuracy of the threads being cut, while not producing hazardous fumes.

NOTE

It may be possible to cut threads without using cutting oil; however, the quality of the threads is often so poor that they will not pass inspection and would fail if placed in service. In addition, the tap or die being used will be damaged and may not be able to be used to cut quality threads later, even if oil is used the next time.

Using Thread Cutting Oils

When taps or dies are being used manually, cutting oil can be applied to the outside surface of the pipe or rod being threaded, or it can be dripped into the hole to be tapped. Additional applications of cutting oil may need to be applied from time to time as the threads are being cut. It is not possible to have too much thread cutting oil, so apply it liberally.

When threads are being cut using a power-threaded or automatic threading machine, the thread cutting oil must be applied as continuously as possible because these machines need it for both cooling and lubrication of the cutter.

Recycling Thread Cutting Oil

Unlike most other lubricating oils, such as those used in engines or on machine bearings, thread-cutting oil can be reused over and over again. Once it is caught in a drip pan or bucket, it can be filtered to remove metal chips and then reused. A drip pan with a built-in filter and a pump for recycling cutting oil are shown in **Figure 20-9A**.

Disposing of Used Thread Cutting Oil

Used cutting oil is considered a hazardous waste like any other oil, so it must be disposed of in accordance with local, state, and national environmental protection laws. Check with city, county, or state offices for instructions on the legal way of disposing of cutting oil.

Disposing of Oil-Soaked Rags

Occupational Safety and Health Administration (OSHA) regulation 1926.252(e) states, "All solvent waste, oily rags, and flammable liquids shall be kept in fire-resistant, covered containers until removed from worksite." An approved metal "Oily Waste" safety disposal can must be used for storing oil-soaked rags until they can be removed from the premises. Approved cans will be labeled.

PIPE THREADING

Pipe threads can be cut manually with a power hand tool or with a pipe-threading machine, **Figures 20-9A and B**. Manually cutting pipe threads is most often used on small jobs or when smaller-diameter pipes are being worked. Power hand-threading tools or pipe-threading machines can be used on small jobs, but they are often required on large jobs to keep up with the production schedule.

Most pipe-cutting dies are ground at an angle to cut mild steel pipe. Although the thread pitch is the same for different materials, the threading dies may need to be ground to a different angle for cutting other materials, such as aluminum, stainless steel, brass, copper, or plastic pipe. Refer to the equipment manufacturer's technical data sheet tables for specific angle requirements for other materials being threaded.

Work-hardened or thermal-hardened pipe must be annealed before threading. Hardened areas can result in the threads tearing, or these hardened areas may break the cutting teeth on the dies. Work hardening can occur on the end of a pipe if it is cut with a roller cutter. Thermal hardening can occur if the pipe is flame cut, welded on, or overheated with an abrasive cutoff wheel.

The outside edge of the pipe should be chamfered to help start the threading dies, **Figure 20-10**. If the pipe was cut using a roller-type pipe cutting tool, it may have a ridge, **Figure 20-11**. This ridge must be removed before starting the threading process.

NOTE

Using a sharp cutting wheel, less cutting pressure, and lubricating the cutting wheel will help to reduce the size of the ridge.

Storing Taps and Dies

It is important to clean, oil, and properly store tap and die cutters after each use. These tools do not have any protective coating to prevent rusting. Rust can form quickly on exposed steel tool surfaces. The sharp cutting edges will be quickly dulled if allowed to rust. Once the sharp cutting edges are damaged, the tap or die cutters must be sharpened or replaced.

Use a clean rag with a few drops of oil to wipe down all the surfaces and remove all metal chips. Once cutters have been cleaned and lightly oiled, they can be returned to their storage case, **Figure 20-12**, or wrapped in a small, lightly oiled rag.

FIGURE 20-9 (A) Pipe threader with automatic oiler, (B) portable pipe threader, (C) pipe stand, (D) pipe support, (E) pipe reamer.

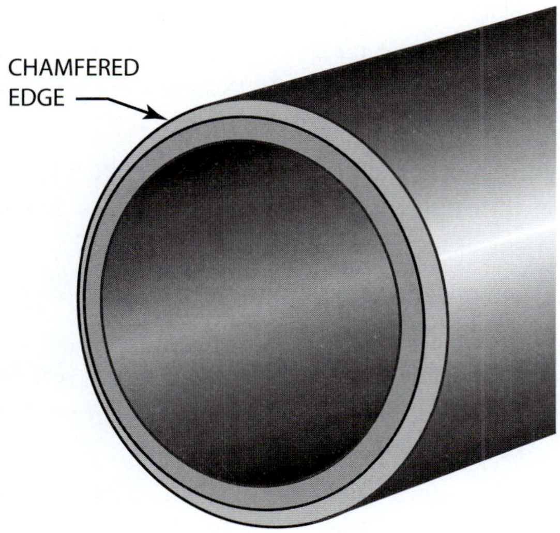

FIGURE 20-10 Chamfering the pipe makes it easier to start the thread die.

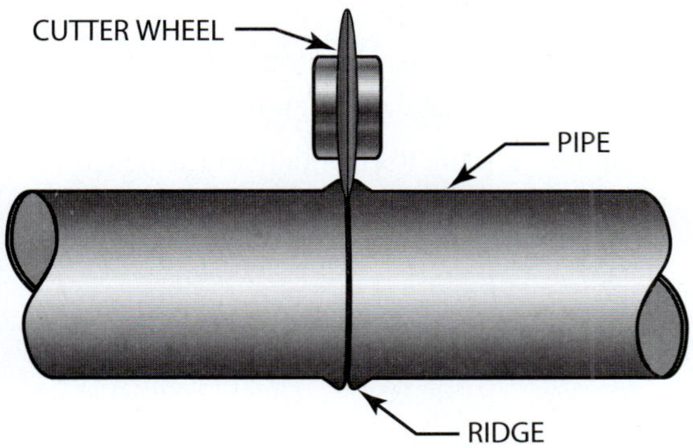

FIGURE 20-11 The ridge left by a pipe cutter must be removed before the pipe is threaded.

FIGURE 20-12 Hand pipe threading kit.

NOTE

Be careful not to touch the cutter with your fingers once it has been cleaned and oiled. Your fingertips contain salts; and if you touch the cutter's surface after it has been cleaned, you may find your fingerprint rusted into it the next time you use it.

PRACTICE 20-1

Manually Threading ½-in. (13-mm) Diameter Mild Steel Pipe

Using a pipe cutting die properly set up according to the manufacturer's specifications, pipe-thread cutting oil, rags, a drip pan or bucket, a pipe vise or bench vise, a 6-in. (150-mm) long or longer section of ½-in. (13-mm) diameter mild steel pipe, and all the required personal protective equipment (PPE), you are going to thread the pipe.

1. Clamp the pipe horizontally in a pipe vise or bench vise with the end to be threaded tilted slightly downward to prevent cutting oil from running back through the pipe and dripping on the floor.
2. Place a drip pan or bucket on the floor under the end of the pipe to catch the metal chips and oil.
3. Check the pipe-threading tool to see that it is free of chips of metal and does not have any broken teeth.
4. Using a small paintbrush or oil can, apply thread-cutting oil on the outside of the pipe.
5. The bushings of the threading tool should slide easily over the end of the pipe, centering the pipe in the thread-cutting dies.
6. Apply pressure against the head of the pipe thread-cutting tool as you rotate the handle clockwise, **Figure 20-13**. Once the teeth on the pipe-cutting

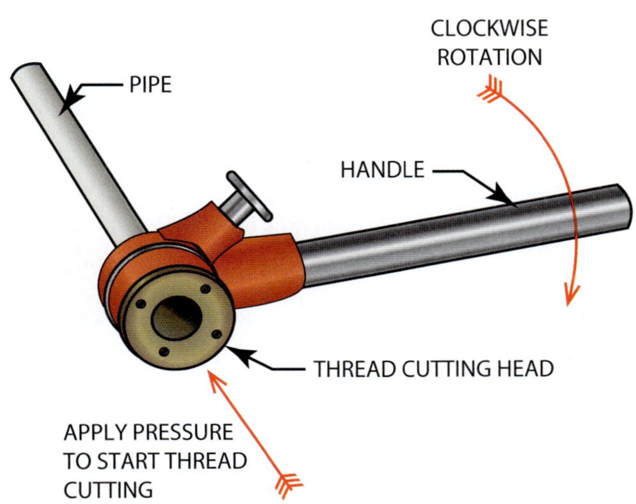

FIGURE 20-13 Parts of a hand pipe threading tool.

tool begin to cut into the pipe, you can reduce or stop applying pressure to the head. The helix of the threads will pull the cutter onto the pipe.

7. Keep turning the cutting head slowly. From time to time, stop, add more thread-cutting oil, and turn the cutter slightly counterclockwise to break up the chips.
8. Enough threads have been cut when the end of the pipe is even with the backside of the cutting head.
9. Turn the cutting tool counterclockwise to remove it from the pipe.
10. Use a rag to wipe off the cutting oil and any loose chips of metal.
11. Examine the threads to see that they are even and free from tears.
12. Use a thread gauge or threaded fitting to check the fit of the threads.
13. Clean up your work area and put away all the tools and supplies. ◆

PRACTICE 20-2

Threading a Hole

Using a drill press, a **tap drill**, which is a drill bit that is sized for the tap as being equal to the minor diameter of the threads to be cut (as shown in **Table 20-2**), thread-cutting oil, a ¼-in.-thick mild steel plate, rags, a center punch, a ball peen hammer, a drip pan or bucket, a drill press vise, a bench vise, and all the required PPE, you are going to thread a drilled hole.

Tap Drill Chart				
Bolt or Screw Size	NC Threads/in.	Drill Size	NF Threads/in.	Drill Size
1/4-in.	20	7	24	4
5/16-in.	18	F	24	I
3/8-in.	16	5/16	24	21/64
7/16-in.	14	3/8	20	25/64
1/2-in.	13	29/64	20	29/64
9/16-in.	12	31/64	18	33/64
5/8-in.	11	17/32	18	37/64
3/4-in.	10	21/32	16	11/16
7/8-in.	9	49/64	13	13/64
1-in.	8	7/8	12	15/16

TABLE 20-2 Tap Drill Chart

1. Clamp the plate horizontally in a drill press vise.
2. Use the punch and hammer to mark the spot where the plate is to be drilled.
3. Place the drip pan or bucket under the drill press to catch the metal chips and excessive cutting oil that may drip off the plate as it is being drilled.
4. Apply cutting oil to the top of the plate. Cutting oil will let the drill bit last longer and allow it to cut a smoother hole.
5. Select the tap drill size for the tap that will be used.
6. Following all the drill press manufacturer's safety rules and wearing all required PPE, begin drilling the tap hole.
7. Add cutting oil to the plate from time to time as needed as the hole is being drilled.
8. When the drill bit begins to break through the plate, reduce your downward pressure on the drill bit to avoid having it grab and possibly break.
9. Remove the plate from the drill press vise, and wipe off the metal chips and excessive cutting oil.
10. Clamp the plate horizontally in the bench vise.
11. Place the drip pan or bucket under the hole to be tapped to catch the metal chips and excessive cutting oil.
12. Place the tap in the T-handle.
13. Apply cutting oil to the top of the plate.
14. Place the end of the tap in the hole and hold it vertically; the end of the tap is tapered to make it easier to start the tapping process, **Figure 20-14**.
15. Apply a downward force on the T-handle, and begin turning it clockwise.

FIGURE 20-14 Nomenclature of a tap.

NOTE

The T-handle will have a tendency to wobble as you begin cutting the threads, so it is important that you hold it vertical and steady.

16. Once the tap begins to cut threads, the downward pressure can be decreased or stopped since the thread helix will pull the tap into the hole.
17. Stop turning the tap clockwise every ½- to ¾-turn and back it up to break up the chips so they can fall free through the chip flute in the tap, **Figure 20-15**. Also, add some cutting oil to lubricate the tap and wash out some of the metal chips.

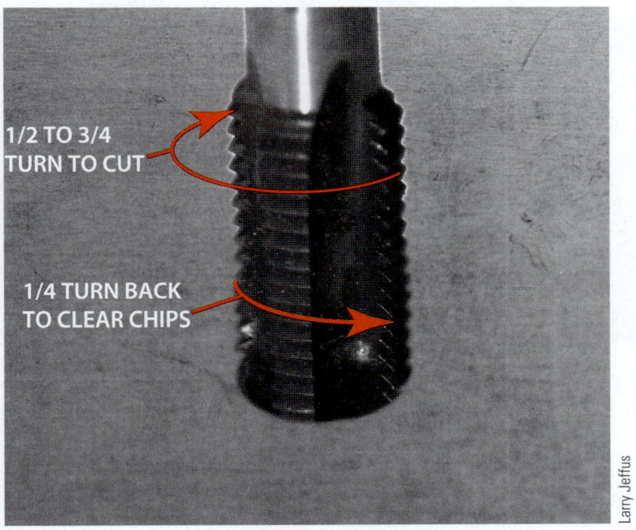

FIGURE 20-15 Proper technique for threading a hole.

18. Once the tap has cut the threads completely through the ¼-in. (6 mm)-thick plate, back it out.
19. Wipe off the metal chips and cutting oil.
20. Examine the threads to see that they are even and free from tears.
21. Check the threads by screwing a bolt in to see that it fits.
22. Clean up your work area and put away all the tools and supplies. ◆

TIGHTENING THREADS

Often, mechanics wrongly believe that tighter is better. Overtightening threads can result in the following types of damage:

- **Stripping**—Occurs when the threads are stressed beyond their shear strength and are torn loose.
- **Warping**—Occurs when a nut or pipe is deformed out of round by the force being applied to it by a wrench.
- **Galling**—Occurs when the surfaces of the threads are torn away by excessive pressure resulting from overtightening. Galling may occur as threads are being stripped.

Torque Wrenches

The most accurate way of properly tightening bolts and screws is to use a torque wrench. Mechanics have been using them for years. When bolts and screws are tightened, they are put in tension and actually stretch slightly. This tensional force works against the mechanical forces trying to pull the parts apart. When properly tightened, the bolts or screws will maintain a tight seal under normal operating pressure.

Although pipe torque wrenches are available, they are not commonly found on most piping jobs.

Pipe Dope and Teflon Tape

Either pipe dope or Teflon tape should be used on pipe threads before they are assembled. Both work well, and some plumbers use both on some medium- or high-pressure systems; however, most jobs have written procedures that will specify which one is to be used on that job.

The primary purpose of these dope and tape products is to lubricate the threads to make it easier to tighten the fitting, reducing leaks. On many fittings, especially brass and copper ones, they help to reduce the possibility of galling. Contrary to popular belief, dope and tape are not designed to seal a joint.

When applying pipe dope or Teflon tape, leave the first one or two threads uncovered; this will help prevent the material from getting into the piping system. When using tape, start wrapping the threads at the top and work down toward the end so that the edges of each wrap cover the first. This will help to keep the tape in place as the fitting is assembled. The number of wraps of tape depends on the thickness of the tape. In most cases, three wraps in the direction of the thread tightening is adequate if all but the first one or two threads are covered.

Tightening Pipe Fittings

Start by screwing the fitting on as far as possible by hand. Next, use a pipe wrench to turn the fitting two complete revolutions. Because of the tapered threads on a pipe fitting, no additional tightening should be required for a leak-free fitting.

FIGURE 20-16 Examples of the use of threaded connectors in a piping system.

Summary

Many welded piping systems may have components such as pumps and valves that will be connected using threaded fittings or are connected to the system with a bolted-together pipe flange, **Figure 20-16**. In these cases, the welder can be required to make these connections to complete the job. Therefore, it is important to know how to properly thread pipe and make leak-free connections. In the case of flanged fittings, proper tightening is required to prevent leaks.

Review Questions

1. What are the two standard bolt and screw thread types in the American Standard?
2. What measuring system does the M-series screw standard use?
3. What is the primary purpose of bolts and nuts?
4. What is the primary purpose of power threads?
5. What is the primary purpose of pipe threads?
6. List two advantages of UNF threads.
7. Give examples of tools or machines that have power threads.
8. What are the three parts of a thread groove?
9. What is the term referring to the distance from any point on one thread to the same point on the next thread?
10. What is the angle formed by the flank or sides of the thread called?
11. How is the diameter measured differently for bolts and machine screws versus pipes?

(Continued)

12. What three terms are used to describe the dimensions of a bolt?
13. What is the main difference between a bolt and a machine screw?
14. What is the purpose of cutting oil used for hand-threading operations?
15. What might happen if oil is not used when cutting threads?
16. Where should you check to find out how to dispose of used cutting oil?
17. If a pipe is work-hardened or thermal-hardened, what must be done before threading?
18. What can be done when cutting pipe to reduce the size of the ridge that forms on the cut edge?
19. What should be done before storing tap and die cutters?
20. As you are cutting pipe thread, when can you reduce applying pressure on the head of the thread-cutting tool?
21. When threading a hole, why should you stop turning the tap clockwise every ½- to ¾-turn and back it up?
22. What can happen if threads are overtightened?
23. What is the most accurate way of properly tightening bolts and screws?
24. Why should pipe dope or Teflon tape be used on pipe threads before they are assembled?

Appendix

I. Student Welding Report and/or SENSE Record

II. AWS SENSE Drawing Detail Form

III. AWS SENSE Workmanship Sample Drawing Form

Appendix I

Student Welding Report and/or SENSE Record

Student Name:______________________________ Date:________________ Class:____________

Practice #:______________, Is this an AWS SENSE Workmanship Sample: Yes___ No ___, AWS SENSE Drawing #: ____________

Welding Parameters

Process(s): ______________ Volts: _____, Amps: _____, or WIRE FEED SPEED for GMAW/FCAW: ________, Shielding Gas(s):_______

Filler Metal AWS #: ____________, Diameter: __________, Shielding Gas CFH: __________, GTAW Tungsten Type:________

Base Metal

Type of Metal:______________, Plate Thickness: ______, Pipe Diameter: ______, Schedule: ______, Welding Position: ______

Metal Preparation

Joint Type(s):________________Weld Grooves Angle: ______________, Root Face: _________, Root Gap: ________

VISUAL INSPECTION REPORT

Criteria According to AWS QC-10 and AWS QC-11

Cracks: Yes ____ No ____, Incomplete Fusion: Yes ____ No ____, Incomplete Joint Penetration: Yes ____ No ____, Overlap: Yes ____ No ____

Undercut: Yes __ No __, Depth 1.___ 2.___ 3.___ 4.___, Acceptable Yes __ No __, Porosity: 1.___ 2.___ 3.___ 4.___, Acceptable Yes __ No __

BEND TEST REPORT

Face Bend	Indication Measurement	Pass Fail	Root Bend	Indication Measurement	Pass Fail
1.			2.		
3.			4.		
Side Bend	Indication Measurement	Pass Fail	Side Bend	Indication Measurement	Pass Fail
1.			2.		
3.			4.		

WELD TEST RESULTS

Weld: Pass ______ Fail ______ Grade:________ Date:____________

Student Signature:______________________________ Instructor Signature: ______________________________

AWS SENSE ACHIEVEMENT RECORD

Module	Required	Topic	Grade	Date	Instructor Initials
1.	Yes	Occupational Orientation			
2.	Yes	Safety and Health of Welders (100% Grade Required)			
3.	Yes	Drawing and Welding Symbol Interpretation			
8.	Yes	Thermal Cutting Processes (Units 1 & 3 Minimum)			
9.	Yes	Welding Inspection and Testing			
Plus One or More of the Following					
4.		SMAW			
5.		GMAW			
6.		FCAW			
7.		GTAW			

Notes: ____

____

____

Appendix II

AWS SENSE Drawing Detail Form

INCH	MM
1/16	1.6
1/8	3.2
1/4	6.4
1/2	12.7
1	25.4

REVISIONS

ZONE	REV	DESCRIPTION	DATE	APPROVED

Notes:

TOLERANCES (Unless Otherwise Specified) DO NOT SCALE DRAWING Fractions: ±1/16" Angles: + 10°,-5"		Performance Qualification Test Information		
DR BY:	DATE:	SIZE	QC NO:	DWG NO:	REV:
CHK BY:	DATE:	APPROVED:	DATE:	SHEET: 1 of __

Appendix III

AWS SENSE Workmanship Sample Drawing Form

NOTES:

ID	QTY	SIZE	METRIC CONVERSION

Performance Qualification Test Information		
DATE:	SCALE:	DWG #:
DR BY:	Tolerances: (Unless otherwise specified) DRAWING NOT TO SCALE Fractions: ± 1/16″ Angles: + 10°, -5°	
APP BY:		

Glossary/Glosario

The terms and definitions in this glossary are extracted from the American Welding Society publication AWS A3.0-80 Welding Terms and Definitions. The terms with an asterisk are from a source other than the American Welding Society. Note: The English term and definition are given first, followed by the same term and definition in Spanish.

A

air carbon arc cutting (CAC-A). A carbon arc cutting process variation removing molten metal with a jet of air.
arco de carbón con aire (CAC-A). Un proceso de cortar con arco de carbón variante que quita el metal derretido con un chorro de aire.

Air carbon arc gouging. It is used to remove welds during repair and replacement pipe work because it is not adversely affected by surface conditions such as rust or corrosion.
arco de carbón con gubia. Se utiliza para eliminar las soldaduras durante la reparación y reemplazo de tubería, ya que no se ve afectada por las condiciones de superficie, tales como la oxidación o corrosión.

***alphabet of lines.** Lines are the language of drawing; they are used to represent various parts of the object being illustrated. The various line types are collectively known as the Alphabet of Lines.
alfabeto de líneas. Las líneas son el idioma del dibujo; se usan para representar las diferentes partes de un objeto que se ilustra. El conjunto de los distintos tipos de líneas se conoce como Alfabeto de Líneas.

American National Standards Institute (ANSI). It is a voluntary organization that oversees the creation of norms and standards that serve as guidelines for thousands of industries worldwide.
Instituto Nacional Estadounidense de Estándares. Es una organización voluntaria que supervisa la creación de normas y estándares que sirven como directrices para miles de industrias en todo el mundo.

***American Welding Society. (AWS).** Organization that promotes the art and science of welding and that publishes international codes and standards.
Sociedad Americana de Soldadores. (AWS). Organización que promueve el arte y la ciencia de la soldadura y que publica códigos y estándares internacionales.

Arc length. This is the length from the tip of the welding electrode to the adjacent surface of the weld pool.
Longitud del arco. Esta es la longitud desde la punta del electrodo de soldadura a la superficie adyacente del baño de soldadura.

Arrow side symbol. This refers to the area below the reference line and any symbol or information appearing below the reference line specifies a weld to be made on the same side of the joint that the arrow touches.
El símbolo del lado de la flecha. Esto se refiere a la zona por debajo de la línea de referencia y cualquier símbolo o información que aparece debajo de la línea de referencia especifica una soldadura que se hará en el mismo lado de la junta que la flecha toca.

As-cast grain structure. It describes crystals that grow in random directions as molten metal solidifies, so there is no real orientation.
Estructure de Grano Como-moldado. Se describe cristales que crecen en direcciones aleatorias como solidifica el metal fundido, así que no hay orientación real.

***atomic hydrogen.** A single free, unbounded hydrogen atom (H) usually formed when molecular hydrogen is exposed to an arc.
hidrógeno atómico. Un solo átomo de hidrógeno (H) libre, que normalmente se forma cuando se expone hidrógeno molecular a un arco.

***automatic welding.** Welding with equipment that requires only occasional or no observation of the welding and no manual adjustment of the equipment controls. Variations of this term are *automatic brazing, automatic soldering, automatic thermal cutting,* and *automatic thermal spraying.*
soldadura automática. Soldadura con equipos que requieren ocasional o ninguna observación, y ningun ajuste manual de los controles de los equipos. Variaciones de éste término son *soldadura fuerte automática, soldadura automática, corte termal automático, rociadura termal automático.*

***axial spray transfer.** Method of metal transfer used by GMAW and FCAW processes.
La transmisión axial arco-spray. Método de transferencia de metal utilizado por los procesos GMAW y FCAW.

B

backhand welding. A welding technique in which the welding torch or gun is directed opposite to the progress of welding. Sometimes referred to as the "pull gun technique" in GMAW and FCAW.
soldadura en revés. Una técnica de soldar la cual el soplete o pistola es guiada en la dirección contraria al adelantamiento de la soldadura. A veces se refiere como una tecnica de "estirar la pistola" en GMAW y FCAW.

benching. It is a way of excavating the side walls of a trench or excavation to prevent cave-ins and looks much like oversized steps.

Bancando. Es una forma de excavar las paredes laterales de una zanja o excavación para evitar derrumbes y se parece mucho a pasos de gran tamaño.

bevel. An angular type of edge preparation.
bisel. Una preparación de tipo angular con filo.

bevel angle. The angle formed between the prepared edge of a member and a plane perpendicular to the surface of the member.
ángulo del bisel. El ángulo formado entre el corte preparado de un miembro y la plana perpendicular a la superficie del miembro.

blueprint. It is a term still used today to refer to mechanical drawings, but originally related to a process once used to reproduce mechanical and architectural drawings. (see Mechanical Drawings)
El Plan Maestro. (anteproyecto) Es un término que todavía se utiliza para referirse a los dibujos mecánicos, pero originalmente relacionada con un proceso, una vez utilizado para reproducir dibujos mecánicos y arquitectónicos. (ver dibujos mecánicos)

branch (piping). It is a section of pipe that joins the primary pipe run.
Tubería de derivación. Se trata de una sección de la tubería que une el conducto de tubería primaria.

brittleness. It is the ease with which a pipe will crack or break apart without any noticeable bending.
fragilidad. Es la facilidad con la que un tubo se agrietao se rompen sin ninguna flexión notable

C

cave-ins. These are the sudden failure of the wall of an excavation and are one of the most serious safety hazards to workers in excavations.
derrumbes. Estos son los fallo repentino de la pared de una excavación y son uno de los riesgos de seguridad más graves para los trabajadores de las excavaciones.

chamfer. It is an angular cut around the end of a pipe that leaves a small flat surface on the inside edge called the root face. (see *Bevel*)
Chaflán. Se trata de un corte angular alrededor del extremo de un tubo que sale de una pequeña superficie plana en el borde interior llamado la cara de la raíz. (ver Bevel)

chromium-carbide precipitation, often called carbide precipitation, occurs in some stainless steels during welding when carbon combines with chromium. This reduces the chromium's ability to protect the pipe from corrosion. Often called *carbide precipitation*.
La precipitación de carburo de cromo. a menudo llamado precipitación de carburo, se produce en algunos aceros inoxidables durante la soldadura cuando el carbono se combina con cromo. Esto reduce la capacidad del cromo para proteger el tubo de la corrosión. A menudo llamada la precipitación de carburo.

clustered porosity. It is a small grouping of porosity usually caused by improper starting techniques.
La porosidad agrupada. Se trata de una pequeña agrupación de porosidad causada generalmente por técnicas de partida incorrectos.

coalescence. The growing together or growth into one body of the materials being welded.
coelescencia. El crecimiento o desarrollo de un cuerpo de los materiales los cuales se están soldando.

codes (American Welding Society). These are voluntary consensus standards that have been developed by the AWS in accordance with the rules of the American National Standards Institute (ANSI)*. They are not a law but often are adopted as a part of an ordinance or law by local, state, or national governments and agencies. (see *Standards*)
códigos (Sociedad Americana de Soldadura). Estas son normas voluntarias de consenso que han sido desarrollados por la AWS, de conformidad con las normas de la American National Standards Institute (ANSI) *. Ellos no son una ley, pero a menudo se adoptan como parte de una ordenanza o ley, estatales o gobiernos y agencias nacionales locales. (ver Normas)

cold crack. A crack occuring in a metal at or near ambient temperatures. Cold cracks can occur in base metal, heat-affected zones, and weld metal zones.
quebradura en frío. Quebradura que ocurre en un metal a temperatura ambiente o a temperaturas cercanas a ésta. Las quebraduras en frío pueden ocurrir en base de metal, zonas afectadas por el calor y zonas de metal de soldadura.

cold lap. The process in which is when molten weld metal is deposited on top of the base metal without fusing the two together.
soldadura de recubrimiento en frío. El proceso en el que es cuando el metal de soldadura fundido se deposita en la parte superior del metal de base sin la fusión de los dos juntos.

collection piping systems. Systems smaller pipes joining larger pipes allowing material to combine and flow to its destination.
sistemas de tuberías de colección. Sistemas de tuberías más pequeñas montaje de los tubos más grandes permitiendo que el material para combinar y el flujo a su destino.

complete joint penetration. When the weld extends completely through the thickness of the joint.
completar la penetración conjunta. Cuando la soldadura se extiende completamente a través del espesor de la junta

compressive strength. The property of a material to resist being crushed.
fuerza compresiva. La propiedad de un material para resistir ser aplastado.

construction pipe welding jobs. Jobs that require a higher degree of welding skills and may require the welder to travel away from their home area. (see *Manufacturing pipe welding jobs* and *Pipeline welding jobs*)

trabajos de soldadura de tuberías de construcción. Los trabajos que requieren un mayor grado de habilidades de soldadura y pueden requerir el soldador que viajar lejos de su zona de origen. (ver Fabricación trabajos de soldadura de tuberías y trabajos de soldadura de tuberías)

consumable insert. Filler metal placed at the joint root before welding and intended to be completely fused into the root of the joint and become part of the weld.
inserción consumible. Metal de relleno antepuesto que se funde completamente en la raíz de la junta y se hace parte de la soldadura.

corrosion. The slow removal of a pipe's interior or exterior surface due to oxidation.
la corrosión. La eliminación lenta de la superficie interior o exterior de un tubo debido a la oxidación.

***cover pass.** The last layer of weld beads on a multipass weld. The final bead should be uniform in width and reinforcement, not excessively wide, and free of any visual defects.
pasada para cubrir. La última capa de cordones soldadura de pasadas múltiples. La pasada final debe ser uniforme en anchura y refuerzo, no excesivamente ancha, y libre de defectos visuales.

cracks. The result of internal stresses within the weld or surrounding metal that exceed the ultimate strength of the metal. Not all cracks are weld discontinuities, but they may occur as a result of welding.
grietas. El resultado de las tensiones internas dentro de la soldadura o metal circundante que exceden la resistencia a la rotura del metal. No todas las grietas son discontinuidades de soldadura, pero pueden ocurrir como resultado de la soldadura.

crater crack. A crack initiated and localized within a crater.
grieta de crater. Una grieta en el crater del cordón de soldar.

Crest. The top or peak of the threads.
Cresta. La parte superior o máximo de las roscas.

crevice corrosion. Oxidation that occurs in the small space (crevice) between two pieces of metal as the result of moisture being trapped in the small space.
corrosión de grieta. Oxidación que ocurre en espacios pequeños (grietas) entre dos piezas de metal como resultado de que la humedad quedara atrapada en el espacio pequeño.

cutting oils. Oils made from petroleum distillates or synthetic or semi-synthetic compounds and are formulated to reduce the wear and heat buildup on taps and dies to improve the accuracy of the threads being cut.
aceites de corte. Aceites hechos de destilados de petróleo o compuestos sintéticos o semi-sintéticos y están formulados para reducir el desgaste y el calor acumulación en los grifos y muere para mejorar la precisión de los hilos que se corta.

cutting tip. The part of an oxygen cutting torch from which the gases issue.
punta para cortar. Esa parte de la antorcha para cortar con oxígeno por donde salen los gases.

cylindrical porosity. Tube-shaped porosity. Cylindrical porosity is often called wormhole because it can travel inside the weld as the weld moves along the weld joint.
porosidad cilíndrica. La porosidad en forma de tubo. Porosidad cilíndrico es a menudo llamado agujero de gusano, ya que puede viajar dentro de la soldadura como los movimientos de soldadura a lo largo de la unión soldada.

D

decimal fraction. A number that is smaller than 1.
fracción decimal. Un número que es menos de 1.

defect. A discontinuity or discontinuities that by nature or accumulated effect (for example, total crack length) render a part or product unable to meet minimum applicable acceptance standards or specifications. This term designates rejectability and flaw. (See also *discontinuity* and *flaw*)
defecto. Una desunión o desuniónes que por la naturaleza o efectos acumulados (por ejemplo, distancia total de una grieta) hace que una parte o producto no esté de acuerdo con las normas o especificaciones mínimas para aceptarse. Este término designado resectabilidad y falta. (Vea también *discontinuidad* y *falta*)

delaminations. Laminates that have separated and now have a gap between the layers of pipe.
deslaminaciones. Los laminados que se han separado y ahora tienen un espacio entre las capas de la tubería.

deposition rate. The weight of material deposited in a unit of time. It is usually expressed as kilograms per hour (kg/hr) (pounds per hour [lb/h]).
relación de deposición. El peso del material depositado en una unidad de tiempo. Es regularmente expresado en kilogramos por hora (kg/hora) (libras por hora [lb/hora]).

Diamétre Nominal (DN). The diameter of pipe, which can vary slightly due to many factors, therefore, the actual dimension for a pipe diameter may vary.
Diamétro Nominal (DN). El diámetro de la tubería, que puede variar ligeramente debido a muchos factores, por lo tanto, la dimensión real de un diámetro de la tubería puede variar.

***dimensioning.** The measurements of an object, such as its length, width, and height, or the measurements for locating such things as parts, holes, and surfaces.
acotación. Las medidas de un objeto, tal como su longitud, ancho, y altura, o las medidas para ubicar cosas como piezas, agujeros o superficies.

dimensioning tolerances. The allowable deviation in accuracy or precision between the measurement specified and the part as laid out or produced.
tolerancias de acotación. La desviación permitida en la exactitud y precisión entre la cota indicada y la parte tal como se establece o se produce.

discontinuity. An interruption of the typical structure of a material, such as a lack of homogeneity in its mechanical,

metallurgical, or physical characteristics. A discontinuity is not necessarily a defect. (See also *defect* and *flaw*)
discontinuidad. Una interrupción de la estructura típica de un material, el que falta de homogenidad en sus caracteristicas mecánicas, metalúrgicas, o fisica. (Vea también *defecto* y *falta*)

distribution piping systems. Systems with smaller individual pipes connected to larger mains as a way of delivering liquids and gases.
sistemas de tuberías de distribución. Los sistemas con tubos individuales más pequeñas conectadas a la red eléctrica más grandes como una forma de entrega de líquidos y gases.

drag angle. The travel angle when the electrode is pointing in a direction opposite to the progression of welding. This angle can also be used to partially define the position of guns, torches, rods, and beams.
ángulo del tiro. El ángulo de avance cuando el electrodo está apuntando en una dirección opuesta del progreso de la soldadura. Este ángulo también se puede usar para parcialmente definir la posición de pistolas, antorchas, varillas, y rayos.

drawing scale. Used to define the ratio between the drawing size and the size of the actual part the drawing represents.
escala de dibujo. Se utiliza para definir la relación entre el tamaño del dibujo y el tamaño de la parte real el dibujo representa.

***dross.** The material expelled from the plasma arc and oxygen assist laser cutting processes, which contains 40% or more of unox-idized base metal.
escoria. Material expelido en los procesos de corte de arco de plasma y de corte láser asistido con oxígeno que contiene 40% o más de base de metal no oxidada.

***dual shield.** FCA welding process that uses both the flux core and an external shielding gas to protect the molten weld pool. See self-shielding.
protección doble. Proceso de soldadura FCA que utiliza un fundente de núcleo y protección externa de gas para proteger la soldadura derretida. Ver autoprotección.

***ductility.** As applied to a soldered or brazed joint, it is the ability of the joint to bend without failing.
ductilidad. Como aplicada a junta de soldadura fuerte o soldadura blanda, es la abilidad de la junta de doblarse sin fallar.

E

electrode extension (GMAW, FCAW, SAW). The length of unmelted electrode extending beyond the end of the contact tube during welding.
extensión del electrodo (GMAW, FCAW, SAW). La distancia de extensión del electrodo que no está derretido más allá de la punta del tubo de contacto durante la soldadura.

***electrode tip.** The end of an electrode where the arc jumps from the electrode to the work.
punta del electrodo. Extremo de un electrodo donde el arco salta desde el electrodo hasta la zona de la pieza de trabajo.

equation. A mathematical statement in which both sides are equal to each other; for example, 2X = 1Y.
ecuación. Un enunciado matemático en el que ambos lados son iguales entre sí; por ejemplo, 2X = 1Y.

Excavation. A man-made cut, cavity, trench, or depression in the earth's surface formed by earth removal.
Excavación. Un corte, cavidad, zanja o depresión hecha por el hombre en la superficie de la tierra formadas por remoción de tierra.

exfoliation corrosion. A process that occurs within the layers of metal causing pieces of the surface to become loose and possibly fall off.
la corrosión por exfoliación. Un proceso que se produce dentro de las capas de metales causando piezas de la superficie que se afloje y posiblemente caiga.

extension ladder. A straight ladder consisting of two or more sections that can be easily extended or lengthened and must be leaned against a stable surface when used. (see *Straight ladders*)
escalera de extensión. Una escalera recta que consiste en dos o más secciones que se puede ampliar fácilmente o alargar y debe ser apoyaban contra una superficie estable cuando se utiliza. (ver escaleras rectas)

F

facing (piping). A process where the edges of the pipe are cut mechanically so that the ends are square.
Paramento. Un proceso en el que los bordes de la tubería se cortan mecánicamente para que los extremos son cuadradas.

fast fill (welding electrodes). The rod's ability to deposit metal quickly.
Lleno rápido (electrodos de soldadura). La capacidad de la varilla para depositar metal rápidamente

***fast-freezing electrode.** An electrode whose flux forms a high-temperature slag that solidifies before the weld metal solidifies, thus holding the molten metal in place. This is an advantage for vertical, horizontal, and overhead welding positions.
electrodo de congelación rápida. Un electrodo cuyo flujo forma una escoria a temperaturas altas que se puede solidificar antes de que el metal de soldadura se pueda solidificar, asi detiene el metal derretido en su lugar. Está es una ventaja en soldaduras de posiciones vertical, horizontal y sobrecabeza.

fire watch. An individual that has been trained in the proper way to sound a fire alarm and in the use of an appropriate fire extinguisher suitable for the type of accidental fire that might occur as the result of welding or cutting in an area.
guardia contra incendios. Un individuo que ha sido entrenado en la forma correcta para que suene una alarma de incendio y en el uso de un extintor de incendios adecuado adecuado para el tipo de incendio accidental que pueda ocurrir como resultado de la soldadura o de corte en un área.

fit for service. A very high likelihood that the weld will never fail during the expected life of the piping system.
aptos para el servicio. Una muy alta probabilidad de que la soldadura nunca fallará durante la vida esperada del sistema de tuberías.

flaw. An undesirable discontinuity.
falta. Una discontinuidad indeseable.

formula. A mathematical statement of the relationship of items. It also defines how one cell of data relates to another cell of data, for example, wt. = [(l" × w" × t") ÷ 1728] × wt./ft.
fórmula. Un enunciado matemático de la relación de artículos. También define cómo una célula de datos se refieren a otra célula de datos, por ejemplo, en peso. = [(L 'x w' x t ') ÷ 1728] × peso. / Ft.

fractions. Two or more numbers used to express a unit smaller than one. Examples of fractions are 1/4, 1/2, 7/16, 4 5/8, and 10 1/2.
fracciones. Dos o más números usados para expresar una unidad más pequeña de uno. Ejemplos de fracciones son 1/4, 1/2, 7/16, 5/8 4, y 10 1/2

freehand (welding). A GTA welding technique where the GTA torch is held so that the tungsten and nozzle are both positioned just above the weld.
soldadura a mano libre. Una técnica de soldadura GTA donde se celebra la antorcha GTA para que el tungsteno y la boquilla están ambos situados justo por encima de la soldadura.

G

galvanic corrosion. A process that occurs between two different types of metal that are placed together or within the same corrosive environment.
corrosión galvánica. Un proceso que se produce entre dos tipos diferentes de metal que se colocan junto o dentro del mismo entorno corrosivo.

gouging. The forming of a bevel or groove by material removal.
escopleando con gubia. Formando un bisel o ranura removiendo el material.

groove angle. The total included angle of the groove between parts to be joined by a groove weld.
ángulo de ranura. El ángulo total incluido de la ranura entre partes para unirse por una soldadura de ranura.

groove face. The surface of a joint member included in the groove.
cara de ranura. La superficie de un miembro de una junta incluido en la ranura.

***GFCI.** Ground fault circuit interrupters are fast-acting circuit breakers that shut off the power to an electrical circuit when they detect a small imbalance in the circuit's electrical flow.
ICFCT. Interruptor del circuito de fallos de conexión a tierra. Tipo de interruptor de circuito de acción rápida que corta el suministro eléctrico a un circuito cuando detecta un pequeño desequilibrio en el flujo eléctrico del circuito

gun angle. The angle between the centerline of the gun and the weld as it relates to the direction of travel.
ángulo de pistola. El ángulo entre la línea central de la pistola y la soldadura que se refiere a la dirección de desplazamiento.

H

hardness. Resistance to penetration.
dureza. Resistencia a la penetración.

header (piping). A large pipe that a number of branch lines or pipes are connected to. A header may feed the branch lines or be a collector.
colector de tuberías. Una gran tubería que una serie de ramales o tuberías están conectadas a. Un encabezado puede alimentar a los ramales o ser un coleccionista.

Heat. The quantity of thermal energy.
Calor. La cantidad de energía térmica.

heat-affected zone. The portion of the base metal whose mechanical properties or microstructure has been altered by the heat of welding, brazing, soldering, or thermal cutting.
zona afectada por el calor. La porción del metal base cuya propiedad mecánica o microestructura ha sido alterada por el calor de soldadura, soldadura fuerte, soldadura blanda, o corte termal.

***high-frequency alternating current.** Electric current that changes polarity at a rate higher than 3 million cycles a second (3 MHz).
corriente alterna de alta frecuencia. Corriente eléctrica que cambia la polaridad a una velocidad de 3 millones de ciclos por segundo (3 MHz).

high-low (piping). A situation where the internal surface of the material does not line up. It results in the surfaces of the pipe not being in the same plane.
Tubería de alta-baja. Una situación en la que la superficie interna del material no se alinea. Es el resultado de las superficies de la tubería no estar en el mismo plano.

high pressure piping systems. Systems categorized by both the pressure and temperature that they are expected to withstand during service.
sistemas de tuberías de alta presión. Sistemas clasificados por tanto la presión y la temperatura que se espera que soportar durante el servicio.

hot crack. A crack occurring in a metal during solidification or at elevated temperatures. Hot cracks can occur in both heat-affected (HAZ) and weld metal (WMZ) zones.
grieta caliente. Una grieta formada a temperaturas cerca de la terminación de la solidificación.

***hot pass.** The welding electrode is passed over the root pass at a higher-than-normal amperage setting and travel rate to reshape an irregular bead and turn out trapped slag. A small amount of metal is deposited during the hot pass so the weld bead is convex, promoting easier cleaning.

pasada caliente. El electrodo de soldadura se pasa sobre la pasada de raíz poniendo el amperaje más alto que lo normal y proporción de avance para reformar un cordón irregular y sacar la escoria atrapada. Una cantidad pequeña de metal es depositada durante la pasada caliente para que el cordón soldado sea convexo, promoviendo más fácil la limpieza.

hydrogen embrittlement. The delayed cracking in steel that may occur hours, days, or weeks following welding. It is a result of hydrogen atoms that dissolved in the molten weld pool during welding.
fragilidad causada por el hidrógeno. El fisuramiento retardado en el acero que puede ocurrir horas, días o semanas después de la soldadura. Es el resultado de la disolución de átomos de hidrógeno en el charco de soldadura derretido durante la soldadura.

I

Inclusions. Foreign objects not melted by the weld arc that are left in the weld groove as weld metal is deposited over them. (see *Coalescence*)
Inclusiones. Objetos extraños no se derritieron por el arco de soldadura que se queda en el surco de soldadura como metal de soldadura se deposita sobre ellos. (ver coalescencia)

incomplete fusion. A weld discontinuity in which fusion did not occur between weld metal and fusion faces or adjoining weld beads.
fusión incompleta. Una discontinuidad en la soldadura en la cual no ocurrió fusión entre el metal soldado y caras de fusión o cordones soldados inmediatos.

incomplete joint penetration. Joint penetration that is unintentionally less than the thickness of the weld joint.
penetración de junta incompleta. Penetración de la junta que no es intencionalmente menos de lo grueso de la junta de soldar.

***induction.** The transfer of heat obtained from the resistance of the work pieces to the flow of induced high frequency welding current.
inducción. Transferencia del calor obtenido de la resistencia de las piezas de trabajo al flujo de corriente inducida de soldadura de alta frecuencia.

inert gas. A gas that does not react chemically with materials.
gas inerte. Un gas que normalmente no se combina químicamente con materiales.

interpass cold lap. Areas with a lack of fusion between the filler metal and previously deposited weld metal.
Lap de interpase en frío. Las áreas con una falta de fusión entre el metal de aporte y el metal de soldadura depositado previamente

interpass temperature. In a multipass weld, the temperature of the weld area between weld passes.
temperatura de pasada interna. En una soldadura de pasadas multiples, la temperatura en la área de la soldadura entre pasadas de soldaduras.

***ionized gas.** A gas that is heated to a point where it becomes conductive. See plasma.
gas ionizado. Gas que pasa a ser conductor una vez que se lo calienta a cierto punto. Ver plasma.

isometric drawings. Pictorial (picture-like) drawings that use lines drawn at 30° angles to the right and left of vertical so that the top, right, and left sides can all be shown in a single drawing.
Dibujos isométricos. dibujos pictoriales (ilustración) que utilizan las líneas dibujadas en ángulos de 30 ° a la derecha e izquierda de la vertical, de manera que la parte superior, a la derecha, e izquierda pueden todos ser mostrado en un solo dibujo.

J

joint geometry. The shape, dimensions, and configuration of a joint prior to welding.
geometría de junta. La figura y dimensión de una junta en sección transversa antes de soldarse.

joint root. The portion of a joint to be welded where the members approach closest to each other. In cross section, the joint root may be either a point, a line, or an area.
raíz de junta. Esa porción de una junta que está para soldarse donde los miembros están más cercanos uno del otro. En la sección transversa, la raíz de la junta puede ser una punta, una línea, o una área.

K

kerf. The width of the cut produced during a cutting process. Refer to drawing for drag.
cortadura. La anchura del corte producido durante un proceso de cortar. Refiérase al dibujo de tiro.

keyhole. The small opening at the leading edge of a weld pool caused by the arc melting away the base metal.
ojo de la cerradura. La pequeña abertura en la franja de un baño de fusión causada por el arco derritiendo el metal base.

***kindling point.** The lowest temperature at which a material will burn.
punto de ignición. La temperatura más baja la cual un material se puede quemar.

L

lack of sidewall fusion. The lack of fusion between the weld metal and the joint face.
falta de fusión pared lateral. La falta de fusión entre el metal de soldadura y la cara de unión.

lamellar tear. A subsurface terrace and steplike crack in the base metal with a basic orientation parallel to the wrought surface caused by tensile stresses in the through-thickness direction of the base metals weakened by the presence of small, dispersed, planar-shaped, nonmetallic inclusions parallel to the metal surface.
rasgadura laminar. Una terraza subsuperficie y una grieta como un escalón en el metal base con una orientación paralela a la superficie forjada. Es causada por tensión en la dirección de lo

grueso-continuo de los metales de base debilitados por la presencia de pequeños, dispersados, formados como plano, inclusiones no metálicas paralelas a la superficie del metal.

laminations. Layers of nonmetallic contaminations such as slag and oxides that are trapped within the steel as it is produced and are formed into layers during the rolling and forming process.
laminaciones. Capas de contaminaciones no metálicos tales como escoria y óxidos que se encuentran atrapados en el acero, ya que se produce y se forman en capas durante el proceso de laminación y conformación.

lateral walking or walking. A term used to describe a pipe as it moves horizontally as it is rolled because of unevenness in the pipe surface or misaligned rollers.
caminando lateral o caminando. Un término utilizado para describir un tubo de medida que se mueve horizontalmente, como que se enrolla a causa de irregularidades en la superficie del tubo o los rodillos desalineados.

light-duty service pipes. Pipes used for bicycle stands, agricultural fences and gates, and art sculptures.
tuberías de servicio ligeros. Tubos utilizados para stands de bicicletas, vallas agrícolas y puertas, y esculturas de arte.

linear porosity. See *Piping porosity.*
Porosidad lineal. Ve Porosidad de tubería

linear slag inclusions. Long strings of slag that lie along one side or both sides of previous weld pass.
inclusiones de escoria lineales. Las cadenas largas de escoria que se encuentran a lo largo de un lado o ambos lados de pasada de soldadura anterior.

liquid penetrant. A technique that uses a solution that is applied to the area that is to be inspected. After a period of time, it is wiped off the surface, and a developer is applied that will show any surface cracks.
líquidos penetrantes. Una técnica que utiliza una solución que se aplica a la zona que va a ser inspeccionado. Después de un período de tiempo, se limpia la superficie, y un revelador se aplica que mostrará ningún grietas en la superficie.

localized corrosion. A type of corrosion that does not occur evenly across the metal surface leaving it covered with varying sizes of pits or dimples. Sometimes called *pitting corrosion.*
la corrosión localizada. Un tipo de corrosión que no se produce de forma homogénea en la superficie del metal dejándolo cubierto con tamaños de hoyos o depresiones variable. A veces llamado la corrosión por picadura.

low-pressure piping systems. Systems that may be used for collector systems such as wastewater systems, agricultural irrigation systems, and building sprinkler systems.
sistemas de tuberías de baja presión. Sistemas que pueden utilizarse para sistemas de colección, como los sistemas de aguas residuales, sistemas de riego agrícolas y sistemas de construcción de rociadores.

M

machine welding. Welding with equipment that performs the welding operation under the constant observation and control of a welding operator. The equipment may or may not perform the loading and unloading of the work.
máquina para soldadura. Soldadura con equipo que ejecutan la operación de soldadura bajo la observación constante de un operador de soldadura. El equipo pueda o no ejecutar el cargar o descargar del trabajo.

magnetic particle process. A process that uses a magnetic field that is set up around the area to be inspected.
proceso de partículas magnéticas. Un proceso que utiliza un campo magnético que se crea alrededor de la zona a inspeccionar.

major diameter. The dimension that is used to identify the size of a bolt and nut.
diámetro mayor. La dimensión que se utiliza para identificar el tamaño de un perno y la tuerca.

manufacturing pipe welding jobs. Jobs that require a higher degree of welding skill but may not require welders to travel away from their home area. (see *Construction pipe welding jobs* and *Pipeline welding jobs*)
fabricación trabajos de soldadura de tubería. Los trabajos que requieren un mayor grado de habilidad de soldadura, pero puede no requerir soldadores que viajar lejos de su zona de origen. (ver trabajos de soldadura de tuberías de construcción y trabajos de soldadura de tuberías)

material takeoff. The process of analyzing drawings, material specifications, and other design documents to determine all the materials that will be needed to fabricate the system. Sometimes referred to as just one word, *takeoff.*
despegue material. El proceso de análisis de dibujos, especificaciones de materiales y otros documentos de diseño para determinar todas las materias que serán necesarios para fabricar el sistema. A veces se refiere como una sola palabra, despegue.

mechanical drawings. An accurate way of convey the requirements and specifications needed to correctly produce the requested item.
dibujos mecánicos. Una manera precisa de transmitir los requisitos y especificaciones necesarias para producir correctamente el elemento solicitado.

***mechanical testing.** It is often a destructive testing of weld specimens to measure strength and other properties. The tests are made on specimens that duplicate the material and weld procedures required for the job.
*** pruebas mecánicas.** A menudo es una prueba destructiva de especímenes de soldadura para medir la resistencia y otras propiedades. Las pruebas se realizan sobre muestras que duplican los procedimientos y materiales de soldadura requeridos para el trabajo.

medium-pressure piping systems. Systems that may be used for water supplies, compressed air, and residential gas distribution.

sistemas de tubería de media presión. Los sistemas que pueden ser utilizados para el suministro de agua, aire comprimido, y distribución de gas residencial.

medium structural service pipes. Pipes used for signposts, railroad crossing signals, Figure 1-20, and truck brush guards.
tuberías de servicio estructurales medianas. Tubos utilizados para la señalización, señales de cruce de ferrocarril, la figura 1-20, y los guardias de pincel camión.

metric system. Abbreviated as SI, comes from the French term Le System International d' Unités. This system is made up of seven base units which include units for length, temperature, weight, etc.
sistema métrico. Abreviado como SI, proviene del término francés Le Sistema Internacional d 'une. Este sistema se compone de unidades de siete bases, que incluyen unidades de longitud, temperatura, peso, etc.

minor diameter. The distance between the thread roots.
diámetro menor. La distancia entre las raíces de rosca.

mismatch (piping). A situation where the inside surfaces of the pipe do not line up causing high low on the alignment of the pipe.
desajuste (tuberías). Una situación en la que las superficies interiores de la tubería no se alinean causando gran baja en la alineación de la tubería.

mixed units. Measurements containing numbers that are expressed in two or more different units.
unidades mixtas. Las mediciones que contienen números que se expresan en dos o más unidades diferentes.

molecular hydrogen. A bonded pair of hydrogen atoms (H_2). This is the configuration that all hydrogen atoms try to form.
hidrógeno molecular. Un par de átomos de hidrógeno (H_2) unidos. Ésta es la configuración que tratan de formar todos los átomos de hidrógeno.

Material Specification Data Sheets (MSDS). Information provided upon request by all material manufacturers that contain detailed information regarding possible hazards resulting from the use of their products.
Hojas de material de especificación de datos (MSDS). La información proporcionada a petición de todos los fabricantes de materiales que contienen información detallada sobre los posibles riesgos derivados de la utilización de sus productos.

N

National Institution for Occupational Safety and Health (NIOSH). An institution that provides scientific research into solutions to reduce the risks of injury and death in industries such as construction. They partner with OSHA to help provide the maximum level of safety for workers. (see *Occupational Safety and Health Administration OSHA*)
Instituto Nacional para la Seguridad y Salud Ocupacional (NIOSH). Una institución que proporciona la investigación científica en soluciones para reducir los riesgos de lesiones y muerte en industrias como la construcción. Se asocian con OSHA para ayudar a proporcionar el máximo nivel de seguridad para los trabajadores. (ver Seguridad y Salud Ocupacional OSHA)

nominal pipe size. The term used to express the diameter when pipe is specified based on inches or schedule.
tamaño nominal de la tubería. El término utilizado para expresar el diámetro cuando se especifica tubería basado en pulgadas o el horario.

***nondestructive testing.** Methods that do not alter or damage the weld being examined; used to locate both surface and internal defects. Methods include visual inspection, penetrant inspection, magnetic particle inspection, radiographic inspection, and ultrasonic inspection.
pruebas no destructivas. Métodos que no alteran ni dañan la soldadura que se está examinando. Se usa para encontrar ambos defectos internos y de superficie. Incluye métodos como inspección visual, inspección penetrante, inspección de partículas magnéticas, inspección de radiografía, inspección ultrasónica.

nonmetallic inclusions. The result of slag from previous weld passes being trapped beneath the next weld pass.
inclusiones no metálicas. El resultado de la escoria de soldadura anterior pasa de ser atrapado debajo de la siguiente pasada de soldadura.

notch sensitive metals. The ones that are more susceptible to stress riser failure.
muescas en metales sensibles. Los que son más susceptibles a la insuficiencia elevador de tensión.

O

Occupational Safety and Health Administration (OSHA). is a federal agency tasked with the job of enforcing safety and health legislation.
Administración de Seguridad y Salud Ocupacional (OSHA). es una agencia federal encargada de la tarea de hacer cumplir la legislación de seguridad y salud.

offset (pipe alignment). Refers to the difference between the surface of one pipe end as it matches up to the adjacent pipe end.
No al ras. (alineación de la tubería). Se refiere a la diferencia entre la superficie de un extremo de tubo, ya que coincide hasta el extremo del tubo adyacente.

orange peel. A method of closing the end of a pipe by cutting sections on the end of a pipe that look much like the sections of an orange, then bringing them together to form an end cap.
piel de naranja. Un método de cerrar el extremo de un tubo cortando secciones en el extremo de un tubo que se parece mucho a las secciones de una naranja, y luego unirlos para formar una tapa de extremo.

orbiting pipe welding. A technique using a machine, sometimes called a "bug," that moves under its own power around the pipe as the welding head moves back and forth across the weld groove.
orbitando soldadura de tuberías. Una técnica que utiliza una máquina, a veces llamado un 'bug', que se mueve por sus propios medios alrededor de la tubería como el cabezal de soldadura se mueve hacia atrás y adelante a través de la ranura de soldadura.

orifice. The hole in the constricting nozzle of the plasma arc torch or plasma spraying gun through which the arc plasma passes. Also known as *constricting orifice (plasma)*.
orificio. El agujero de la boquilla de constricción de la antorcha de arco de plasma o pistola de pulverización de plasma a través del cual pasa el arco de plasma. También conocido como orificio de constricción (plasma).

orthographic drawings. One or more two-dimensional drawings called views that are drawn as if you were looking directly at one side of the object, i.e., front, top, right side, etc. *See also* Mechanical drawings.
dibujos ortográficos. Uno o más dibujos bidimensionales llamados puntos de vista que se dibujan como si estuviera mirando directamente a un lado del objeto, es decir, frente, parte superior, lado derecho, etc. Ver también dibujos mecánicos.

other side symbol. The area above the reference line and any symbol or information appearing above the reference line specifies a weld to be made on the opposite side of the joint that the arrow touches.
símbolo de otro lado. El área por encima de la línea de referencia y cualquier símbolo o información que aparece por encima de la línea de referencia especifica una soldadura que se hará en el lado opuesto de la articulación que toca la flecha.

outside diameter. The outside measurement of the diameter of a pipe.
diámetro exterior. La medición fuera del diámetro de un tubo.

overlap. The protrusion of weld metal beyond the toe, face, or root of the weld; in resistance seam welding, the area in the preceding weld remelted by the succeeding weld.
traslapo. El metal de la soldadura que sobresale más allá del pie, cara, o de la raíz de una soldadura; en soldaduras de costuras por resistencia, la área de la soldadura anterior se rederrite por la soldadura subsiguiente.

P

Personal Protective Equipment (PPE). Items that provide protection to an individual such as clothing, gloves, goggles, helmets, respirators, harnesses etc. designed to prevent injury or death.
Equipo de Protección Personal (EPP). Los productos que ofrecen protección a un individuo como ropa, guantes, gafas, cascos, mascarillas, arneses, etc. diseñados para evitar lesiones o la muerte.

perspective drawings. These drawings are the most picture-like because the lines converge at points on the horizon line just like railroad tracks appear to converge at the horizon.
dibujos en perspectiva. Estos dibujos son los más foto-como porque las líneas convergen en puntos sobre la línea del horizonte al igual que las vías del tren parecen converger en el horizonte.

pilot arc. A low-current arc between the electrode and the constricting nozzle of the plasma arc torch to ionize the gas and facilitate the start of the welding arc.
piloto del arco. Un arco de corriente baja en medio del electrodo y la boquilla constreñida de la antorcha de arco de plasma para ionizar el gas y facilitar el arranque del arco para soldar.

pinch effect. The second stage of the short circuit transfer process and is the force that squeezes the molten end of the electrode off so the droplet can separate to restart the arc.
Efecto de pinzar. La segunda etapa del proceso de transferencia de corto circuito y es la fuerza que aprieta el extremo fundido del electrodo fuera por lo que la gotita puede separar para reiniciar el arco.

pipe grades. Standards that further define the manufacturing process and chemical composition that would give each grade its unique properties.
grados de tubería. Normas que definen aún más el proceso de fabricación y composición química que daría a cada grado sus propiedades únicas.

pipe standard. A well-defined set of specifications covering the manufacturing process and chemical composition of pipes and fittings that fall within each standard.
estándar de tubería. Un conjunto bien definido de especificaciones que cubren el proceso de fabricación y composición química de tuberías y accesorios que caen dentro de cada norma.

pipeline welding jobs. Jobs that require a higher degree of welding skills and will require the welder to travel away from their home area. See Manufacturing pipe welding jobs and Construction pipe welding jobs.
trabajos de soldadura de tuberías. Los trabajos que requieren un mayor grado de habilidades de soldadura y requerirán el soldador que viajar lejos de su zona de origen. Ver fabricación trabajos de soldadura de tuberías y trabajos de soldadura de tuberías de construcción.

piping porosity (or wormhole). Most often caused by contamination at the root. This porosity is unique because its formation depends on the gas escaping from the weld pool at the same rate as the pool is solidifying. Also known as *linear porosity*.
porosidad de tuberías (o agujero de gusano). Muy a menudo causada por la contaminación en la raíz. Esta porosidad es único porque su formación depende del gas que se escapa de la piscina de soldadura en la misma tasa que la piscina está solidificando. También conocida como la porosidad lineal.

pitch. The angular rotation of a moving body about an axis perpendicular to its direction of motion and in the same plane as its top side.

grado de inclinación. La rotación angular de un cuerpo en movimiento alrededor de un eje perpendicular a su dirección y en el mismo plano como el del lado de arriba.

pitch diameter. The measurement at the center line between the major and minor diameters.
diámetro de paso. La medida en la línea de centro entre los diámetros mayor y menor.

pitting corrosion. A localized corrosion that does not occur evenly across the metal surface leaving it covered with varying sizes of pits or dimples.
la corrosión por picaduras. A la corrosión localizada que no se produce de forma homogénea en la superficie del metal dejándolo cubierto con tamaños de hoyos o depresiones variable.

Planned Maintenance (PM). A schedule of routine tasks that manufacturers recommend for the upkeep of their equipment as a way of extending their service life. Also known as *preventative maintenance.*
Mantenimiento planeado (MP). Un calendario de tareas de rutina que los fabricantes recomiendan para el mantenimiento de sus equipos como una forma de extender su vida útil. También conocido como el mantenimiento preventivo.

plasma arc cutting (PAC). An arc cutting process employing a constricted arc and removes the molten metal with a high-velocity jet of ionized gas issuing from the constricting orifice.
cortes con arco de plasma (PAC). Un proceso de cortar con el arco que usa un arco constreñido y quita el metal derretido con un chorro de alta velocidad de gas ionizado que sale de la orifice constringente.

***plasma arc gouging.** See *plasma arc cutting.*
gubiadura con arco de plasma. Ver *cortes con arco de plasma.*

porosity. Cavity-type discontinuities formed by gas entrapment during solidification or in a thermal spray deposit.
porosidad. Un tipo de cavidad de desuniones formadas por gas atrapado durante la solidificación o en un deposito rociado termal.

postheating. The application of heat to an assembly after welding, brazing, soldering, thermal spraying, or thermal cutting.
poscalentamiento. La aplicación de calor a una asamblea después de la soldadura, soldadura fuerte, soldadura blanda, rociado termal o corte termal.

Powered Air-Purifying Respirators (PAPR). Air-purifying respirators that use a blower to force the ambient air through air-purifying elements to the inlet covering.
Los respiradores purificadores de aire (PAPR). Respiradores purificadores de aire que utilizan un ventilador para forzar el aire ambiente a través de elementos purificadores de aire a la cubierta de entrada.

preheat flames (OF cutting). The flames surrounding the cutting orifice that heat the metal surface to its kindling point. (see *Kindling point*)
llamas de precalentamiento (de corte). Las llamas que rodean el orificio de corte que calentar la superficie de metal a su punto de leña. (véase el punto de quemar)

***preheat holes.** The cutting tip has a central hole through which the oxygen flows. Surrounding this central hole are a number of other holes called preheat holes. The differences in the type or number of preheat holes determine the type of fuel gas to be used in the tip.
agujeros para precalentamiento. La boquilla para cortar tiene un agujero central por donde corre el oxígeno. Rodeando este agujero central hay un numero de otros agujeros que se llaman agujeros para precalentar. Las diferencias en el tipo o número de agujeros percalentados determina el tipo de gas combustible que se usará en la boquilla.

preheating. The application of heat to the base metal immediately before welding, brazing, soldering, thermal spraying, or cutting.
precalentamiento. La aplicación de calor al metal base inmediatamente antes de la soldadura, soldadura fuerte, soldadura blanda, rociado termal o cortes.

process piping systems. Systems used to convey various materials for distribution throughout a plant or facility.
sistemas de tuberías de proceso. Sistemas utilizados para transportar diversos materiales para su distribución a lo largo de una planta o instalación.

profile angle. The angle formed by the thread face or sides of the thread.
ángulo de perfil. El ángulo formado por la cara de rosca o en los lados de la rosca.

***pulse-arc metal transfer.** A gas metal arc welding process where the current pulses from a level below the transition current to a level above the transition current to achieve a controlled spray transfer at lower average currents; spray transfer occurs at the higher current level.
***transferencia de metal pulso de arco.** Un proceso de soldadura por arco metálico con gas, donde los pulsos de corriente a partir de un nivel por debajo de la corriente de transición a un nivel por encima de la corriente de transición para lograr una transferencia por pulverización controlada a corrientes medio inferior; transferencia por pulverización se produce en el nivel actual más alta

push angle. The travel angle when the electrode is pointing in the direction of weld progression. This angle can also be used to partially define the position of guns, torches, rods, and beams.
ángulo de empuje. El ángulo de avance cuando el electrodo apunta en la dirección en que la soldadura progresa. Este ángulo también puede ser usado para parcialmente definir la posición de pistolas, antorchas, varillas, y rayos.

R

rise (piping). A term given to the vertical distance at the center point of bends of two pieces of pipe that are parallel to each other. This is to transition the pipe from a lower level to a higher level.
Distancia vertical(de tuberías). Un término dado a la distancia vertical en el punto central de curvas de dos piezas de

tubería que son paralelos entre sí. Esta es la transición de la tubería de un nivel inferior a un nivel superior.

root cracks. Cracks in the weld or heat-affected zone (HAZ) occurring at the root of a weld.
grietas de raíz. Las grietas en la soldadura o la zona afectada por el calor (HAZ) que se produce en la raíz de una soldadura.

root face. The portion of the groove face adjacent to the root of the joint.
cara de raíz. La porción de la cara de la ranura adyacente a la raíz de la junta.

root opening. The separation between the members to be joined at the root of the joint.
abertura de raíz. La separación entre los miembros que están para unirse a la raíz de la junta.

***root pass.** The first weld of a multipass weld. The root pass fuses the two pieces together and establishes the depth of weld metal penetration.
pasada de raíz. La primera soldadura de una soldadura de pasadas múltiples. La pasada de raíz funde las dos piezas juntas y establece la profundidad de la penetración del metal soldado.

run (piping). The term given to the horizontal distance at the center point of bends of two pipes that are parallel to each other and running in the horizontal position.
Distancia lineal (de tuberías). El término dado a la distancia horizontal en el punto central de curvas de dos tubos que son paralelas entre sí y que se ejecutan en la posición horizontal.

S

saddle in (piping). Refers to how the branch pipe fits up with the header. This is where the branch pipe is beveled so that it projects inside the header pipe. The bevel on the branch pipe rests on the bevel inside of the header pipe.
ensillar (tuberías). Se refiere a cómo el tubo de derivación se ajusta con la cabecera. Aquí es donde el ramal de tubería está biselado para que se proyecta en el interior del tubo colector. El bisel en el ramal de tubería se apoya en el bisel interior del tubo colector.

saddle on (piping). Refers to how the branch pipe fits up with the header. The saddle-on technique is when the through thickness of the branch pipe rests on the outside diameter of the header pipe. The branch pipe is beveled inward so that the bevel rests on the outside of the header pipe.
Montura (tuberías). Se refiere a cómo el tubo de derivación se ajusta con la cabecera. La técnica de silla de montar en es cuando el espesor a través de la tubería de ramificación se basa en el diámetro exterior de la tubería de cabecera. El tubo de derivación está biselada hacia el interior de modo que el bisel se apoya sobre el exterior del tubo colector.

Safety Data Sheets (SDSs). Formerly Material Specification Data Sheets (MSDSs), SDSs include information such as the properties of each chemical, such as the hazardous chemical information associated with the use and handling of the product; the physical, health, and environmental health hazards; protective measures; and safety precautions for handling, storing, and transporting the chemical.
Fichas de Datos de Seguridad (FDS). Anteriormente hojas de material de especificación de datos (MSDS), SDS incluyen información como las propiedades de cada producto químico, como por ejemplo la información de producto químico peligroso asociado con el uso y manejo del producto; la física, la salud y los riesgos de salud ambiental; medidas de protección; y medidas de seguridad para el manejo, almacenamiento y transporte de la sustancia química.

scattered inclusions. Inclusions that resemble porosity but, unlike porosity, they are generally not spherical. These inclusions can also result from inadequate removal of earlier slag deposits and poor manipulation of the arc. Additionally, heavy mill scale or rust serves as their source, or they can result from unfused pieces of damaged electrode coatings falling into the weld. In radiographs some detail will appear.
inclusiones dispersas. Inclusiones que se asemejan a la porosidad, pero, a diferencia de porosidad, por lo general no son esféricas. Estas inclusiones pueden también resultar de la eliminación inadecuada de los depósitos de escoria anteriores y la mala manipulación del arco. Además, cascarilla de laminación pesada o herrumbre sirve como su fuente, o pueden ser el resultado de piezas no fusionadas de los recubrimientos de electrodos dañados que caen en la soldadura. En las radiografías aparecerá algún detalle.

scope. The section of a Welding Procedure Specification that describes all of the parameters regarding pipe material type, size, schedule, etc. that a particular code or quality manual applies to. It will define the requirements for fabricating a particular type of pipe joint. The "Scope" portion of the code outlines what is covered by the code. Most codes have five or more sections that cover terms and definitions, specifications, qualification, design and workmanship, and inspection.
ámbito de aplicación. La sección de una especificación del procedimiento de soldadura que se describen todos los parámetros relacionados con tubería tipo de material, tamaño, horario, etc., que un código particular o manual de calidad se aplica a. Se definirá los requisitos para la fabricación de un determinado tipo de junta de tubería. La parte de 'Alcance' del código resume lo que está cubierto por el código. La mayoría de los códigos tienen cinco o más secciones que cubren los términos y definiciones, especificaciones, cualificación, diseño y mano de obra, y la inspección.

seamed pipe. A pipe manufactured from sheets or coils of flat stock that is formed into a tube that is sealed with a longitudinal weld along the full length of the pipe. *See also* Seamless pipe.
tubería con costura. Un tubo fabricado a partir de hojas o rollos de material plano que se forma en un tubo que se sella con una soldadura longitudinal a lo largo de toda la longitud de la tubería. Ver también tubos sin costura.

seamless pipe. A pipe formed from a solid piece of metal so there is no need for a seam. *See also* Seamed pipe.
tubo sin costura. Un tubo formado a partir de una pieza sólida de metal lo que no hay necesidad de una costura. Ver tubería con costura también.

Self-Contained Breathing Apparatus (SCBA). Atmosphere-supplying respirators for which the breathing air source is designed to be carried by the user.
Equipo de respiración autónoma (SCBA). Respiradores suplidores de atmósfera para el que la fuente de aire de respiración está diseñado para ser llevadas por el usuario.

***self-shielding.** As it relates to FCA welding, it refers to electrodes that contain enough fluxing agents inside the electrode to provide complete weld protection without the need to provide a shielding gas. (See *dual shield*)
autoprotección. En relación a la soldadura FCA. Se refiere a los electrodos que contienen suficiente agentes fundentes en su interior como para proporcionar protección completa de la soldadura sin la necesidad de usar gas de protección.

Schools Excelling through National Skills Education, SENSE (AWS). A set of standards and guidelines for welding education programs. It provides a low-cost internationally recognized welding student certifying program.
Escuelas Sobresalientes a través de Habilidades de Educación Nacional, SENSE (AWS). Un conjunto de normas y directrices para la soldadura de los programas de educación. Proporciona un bajo costo programa de soldadura de certificación estudiante reconocido internacionalmente.

***shear strength.** As applied to a soldered or brazed joint, it is the ability of the joint to withstand a force applied parallel to the joint.
fuerza cizallada. Asi como es aplicada a una junta de soldadura fuerte o soldadura blanda, es la habilidad de la junta de resistir una fuerza aplicada al paralelo de la junta.

Shoring. Any method of reinforcing or securing the walls of a trench or excavation to prevent cave-ins by bracing them with one of several types of materials.
Apuntalamiento. Cualquier método de refuerzo o asegurar las paredes de una zanja o excavación para evitar derrumbes por refuerzos con uno de varios tipos de materiales.

short-circuit metal transfer. A method of metal transfer that has a low heat input that allows for welding thin materials.
corto circuito de transferencia de metal. Un método de transferencia de metal que tiene una entrada de bajo calor que permite para la soldadura de materiales delgados.

shutoff valves. Types of valves that can only be used in the full open or closed positions. (see *Throttling valves*)
válvulas de cierre. Tipos de válvulas que sólo se pueden utilizar en las posiciones completos abierta o cerrada. (ver válvulas de estrangulamiento)

slag (arc welding). The nonmetallic product resulting from the mutual dissolution of flux and nonmetallic impurities in some welding and brazing processes.
escoria (soldadura de arco). El producto no metálico resultante de la disolución mutua de impurezas de flujo y no metálicos en algunos procesos de soldadura y la soldadura fuerte.

slag (thermal cutting). The metallic product resulting from the removal of the material as the kerf is formed.
escoria (corte térmico). Se forma el producto metálico resultante de la eliminación del material como la entalladura.

slag inclusion. Nonmetallic solid material entrapped in weld metal or between weld metal and base metal.
inclusion de escoria. Material sólido no metálico atrapado en el metal de soldadura o entre el metal de soldadura y el metal base.

Sloping. The tapering back of the sides of a trench or excavation to an angle that will prevent a cave-in.
Inclinación. La parte posterior se estrecha de los lados de una zanja o excavación de un ángulo que evite un derrumbe.

spherical porosity. A form of porosity that is either spherical porosity that is ball-shaped or cylindrical porosity that is tube-shaped.
porosidad esférica. Una forma de porosidad que es ya sea la porosidad esférica que es en forma de bola o porosidad que es cilíndrica en forma de tubo.

Standard System (measurement). A measurement system sometimes referred to as the English system and is the most commonly used system in the U.S. today for measuring liquids, weight, length, area, volume, mass, and temperature.
Sistema estándar (medición). Un sistema de medición refiere a veces como el sistema de Inglés y es el sistema más utilizado en los EE.UU. de hoy para líquidos de medida, peso, longitud, área, volumen, masa y temperatura.

standards (welding). Very detailed and specific requirements on how welds are to be made, including even the smallest details. For example, a standard would include specific tolerances for the groove, electrode type and size, number of weld passes etc. (see *Codes*)
normas o estándares (soldadura). Requisitos muy detallados y específicos sobre cómo soldaduras deben realizarse, incluyendo los detalles más pequeños. Por ejemplo, un estándar incluiría tolerancias específicas para la ranura, del tipo de electrodo y el tamaño, número de pasadas de soldadura (ver los códigos etc.)

standoff distance. The distance between a nozzle and the workpiece.
distancia de alejamiento. La distancia entre la boquilla y la pieza de trabajo.

stepladders. Ladders that can be opened so they are self-supporting. (see *Straight ladders* and *Extension ladders*)
escaleras de tijera. Las escaleras que se pueden abrir para que sean autosuficientes. (ver escaleras rectas y escaleras de extensión)

straight ladders. Ladders that are not self-supporting so they must be leaned against a stable surface. Straight ladders are available as fixed length ladders or extension ladders. *See also* Stepladders and Extension ladders.

escaleras rectas. Las escaleras que no son autosuficientes por lo que deben ser apoyados en una superficie estable. Escaleras rectas están disponibles como escaleras de longitud fija o escaleras de extensión. Véase también escaleras de tijera y escaleras de extensión.

Stress riser (welding) is a point of stress concentration usually caused by an abrupt change in the thickness or shape of a material or the presence of a sharp groove or crack on the material's surface.
Elevador de tensión (soldadura) es un punto de concentración de esfuerzos causada generalmente por un cambio brusco en el espesor o la forma de un material o la presencia de una ranura agudo o grieta en la superficie del material.

stringer bead. A type of weld bead made without appreciable weaving motion. (See also *weave bead*)
cordón encordador. Un tipo de cordón de soldadura sin movimiento del tejido apreciable. (Vea también *cordón tejido)*

suck back. The concave root surface of a weld.
El retroceso. La superficie de la raíz cóncava de una soldadura.

Supplied-Air Respirators (SAR). Atmosphere-supplying respirators that have air piped in through a flexible hose from a large central air supply. Also known as *airline respirators.*
Los respiradores con suministro de aire (SAR). Respiradores suplidores de atmósfera que tienen aire canalizado a través de una manguera flexible de un gran suministro de aire central. También conocido como respiradores de línea.

T

tack weld. A weld made to hold parts of a weldment in proper alignment until the final welds are made.
soldadura de puntos aislados. Una soldadura hecha para detener las partes en su propio alineamiento hasta que se hagan las soldaduras finales.

take outs. The distance that an elbow extends the centerline past the end of a piece of pipe.
El exceso de codo . La distancia que un codo se extiende la línea central más allá del extremo de una pieza de tubería.

tap drill. A drill bit that is used to drill a hole that will be threaded using a tap. The tap drill size is usually equal to the minor diameter of the threads that will be cut.
Taladro de rosca. Una broca que se utiliza para perforar un agujero que se enrosca el uso de un grifo. El tamaño de perforación del grifo es generalmente igual al diámetro menor de la rosca que se cortan.

Temperature. The level of thermal activity.
La temperatura. El nivel de actividad térmica.

***tensile strength.** As applied to a brazed or soldered joint, the ability of the joint to withstand being pulled apart.
resistencia a la tensión. Como es aplicada a una junta de soldadura fuerte o soldadura blanda, la capacidad de una junta que resista ser estirada hasta que se rompa en dos pedazos.

thermal conductivity. A factor that affects the degree of distortion experienced during welding.
conductividad térmica. Un factor que afecta el grado de distorsión experimentó durante la soldadura.

thermal expansion. The change in volume that metal has as the result of the metal being heated.
expansión termal. El cambio en el volumen de metal que tiene como resultado de que el metal se calienta.

thread face. Term that refers to the side of the thread. Also known as *thread flank.*
cara de hilo. Término que se refiere a la parte de la rosca. También conocido como flanco de hilo.

Threads. Uniformly shaped and spaced ridges around the outside or inside of an object that are used to hold parts together, transfer power, or adjust parts. There are a number of different shapes and sizes of threads for different applications.
Hilos. Uniformemente crestas formadas y espaciadas alrededor del exterior o en el interior de un objeto que se utiliza para mantener las piezas juntas, transferir el poder, o ajustar las piezas. Hay un número de diferentes formas y tamaños de hilos para diferentes aplicaciones.

throat cracks. Longitudinal cracks that appear in the face of the weld that are open to the surface and are easily detected.
grietas de garganta. Grietas longitudinales que aparecen en la cara de la soldadura que están abiertos a la superficie y se detectan fácilmente.

throttling valves. Types of valves that can be used to control the flow in the pipe. Throttling valves can also be used as shutoff valves. (see *Shutoff valves)*
Válvulas reguladoras. Tipos de válvulas que se pueden utilizar para controlar el flujo en la tubería. Válvulas reguladoras también pueden utilizarse como válvulas de cierre. (ver válvulas de cierre)

toe cracks. Cracks in the base metal occurring at the toe of a weld.
grietas dedo del pie. Las grietas en el metal de base se produce en la punta de una soldadura.

torsional strength. The property of a material to withstand a twisting force.
resistencia a la torsión. La propiedad de un material para resistir una fuerza de torsión.

Toughness. The property that allows a pipe to withstand forces, sudden shock, or bends without fracturing.
Dureza. La propiedad que permite una tubería para resistir fuerzas, golpe repentino, o se dobla sin romperse.

transportation piping systems. Systems used to move various products throughout the community, state, or nation. Sometimes referred to as *cross-country pipelines.*
sistemas de tuberías de transporte. Sistemas utilizados para mover productos en toda la comunidad, estado o nación. A veces se refiere como tuberías de esquí de fondo.

travel. The diagonal distance at the center point of bends of two pipes that are running parallel to each other.
Distancia diagonal. La distancia diagonal en el punto central de curvas de dos tubos que se están ejecutando en paralelo entre sí.

trench. A narrow underground excavation that is deeper than it is wide, and no wider than 15 feet (4.5 meters).
zanja. Una excavación estrecha subterráneo que es más profundo que ancho, y no hay más ancho de 15 pies (4.5 metros).

U

ultimate strength. A measure of the load that breaks a specimen.
resistencia a la rotura. Una medida de la carga que rompe un espécimen.

underbead cracks. Cracks in the heat-affected zone, generally not extending to the surface of the base metal.
Las fisuras. Las grietas en la zona afectada por el calor, por lo general no se extiende a la superficie del metal base.

underfill. A depression on the face of the weld or root surface extending below the surface of the adjacent base metal.
faltante de material. Una depresión en la cara de la soldadura o la superficie de la raíz extendiéndose más abajo de la superficie del adyacente metal base.

Underwriters Laboratories (UL). An international safety laboratory that provides safety testing standards for almost any item used residentially or commercially as a way of protecting individuals and property.
Laboritorios Underwriters (LU). Un laboratorio de seguridad internacional que proporciona normas de ensayo de seguridad para casi cualquier elemento utilizado residencialmente o comercialmente como una forma de proteger a las personas y los bienes.

uniform or general corrosion. The most common type of corrosion. It is the oxide layer you see uniformly covering the entire surface of an exposed plate.
la corrosión general o uniforme. El tipo más común de corrosión. Es la capa de óxido que se ve que cubre uniformemente toda la superficie de una placa expuesta.

uniformly scattered porosity. A situation frequently caused by poor welding techniques or faulty materials.
porosidad dispersada uniformemente. Una situación causada con frecuencia por las técnicas de soldadura pobres o materiales defectuosos.

W

***wagon tracks.** A pattern of trapped slag inclusions in the weld that show up as discontinuities in X-rays of the weld.
huellas de carreta. Una muestra de inclusiones de escoria atrapadas en la soldadura que enseña que hay discontinuidades en los rayos-x de la soldadura.

walking the cup (welding). A method of controlling the arc length when making a GTA weld by resting the cup on the work piece. Also known as *cup walking.*
caminando la copa (soldadura). Un método para controlar la longitud del arco al hacer una soldadura GTA por el reposo de la taza sobre la pieza de trabajo. También conocido como vaso caminar.

washing (thermal cutting). The process of removing the surface of a material over a large area.
lavando (corte térmico). El proceso de eliminación de la superficie de un material sobre un área grande.

weld toe. The junction of the weld face and the base metal.
pie de la soldadura. La unión de la cara de la soldadura y el metal base.

Welding Procedure Specification (WPS). A document providing in detail the required variables for specific application to ensure repeatability by properly trained welders and welding operators.
calificación de procedimiento de soldadura (WPA). Un documento que provee en detalle los variables requeridos para la aplicación especifica para asegurar la habilidad de repetir el procedimiento por soldadores y operadores que estén propiamente preparados.

whole numbers. Numbers used to express units in increments of 1. They can be divided evenly by the number 1. Examples of whole numbers are 1, 2, 3, 4, 5, 6, 7, 8, 9, 10.
números enteros. Números utilizadas para expresar unidades en incrementos de 1. Se pueden dividir de manera uniforme por el número 1. Ejemplos de números enteros son 1, 2, 3, 4, 5, 6, 7, 8, 9, 10.

work angle (pipe). The angle less than 90° between a line perpendicular to the cylindrical pipe surface at the point of intersection of the weld axis and the extension of the electrode axis, and a plane determined by the electrode axis and a line tangent to the pipe at the same point. In a T-joint, the line is perpendicular to the non-butting member. This angle can also be used to partially define the position of guns, torches, rods, and beams.
ángulo de trabajo (tubo). El ángulo menos de 90° entre una línea, la cual es perpendicular a la superficie de un tubo cilíndrico al punto de intersección del eje de la soldadura y la extensión del eje del electrodo, y un plano determinado por el eje del electrodo y una línea tangente al tubo al mismo punto. En una junta-T, la línea es perpendicular a un miembro que no topa. Este ángulo puede también usarse para definir parcialmente la posición de pistolas, antorchas, varillas, y rayos.

Wormhole. see *Piping porosity*
Hoyo de gusano. ve la porosidad de tubería

Y

yield point. The point during tensile loading when the pipe's metal stops stretching and begins to be permanently made longer by deforming.
límite de elasticidad. El momento de carga de tracción cuando el metal de la tubería deja de estiramiento y comienza a ser permanentemente hace ya deformando.

yield strength. The amount of strain needed to permanently deform a test specimen.
límite elástico. La cantidad de tensión necesaria para deformar permanentemente una muestra de ensayo.

*** Courtesy of the American Welding Society**

Index

A

D

E

H

I

J

Q

R

S

U

V

W

Y

Z